D0765543

Hay Library
Western Wyoming Community College

FIBER OPTIC
COMMUNICATIONS

DISCARDED

621.3692 K960F 2014
Shiva Kumar & M. Jamal Deen
Fiber Optic Communications

FIBER OPTIC COMMUNICATIONS
FUNDAMENTALS AND APPLICATIONS

Shiva Kumar and M. Jamal Deen

Department of Electrical and Computer Engineering, McMaster University, Canada

This edition first published 2014
© 2014 John Wiley & Sons, Ltd

Registered office

John Wiley & Sons Ltd, The Atrium, Southern Gate, Chichester, West Sussex, PO19 8SQ, United Kingdom

For details of our global editorial offices, for customer services and for information about how to apply for permission to reuse the copyright material in this book please see our website at www.wiley.com.

The right of the author to be identified as the author of this work has been asserted in accordance with the Copyright, Designs and Patents Act 1988.

All rights reserved. No part of this publication may be reproduced, stored in a retrieval system, or transmitted, in any form or by any means, electronic, mechanical, photocopying, recording or otherwise, except as permitted by the UK Copyright, Designs and Patents Act 1988, without the prior permission of the publisher.

Wiley also publishes its books in a variety of electronic formats. Some content that appears in print may not be available in electronic books.

Designations used by companies to distinguish their products are often claimed as trademarks. All brand names and product names used in this book are trade names, service marks, trademarks or registered trademarks of their respective owners. The publisher is not associated with any product or vendor mentioned in this book.

Limit of Liability/Disclaimer of Warranty: While the publisher and author have used their best efforts in preparing this book, they make no representations or warranties with respect to the accuracy or completeness of the contents of this book and specifically disclaim any implied warranties of merchantability or fitness for a particular purpose. It is sold on the understanding that the publisher is not engaged in rendering professional services and neither the publisher nor the author shall be liable for damages arising herefrom. If professional advice or other expert assistance is required, the services of a competent professional should be sought

The advice and strategies contained herein may not be suitable for every situation. In view of ongoing research, equipment modifications, changes in governmental regulations, and the constant flow of information relating to the use of experimental reagents, equipment, and devices, the reader is urged to review and evaluate the information provided in the package insert or instructions for each chemical, piece of equipment, reagent, or device for, among other things, any changes in the instructions or indication of usage and for added warnings and precautions. The fact that an organization or Website is referred to in this work as a citation and/or a potential source of further information does not mean that the author or the publisher endorses the information the organization or Website may provide or recommendations it may make. Further, readers should be aware that Internet Websites listed in this work may have changed or disappeared between when this work was written and when it is read. No warranty may be created or extended by any promotional statements for this work. Neither the publisher nor the author shall be liable for any damages arising herefrom.

Library of Congress Cataloging-in-Publication Data

Kumar, Shiva, Dr.
  Fiber optic communications : fundamentals and applications / Shiva Kumar and M. Jamal Deen.
    pages cm
  Includes bibliographical references and index.
  ISBN 978-0-470-51867-0 (cloth)
1. Optical fiber communication. 2. Fiber optics. I. Deen, M. Jamal. II. Title.
  TK5103.592.F52K816 2014
  621.36'92–dc23

                                                                    2013043803

A catalogue record for this book is available from the British Library.

ISBN: 9780470518670

Set in 10/12pt TimesLTStd by Laserwords Private Limited, Chennai, India

MJD

To my late parents, Mohamed and Zabeeda Deen

SK

To my late parents, Saraswathi and Narasinga Rao

Contents

---

[*] Advanced material which may need additional explanation for undergraduate readers

Preface

The field of fiber-optic communications has advanced significantly over the last three decades. In the early days, most of the fiber's usable bandwidth was significantly under-utilized as the transmission capacity was quite low and hence, there was no need to apply techniques developed in non-optical communication systems to improve the spectral efficiency. However, with the recent revival of coherent detection, high spectral efficiency can be realized using advanced modulation formats.

This book grew out of our notes for undergraduate and graduate courses on fiber-optic communications. Chapters 1 to 6 discuss, in depth, the physics and engineering applications of photonic and optoelectronic devices used in fiber-optic communication systems. Chapters 7 to 11 focus on transmission system design, various propagation impairments, and how to mitigate them.

Chapters 1 to 7 are intended for undergraduate students at the senior level or for an introductory graduate course. The sections with asterisks may be omitted for undergraduate teaching or they may be covered qualitatively without the rigorous analysis provided. Chapters 8 to 11 are intended for an advanced course on fiber-optic systems at the graduate level and also for researchers working in the field of fiber-optic communications. Throughout the book, most of the important results are obtained by first principles rather than citing research articles. Each chapter has many worked problems to help students understand and reinforce the concepts.

Optical communication is an interdisciplinary field that combines photonic/optoelectronic devices and communication systems. The study of photonic devices requires a background in electromagnetics. Therefore, Chapter 1 is devoted to a review of electromagnetics and optics. The rigorous analysis of fiber modes in Chapter 2 would not be possible without understanding the Maxwell equations reviewed in Chapter 1. Chapter 2 introduces students to optical fibers. The initial sections deal with the qualitative understanding of light propagation in fibers using ray optics theory, and in later sections an analysis of fiber modes using wave theory is carried out. The fiber is modeled as a linear system with a transfer function, which enables students to interpret fiber chromatic dispersion and polarization mode dispersion as some kind of filter.

Two main components of an optical transmitter are the optical source, such as a laser, and the optical modulator, and these components are discussed in Chapters 3 and 4, respectively. After introducing the basic concepts, such as spontaneous and stimulated emission, various types of semiconductor laser structures are covered in Chapter 3. Chapter 4 deals with advanced modulation formats and different types of optical modulators that convert electrical data into optical data. Chapter 5 deals with the reverse process – conversion of optical data into electrical data. The basic principles of photodetection are discussed. This is followed by a detailed description of common types of photodetectors. Then, direct detection and coherent detection receivers are covered in detail. Chapter 6 is devoted to the study of optical amplifiers. The physical principles underlying the amplifying action and the system impact of amplifier noise are covered in Chapter 6.

In Chapters 7 and 8, the photonics and optoelectronics devices discussed so far are put together to form a fiber-optic transmission system. Performance degradations due to fiber loss, fiber dispersion, optical amplifier noise, and receiver noise are discussed in detail in Chapter 7. Scaling laws and engineering rules for fiber-optic transmission design are also provided. Performance analysis of various modulation formats with direct detection and coherent detection is carried out in Chapter 8.

To utilize the full bandwidth of the fiber channel, typically, channels are multiplexed in time, polarization and frequency domains, which is the topic covered in Chapter 9. So far the fiber-optic system has been treated as a linear system, but in reality it is a nonlinear system due to nonlinear effects such as the Kerr effect and Raman effect. The origin and impact of fiber nonlinear effects are covered in detail in Chapter 10.

The last chapter is devoted to the study of digital signal processing (DSP) for fiber communication systems, which has drawn significant research interest recently. Rapid advances in DSP have greatly simplified the coherent detection receiver architecture – phase and polarization alignment can be done in the electrical domain using DSP instead of using analog optical phase-locked loop and polarization controllers. In addition, fiber chromatic dispersion, polarization mode dispersion and even fiber nonlinear effects to some extent can be compensated for using DSP. About a decade ago, these effects were considered detrimental. Different types of algorithm to compensate for laser phase noise, chromatic dispersion, polarization mode dispersion and fiber nonlinear impairments are discussed in this chapter.

Supplementary material including PowerPoint slides and MATLAB coding can be found by following the related websites link from the book home page at http://eu.wiley.com/WileyCDA/WileyTitle/productCd -0470518677.html.

Acknowledgments

MJD sincerely acknowledges several previous doctoral students: CLF Ma, Serguei An, Yegao Xiao, Yasser El-batawy, Yasaman Ardershirpour, Naser Faramarzpour and Munir Eldesouki, as well as Dr. Ognian Marinov, for their generous assistance and support. He is also thankful to his wife Meena as well as their sons, Arif, Imran and Tariq, for their love, support and understanding over the years.

SK would like to thank his former and current research students, P. Zhang, D. Yang, M. Malekiha, S.N. Shahi and J. Shao, for reading various chapters and assisting with the manuscript. He would also like to thank Professor M. Karlsson and Dr. S. Burtsev for making helpful suggestions on several chapters. Finally, he owes a debt of gratitude to his wife Geetha as well as their children Samarth, Soujanya and Shashank for their love, patience and understanding.

1

Electromagnetics and Optics

1.1 Introduction

In this chapter, we will review the basics of electromagnetics and optics. We will briefly discuss various laws of electromagnetics leading to Maxwell's equations. Maxwell's equations will be used to derive the wave equation, which forms the basis for the study of optical fibers in Chapter 2. We will study elementary concepts in optics such as reflection, refraction, and group velocity. The results derived in this chapter will be used throughout the book.

1.2 Coulomb's Law and Electric Field Intensity

In 1783, Coulomb showed experimentally that the force between two charges separated in free space or vacuum is directly proportional to the product of the charges and inversely proportional to the square of the distance between them. The force is repulsive if the charges are alike in sign, and attractive if they are of opposite sign, and it acts along the straight line connecting the charges. Suppose the charge q_1 is at the origin and q_2 is at a distance r as shown in Fig. 1.1. According to Coulomb's law, the force F_2 on the charge q_2 is

$$\mathbf{F}_2 = \frac{q_1 q_2}{4\pi \epsilon r^2}\mathbf{r}, \tag{1.1}$$

where $\mathbf{r}$ is a unit vector in the direction of r and ϵ is called the *permittivity* that depends on the medium in which the charges are placed. For free space, the permittivity is given by

$$\epsilon_0 = 8.854 \times 10^{-12}\,\mathrm{C^2/Nm^2}. \tag{1.2}$$

For a dielectric medium, the permittivity ϵ is larger than ϵ_0. The ratio of the permittivity of a medium to the permittivity of free space is called the relative permittivity, ϵ_r,

$$\frac{\epsilon}{\epsilon_0} = \epsilon_r. \tag{1.3}$$

It would be convenient if we could find the force on a test charge located at any point in space due to a given charge q_1. This can be done by taking the test charge q_2 to be a unit positive charge. From Eq. (1.1), the force on the test charge is

$$\mathbf{E} = \mathbf{F}_2 = \frac{q_1}{4\pi \epsilon r^2}\mathbf{r}. \tag{1.4}$$

Fiber Optic Communications: Fundamentals and Applications, First Edition. Shiva Kumar and M. Jamal Deen.
© 2014 John Wiley & Sons, Ltd. Published 2014 by John Wiley & Sons, Ltd.

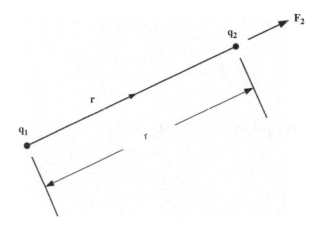

Figure 1.1 Force of attraction or repulsion between charges.

The electric field intensity is defined as the force on a positive unit charge and is given by Eq. (1.4). The electric field intensity is a function only of the charge q_1 and the distance between the test charge and q_1.

    For historical reasons, the product of electric field intensity and permittivity is defined as the electric flux density **D**,

$$\mathbf{D} = \epsilon\mathbf{E} = \frac{q_1}{4\pi r^2}\mathbf{r}. \tag{1.5}$$

The electric flux density is a vector with its direction the same as the electric field intensity. Imagine a sphere S of radius r around the charge q_1 as shown in Fig. 1.2. Consider an incremental area ΔS on the sphere. The electric flux crossing this surface is defined as the product of the normal component of **D** and the area ΔS.

$$\text{Flux crossing} \quad \Delta S = \Delta\psi = D_n\Delta S, \tag{1.6}$$

where D_n is the normal component of **D**. The direction of the electric flux density is normal to the surface of the sphere and therefore, from Eq. (1.5), we obtain $D_n = q_1/4\pi r^2$. If we add the differential contributions to the flux from all the incremental surfaces of the sphere, we obtain the total electric flux passing through the sphere,

$$\psi = \int d\psi = \oint_S D_n dS. \tag{1.7}$$

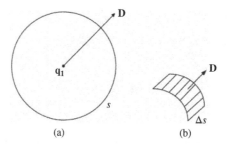

(a)   (b)

Figure 1.2   (a) Electric flux density on the surface of the sphere. (b) The incremental surface ΔS on the sphere.

Since the electric flux density D_n given by Eq. (1.5) is the same at all points on the surface of the sphere, the total electric flux is simply the product of D_n and the surface area of the sphere $4\pi r^2$,

$$\psi = \oint_S D_n dS = \frac{q_1}{4\pi r^2} \times \text{surface area} = q_1. \tag{1.8}$$

Thus, the total electric flux passing through a sphere is equal to the charge enclosed by the sphere. This is known as *Gauss's law*. Although we considered the flux crossing a sphere, Eq. (1.8) holds true for any arbitrary closed surface. This is because the surface element ΔS of an arbitrary surface may not be perpendicular to the direction of **D** given by Eq. (1.5) and the projection of the surface element of an arbitrary closed surface in a direction normal to D is the same as the surface element of a sphere. From Eq. (1.8), we see that the total flux crossing the sphere is independent of the radius. This is because the electric flux density is inversely proportional to the square of the radius while the surface area of the sphere is directly proportional to the square of the radius and therefore, the total flux crossing a sphere is the same no matter what its radius is.

So far, we have assumed that the charge is located at a point. Next, let us consider the case when the charge is distributed in a region. The volume charge density is defined as the ratio of the charge q and the volume element ΔV occupied by the charge as it shrinks to zero,

$$\rho = \lim_{\Delta V \to 0} \frac{q}{\Delta V}. \tag{1.9}$$

Dividing Eq. (1.8) by ΔV where ΔV is the volume of the surface S and letting this volume shrink to zero, we obtain

$$\lim_{\Delta V \to 0} \frac{\oint_S D_n dS}{\Delta V} = \rho. \tag{1.10}$$

The left-hand side of Eq. (1.10) is called the *divergence* of **D** and is written as

$$\text{div } \mathbf{D} = \nabla \cdot \mathbf{D} = \lim_{\Delta V \to 0} \frac{\oint_S D_n dS}{\Delta V}; \tag{1.11}$$

Eq. (1.11) can be written as

$$\text{div } \mathbf{D} = \rho. \tag{1.12}$$

The above equation is called the *differential form of Gauss's law* and it is the first of Maxwell's four equations. The physical interpretation of Eq. (1.12) is as follows. Suppose a gunman is firing bullets in all directions, as shown in Fig. 1.3 [1]. Imagine a surface S_1 that does not enclose the gunman. The net outflow of the bullets through the surface S_1 is zero, since the number of bullets entering this surface is the same as the number of bullets leaving the surface. In other words, there is no *source* or *sink* of bullets in the region S_1. In this case, we say that the divergence is zero. Imagine a surface S_2 that encloses the gunman. There is a net outflow of bullets since the gunman is the *source* of bullets and lies within the surface S_2, so the divergence is not zero. Similarly, if we imagine a closed surface in a region that encloses charges with charge density ρ, the divergence is not zero and is given by Eq. (1.12). In a closed surface that does not enclose charges, the divergence is zero.

1.3 Ampere's Law and Magnetic Field Intensity

Consider a conductor carrying a direct current I. If we bring a magnetic compass near the conductor, it will orient in the direction shown in Fig. 1.4(a). This indicates that the magnetic needle experiences the magnetic field produced by the current. The magnetic field intensity **H** is defined as the force experienced by an isolated

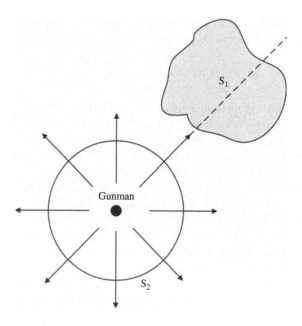

Figure 1.3   Divergence of bullet flow.

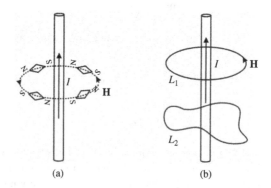

Figure 1.4   (a) Direct current-induced constant magnetic field. (b) Ampere's circuital law.

unit positive magnetic charge (note that an isolated magnetic charge q_m does not exist without an associated $-q_m$), just like the electric field intensity $\mathbf{E}$ is defined as the force experienced by a unit positive electric charge.

Consider a closed path L_1 or L_2 around the current-carrying conductor, as shown in Fig. 1.4(b). Ampere's circuital law states that the line integral of $\mathbf{H}$ about any closed path is equal to the direct current enclosed by that path,

$$\oint_{L_1} \mathbf{H} \cdot d\mathbf{L} = \oint_{L_2} \mathbf{H} \cdot d\mathbf{L} = I. \tag{1.13}$$

The above equation indicates that the sum of the components of $\mathbf{H}$ that are parallel to the tangent of a closed curve *times* the differential path length is equal to the current enclosed by this curve. If the closed path is a circle (L_1) of radius r, due to circular symmetry, the magnitude of $\mathbf{H}$ is constant at any point on L_1 and its direction is shown in Fig. 1.4(b). From Eq. (1.13), we obtain

$$\oint_{L_1} \mathbf{H} \cdot d\mathbf{L} = H \times \text{circumference} = I \qquad (1.14)$$

or

$$H = \frac{I}{2\pi r}. \qquad (1.15)$$

Thus, the magnitude of the magnetic field intensity at a point is inversely proportional to its distance from the conductor. Suppose the current is flowing in the z-direction. The z-component of the current density J_z may be defined as the ratio of the incremental current ΔI passing through an elemental surface area $\Delta S = \Delta X \Delta Y$ perpendicular to the direction of the current flow as the surface ΔS shrinks to zero,

$$J_z = \lim_{\Delta S \to 0} \frac{\Delta I}{\Delta S}. \qquad (1.16)$$

The current density $\mathbf{J}$ is a vector with its direction given by the direction of the current. If $\mathbf{J}$ is not perpendicular to the surface ΔS, we need to find the component J_n that is perpendicular to the surface by taking the dot product

$$J_n = \mathbf{J} \cdot \mathbf{n}, \qquad (1.17)$$

where $\mathbf{n}$ is a unit vector normal to the surface ΔS. By defining a vector $\Delta \mathbf{S} = \Delta S \mathbf{n}$, we have

$$J_n \Delta S = \mathbf{J} \cdot \Delta \mathbf{S} \qquad (1.18)$$

and the incremental current ΔI is given by

$$\Delta I = \mathbf{J} \cdot \Delta \mathbf{S}. \qquad (1.19)$$

The total current flowing through a surface S is obtained by integrating,

$$I = \int_S \mathbf{J} \cdot d\mathbf{S}. \qquad (1.20)$$

Using Eq. (1.20) in Eq. (1.13), we obtain

$$\oint_{L1} \mathbf{H} \cdot d\mathbf{L} = \int_S \mathbf{J} \cdot d\mathbf{S}, \qquad (1.21)$$

where S is the surface whose perimeter is the closed path L_1.

In analogy with the definition of electric flux density, magnetic flux density is defined as

$$\mathbf{B} = \mu \mathbf{H}, \qquad (1.22)$$

where μ is called the *permeability*. In free space, the permeability has a value

$$\mu_0 = 4\pi \times 10^{-7} \, \text{N/A}^2. \qquad (1.23)$$

In general, the permeability of a medium μ is written as a product of the permeability of free space μ_0 and a constant that depends on the medium. This constant is called the relative permeability μ_r,

$$\mu = \mu_0 \mu_r. \qquad (1.24)$$

The magnetic flux crossing a surface S can be obtained by integrating the normal component of magnetic flux density,

$$\psi_m = \int_S B_n dS. \tag{1.25}$$

If we use Gauss's law for the magnetic field, the normal component of the magnetic flux density integrated over a closed surface should be equal to the magnetic charge enclosed. However, no isolated magnetic charge has ever been discovered. In the case of an electric field, the flux lines start from or terminate on electric charges. In contrast, magnetic flux lines are closed and do not emerge from or terminate on magnetic charges. Therefore,

$$\psi_m = \int_S B_n dS = 0 \tag{1.26}$$

and in analogy with the differential form of Gauss's law for an electric field, we have

$$\text{div } \mathbf{B} = 0. \tag{1.27}$$

The above equation is one of Maxwell's four equations.

1.4 Faraday's Law

Consider an iron core with copper windings connected to a voltmeter, as shown in Fig. 1.5. If you bring a bar magnet close to the core, you will see a deflection in the voltmeter. If you stop moving the magnet, there will be no current through the voltmeter. If you move the magnet away from the conductor, the deflection of the voltmeter will be in the opposite direction. The same results can be obtained if the core is moving and the magnet is stationary. Faraday carried out an experiment similar to the one shown in Fig. 1.5 and from his experiments, he concluded that the time-varying magnetic field produces an electromotive force which is responsible for a current in a closed circuit. An electromotive force (e.m.f.) is simply the electric field intensity integrated over the length of the conductor or in other words, it is the voltage developed. In the absence of electric field intensity, electrons move randomly in all directions with a zero net current in any direction. Because of the electric field intensity (which is the force experienced by a unit electric charge) due to a time-varying magnetic field, electrons are forced to move in a particular direction leading to current.

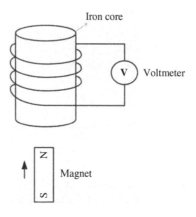

Figure 1.5   Generation of e.m.f. by moving a magnet.

Faraday's law is stated as

$$\text{e.m.f.} = -\frac{d\psi_m}{dt}, \tag{1.28}$$

where e.m.f. is the electromotive force about a closed path L (that includes a conductor and connections to a voltmeter), ψ_m is the magnetic flux crossing the surface S whose perimeter is the closed path L, and $d\psi_m/dt$ is the time rate of change of this flux. Since e.m.f. is an integrated electric field intensity, it can be expressed as

$$\text{e.m.f.} = \oint_L \mathbf{E} \cdot d\mathbf{l}. \tag{1.29}$$

The magnetic flux crossing the surface S is equal to the sum of the normal component of the magnetic flux density at the surface *times* the elemental surface area dS,

$$\psi_m = \int_S B_n dS = \int_S \mathbf{B} \cdot d\mathbf{S}, \tag{1.30}$$

where $d\mathbf{S}$ is a vector with magnitude dS and direction normal to the surface. Using Eqs. (1.29) and (1.30) in Eq. (1.28), we obtain

$$\oint_L \mathbf{E} \cdot d\mathbf{l} = -\frac{d}{dt}\int_S \mathbf{B} \cdot d\mathbf{S}$$
$$= -\int_S \frac{\partial \mathbf{B}}{\partial t} \cdot d\mathbf{S}. \tag{1.31}$$

In Eq. (1.31), we have assumed that the path is stationary and the magnetic flux density is changing with time; therefore the elemental surface area is not time dependent, allowing us to take the partial derivative under the integral sign. In Eq. (1.31), we have a line integral on the left-hand side and a surface integral on the right-hand side. In vector calculus, a line integral could be replaced by a surface integral using Stokes's theorem,

$$\oint_L \mathbf{E} \cdot d\mathbf{l} = \int_S (\nabla \times \mathbf{E}) \cdot d\mathbf{S} \tag{1.32}$$

to obtain

$$\int_S \left[\nabla \times \mathbf{E} + \frac{\partial \mathbf{B}}{\partial t} \right] \cdot d\mathbf{S} = 0. \tag{1.33}$$

Eq. (1.33) is valid for any surface whose perimeter is a closed path. It holds true for any arbitrary surface only if the integrand vanishes, i.e.,

$$\nabla \times \mathbf{E} = -\frac{\partial \mathbf{B}}{\partial t}. \tag{1.34}$$

The above equation is Faraday's law in the differential form and is one of Maxwell's four equations.

1.4.1   Meaning of Curl

The curl of a vector $\mathbf{A}$ is defined as

$$\text{curl } \mathbf{A} = \nabla \times \mathbf{A} = F_x \mathbf{x} + F_y \mathbf{y} + F_z \mathbf{z} \tag{1.35}$$

where

$$F_x = \frac{\partial A_z}{\partial y} - \frac{\partial A_y}{\partial z}, \tag{1.36}$$

$$F_y = \frac{\partial A_x}{\partial z} - \frac{\partial A_z}{\partial x}, \tag{1.37}$$

$$F_z = \frac{\partial A_y}{\partial x} - \frac{\partial A_x}{\partial y}. \tag{1.38}$$

Consider a vector **A** with only an x-component. The z-component of the curl of **A** is

$$F_z = -\frac{\partial A_x}{\partial y}. \tag{1.39}$$

Skilling [2] suggests the use of a paddle wheel to measure the curl of a vector. As an example, consider the water flow in a river as shown in Fig. 1.6(a). Suppose the velocity of water (A_x) increases as we go from the bottom of the river to the surface. The length of the arrow in Fig. 1.6(a) represents the magnitude of the water velocity. If we place a paddle wheel with its axis perpendicular to the paper, it will turn clockwise since the upper paddle experiences more force than the lower paddle (Fig. 1.6(b)). In this case, we say that curl exists along the axis of the paddle wheel in the direction of an inward normal to the surface of the page (z-direction). A larger speed of the paddle means a larger value of the curl.

Suppose the velocity of water is the same at all depths, as shown in Fig. 1.7. In this case the paddle wheel will not turn, which means there is no curl in the direction of the axis of the paddle wheel. From Eq. (1.39), we find that the z-component of the curl is zero if the water velocity A_x does not change as a function of depth y.

Eq. (1.34) can be understood as follows. Suppose the x-component of the electric field intensity E_x is changing as a function of y in a conductor, as shown in Fig. 1.8. This implies that there is a curl perpendicular to the page. From Eq. (1.34), we see that this should be equal to the time derivative of the magnetic field intensity

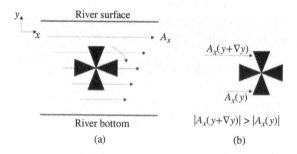

Figure 1.6   Clockwise movement of the paddle when the velocity of water increases from the bottom to the surface of a river.

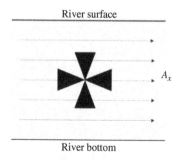

Figure 1.7   Velocity of water constant at all depths. The paddle wheel does not rotate in this case.

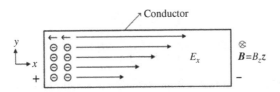

Figure 1.8   Induced electric field due to the time-varying magnetic field perpendicular to the page.

in the z-direction. In other words, the time-varying magnetic field in the z-direction induces an electric field intensity as shown in Fig. 1.8. The electrons in the conductor move in a direction opposite to E_x (Coulomb's law), leading to the current in the conductor if the circuit is closed.

1.4.2   Ampere's Law in Differential Form

From Eq. (1.21), we have

$$\oint_{L_1} \mathbf{H} \cdot d\mathbf{l} = \int_S \mathbf{J} \cdot d\mathbf{S}. \tag{1.40}$$

Using Stokes's theorem (Eq. (1.32)), Eq. (1.40) may be rewritten as

$$\int_S (\nabla \times \mathbf{H}) \cdot d\mathbf{S} = \int_S \mathbf{J} \cdot d\mathbf{S} \tag{1.41}$$

or

$$\nabla \times \mathbf{H} = \mathbf{J}. \tag{1.42}$$

The above equation is the differential form of Ampere's circuital law and it is one of Maxwell's four equations for the case of current and electric field intensity not changing with time. Eq. (1.40) holds true only under non-time-varying conditions. From Faraday's law (Eq. (1.34)), we see that if the magnetic field changes with time, it produces an electric field. Owing to symmetry, we might expect that the time-changing electric field produces a magnetic field. However, comparing Eqs. (1.34) and (1.42), we find that the term corresponding to a time-varying electric field is missing in Eq. (1.42). Maxwell proposed adding a term to the right-hand side of Eq. (1.42) so that a time-changing electric field produces a magnetic field. With this modification, Ampere's circuital law becomes

$$\nabla \times \mathbf{H} = \mathbf{J} + \frac{\partial \mathbf{D}}{\partial t}. \tag{1.43}$$

In the absence of the second term on the right-hand side of Eq. (1.43), it can be shown that the law of conservation of charges is violated (see Exercise 1.4). The second term is known as the displacement current density.

1.5   Maxwell's Equations

Combining Eqs. (1.12), (1.27), (1.34) and (1.43), we obtain

$$\text{div } \mathbf{D} = \rho, \tag{1.44}$$

$$\text{div } \mathbf{B} = 0, \tag{1.45}$$

where c is the velocity of light in free space. Before Maxwell's time, electrostatics, magnetostatics, and optics were unrelated. Maxwell unified these three fields and showed that the light wave is actually an electromagnetic wave with velocity given by Eq. (1.63).

1.5.4 Propagation in a Dielectric Medium

Similar to Eq. (1.63), the velocity of light in a medium can be written as

$$v = \frac{1}{\sqrt{\mu\epsilon}}, \tag{1.64}$$

where $\mu = \mu_0\mu_r$ and $\epsilon = \epsilon_0\epsilon_r$. Therefore,

$$v = \frac{1}{\sqrt{\mu_0\epsilon_0\mu_r\epsilon_r}}. \tag{1.65}$$

Using Eq. (1.64) in Eq. (1.65), we have

$$v = \frac{c}{\sqrt{\mu_r\epsilon_r}}. \tag{1.66}$$

For dielectrics, $\mu_r = 1$ and the velocity of light in a dielectric medium can be written as

$$v = \frac{c}{\sqrt{\epsilon_r}} = \frac{c}{n}, \tag{1.67}$$

where $n = \sqrt{\epsilon_r}$ is called the refractive index of the medium. The refractive index of a medium is greater than 1 and the velocity of light in a medium is less than that in free space.

1.6   1-Dimensional Wave Equation

Using Eq. (1.64) in Eq. (1.62), we obtain

$$\frac{\partial^2 E_x}{\partial z^2} = \frac{1}{v^2}\frac{\partial^2 E_x}{\partial t^2}. \tag{1.68}$$

Elimination of E_x from Eqs. (1.55) and (1.58) leads to the same equation for H_y,

$$\frac{\partial^2 H_y}{\partial z^2} = \frac{1}{v^2}\frac{\partial H_y}{\partial t^2}. \tag{1.69}$$

To solve Eq. (1.68), let us try a trial solution of the form

$$E_x(t, z) = f(t + \alpha z), \tag{1.70}$$

where f is an arbitrary function of $t + \alpha z$. Let

$$u = t + \alpha z, \tag{1.71}$$

$$\frac{\partial u}{\partial z} = \alpha, \quad \frac{\partial u}{\partial t} = 1, \tag{1.72}$$

$$\frac{\partial E_x}{\partial z} = \frac{\partial E_x}{\partial u}\frac{\partial u}{\partial z} = \frac{\partial E_x}{\partial u}\alpha, \tag{1.73}$$

$$\frac{\partial^2 E_x}{\partial z^2} = \frac{\partial^2 E_x}{\partial u^2}\alpha^2, \tag{1.74}$$

$$\frac{\partial^2 E_x}{\partial t^2} = \frac{\partial^2 E_x}{\partial u^2}. \tag{1.75}$$

Using Eqs. (1.74) and (1.75) in Eq. (1.68), we obtain

$$\alpha^2 \frac{\partial^2 E_x}{\partial u^2} = \frac{1}{v^2}\frac{\partial^2 E_x}{\partial u^2}. \tag{1.76}$$

Therefore,

$$\alpha = \pm\frac{1}{v}, \tag{1.77}$$

$$E_x = f\left(t + \frac{z}{v}\right) \quad \text{or} \quad E_x = f\left(t - \frac{z}{v}\right). \tag{1.78}$$

The negative sign implies a forward-propagating wave and the positive sign indicates a backward-propagating wave. Note that f is an arbitrary function and it is determined by the initial conditions as illustrated by the following examples.

Example 1.1

Turn on a flash light for 1 ns then turn it off. You will generate a pulse as shown in Fig. 1.9 at the flash light ($z = 0$) (see Fig. 1.10). The electric field intensity oscillates at light frequencies and the rectangular shape shown in Fig. 1.9 is actually the absolute field envelope. Let us ignore the fast oscillations in this example and write the field (which is actually the field envelope[1]) at $z = 0$ as

$$E_x(t, 0) = f(t) = A_0 \operatorname{rect}\left(\frac{t}{T_0}\right), \tag{1.79}$$

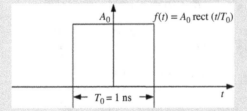

Figure 1.9 Electrical field $E_x(t, 0)$ at the flash light.

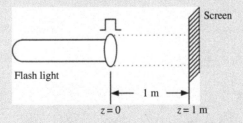

Figure 1.10 The propagation of the light pulse generated at the flash light.

---

[1] It can be shown that the field envelope also satisfies the wave equation.

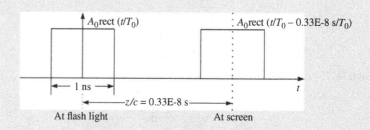

Figure 1.11   The electric field envelopes at the flash light and at the screen.

where

$$\text{rect}(x) = \begin{cases} 1, & \text{if} \quad |x| < 1/2 \\ 0, & \text{otherwise} \end{cases} \tag{1.80}$$

and $T_0 = 1$ ms. The speed of light in free space $v = c \simeq 3 \times 10^8$ m/s. Therefore, it takes 0.33×10^{-8} s to get the light pulse on the screen. At $z = 1$ m (see Fig. 1.11),

$$E_x(t,z) = f\left(t - \frac{z}{v}\right) = A_0 \, \text{rect}\left(\frac{t - 0.33 \times 10^{-8}}{T_0}\right). \tag{1.81}$$

Example 1.2

A laser shown in Fig. 1.12 operates at 191 THz. Under ideal conditions and ignoring transverse distributions, the laser output may be written as

$$E_x(t,0) = f(t) = A_0 \cos(2\pi f_0 t), \tag{1.82}$$

where $f_0 = 191$ THz. The laser output arrives at the screen after 0.33×10^{-8} s (see Fig. 1.12). The electric field intensity at the screen may be written as

$$\begin{aligned} E_x(t,z) &= f\left(t - \frac{z}{v}\right) \\ &= A \cos\left[2\pi f_0\left(t - \frac{z}{v}\right)\right] \\ &= A \cos\left[2\pi f_0(t - 0.33 \times 10^{-8})\right]. \end{aligned} \tag{1.83}$$

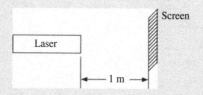

Figure 1.12   The propagation of laser output in free space.

Example 1.3

The laser output is reflected by a mirror and it propagates in a backward direction as shown in Fig. 1.13. In Eq. (1.78), the positive sign corresponds to a backward-propagating wave. Suppose that at the mirror, the electromagnetic wave undergoes a phase shift of ϕ.[2] The backward-propagating wave can be described by (see Eq. (1.78))

$$E_{x-} = A\cos\left[2\pi f_0(t + z/v) + \phi\right]. \tag{1.84}$$

The forward-propagating wave is described by (see Eq. (1.83))

$$E_{x+} = A\cos\left[2\pi f_0(t - z/v)\right]. \tag{1.85}$$

The total field is given by

$$E_x = E_{x+} + E_{x-}. \tag{1.86}$$

Figure 1.13   Reflection of the laser output by a mirror.

1.6.1   1-Dimensional Plane Wave

The output of the laser in Example 1.2 propagates as a *plane wave*, as given by Eq. (1.83). A plane wave can be written in any of the following forms:

$$
\begin{aligned}
E_x(t,z) &= E_{x0}\cos\left[2\pi f\left(t - \frac{z}{v}\right)\right] \\
&= E_{x0}\cos\left[2\pi ft - \frac{2\pi}{\lambda}z\right] \\
&= E_{x0}\cos\left(\omega t - kz\right),
\end{aligned}
\tag{1.87}
$$

where v is the velocity of light in the medium, f is the frequency, $\lambda = v/f$ is the wavelength, $\omega = 2\pi f$ is the angular frequency, $k = 2\pi/\lambda$ is the wavenumber, and k is also called the propagation constant. Frequency and wavelength are related by

$$v = f\lambda, \tag{1.88}$$

or equivalently

$$v = \frac{\omega}{k}. \tag{1.89}$$

Since H_y also satisfies the wave equation (Eq. (1.69)), it can be written as

$$H_y = H_{y0}\cos\left(\omega t - kz\right). \tag{1.90}$$

From Eq. (1.58), we have

$$\frac{\partial H_y}{\partial z} = -\epsilon\frac{\partial E_x}{\partial t}. \tag{1.91}$$

---

[2] If the mirror is a perfect conductor, $\phi = \pi$.

Using Eq. (1.87) in Eq. (1.91), we obtain

$$\frac{\partial H_y}{\partial z} = \epsilon \omega E_{x0} \sin (\omega t - kz). \tag{1.92}$$

Integrating Eq. (1.92) with respect to z,

$$H_y = \frac{\epsilon E_{x0} \omega}{k} \cos (\omega t - kz) + D, \tag{1.93}$$

where D is a constant of integration and could depend on t. Comparing Eqs. (1.90) and (1.93), we see that D is zero and using Eq. (1.89) we find

$$\frac{E_{x0}}{H_{y0}} = \frac{1}{\epsilon v} = \eta, \tag{1.94}$$

where η is the intrinsic impedance of the dielectric medium. For free space, $\eta = 376.47\,\text{Ohms}$. Note that E_x and H_y are independent of x and y. In other words, at time t, the phase $\omega t - kz$ is constant in a transverse plane described by $z = $ constant and therefore, they are called plane waves.

1.6.2   Complex Notation

It is often convenient to use complex notation for electric and magnetic fields in the following forms:

$$\tilde{E}_x = E_{x0} e^{i(\omega t - kz)} \;\; \text{or} \;\; \tilde{E}_x = E_{x0} e^{-i(\omega t - kz)} \tag{1.95}$$

and

$$\tilde{H}_y = H_{y0} e^{i(\omega t - kz)} \;\; \text{or} \;\; \tilde{H}_y = H_{y0} e^{-i(\omega t - kz)}. \tag{1.96}$$

This is known as an analytic representation. The actual electric and magnetic fields can be obtained by

$$E_x = \text{Re}\left[\tilde{E}_x\right] = E_{x0} \cos (\omega t - kz) \tag{1.97}$$

and

$$H_y = \text{Re}\left[\tilde{H}_y\right] = H_{y0} \cos (\omega t - kz). \tag{1.98}$$

In reality, the electric and magnetic fields are not complex, but we represent them in the complex forms of Eqs. (1.95) and (1.96) with the understanding that the real parts of the complex fields correspond to the actual electric and magnetic fields. This representation leads to mathematical simplifications. For example, differentiation of a complex exponential function is the complex exponential function multiplied by some constant. In the analytic representation, superposition of two eletromagnetic fields corresponds to addition of two complex fields. However, care should be exercised when we take the product of two electromagnetic fields as encountered in nonlinear optics. For example, consider the product of two electrical fields given by

$$E_{xn} = A_n \cos (\omega_n t - k_n z), \quad n = 1, 2 \tag{1.99}$$

$$E_{x1} E_{x2} = \frac{A_1 A_2}{2} \cos [(\omega_1 + \omega_2)t - (k_1 + k_2)z]$$
$$+ \cos [(\omega_1 - \omega_2)t - (k_1 - k_2)z]. \tag{1.100}$$

The product of the electromagnetic fields in the complex forms is

$$\tilde{E}_{x1} \tilde{E}_{x2} = A_1 A_2 \exp [i(\omega_1 + \omega_2)t - i(k_1 + k_2)z]. \tag{1.101}$$

If we take the real part of Eq. (1.101), we find

$$\text{Re}\left[\tilde{E}_{x1}\tilde{E}_{x1}\right] = A_1 A_2 \cos\left[(\omega_1 + \omega_2)t - (k_1 + k_2)z\right]$$

$$\neq E_{x1}E_{x2}. \tag{1.102}$$

In this case, we should use the real form of electromagnetic fields. In the rest of this book we sometimes omit ~ and use $E_x(H_y)$ to represent a complex electric (magnetic) field with the understanding that the real part is the actual field.

1.7 Power Flow and Poynting Vector

Consider an electromagnetic wave propagating in a region V with the cross-sectional area A as shown in Fig. 1.14. The propagation of a plane electromagnetic wave in the source-free region is governed by Eqs. (1.58) and (1.55),

$$\epsilon\frac{\partial E_x}{\partial t} = -\frac{\partial H_y}{\partial z} \tag{1.103}$$

$$\mu\frac{\partial H_y}{\partial t} = -\frac{\partial E_x}{\partial z}. \tag{1.104}$$

Multiplying Eq. (1.103) by E_x and noting that

$$\frac{\partial E_x^2}{\partial t} = 2E_x\frac{\partial E_x}{\partial t}, \tag{1.105}$$

we obtain

$$\frac{\epsilon}{2}\frac{\partial E_x^2}{\partial t} = -E_x\frac{\partial H_y}{\partial z}. \tag{1.106}$$

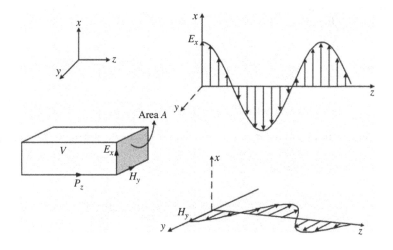

Figure 1.14   Electromagnetic wave propagation in a volume V with cross-sectional area A.

Similarly, multiplying Eq. (1.104) by H_y, we have

$$\frac{\mu}{2}\frac{\partial H_y^2}{\partial t} = -H_y\frac{\partial E_x}{\partial z}. \tag{1.107}$$

Adding Eqs. (1.107) and (1.106) and integrating over the volume V, we obtain

$$\frac{\partial}{\partial t}\int_V\left[\frac{\epsilon E_x^2}{2} + \frac{\mu H_y^2}{2}\right]dV = -A\int_0^L\left[E_x\frac{\partial H_y}{\partial z} + H_y\frac{\partial E_x}{\partial z}\right]dz. \tag{1.108}$$

On the right-hand side of Eq. (1.108), integration over the transverse plane yields the area A since E_x and H_y are functions of z only. Eq. (1.108) can be rewritten as

$$\frac{\partial}{\partial t}\int_V\left[\frac{\epsilon E_x^2}{2} + \frac{\mu H_y^2}{2}\right]dV = -A\int_0^L\frac{\partial}{\partial z}\left[E_xH_y\right]dz = -AE_xH_y\bigg|_0^L. \tag{1.109}$$

The terms $\epsilon E_x^2/2$ and $\mu H_y^2/2$ represent the *energy densities* of the electric field and the magnetic field, respectively. The left-hand side of Eq. (1.109) can be interpreted as the power crossing the area A and therefore, E_xH_y is the power per unit area or the *power density* measured in watts per square meter (W/m^2). We define a Poynting vector $\mathcal{P}$ as

$$\mathcal{P} = \mathbf{E}\times\mathbf{H}. \tag{1.110}$$

The z-component of the Poynting vector is

$$P_z = E_xH_y. \tag{1.111}$$

The direction of the Poynting vector is normal to both $\mathbf{E}$ and $\mathbf{H}$, and is in fact the direction of power flow.

In Eq. (1.109), integrating the energy density over volume leads to energy $\mathcal{E}$ and, therefore, it can be rewritten as

$$\frac{1}{A}\frac{d\mathcal{E}}{dt} = P_z(0) - P_z(L). \tag{1.112}$$

The left-hand side of (1.112) represents the rate of change of energy per unit area and therefore, P_z has the dimension of power per unit area or power density. For light waves, the power density is also known as the *optical intensity*. Eq. (1.112) states that the difference in the power entering the cross-section A and the power leaving the cross-section A is equal to the rate of change of energy in the volume V. The plane-wave solutions for E_x and H_y are given by Eqs. (1.87) and (1.90),

$$E_x = E_{x0}\cos(\omega t - kz), \tag{1.113}$$

$$H_y = H_{y0}\cos(\omega t - kz), \tag{1.114}$$

$$P_z = \frac{E_{x0}^2}{\eta}\cos^2(\omega t - kz). \tag{1.115}$$

The average power density may be found by integrating it over one cycle and dividing by the period $T = 1/f$,

$$P_z^{av} = \frac{1}{T}\frac{E_{x0}^2}{\eta}\int_0^T\cos^2(\omega t - kz)dt, \tag{1.116}$$

$$= \frac{1}{T}\frac{E_{x0}^2}{\eta}\int_0^T\frac{1 + \cos[2(\omega t - kz)]}{2}dt \tag{1.117}$$

$$= \frac{E_{x0}^2}{2\eta}. \tag{1.118}$$

The integral of the cosine function over one period is zero and, therefore, the second term of Eq. (1.118) does not contribute after the integration. The average power density $\mathcal{P}_z^{av}$ is proportional to the square of the electric field amplitude. Using complex notation, Eq. (1.111) can be written as

$$\mathcal{P}_z = \mathrm{Re}\left[\tilde{E}_x\right]\mathrm{Re}\left[\tilde{H}_y\right] \tag{1.119}$$

$$= \frac{1}{\eta}\mathrm{Re}\left[\tilde{E}_x\right]\mathrm{Re}\left[\tilde{E}_x\right] = \frac{1}{\eta}\left[\frac{\tilde{E}_x + \tilde{E}_x^*}{2}\right]\left[\frac{\tilde{E}_x + \tilde{E}_x^*}{2}\right]. \tag{1.120}$$

The right-hand side of Eq. (1.120) contains product terms such as $\tilde{E}_x^2$ and $\tilde{E}_x^{*2}$. The average of E_x^2 and E_x^{*2} over the period T is zero, since they are sinusoids with no d.c. component. Therefore, the average power density is given by

$$\mathcal{P}_z^{av} = \frac{1}{2\eta T}\int_0^T |\tilde{E}_x|^2 dt = \frac{|\tilde{E}_x|^2}{2\eta}, \tag{1.121}$$

since $|\tilde{E}_x|^2$ is a constant for the plane wave. Thus, we see that, in complex notation, the average power density is proportional to the absolute square of the field amplitude.

Example 1.4

Two monochromatic waves are superposed to obtain

$$\tilde{E}_x = A_1 \exp\left[i(\omega_1 t - k_1 z)\right] + A_2 \exp\left[i(\omega_2 t - k_2 z)\right]. \tag{1.122}$$

Find the average power density of the combined wave.

Solution:
From Eq. (1.121), we have

$$\mathcal{P}_z^{av} = \frac{1}{2\eta T}\int_0^T |\tilde{E}_x|^2 dt$$

$$= \frac{1}{2\eta T}\left\{ T|A_1|^2 + T|A_2|^2 + A_1 A_2^\star \int_0^T \exp\left[i(\omega_1 - \omega_2)t - i(k_1 - k_2)z\right]dt \right.$$

$$\left. + A_2 A_1^\star \int_0^T \exp\left[-i(\omega_1 - \omega_2) + i(k_1 - k_2)z\right] \right\} dt. \tag{1.123}$$

Since integrals of sinusoids over the period T are zero, the last two terms in Eq. (1.123) do not contribute, which leads to

$$\mathcal{P}_z^{av} = \frac{|A_1|^2 + |A_2|^2}{2\eta}. \tag{1.124}$$

Thus, the average power density is the sum of absolute squares of the amplitudes of monochromatic waves.

1.8   3-Dimensional Wave Equation

From Maxwell's equations, the following wave equation could be derived (see Exercise 1.6):

$$\frac{\partial^2 \psi}{\partial x^2} + \frac{\partial^2 \psi}{\partial y^2} + \frac{\partial^2 \psi}{\partial z^2} - \frac{1}{v^2}\frac{\partial^2 \psi}{\partial t^2} = 0, \tag{1.125}$$

where ψ is any one of the components $E_x, E_y, E_z, H_x, H_y, H_z$. As before, let us try a trial solution of the form

$$\psi = f(t - \alpha_x x - \alpha_y y - \alpha_z z).$$ (1.126)

Proceeding as in Section 1.6, we find that

$$\alpha_x^2 + \alpha_y^2 + \alpha_z^2 = \frac{1}{v^2}.$$ (1.127)

If we choose the function to be a cosine function, we obtain a 3-dimensional plane wave described by

$$\psi = \psi_0 \cos\left[\omega\left(t - \alpha_x x - \alpha_y y - \alpha_z z\right)\right]$$ (1.128)

$$= \psi_0 \cos\left(\omega t - k_x x - k_y y - k_z z\right),$$ (1.129)

where $k_r = \omega \alpha_r$, $r = x, y, z$. Define a vector $\mathbf{k} = k_x \mathbf{x} + k_y \mathbf{y} + k_z \mathbf{z}$. $\mathbf{k}$ is known as a *wave vector*. Eq. (1.127) becomes

$$\frac{\omega^2}{k^2} = v^2 \quad \text{or} \quad \frac{\omega}{k} = \pm v,$$ (1.130)

where k is the magnitude of the vector $\mathbf{k}$,

$$k = \sqrt{k_x^2 + k_y^2 + k_z^2}.$$ (1.131)

k is also known as the *wavenumber*. The angular frequency ω is determined by the light source, such as a laser or light-emitting diode (LED). In a linear medium, the frequency of the launched electromagnetic wave can not be changed. The frequency of the plane wave propagating in a medium of refractive index n is the same as that of the source, although the wavelength in the medium decreases by a factor n. For given angular frequency ω, the wavenumber in a medium of refractive index n can be determined by

$$k = \frac{\omega}{v} = \frac{\omega n}{c} = \frac{2\pi n}{\lambda_0},$$ (1.132)

where $\lambda_0 = c/f$ is the free-space wavelength. For free space, $n = 1$ and the wavenumber is

$$k_0 = \frac{2\pi}{\lambda_0}.$$ (1.133)

The wavelength λ_m in a medium of refractive index n can be defined by

$$k = \frac{2\pi}{\lambda_m}.$$ (1.134)

Comparing (1.132) and (1.134), it follows that

$$\lambda_m = \frac{\lambda_0}{n}.$$ (1.135)

Example 1.5

Consider a plane wave propagating in the x–z plane making an angle of $30°$ with the z-axis. This plane wave may be described by

$$\psi = \psi_0 \cos\left(\omega t - k_x x - k_z z\right).$$ (1.136)

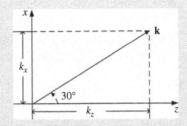

Figure 1.15 A plane wave propagates at angle 30° with the z-axis.

The wave vector $\mathbf{k} = k_x\mathbf{x} + k_z\mathbf{z}$. From Fig. 1.15, $k_x = k\cos 60^\circ = k/2$ and $k_z = k\cos 30^\circ = k\sqrt{3}/2$. Eq. (1.136) may be written as

$$\psi = \psi_0 \cos\left[\omega t - k\left(\frac{1}{2}x + \frac{\sqrt{3}}{2}z\right)\right]. \tag{1.137}$$

1.9   Reflection and Refraction

Reflection and refraction occur when light enters into a new medium with a different refractive index. Consider a ray incident on the mirror MM$'$, as shown in Fig. 1.16. According to the law of reflection, the angle of reflection ϕ_r is equal to the angle of incidence ϕ_i,

$$\phi_i = \phi_r.$$

The above result can be proved from Maxwell's equations with appropriate boundary conditions. Instead, let us use *Fermat's principle* to prove it. There are an infinite number of paths to go from point A to point B after striking the mirror. Fermat's principle can be stated loosely as follows: out of the infinite number of paths to go from point A to point B, light chooses the path that takes the shortest transit time. In Fig. 1.17, light could choose AC$'$B, AC$''$B, AC$'''$B, or any other path. But it chooses the path AC$'$B, for which $\phi_i = \phi_r$. Draw the line M$'$B$' = $ BM$'$ so that BC$' = $ C$'$B$'$, BC$'' = $ C$''$B$'$, and so on. If AC$'$B$'$ is a straight line, it would be the shortest of all the paths connecting A and B$'$. Since AC$'$B($=$ AC$'$B$'$), it would be the shortest path to go from A to B after striking the mirror and therefore, according to Fermat's principle, light chooses the path AC$'$B which takes the shortest time. To prove that $\phi_i = \phi_r$, consider the point C$'$. Adding up all the angles at C$'$, we find

$$\phi_i + \phi_r + 2(\pi/2 - \phi_r) = 2\pi \tag{1.138}$$

or

$$\phi_i = \phi_r. \tag{1.139}$$

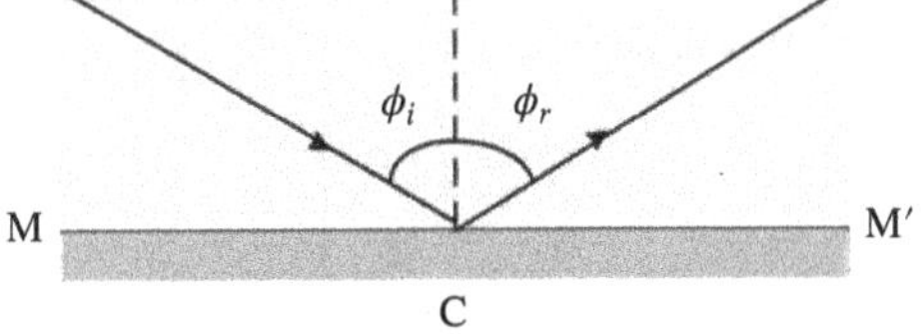

Figure 1.16 Reflection of a light wave incident on a mirror.

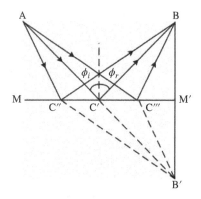

Figure 1.17 Illustration of Fermat's principle.

1.9.1 Refraction

In a medium with constant refractive index, light travels in a straight line. But as the light travels from a rarer medium to a denser medium, it bends toward the normal to the interface, as shown in Fig. 1.18. This phenomenon is called *refraction*, and it can be explained using Fermat's principle. Since the speeds of light in two media are different, the path which takes the shortest time to reach B from A may not be a straight line AB. Feynmann *et al.* [1] give the following analogy: suppose there is a little girl drowning in the sea at point B and screaming for help as illustrated in Fig. 1.19. You are at point A on the land. Obviously, the paths AC_2B and AC_3B take a longer time. You could choose the straight-line path AC_1B. But since running takes less time than swimming, it is advantageous to travel a slightly longer distance on land than sea. Therefore, the path AC_0B would take a shorter time than AC_1B. Similarly, in the case of light propagating from a rare medium to a dense medium (Fig. 1.20), light travels faster in the rare medium and therefore, the path AC_0B may take a shorter time than AC_1B. This explains why light bends toward the normal. To obtain a relation between the angle of incidence ϕ_1 and the angle of refraction ϕ_2, let us consider the time taken by light to go from A to B via several paths:

$$t_x = \frac{n_1 AC_x}{c} + \frac{n_2 C_x B}{c}, \quad x = 0, 1, 2, \ldots \tag{1.140}$$

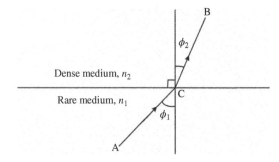

Figure 1.18 Refraction of a plane wave incident at the interface of two dielectrics.

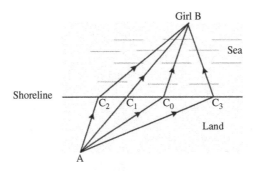

Figure 1.19 Different paths to connect A and B.

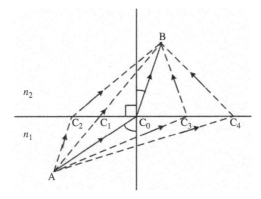

Figure 1.20 Illustration of Fermat's principle for the case of refraction.

From Fig. 1.21, we have

$$AD = x, \quad C_x D = y, \quad AC_x = \sqrt{x^2 + y^2}, \tag{1.141}$$

$$BE = AF - x, \quad BC_x = \sqrt{(AF - x)^2 + BG^2}. \tag{1.142}$$

Substituting this in Eq. (1.140), we find

$$t_x = \frac{n_1 \sqrt{x^2 + y^2}}{c} + \frac{n_2 \sqrt{(AF - x)^2 + BG^2}}{c}. \tag{1.143}$$

Note that AF, BG, and y are constants as x changes. Therefore, to find the path that takes the least time, we differentiate t_x with respect to x and set it to zero,

$$\frac{dt_x}{dx} = \frac{n_1 x}{\sqrt{x^2 + y^2}} - \frac{n_2(AF - x)}{\sqrt{(AF - x)^2 + BG^2}} = 0. \tag{1.144}$$

From Fig. 1.21, we have

$$\frac{x}{\sqrt{x^2 + y^2}} = \sin \phi_1, \quad \frac{AF - x}{\sqrt{(AF - x)^2 + BG^2}} = \sin \phi_2. \tag{1.145}$$

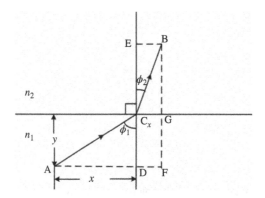

Figure 1.21   Refraction of a light wave.

Therefore, Eq. (1.144) becomes

$$n_1 \sin \phi_1 = n_2 \sin \phi_2. \tag{1.146}$$

This is called *Snell's law*. If $n_2 > n_1$, $\sin \phi_1 > \sin \phi_2$ and $\phi_1 > \phi_2$. This explains why light bends toward the normal in a denser medium, as shown in Fig. 1.18.

When $n_1 > n_2$, from Eq. (1.146), we have $\phi_2 > \phi_1$. As the angle of incidence ϕ_1 increases, the angle of refraction ϕ_2 increases too. For a particular angle, $\phi_1 = \phi_c$, ϕ_2 becomes $\pi/2$,

$$n_1 \sin \phi_c = n_2 \sin \pi/2 \tag{1.147}$$

or

$$\sin \phi_c = n_2/n_1. \tag{1.148}$$

The angle ϕ_c is called the *critical angle*. If the angle of incidence is increased beyond the critical angle, the incident optical ray is reflected completely as shown in Fig. 1.22. This is called *total internal reflection* (TIR), and it plays an important role in the propagation of light in optical fibers.

Note that the statement that light chooses the path that takes the least time is not strictly correct. In Fig. 1.16, the time to go from A to B directly (without passing through the mirror) is the shortest and we may wonder why light should go through the mirror. However, if we put the constraint that light has to pass through the mirror, the shortest path would be ACB and light indeed takes that path. In reality, light takes the direct path

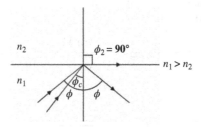

Figure 1.22   Total internal reflection when $\phi > \phi_c$.

AB as well as ACB. A more precise statement of Fermat's principle is that light chooses a path for which the transit time is an *extremum*. In fact, there could be several paths satisfying the condition of extremum and light chooses all those paths. By extremum, we mean there could be many neighboring paths and the change of time of flight with a small change in the path length is zero to first order.

Example 1.6

The critical angle for the glass−air interface is 0.7297 rad. Find the refractive index of glass.

Solution:
The refractive index of air is close to unity. From Eq. (1.148), we have

$$\sin \phi_c = n_2/n_1. \tag{1.149}$$

With $n_2 = 1$, the refractive index of glass, n_1 is

$$n_1 = 1/\sin \phi_c$$
$$= 1.5. \tag{1.150}$$

Example 1.7

The output of a laser operating at 190 THz is incident on a dielectric medium of refractive index 1.45. Calculate (a) the speed of light, (b) the wavelength in the medium, and (c) the wavenumber in the medium.

Solution:
(a) The speed of light in the medium is given by

$$v = \frac{c}{n} \tag{1.151}$$

where $c = 3 \times 10^8$ m/s, $n = 1.45$, so

$$v = \frac{3 \times 10^8 \text{ m/s}}{1.45} = 2.069 \times 10^8 \text{ m/s}. \tag{1.152}$$

(b) We have

$$\text{speed} = \text{frequency} \times \text{wavelength}$$
$$v = f \lambda_m \tag{1.153}$$

where $f = 190$ THz, $v = 2.069 \times 10^8$ m/s, so

$$\lambda_m = \frac{2.069 \times 10^8}{190 \times 10^{12}} \text{ m} = 1.0889 \text{ μm}. \tag{1.154}$$

(c) The wavenumber in the medium is

$$k = \frac{2\pi}{\lambda_m} = \frac{2\pi}{1.0889 \times 10^{-6}} = 5.77 \times 10^6 \text{ m}^{-1}. \tag{1.155}$$

Example 1.8

The output of the laser of Example 1.7 is incident on a dielectric slab with an angle of incidence $= \pi/3$, as shown in Fig. 1.23. (a) Calculate the magnitude of the wave vector of the refracted wave and (b) calculate the x-component and z-component of the wave vector. The other parameters are the same as in Example 1.7.

Solution:
Using Snell's law, we have

$$n_1 \sin \phi_1 = n_2 \sin \phi_2. \tag{1.156}$$

For air $n_1 \approx 1$, for the slab $n_2 = 1.45$, $\phi_1 = \pi/3$. So,

$$\phi_2 = \sin^{-1} \left\{ \frac{\sin (\pi/3)}{1.45} \right\} = 0.6401 \text{ rad.} \tag{1.157}$$

The electric field intensity in the dielectric medium can be written as

$$E_y = A \cos (\omega t - k_x x - k_z z). \tag{1.158}$$

(a) The magnitude of the wave vector is the same as the wavenumber, k. It is given by

$$|\mathbf{k}| = k = \frac{2\pi}{\lambda_m} = 5.77 \times 10^6 \text{ m}^{-1}. \tag{1.159}$$

(b) The z-component of the wave vector is

$$k_z = k \cos (\phi_2) = 5.77 \times 10^6 \times \cos (0.6401) \text{ m}^{-1} = 4.62 \times 10^6 \text{ m}^{-1}. \tag{1.160}$$

The x-component of the wave vector is

$$k_x = k \sin (\phi_2) = 5.77 \times 10^6 \times \sin (0.6401) \text{ m}^{-1} = 3.44 \times 10^6 \text{ m}^{-1}. \tag{1.161}$$

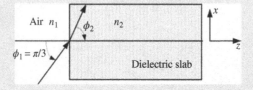

Figure 1.23   Reflection of light at air–dielectric interface.

1.10   Phase Velocity and Group Velocity

Consider the superposition of two monochromatic electromagnetic waves of frequencies $\omega_0 + \Delta\omega/2$ and $\omega_0 - \Delta\omega/2$ as shown in Fig. 1.24. Let $\Delta\omega \ll \omega_0$. The total electric field intensity can be written as

$$E = E_1 + E_2. \tag{1.162}$$

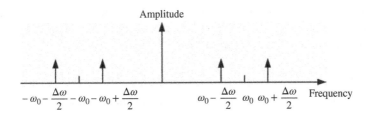

Figure 1.24 The spectrum when two monochromatic waves are superposed.

Let the electric field intensity of these waves be

$$E_1 = \cos\left[(\omega_0 - \Delta\omega/2)t - (k - \Delta k/2)z\right],$$ (1.163)

$$E_2 = \cos\left[(\omega_0 + \Delta\omega/2)t - (k + \Delta k/2)z\right].$$ (1.164)

Using the formula

$$\cos C + \cos D = 2\cos\left(\frac{C+D}{2}\right)\cos\left(\frac{C-D}{2}\right),$$

Eq. (1.162) can be written as

$$E = \underbrace{2\cos\left(\Delta\omega t - \Delta k z\right)}_{\text{field envelope}}\underbrace{\cos\left(\omega_0 t - k_0 z\right)}_{\text{carrier}}.$$ (1.165)

Eq. (1.165) represents the modulation of an optical carrier of frequency ω_0 by a sinusoid of frequency $\Delta\omega$. Fig. 1.25 shows the total electric field intensity at $z = 0$. The broken line shows the field envelope and the solid line shows rapid oscillations due to the optical carrier. We have seen before that

$$v_{ph} = \frac{\omega_0}{k_0}$$

is the velocity of the carrier. It is called the *phase velocity*. Similarly, from Eq. (1.165), the speed with which the envelope moves is given by

$$v_g = \frac{\Delta\omega}{\Delta k}$$ (1.166)

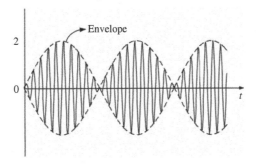

Figure 1.25 Superposition of two monochromatic electromagnetic waves. The broken lines and solid lines show the field envelope and optical carrier, respectively.

where v_g is called the *group velocity*. Even if the number of monochromatic waves traveling together is more than two, an equation similar to Eq. (1.165) can be derived. In general, the speed of the envelope (group velocity) could be different from that of the carrier. However, in free space,

$$v_g = v_{ph} = c.$$

The above result can be proved as follows. In free space, the velocity of light is independent of frequency,

$$\frac{\omega_1}{k_1} = \frac{\omega_2}{k_2} = c = v_{ph}. \tag{1.167}$$

Let

$$\omega_1 = \omega_0 - \frac{\Delta\omega}{2}, \qquad k_1 = k_0 - \frac{\Delta k}{2}, \tag{1.168}$$

$$\omega_2 = \omega_0 + \frac{\Delta\omega}{2}, \qquad k_2 = k_0 + \frac{\Delta k}{2}. \tag{1.169}$$

From Eqs. (1.168) and (1.169), we obtain

$$\frac{\omega_2 - \omega_1}{k_2 - k_1} = \frac{\Delta\omega}{\Delta k} = v_g. \tag{1.170}$$

From Eq. (1.167), we have

$$\omega_1 = ck_1,$$

$$\omega_2 = ck_2,$$

$$\omega_1 - \omega_2 = c(k_1 - k_2). \tag{1.171}$$

Using Eqs. (1.170) and (1.171), we obtain

$$\frac{\omega_1 - \omega_2}{k_1 - k_2} = c = v_g. \tag{1.172}$$

In a dielectric medium, the velocity of light v_{ph} could be different at different frequencies. In general,

$$\frac{\omega_1}{k_1} \neq \frac{\omega_1}{k_2}. \tag{1.173}$$

In other words, the phase velocity v_{ph} is a function of frequency,

$$v_{ph} = v_{ph}(\omega), \tag{1.174}$$

$$k = \frac{\omega}{v_{ph}(\omega)} = k(\omega). \tag{1.175}$$

In the case of two sinusoidal waves, the group speed is given by Eq. (1.166),

$$v_g = \frac{\Delta\omega}{\Delta k}. \tag{1.176}$$

In general, for an arbitrary cluster of waves, the group speed is defined as

$$v_g = \lim_{\Delta k \to 0} \frac{\Delta\omega}{\Delta k} = \frac{d\omega}{dk}. \tag{1.177}$$

Sometimes it is useful to define the *inverse group speed* β_1 as

$$\beta_1 = \frac{1}{v_g} = \frac{dk}{d\omega}. \tag{1.178}$$

β_1 could depend on frequency. If β_1 changes with frequency in a medium, it is called a *dispersive medium*. Optical fiber is an example of a dispersive medium, which will be discussed in detail in Chapter 2. If the refractive index changes with frequency, β_1 becomes frequency dependent. Since

$$k(\omega) = \frac{\omega n(\omega)}{c}, \tag{1.179}$$

from Eq. (1.178) it follows that

$$\beta_1(\omega) = \frac{n(\omega)}{c} + \frac{\omega}{c}\frac{dn(\omega)}{d\omega}. \tag{1.180}$$

Another example of a dispersive medium is a prism, in which the refractive index is different for different frequency components. Consider a white light incident on the prism, as shown in Fig. 1.26. Using Snell's law for the air–glass interface on the left, we find

$$\phi_2(\omega) = \sin^{-1}\left(\frac{\sin \phi_1}{n_2(\omega)}\right) \tag{1.181}$$

where $n_2(\omega)$ is the refractive index of the prism. Thus, different frequency components of a white light travel at different angles, as shown in Fig. 1.26. Because of the material dispersion of the prism, a white light is spread into a rainbow of colors.

Next, let us consider the co-propagation of electromagnetic waves of different angular frequencies in a range $[\omega_1, \omega_2]$ with the mean angular frequency ω_0 as shown in Fig. 1.27. The frequency components near the left edge travel at an inverse speed of $\beta_1(\omega_1)$. If the length of the medium is L, the frequency components corresponding to the left edge would arrive at L after a delay of

$$T_1 = \frac{L}{v_g(\omega_1)} = \beta_1(\omega_1)L.$$

Similarly, the frequency components corresponding to the right edge would arrive at L after a delay of

$$T_2 = \beta_1(\omega_2)L.$$

The delay between the left-edge and the right-edge frequency components is

$$\Delta T = |T_1 - T_2| = L|\beta_1(\omega_1) - \beta_1(\omega_2)|. \tag{1.182}$$

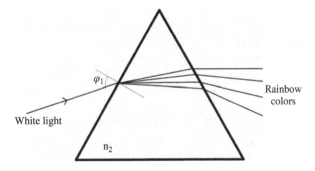

Figure 1.26   Decomposition of white light into its constituent colors.

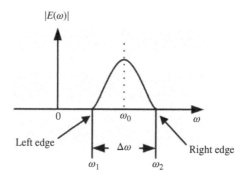

Figure 1.27   The spectrum of an electromagnetic wave.

Differentiating Eq. (1.178), we obtain

$$\frac{d\beta_1}{d\omega} = \frac{d^2k}{d\omega^2} \equiv \beta_2.$$

(1.183)

β_2 is called the *group velocity dispersion* parameter. When $\beta_2 > 0$, the medium is said to exhibit a *normal dispersion*. In the normal-dispersion regime, low-frequency (red-shifted) components travel faster than high-frequency (blue-shifted) components. If $\beta_2 < 0$, the opposite occurs and the medium is said to exhibit an *anomalous dispersion*. Any medium with $\beta_2 = 0$ is non-dispersive. Since

$$\frac{d\beta_1}{d\omega} = \lim_{\Delta\omega \to 0} \frac{\beta_1(\omega_1) - \beta_1(\omega_2)}{\omega_1 - \omega_2} = \beta_2$$

(1.184)

and

$$\beta_1(\omega_1) - \beta_1(\omega_2) \simeq \beta_2 \Delta\omega,$$

(1.185)

using Eq. (1.185) in Eq. (1.182), we obtain

$$\Delta T = L|\beta_2|\Delta\omega.$$

(1.186)

In free space, β_1 is independent of frequency, $\beta_2 = 0$, and, therefore, the delay between left- and right-edge components is zero. This means that the pulse duration at the input ($z = 0$) and output ($z = L$) would be the same. However, in a dispersive medium such as optical fiber, the frequency components near ω_1 could arrive earlier (or later) than those near ω_2, leading to pulse broadening.

Example 1.9

An optical signal of bandwidth $100\,\text{GHz}$ is transmitted over a dispersive medium with $\beta_2 = 10\,\text{ps}^2/\text{km}$. The delay between minimum and maximum frequency components is found to be $3.14\,\text{ps}$. Find the length of the medium.

Solution:

$$\Delta\omega = 2\pi 100\,\text{Grad/s}, \quad \Delta T = 3.14\,\text{ps}, \quad \beta_2 = 10\,\text{ps}^2/\text{km}.$$

(1.187)

Substituting Eq. (1.187) in Eq. (1.186), we find $L = 500\,\text{m}$.

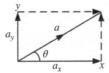

Figure 1.28   The *x*- and *y*-polarization components of a plane wave. The magnitude is $|a| = \sqrt{a_x^2 + a_y^2}$ and the angle is $\theta = \tan^{-1}(a_y/a_x)$.

1.11   Polarization of Light

So far we have assumed that the electric and magnetic fields of a plane wave are along the *x*- and *y*-directions, respectively. In general, an electric field can be in any direction in the *x*–*y* plane. This plane wave propagates in the *z*-direction. The electric field intensity can be written as

$$E = A_x \mathbf{x} + A_y \mathbf{y}, \tag{1.188}$$

$$A_x = a_x \exp\left[i(\omega t - kz) + i\phi_x\right], \tag{1.189}$$

$$A_y = a_y \exp\left[i(\omega t - kz) + i\phi_y\right], \tag{1.190}$$

where a_x and a_y are amplitudes of the *x*- and *y*-polarization components, respectively, and ϕ_x and ϕ_y are the corresponding phases. Using Eqs. (1.189) and (1.190), Eq. (1.188) can be written as

$$E = \mathbf{a} \exp\left[i(\omega t - kz) + i\phi_x\right], \tag{1.191}$$

$$\mathbf{a} = a_x \mathbf{x} + a_y \exp\left(i\Delta\phi\right)\mathbf{y}, \tag{1.192}$$

where $\Delta\phi = \phi_y - \phi_x$. Here, **a** is the complex field envelope vector. If one of the polarization components vanishes ($a_y = 0$, for example), the light is said to be *linearly polarized* in the direction of the other polarization component (the *x*-direction). If $\Delta\phi = 0$ or π, the light wave is also linearly polarized. This is because the magnitude of **a** in this case is $a_x^2 + a_y^2$ and the direction of **a** is determined by $\theta = \pm\tan^{-1}(a_y/a_x)$ with respect to the *x*-axis, as shown in Fig. 1.28. The light wave is linearly polarized at an angle θ with respect to the *x*-axis. A plane wave of angular frequency ω is characterized completely by the complex field envelope vector **a**. It can also be written in the form of a column matrix, known as the *Jones vector*:

$$\mathbf{a} = \begin{bmatrix} a_x \\ a_y \exp\left(i\Delta\phi\right) \end{bmatrix}. \tag{1.193}$$

The above form will be used for the description of polarization mode dispersion in optical fibers.

Exercises

1.1   Two identical charges are separated by 1 mm in vacuum. Each of them experience a repulsive force of 0.225 N. Calculate (a) the amount of charge and (b) the magnitude of electric field intensity at the location of a charge due to the other charge.

(Ans: (a) 5 nC; (b) 4.49×10^7 N/C.)

1.2   The magnetic field intensity at a distance of 1 mm from a long conductor carrying d.c. is 239 A/m. The cross-section of the conductor is 2 mm^2. Calculate (a) the current and (b) the current density.

(Ans: (a) 1.5 A; (b) 7.5 $\times 10^5$ A/m^2.)

1.3   The electric field intensity in a conductor due to a time-varying magnetic field is

$$\mathbf{E} = 6\cos{(0.1y)}\cos{(10^5 t)}\mathbf{x} \text{ V/m} \qquad\qquad (1.194)$$

Calculate the magnetic flux density. Assume that the magnetic flux density is zero at $t = 0$.

(Ans: $\mathbf{B} = -0.6\sin{(0.1y)}\sin{(10^6 t)}\mathbf{z}$ μT.)

1.4   The law of conservation of charges is given by

$$\nabla \cdot \mathbf{J} + \frac{\partial \rho}{\partial t} = 0.$$

Show that Ampere's law given by Eq. (1.42) violates the law of conservation of charges and Maxwell's equation given by Eq. (1.43) is in agreement with the law of conservation of charges.
    Hint: Take the divergence of Eq. (1.42) and use the vector identity

$$\nabla \cdot \nabla \times \mathbf{H} = 0.$$

1.5   The x-component of the electric field intensity of a laser operating at 690 nm is

$$E_x(t, 0) = 3\,\text{rect}\,(t/T_0)\cos{(2\pi f_0 t)} \text{ V/m}, \qquad\qquad (1.195)$$

where $T_0 = 5$ ns. The laser and screen are located at $z = 0$ and $z = 5$ m, respectively. Sketch the field intensities at the laser and the screen in the time and frequency domain.

1.6   Starting from Maxwell's equations (Eqs. (1.48)–(1.51)), prove that the electric field intensity satisfies the wave equation

$$\nabla^2 \mathbf{E} - \frac{1}{c^2}\frac{\partial^2 \mathbf{E}}{\partial t^2} = 0.$$

Hint: Take the curl on both sides of Eq. (1.50) and use the vector identity

$$\nabla \times \nabla \times \mathbf{E} = \nabla(\nabla \cdot \mathbf{E}) - \nabla^2 \mathbf{E}.$$

1.7   Determine the direction of propagation of the following wave:

$$E_x = E_{x0} = \cos\left[\omega\left(t - \frac{\sqrt{3}}{2c}z + \frac{x}{2c}\right)\right].$$

1.8   Show that

$$\Psi = \Psi_0 \exp\left[-\left(\omega t - k_x x - k_y y - k_z z\right)^2\right] \qquad\qquad (1.196)$$

is a solution of the wave equation (1.125) if $\omega^2 = v^2(k_x^2 + k_y^2 + k_z^2)$.
    Hint: Substitute Eq. (1.196) into the wave equation (1.125).

1.9 A light wave of wavelength (free space) 600 nm is incident on a dielectric medium of relative permittivity 2.25. Calculate (a) the speed of light in the medium, (b) the frequency in the medium, (c) the wavelength in the medium, (d) the wavenumber in free space, and (e) the wavenumber in the medium.

(Ans: (a) 2×10^8 m/s; (b) 500 THz; (c) 400 nm; (d) 1.047×10^7 m^{-1}; (e) 1.57×10^7 m^{-1}.)

1.10 State Fermat's principle and explain its applications.

1.11 A light ray propagating in a dielectric medium of index $n = 3.2$ is incident on the dielectric–air interface. (a) Calculate the critical angle; (b) if the angle of incidence is $\pi/4$, will it undergo total internal reflection?

(Ans: (a) 0.317 rad; (b) yes.)

1.12 Consider a plane wave making an angle of $\pi/6$ radians with the mirror, as shown in Fig. 1.29. It undergoes reflection at the mirror and refraction at the glass–air interface. Provide a mathematical expression for the plane wave in the air corresponding to segment CD. Ignore phase shifts and losses due to reflections.

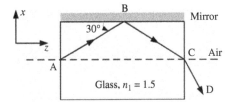

Figure 1.29 Plane-wave reflection at the glass–mirror interface.

1.13 Find the average power density of the superposition of N electromagnetic waves given by

$$E_x = \sum_{n=1}^{N} A_n \exp\left[in(\omega t - kz)\right]. \tag{1.197}$$

1.14 A plane electromagnetic wave of wavelength 400 nm is propagating in a dielectric medium of index $n = 1.5$. The electric field intensity is

$$\mathbf{E}^+ = 2\cos\left(2\pi f_0 t(t - z/v)\right)\mathbf{x} \text{ V/m}. \tag{1.198}$$

(a) Determine the Poynting vector. (b) This wave is reflected by a mirror. Assume that the phase shift due to reflection is π. Determine the Poynting vector for the reflected wave. Ignore losses due to propagation and mirror reflections.

1.15 An experiment is conducted to calculate the group velocity dispersion coefficient of a medium of length 500 m by sending two plane waves of wavelengths 1550 nm and 1550.1 nm. The delay between these frequency components is found to be 3.92 ps. Find $|\beta_2|$. The transit time for the higher-frequency component is found to be less than that for the lower-frequency component. Is the medium normally dispersive?

(Ans: $100 \, \text{ps}^2/\text{km}$. No.)

Further Reading

J.D. Jackson, *Classical Electrodynamics*, 3rd edn. John Wiley & Sons, New York, 1998.
M. Born and E. Wolf, *Principles of Optics*, 7th edn. Cambridge University Press, Cambridge, 2003.
A. Ghatak and K. Thyagarajan, *Optical Electronics*. Cambridge University Press, Cambridge, 1998.
W.H. Hayt, Jr., *Engineering Electromagnetics*, 5th edn. McGraw-Hill, Newyork, 1989.
B.E.A. Saleh and M.C. Teich, *Fundamentals of Photonics*. John Wiley & Sons, Hoboken, NJ, 2007.

References

[1] R.P. Feynmann, R.B. Leighton, and M. Sands, *The Feynman Lectures on Physics*, Vol. 1. Addison-Wesley, New york, 1963.
[2] H.H. Skilling, *Fundamentals of Electric Waves*, 2nd edn. John Wiley & Sons, New York, 1948.

2

Optical Fiber Transmission

2.1 Introduction

Until the mid-1970s, communication systems transmitted information over copper cables or free space. In 1966, Charles Kao and George Hockham working at Standard Telecommunications in the UK proposed that an optical fiber might be used as a means of communication provided the signal loss could be much less than 20 dB/km [1]. They also illustrated that the attenuation in fibers available at that time was caused by impurities which could be removed. At Corning Glass Works, Robert Maurer, Dould Keck, and Peter Schultz worked with fused silica, a material that can be made extremely pure. In 1970, they developed a single-mode fiber with attenuation below 20 dB/km [2]. In 1977, the first optical telecommunication system was installed about 1.5 miles under downtown Chicago and each optical fiber carried the equivalent of 672 voice channels. In 1979, single-mode fibers with a loss of only 0.2 dB/km at 1550 nm were fabricated [3]. The availability of low-loss fibers combined with the advent of semiconductor lasers led to a new era of optical fiber communication. Today, more than 80% of the world's long-distance traffic is carried over optical fiber cable and about 25 million kilometers of optical fiber has been installed worldwide.

This chapter deals with light propagation in optical fibers. Multi-mode and single-mode fibers are discussed using a ray-optics description in Section 2.3. A rigorous solution of the wave equation is derived in Section 2.4, and a wave-optics description of the single-mode and multi-mode fibers is presented. Pulse propagation in single-mode fibers is discussed in Section 2.5. The comparison between single-mode fibers and multi-mode fibers is made in Section 2.6. Section 2.7 focuses on the design of single-mode fibers.

2.2 Fiber Structure

An optical fiber consists of a central core clad with a material of slightly lower refractive index, as shown in Fig. 2.1. If the refractive index of the core is constant, such a fiber is called a *step-index* fiber. Most of the fibers are made from glass, i.e., silica. The refractive index of the core is increased by doping the silica with GeO_2. The cladding is pure silica. A polymer jacket is used to protect the fiber from moisture and abrasion. For short-distance (<1 km) and low-bit-rate ($\sim$ Mb/s) transmission systems, plastic fibers can be used. They are: (i) inexpensive, (ii) flexible, and (iii) easy to install and connect. However, they do not transmit light efficiently because of high absorption. For long-distance and high-bit-rate systems, glass fibers are typically used. Optical fibers have the following advantages over copper cable.

(i) *Bandwidth*: To transmit more bits of information in a given time period, the transmission medium should have a high bandwidth. Typically, the bandwidth is of the order of the carrier frequency. In the case of

Fiber Optic Communications: Fundamentals and Applications, First Edition. Shiva Kumar and M. Jamal Deen.
© 2014 John Wiley & Sons, Ltd. Published 2014 by John Wiley & Sons, Ltd.

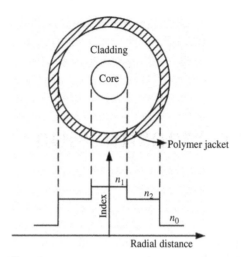

Figure 2.1   Refractive index profile and cross-section of a step-index fiber.

optical signals, the carrier frequency is 200 THz and the bandwidth of the fiber is several THz, whereas the bandwidth of the copper cable is typically several GHz or MHz.

(ii) *Attenuation*: The loss of a silica optical fiber is around 0.2 dB/km, which is much lower than that of copper cable. Because of the lower loss, optical signals can propagate over a longer distance without requiring repeaters.

(iii) *Electromagnetic interference (EMI)*: Optical fibers are not affected by electromagnetic interference. This is because optical fibers are purely dielectric waveguides with no metal parts. In the case of copper cables, electromagnetic noise fields set up conduction currents which interfere with the signal transmission.

2.3   Ray Propagation in Fibers

Consider a step-index fiber with core index n_1 greater than the cladding index n_2. Let ϕ_c be the critical angle. Consider a ray with an angle $\phi > \phi_c$, as shown in Fig. 2.2. This ray undergoes total internal reflection at B. The reflected ray BC undergoes total internal reflection again at C. This process continues till the output end and is called *frustrated total internal reflection*. With this mechanism, light is successfully transmitted from the input end to the output end of the fiber.

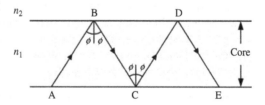

Figure 2.2   Signal propagation in a fiber by frustrated total internal reflection.

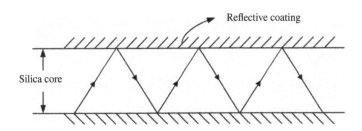

Figure 2.3   Signal propagation in a fiber by repeated normal reflections.

The *power reflection coefficient* may be defined as

$$R_p = \frac{\text{reflected power}}{\text{incident power}}. \tag{2.1}$$

In the case of total internal reflection, $R_p = 1$, which means all the power of the incident ray appears in the reflected ray. That is why it is called total internal reflection (TIR). In contrast, the normal reflection is always accompanied by refraction and $R_p < 1$. Suppose the core of a silica fiber is surrounded by a reflective coating instead of a dielectric cladding, as shown in Fig. 2.3. In this case, the light is guided by the "normal" reflections at the interface. By choosing a coating with high reflectivity, the power loss during each reflection can be minimized. However, significant power is lost after multiple reflections. In contrast, TIR is a more efficient way to transmit an optical signal over a long distance.

2.3.1   Numerical Aperture

Consider a ray which is incident on the fiber input making an angle i as shown in Fig. 2.4. Using Snell's law, we have

$$\sin i = n_1 \sin \theta = n_1 \cos \phi, \tag{2.2}$$

where we have assumed the refractive index of air to be unity. If this ray has to suffer total internal reflection at the core–cladding interface, the angle ϕ should be larger than the critical angle ϕ_c,

$$\phi > \phi_c,$$

$$\sin \phi > \sin \phi_c. \tag{2.3}$$

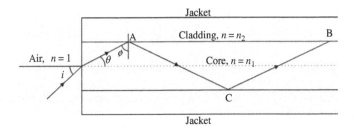

Figure 2.4   Numerical aperture of the fiber. If the incidence angle i is less than the acceptance angle, it undergoes total internal reflection.

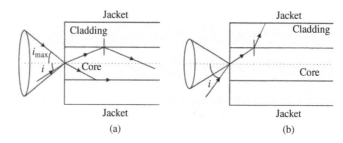

Figure 2.5   (a) If $i \le i_{max}$, light is guided. (b) If $i > i_{max}$, light escapes from the core.

Using Eq. (1.148), we obtain

$$n_1^2 \sin^2 \phi > n_2^2$$

or

$$n_1^2 \cos^2 \phi < n_1^2 - n_2^2. \tag{2.4}$$

Using Eqs. (2.2) and (2.4), it follows that, to have a total internal reflection, we should have the following condition:

$$\sin i < (n_1^2 - n_2^2)^{1/2}. \tag{2.5}$$

If $(n_1^2 - n_2^2)^{1/2} > 1$, total internal reflection occurs for any incidence angle i. But for most of the practical fiber designs, $(n_1^2 - n_2^2)^{1/2} \ll 1$. In this case, as the angle of incidence i increases, ϕ decreases and the light ray could escape the core–cladding interface without total internal reflection. From Eq. (2.5), the maximum value of $\sin i$ for a ray to be guided is given by

$$\sin i_{max} = (n_1^2 - n_2^2)^{1/2}. \tag{2.6}$$

Therefore, the *numerical aperture* (NA) of the fiber is defined as

$$\text{NA} = \sin i_{max} = (n_1^2 - n_2^2)^{1/2}, \tag{2.7}$$

and i_{max} is called the *acceptance angle*. Let us define the relative index difference as

$$\Delta = \frac{n_1 - n_2}{n_1}. \tag{2.8}$$

If the difference between n_1 and n_2 is small, $n_1 + n_2 \approx 2n_1$ and Eq. (2.7) can be approximated as

$$\text{NA} \approx n_1 (2\Delta)^{1/2}. \tag{2.9}$$

Let us construct a cone with the semi-angle being equal to i_{max}, as shown in Fig. 2.5(a). If the incident ray is within the cone ($i < i_{max}$), it will be guided through the fiber. Otherwise, it will escape to the cladding and then to the jacket, as shown in Fig. 2.5(b). From a practical standpoint, it is desirable to have most of the source power launched to the fiber, which requires large NA.

Example 2.1

The core and cladding refractive indices of a multi-mode fiber are 1.47 and 1.45, respectively. Find (a) the numerical aperture, (b) the acceptance angle, and (c) the relative index difference Δ.

Solution:
(a) $n_1 = 1.47$, $n_2 = 1.45$. From Eq. (2.7), we find

$$\text{NA} = (n_1^2 - n_2^2)^{1/2} = 0.2417. \tag{2.10}$$

(b) From Eq. (2.6), the acceptance angle is

$$i_{\max} = \sin^{-1}(\text{NA}) = 0.2441 \, \text{rad}. \tag{2.11}$$

(c) From Eq. (2.8), the refractive index difference Δ is

$$\Delta = \frac{n_1 - n_2}{n_1} = 0.0136. \tag{2.12}$$

2.3.2 Multi-Mode and Single-Mode Fibers

If the index difference $(n_1 - n_2)$ is large or the core radius a is much larger than the wavelength of light, an optical fiber supports multiple guided modes. A guided mode can be imagined as a ray that undergoes total internal reflection. A mathematical description of guided modes is provided in Section 2.4. From ray-optics theory, it follows that total internal reflection occurs for any angle in the interval $[\phi_c, \ \pi/2]$. This implies an infinite number of guided modes. However, from the wave-optics theory, it follows that not all the angles in the interval $[\phi_c, \ \pi/2]$ are permitted. Light guidance occurs only at discrete angles $\{\phi_1, \phi_2, \cdots \}$ in the interval $[\phi_c, \ \pi/2]$, as shown in Fig. 2.6. Each discrete angle in Fig. 2.6 corresponds to a guided mode. Typically, a multi-mode fiber can support thousands of guided modes. As the index difference $(n_1 - n_2)$ becomes very large and/or the core diameter becomes much larger than the wavelength of light, the fiber supports a very large number of modes N which approaches infinity, and total internal reflection occurs for nearly any angle in the interval $[\phi_c, \ \pi/2]$. In this case, the ray-optics theory is valid. As the index difference $(n_1 - n_2)$ becomes smaller and/or the core diameter becomes comparable with the wavelength of light, the number of guided modes decreases. In fact, by the proper design, a fiber could support only one guided mode (in ray-optics language, one ray with a specific angle). Such a fiber is called a *single-mode fiber*, which is of significant importance for high-speed optical communication.

2.3.3 Dispersion in Multi-Mode Fibers

A light pulse launched into a fiber broadens as it propagates down the fiber because of the different times taken by different rays or modes to propagate through the fiber. This is known as *intermodal dispersion*. In Fig. 2.7, the path length of ray 1 is longer than that of ray 3 and, therefore, the fraction of the incident pulse carried by ray 3 arrives earlier than that by ray 1, leading to pulse broadening.

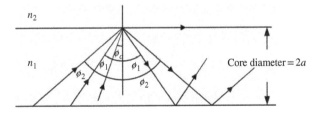

Figure 2.6 When the angle of incidence exceeds ϕ_c, total internal reflection occurs only for certain discrete angles.

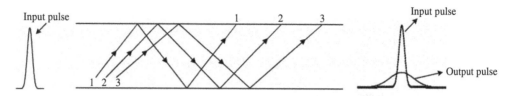

Figure 2.7 Pulse broadens because of the different times taken by different rays to pass through the fiber.

Suppose an impulse is launched to the fiber. Let us now estimate the pulse width at the output end. For a ray making an angle ϕ with the axis (see Fig. 2.8), the distance AB is traversed in time

$$t_{AB} = \frac{AC + CB}{v_1} = \frac{AB}{v_1 \sin \phi}, \tag{2.13}$$

where $v_1 = c/n_1$ is the speed of light in the core. ACB can be imagined as one unit cell. Let the fiber length L be composed of N such unit cells. The time taken by the ray to traverse a fiber length L is

$$t_L = \frac{N(AB)n_1}{c \sin \phi} = \frac{n_1 L}{c \sin \phi}. \tag{2.14}$$

For multi-mode fibers, we assume that all the rays making angles in the interval $[\phi_c, \ \pi/2]$ are present. This is a good approximation if the multi-mode fiber supports many modes. The ray which makes an angle $\phi = \pi/2$ propagates almost along the axis and takes the shortest time. From Eq. (2.14), the time taken by this ray is

$$t_{min} = \frac{n_1 L}{c \sin \pi/2} = \frac{n_1 L}{c}. \tag{2.15}$$

The ray which makes an angle $\phi = \phi_c$ takes the longest time. The time taken by this ray is

$$t_{max} = \frac{n_1 L}{c \sin \phi_c} = \frac{n_1^2 L}{c n_2}. \tag{2.16}$$

The time taken by a ray with angle ϕ in the interval $[\phi_c, \ \pi/2]$ is somewhere in between t_{min} and t_{max}. If an impulse is incident at the input end, it would excite all the rays in the interval $[\phi_c, \ \pi/2]$ and the rays occupy a time interval at the output end of duration ΔT given by

$$\Delta T = t_{max} - t_{min} = \frac{n_1^2 L}{c n_2} - \frac{n_1 L}{c} = \frac{n_1 L}{c} \left(\frac{n_1}{n_2} - 1 \right) = \frac{n_1^2 L \Delta}{c n_2}, \tag{2.17}$$

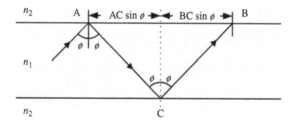

Figure 2.8 A ray undergoing multiple total internal reflections in a multi-mode fiber.

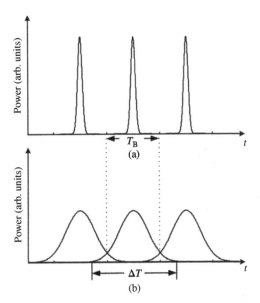

Figure 2.9 The pulse train at (a) fiber input and (b) fiber output. The individual pulses shown here are outputs in the absence of input pulses at the other bit slots.

where Δ is defined in Eq. (2.8). Fig. 2.9(b) shows the power profiles of individual pulses in the absence of other pulses. The pulse width at the output end is ΔT, as shown in Fig. 2.9(b). If the bit rate is B, the interval between bits is given by

$$T_B = \frac{1}{B}. \tag{2.18}$$

To avoid intersymbol interference, the pulse width $\Delta T \leq T_B$. Using Eqs. (2.17) and (2.18), we have

$$BL \leq \frac{cn_2}{n_1^2 \Delta}. \tag{2.19}$$

Eq. (2.19) provides the maximum bit rate–distance product possible for multi-moded fibers. From Eq. (2.19), we see that the product BL can be maximized by decreasing Δ, but from Eq. (2.9), we see that it leads to a reduction in NA, which is undesirable since it lowers the power launched to the fiber. So, there is a trade-off between power coupling efficiency and the maximum achievable bit rate–distance product.

From a practical standpoint, it is desirable to reduce the delay ΔT. From Eq. (2.17), we see that the delay ΔT increases linearly with fiber length L. The quantity $\Delta T / L$ is a measure of intermodal dispersion.

Example 2.2

Consider a multi-mode fiber with $n_1 = 1.46$, $\Delta = 0.01$, and fiber length $L = 1$ km. From Eq. (2.8)

$$n_2 = n_1(1 - \Delta) = 1.4454 \tag{2.20}$$

and

$$\Delta T = \frac{n_1^2 L \Delta}{cn_2} \approx 50\,\text{ns}. \tag{2.21}$$

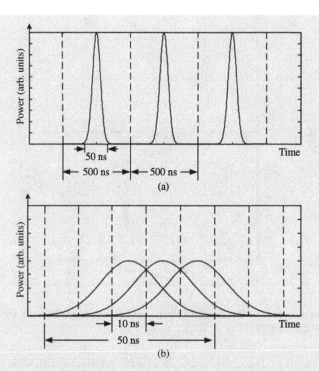

Figure 2.10 A pulse train at the fiber output: (a) bit rate $B = 2$ Mb/s; (b) $B = 100$ Mb/s. The individual pulses shown here are the outputs in the absence of input pulses at the other bit slots.

This implies that an impulse traversing through the fiber becomes a pulse of duration about 50 ns. If $B = 2$ Mb/s, $T_B = 500$ ns and the pulses at the output end are quite resolvable, as shown in Fig. 2.10(a). However, if the bit interval is 10 ns ($B = 100$ Mb/s), the pulses would be absolutely unresolvable at the output end, as shown in Fig. 2.10(b). From Eq. (2.19), the maximum bit rate–distance product is

$$(BL)_{\text{max}} = \frac{cn_2}{n_1^2 \Delta} = 20.3 \,\text{Mb/s km.} \tag{2.22}$$

This implies that the maximum achievable bit rate is 20.3 Mb/s for a system consisting of a 1-km fiber. Note that the power profiles in Fig. 2.10 are those of individual pulses in the absence of other pulses. To find the actual power profiles, the fields of individual pulses should be added and then the power of the combined field should be calculated.

2.3.4   Graded-Index Multi-Mode Fibers

In a step-index multi-mode fiber, the pulse width at the output is given by

$$\Delta T = \frac{n_1^2 L \Delta}{cn_2}. \tag{2.23}$$

To minimize this delay, graded-index multi-mode fibers are used. The refractive index profile, in this case, is given by

$$n(r) = \begin{cases} n_1 \left[1 - \Delta (r/a)^\alpha \right] & \text{if} \quad r < a \\ n_2 = n_1 (1 - \Delta) & \text{otherwise,} \end{cases}$$

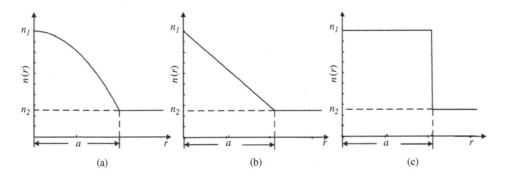

Figure 2.11   (a) Parabolic index profile. (b) Triangular index profile. (c) Step-index fiber.

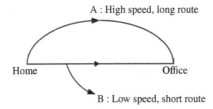

Figure 2.12   Two paths to connect home and office which could take roughly the same time.

where a is the core radius, n_2 is the cladding index, n_1 is the core index at $r = 0$, $\Delta = (n_1 - n_2)/n_1$, and α determines the index profile. When $\alpha = 2$, such a profile is called a *parabolic index profile*, as shown in Fig. 2.11(a). When $\alpha = \infty$, we get a step-index profile. A rigorous mathematical calculation shows that, if $\alpha = 2(1 - \Delta)(\approx 2)$, the pulse width, ΔT, is a minimum and is given by [4, 5]

$$\Delta T = \frac{n_1 \Delta^2 L}{8c}. \tag{2.24}$$

From Eqs. (2.23) and (2.24), we see that the pulse broadening is proportional to Δ in step-index fibers whereas it is proportional to Δ^2 in graded-index fibers (with $\alpha = 2(1 - \Delta)$). Since $\Delta \ll 1$, pulse broadening can be significantly reduced using graded-index fibers.

The reduction in pulse delay can be understood from the following analogy: suppose A takes a highway to go to the office from his home which is faster, but longer distance, as shown in Fig. 2.12. B takes a shorter route but there are many traffic lights, leading to delays. It is possible that A and B arrive at the same time to the office, which implies that there is no delay between the arrival times of A and B. In the case of graded-index fibers, the axial ray is confined mostly to the core center (because it undergoes total internal reflection closer to the center of the core due to the higher refractive index) and travels slowly because of the higher refractive index. The off-axis ray travels faster because it passes through the region of lower refractive index. But it has to travel a longer zig-zag path. So, the arrival time difference between these rays could be smaller. This explains why the delay given by Eq. (2.24) is smaller than that given by Eq. (2.23).

Example 2.3

Compare ΔT for a step-index fiber with that for a parabolic-index fiber. Length $= 1$ km, $n_1 = 1.47$, and $n_2 = 1.45$.

Solution:

$$\Delta = \frac{n_1 - n_2}{n_1} = 0.0136. \tag{2.25}$$

For a step-index fiber, from Eq. (2.23), we find

$$\Delta T = \frac{n_1^2 L \Delta}{c n_2} = 67.58 \text{ ns}. \tag{2.26}$$

For a parabolic-index fiber, from Eq. (2.24), we find

$$\Delta T = \frac{n_1^2 \Delta^2 L}{8c} = 0.1133 \text{ ns}. \tag{2.27}$$

Thus, we see that the intermodal dispersion can be significantly reduced by using a parabolic-index fiber.

2.4   Modes of a Step-Index Optical Fiber*

To understand the electromagnetic field propagation in optical fibers, we should solve Maxwell's equations with the condition that the tangential components of electric and magnetic fields should be continuous at the interface between core and cladding [6, 7]. When the refractive index difference between core and cladding is small, a weakly guiding or scalar wave approximation can be made [8–11] and in this approximation, the electromagnetic field is assumed to be nearly transverse as in the case of free-space propagation. Under this approximation, the one set of modes consists of E_x and H_y components (x-polarized) and the other set of modes consists of E_y and H_x components (y-polarized). These two sets of modes are independent and known as *linearly polarized (LP) modes*. The x- or y- component of the electric field intensity satisfies the scalar wave equation Eq. (1.125),

$$\nabla^2 \psi - \frac{1}{v^2(r)} \frac{\partial^2 \psi}{\partial t^2} = 0, \tag{2.28}$$

where $v(r)$ is the speed of light given by

$$v(r) = \frac{c}{n(r)} \tag{2.29}$$

with

$$n(r) = \begin{cases} n_1 & \text{for } r < a \\ n_2 & \text{for } r \geq a \end{cases}, \tag{2.30}$$

where $a =$ core radius. We assume that $n_1 > n_2$, as shown in Fig. 2.13. In cylindrical coordinates, the Laplacian operator ∇^2 can be written as

$$\nabla^2 \psi = \frac{\partial^2 \psi}{\partial r^2} + \frac{1}{r} \frac{\partial \psi}{\partial r} + \frac{1}{r^2} \frac{\partial^2 \psi}{\partial \phi^2} + \frac{\partial^2 \psi}{\partial z^2}. \tag{2.31}$$

Suppose this fiber is excited with a laser oscillating at angular frequency ω. In a linear dielectric medium, the frequency of the electromagnetic field should be the same as that of the source. Therefore, we look for a solution of Eq. (2.28) in the form

$$\Psi(r, \phi, z, t) = f(r, \phi, z)e^{-i\omega t}. \tag{2.32}$$

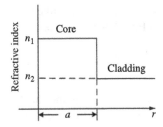

Figure 2.13   Refractive-index profile of a step-index fiber.

Substituting Eq. (2.32) in Eq. (2.28) and using Eqs. (2.29) and (2.31), we obtain

$$\frac{\partial^2 f}{\partial r^2} + \frac{1}{r}\frac{\partial f}{\partial r} + \frac{1}{r^2}\frac{\partial^2 f}{\partial \phi^2} + \frac{\partial^2 f}{\partial z^2} + k_0^2 n^2(r) f = 0, \tag{2.33}$$

where $k_0 = \omega/c = 2\pi/\lambda_0$ is the free-space wavenumber. The above equation is known as the *Helmholtz equation*. We solve Eq. (2.33) by separation of variables:

$$f(r, \phi, z) = R(r)\Phi(\phi)Z(z). \tag{2.34}$$

This technique may not work for all types of partial differential equations. Especially, if the partial differential equation is nonlinear, the method of separation of variables fails. Substituting Eq. (2.34) in Eq. (2.33), we obtain

$$\left(\frac{d^2 R}{dr^2} + \frac{1}{r}\frac{dR}{dr}\right)\Phi Z + \frac{1}{r^2}\frac{d^2\Phi}{d\phi^2}RZ + \frac{d^2 Z}{dz^2}R\Phi + k_0^2 n^2(r)R\Phi Z = 0. \tag{2.35}$$

Dividing Eq. (2.35) by $R\Phi Z$, we obtain

$$\frac{1}{R}\left(\frac{d^2 R}{dr^2} + \frac{1}{r}\frac{dR}{dr}\right) + k_0^2 n^2(r) + \frac{1}{\Phi r^2}\frac{d^2\Phi}{d\phi^2} = -\frac{d^2 Z}{dz^2}\frac{1}{Z}. \tag{2.36}$$

In Eq. (2.34), we assumed that f can be decomposed into three parts R, Φ, and Z which are functions of r, ϕ, and z, respectively. Since the right-hand side of Eq. (2.36) depends only on z while the left-hand side of Eq. (2.36) depends only on R and Φ, they can be equated only if each of them is a constant independent of r, ϕ, and z. Let this constant be β^2:

$$-\frac{1}{Z}\frac{d^2 Z}{dz^2} = \beta^2, \tag{2.37}$$

$$Z(z) = A_1 e^{i\beta z} + A_2 e^{-i\beta z}. \tag{2.38}$$

Using Eq. (2.34) and substituting Eq. (2.38) in Eq. (2.32), we obtain

$$\psi(r, \phi, z, t) = R(r)\Phi(\phi)\left[A_1 e^{-i(\omega t - \beta z)} + A_2 e^{-i(\omega t + \beta z)}\right]. \tag{2.39}$$

The first and second terms represent forward- and backward-propagating waves, respectively. In this section, let us consider only the forward-propagating modes by setting $A_2 = 0$. For example, the laser output is launched to the fiber from the left so that only forward-propagating modes are excited. If the fiber medium has no defect, there would be no reflection occurring within the fiber and the assumption of a forward-propagating mode is valid. From the left-hand side of Eq. (2.36), we obtain

$$\frac{r^2}{R}\left(\frac{d^2R}{dr^2} + \frac{1}{r}\frac{dR}{dr}\right) + r^2\left[k_0^2 n^2(r) - \beta^2\right] = -\frac{1}{\Phi}\frac{d^2\Phi}{d\phi^2}. \tag{2.40}$$

The left-hand side of Eq. (2.40) is a function of r only and the right-hand side is a function of ϕ only. As before, each of these terms should be a constant. Let this constant be m^2:

$$-\frac{1}{\Phi}\frac{d^2\Phi}{d\phi^2} = m^2, \tag{2.41}$$

$$\Phi(\phi) = B_1 e^{im\phi} + B_2 e^{-im\phi}. \tag{2.42}$$

The first and second terms represent the modes propagating in counter-clockwise and clockwise directions, respectively, when m is positive. Let us consider only one set of modes, say counter-clockwise modes, and set $B_2 = 0$. If the initial conditions at the input end of the fiber are such that both types of modes are excited, we can not ignore the second term in Eq. (2.42). In Section 2.4.6, we will study how to combine various modes to satisfy the given initial conditions. Using Eq. (2.41) in Eq. (2.40), we obtain

$$\frac{d^2R}{dr^2} + \frac{1}{r}\frac{dR}{dr} + \left[k_0^2 n^2(r) - \beta^2 - \frac{m^2}{r^2}\right]R = 0. \tag{2.43}$$

Using Eq. (2.30) for $n^2(r)$, we obtain

$$\frac{d^2R}{dr^2} + \frac{1}{r}\frac{dR}{dr} - \frac{m^2}{r^2}R + (k_0^2 n_1^2 - \beta^2)R = 0 \quad r < a, \tag{2.44}$$

$$\frac{d^2R}{dr^2} + \frac{1}{r}\frac{dR}{dr} - \frac{m^2}{r^2}R + (k_0^2 n_2^2 - \beta^2)R = 0 \quad r \geq a. \tag{2.45}$$

Fiber modes can be classified into two types: (i) $k_0^2 n_2^2 < \beta^2 < k_0^2 n_1^2$–these modes are called *guided modes* and (ii) $\beta^2 < k_0^2 n_2^2$–these modes are called *radiation modes*. It can be shown that there exists no mode when $\beta^2 > k_0^2 n_1^2$.

2.4.1   Guided Modes

Since $\beta^2 < k_0^2 n_1^2$, the last term in Eq. (2.44) is positive and the solution of Eq. (2.44) for this case is given by the Bessel functions

$$R(r) = C_1 J_m(\alpha_1 r) + C_2 Y_m(\alpha_1 r), \quad r \leq a, \tag{2.46}$$

where $\alpha_1 = \sqrt{k_0^2 n_1^2 - \beta^2}$, $J_m(\alpha_1 r)$ and $Y_m(\alpha_1 r)$ are the Bessel functions of first kind and second kind, respectively, and are plotted in Fig. 2.14. The solution $Y_m(\alpha_1 r)$ has to be rejected, since it becomes $-\infty$ as $r \to 0$. Therefore,

$$R(r) = C_1 J_m(\alpha_1 r), \quad r \leq a. \tag{2.47}$$

Since $\beta^2 > k_0^2 n_2^2$, the last term in Eq. (2.45) is negative. The solution of Eq. (2.45) is given by the modified Bessel function

$$R(r) = D_1 K_m(\alpha_2 r) + D_2 I_m(\alpha_2 r), \quad r \geq a, \tag{2.48}$$

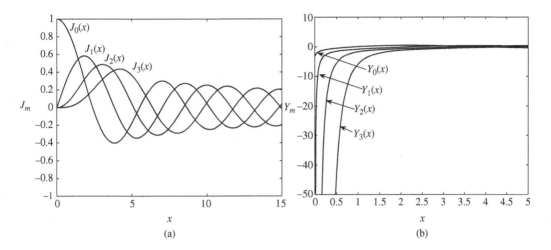

Figure 2.14 Bessel functions of (a) the first kind and (b) the second kind.

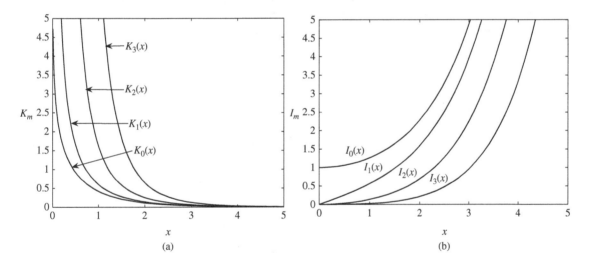

Figure 2.15 Modified Bessel function of (a) the first kind and (b) the second kind.

where $\alpha_2 = \sqrt{\beta^2 - k_0^2 n_2^2}$. $I_m(\alpha_2 r)$ and $K_m(\alpha_2 r)$ are modified Bessel functions of the first and second kind, respectively, and are plotted in Fig. 2.15. The solution $I_m(\alpha_2 r)$ has to be rejected since it becomes infinite as $r \to \infty$. Therefore,

$$R(r) = D_1 K_m(\alpha_2 r), \quad r \geq a. \tag{2.49}$$

Now we make use of the fact that R and dR/dr should be continuous at the core–cladding interface. If dR/dr is not continuous, d^2R/dr^2 will be a Dirac delta function centered at the interface and from Eq. (2.43), we find that d^2R/dr^2 could be a Dirac delta function only if the refractive index at the interface is infinity. Otherwise, Eq. (2.43) will not be satisfied at the interface. Since the refractive index is finite, it follows that dR/dr is continuous at the interface. Similarly, it can be proved that $R(r)$ is continuous at the interface. Continuity of

$R(r)$ and dR/dr at the core–cladding interface leads to the following equations:

$$C_1 J_m(\alpha_1 a) = D_1 K_m(\alpha_2 a),\qquad(2.50)$$

$$C_1 \alpha_1 J'_m(\alpha_1 a) = D_1 \alpha_2 K'_m(\alpha_2 a),\qquad(2.51)$$

where $'$ denotes differentiation with respect to the argument. Dividing Eq. (2.51) by Eq. (2.50), we obtain the following eigenvalue equation:

$$\frac{J'_m(\alpha_1 a)}{J_m(\alpha_1 a)} = \frac{\alpha_2}{\alpha_1}\frac{K'_m(\alpha_2 a)}{K_m(\alpha_2 a)}\qquad(2.52)$$

where

$$\alpha_1 = \sqrt{k_0^2 n_1^2 - \beta^2}\qquad(2.53)$$

and

$$\alpha_2 = \sqrt{\beta^2 - k_0^2 n_2^2}.\qquad(2.54)$$

Note that in Eq. (2.52), the only unknown parameter is the propagation constant β. It is not possible to solve Eq. (2.52) analytically. Eq. (2.52) may be solved numerically to obtain the possible values of β. It would be easier to solve Eq. (2.52) numerically if we avoid differentiations in Eq. (2.52). This can be done using the following identities:

$$\alpha_1 a J'_m(\alpha_1 a) = -m J_m(\alpha_1 a) + \alpha_1 a J_{m-1}(\alpha_1 a),\qquad(2.55)$$

$$\alpha_2 a K'_m(\alpha_2 a) = -m K_m(\alpha_2 a) - \alpha_2 a K_{m-1}(\alpha_2 a).\qquad(2.56)$$

Using Eqs. (2.55) and (2.56) in Eq. (2.52), we obtain

$$\frac{J_{m-1}(\alpha_1 a)}{J_m(\alpha_1 a)} = -\frac{\alpha_2}{\alpha_1}\frac{K_{m-1}(\alpha_2 a)}{K_m(\alpha_2 a)}.\qquad(2.57)$$

The propagation constants β obtained after solving Eq. (2.57) lie in the interval $[k_0 n_2,\ k_0 n_1]$. It is convenient to define the normalized propagation constant

$$b = \frac{\beta^2/k_0^2 - n_2^2}{n_1^2 - n_2^2}\qquad(2.58)$$

so that when $\beta = k_0 n_2$, $b = 0$ and when $\beta = k_0 n_1$, $b = 1$. For any guided mode of a step-index fiber, we have $0 < b < 1$. Eq. (2.57) can be solved for various design parameters such as wavelength λ and core radius a, and the numerically calculated propagation constant β can be plotted as a function of a specific design parameter. Instead, it is more convenient to define the normalized frequency

$$V = a\sqrt{\alpha_1^2 + \alpha_2^2} = k_0 a\sqrt{n_1^2 - n_2^2}$$
$$= \frac{2\pi \bar{f} a}{c}(n_1^2 - n_2^2)^{1/2},\qquad(2.59)$$

where $\bar{f}$ is the mean frequency of the light wave. Using Eq. (2.7), Eq. (2.59) may be rewritten as

$$V = \frac{2\pi a}{\lambda}\text{NA}.\qquad(2.60)$$

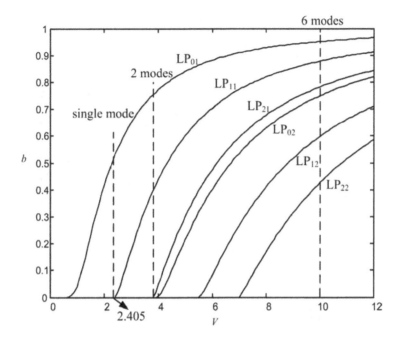

Figure 2.16   Plot of normalized propagation constant b versus normalized frequency V.

The solutions of Eq. (2.57) for a normalized propagation constant b as a function of normalized frequency V give us universal curves as shown in Fig. 2.16.

To solve Eq. (2.57), we first calculate V using Eq. (2.59). Let us first set $m = 0$. Since Eq. (2.57) is an implicit function of b, the left-hand side and right-hand side of Eq. (2.57) are plotted for various values of b in the interval $[0, 1]$. The point of intersection of the curves corresponding to the left-hand side and right-hand side of Eq. (2.57) gives the normalized propagation constant b of the guided mode supported by the fiber. As can be seen from Fig. 2.17, there can be many intersections which means there are several guided-mode solutions. This procedure is repeated for $m = 1, 2, \ldots, M$. For $m = M + 1$, we find that Eq. (2.57) admits no solution. When there is only one intersection ($M = 1$), such a fiber is called a single-mode fiber. The values of b corresponding to the intersections for a particular value of V are shown in Fig. 2.16 by the broken lines. This process is repeated for different values of V.

The advantage of the universal curve shown in Fig. 2.16 is that it can be used for a step-index fiber with arbitrary refractive indices and core radius. The dependence of the propagation constant β on the specific design parameter can be extracted from Fig. 2.16 using Eqs. (2.58) and (2.59).

For many applications, it is required to obtain the frequency dependence of propagation constant β in a single-mode fiber. This information can be obtained from Fig. 2.16. For the given fiber parameters and for the desired range of frequencies, V-parameters are calculated using Eq. (2.59). Using Fig. 2.16, the corresponding normalized propagation constants, b are calculated (corresponding to LP_{01}) and with the help of Eq. (2.58), propagation constants β for this range of frequencies can be calculated. For a specific value of m, Eq. (2.57) has a finite number of solutions and the nth solution is known as the LP_{mn} mode. LP stands for linearly polarized mode. Under the weakly guiding approximation, the electromagnetic field is nearly transverse and each LP mode corresponds to either an x-polarized or a y-polarized mode. For an ideal cylindrical fiber, the propagation constants of the x-polarized LP_{mn} and the y-polarized LP_{mn} are identical. When the refractive

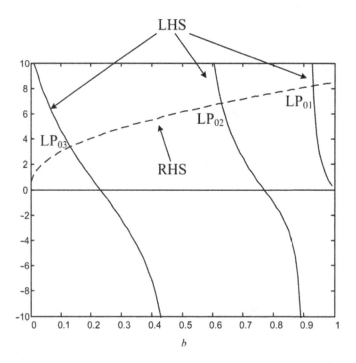

Figure 2.17 Left-hand side (LHS) and right-hand side (RHS) of Eq. (2.57) as a function of b. $V = 8$.

index difference between the core and the cladding is large, a weakly guiding approximation is not valid. Under this condition, E_z and/or H_z could be nonzero [5, 12].

Suppose one of the solutions of Eq. (2.57) is β_{mn}. Using Eqs. (2.47), (2.49), (2.42), and (2.39), the electric field distribution of this guided mode can be written as

$$\psi = \begin{cases} C_1 J_m\left(\alpha_1 r\right) e^{-i(\omega t - \beta_{mn} z - im\phi)} & \text{for } r \leq a \\ D_1 K_m(\alpha_2 r) e^{-i(\omega t - \beta_{mn} z - im\phi)} & \text{for } r \geq a. \end{cases} \tag{2.61}$$

From Eq. (2.50), we have

$$D_1 = \frac{C_1 J_m(\alpha_1 a)}{K_m(\alpha_2 a)}. \tag{2.62}$$

Using Eq. (2.62) in Eq. (2.61), we see that the only unknown parameter in Eq. (2.61) is C_1, which can be determined from the average power carried by this guided mode. In Section 1.7, the average power density carried by an electromagnetic wave in a homogeneous medium is found to be

$$P_z^{\text{av}} = \frac{|\psi|^2}{2\eta}. \tag{2.63}$$

Under the weakly guiding approximation, the field is nearly transverse and Eq. (2.63) may be used to calculate the power. The total power carried by a mode is [8, 9]

$$P_{\text{tot}} = |C_1|^2 \left\{ \int_0^{2\pi} \int_0^a \frac{J_m^2\left(\alpha_1 r\right)}{2\eta_1} r dr d\phi + \int_0^{2\pi} \int_a^\infty \frac{J_m^2(\alpha_1 a)}{2\eta_2 K_m^2(\alpha_2 a)} K_m^2(\alpha_2 r) r dr d\phi \right\}$$

$$= F_{mn} |C_1|^2, \tag{2.64}$$

where η_1 and η_2 are the characteristic impedances of core and cladding, respectively. F_{mn} can be determined after performing the integrations in Eq. (2.64) numerically. Eq. (2.61) can be normalized so that the power carried by this mode is unity,

$$P_{\text{tot}} = 1 \quad \text{or} \quad C_1 = \frac{1}{\sqrt{F_{mn}}} \tag{2.65}$$

and

$$\psi = R_{mn}(r)e^{-i(\omega t - \beta_{mn}z - im\phi)}, \tag{2.66}$$

where

$$R_{mn}(r) = \begin{cases} J_m\left(\alpha_1 r\right) / \sqrt{F_{mn}} & \text{for} \quad r \le a \\ [J_m(\alpha_1 a)/K_m(\alpha_2 a)]K_m(\alpha_2 r)/\sqrt{F_{mn}} & \text{for} \quad r > a \end{cases}. \tag{2.67}$$

Figs. 2.18–2.21 show the optical intensity as a function of radial distance for various LP_{mn} modes. The total number of guided modes M is given by an approximate expression [4, 13],

$$M \cong \frac{V^2}{2}. \tag{2.68}$$

2.4.2 Mode Cutoff

Fig. 2.16 shows the plot of normalized propagation constant b as a function of normalized frequency V obtained by solving Eq. (2.57). From Fig. 2.16, we see that when $V = 10$, there are six possible values of b which means there are six guided modes. From Eq. (2.59), it follows that V is large if the ratio of core radius to wavelength is large or the index difference is large. When $V = 3$, the fiber supports two modes, LP_{01} and LP_{11}. From Fig. 2.16, we see that when $V < 2.4048$, the fiber supports only one mode. This can be proved as follows. Any LP_{mn} mode ceases to exist when $b < 0$ because it then corresponds to a radiation mode, since $\beta < k_0 n_2$ (Eq. (2.58)). For LP_{11} mode, when $\beta = k_0 n_2$, using Eqs. (2.54) and (2.58), we obtain

$$\alpha_2 = 0 \quad \text{and} \quad b = 0. \tag{2.69}$$

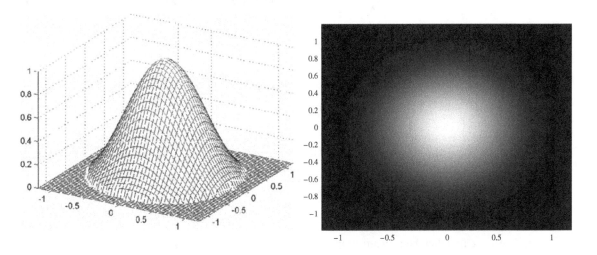

Figure 2.18 Optical field and power distributions of LP_{01} mode. $V = 5$ and $b = 0.84$.

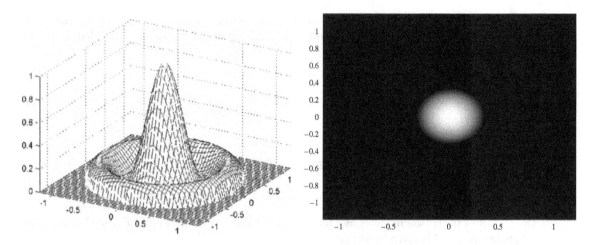

Figure 2.19   Optical field and power distributions of LP_{02} mode. $V = 5.3$ and $b = 0.278$.

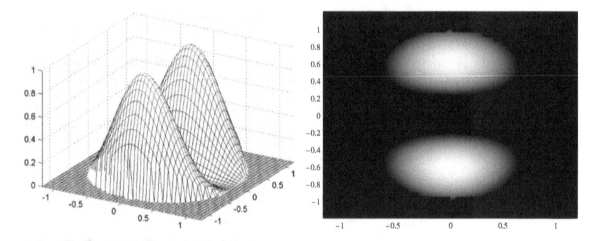

Figure 2.20   Optical field and power distributions of LP_{11} mode. $V = 4$ and $b = 0.44$.

In this case, using Eq. (2.53) in Eq. (2.59), we find $V = \alpha_1 a$ and with $m = 1$, Eq. (2.57) becomes

$$J_0(V) = 0. \tag{2.70}$$

The first zero of the zeroth-order Bessel function (Fig. 2.14) occurs at $V = 2.4048$. Therefore, LP_{11} ceases to exist if $V < 2.4048$. Thus, when $0 < V < 2.4048$, the fiber supports only LP_{01} mode. In other words, to ensure that the fiber is single-moded, V should be smaller than 2.4048.

2.4.3   Effective Index

From Eq. (2.39), we find that the forward-propagating mode can be written as

$$\psi(r, \phi, z, t) = A_1 R(r)\Phi(\phi) \exp\left[-i(\omega t - \beta z)\right]. \tag{2.71}$$

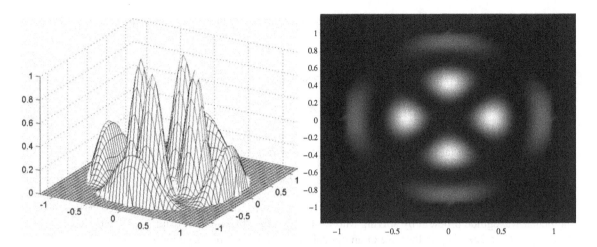

Figure 2.21 Optical field and power distributions of LP_{22} mode. $V = 10.3$ and $b = 0.456$.

In analogy with a 1-dimensional plan wave (Section 1.6.1), the phase speed of this mode can be written as

$$v = \frac{\omega}{\beta}. \tag{2.72}$$

The phase speed of an electromagnetic wave propagating in a uniform medium of refractive index n_{eff} is

$$v = \frac{c}{n_{\text{eff}}}. \tag{2.73}$$

Combining Eqs. (2.72) and (2.73), we find

$$\beta = \frac{\omega}{c} n_{\text{eff}} = k_0 n_{\text{eff}}. \tag{2.74}$$

Thus, n_{eff} can be interpreted as the effective index "seen" by the mode with propagating constant β. Guided modes occur if $k_0 n_2 < \beta < k_0 n_1$ or $n_2 < n_{\text{eff}} < n_1$.

2.4.4   2-Dimensional Planar Waveguide Analogy

Consider a 2-D planar waveguide as shown in Fig. 2.22. Let

$$n(x) = \begin{cases} n_1 & \text{for} \quad |x| < d/2 \\ n_2 & \text{otherwise} \end{cases}$$

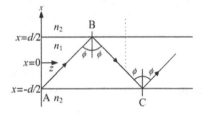

Figure 2.22   A 2-dimensional planar waveguide.

and $n_1 > n_2$. For simplicity, let us ignore the variations with respect to the y-coordinate. The ray AB corresponds to a plane wave,

$$E_{\text{incident}} = Ae^{-i\omega t+i(\alpha x+i\beta z)}. \tag{2.75}$$

There is a total internal reflection at the interface. The reflected ray BC corresponds to

$$E_{\text{ref}} = Ae^{-i\omega t-i\alpha x+i\beta z}. \tag{2.76}$$

Note that the z-component of the wave vector does not change after the reflection, but the x-component reverses its sign. The total field in the waveguide can be written as

$$E = E_{\text{incident}} + E_{\text{ref}} = 2A\cos(\alpha x)e^{-i(\omega t-\beta z)}. \tag{2.77}$$

Thus, incident and reflected plane waves set up a standing wave in the x-direction. The rigorous solution to the planar waveguide problem by solving Maxwell's equation shows that β can take discrete values β_n and $2\pi n_2/\lambda_0 < \beta_n < 2\pi n_1/\lambda_0$. In the case of an optical fiber, $\cos(\alpha x)$ is replaced by the Bessel function and the rest is nearly the same. In Section 2.4 we have found that for single-mode fibers there is only one mode, with field distribution given by

$$\psi = A_1 R_{01}(r)e^{-i\omega t+i\beta_{01}z} \tag{2.78}$$

where β_{01} is the propagation constant obtained by solving Eq. (2.57). Therefore, a guided mode of an optical fiber can be imagined as a standing wave in transverse directions and a propagating wave in the z-direction resulting from the superposition of the ray AB and the reflected ray BC. The propagation constant β_{01} and angle ϕ are related by

$$\beta_{01} = k_0 n_1 \sin\phi. \tag{2.79}$$

the discrete value of the propagation constant β_{01} implies that ϕ can not take arbitrary values in the interval $[\phi_c, \pi/2]$, but only a discrete value as determined from Eqs. (2.57) and (2.79).

2.4.5 Radiation Modes

For radiation modes, $\beta^2 < k_0^2 n_2^2$. In this case, the last terms on the left-hand sides of Eqs. (2.44) and (2.45) are both positive and their solutions are given by Bessel functions,

$$R(r) = \begin{cases} C_1 J_m(\alpha_1 r) & r \le a \\ E_1 J_m(\alpha_2 r) + E_2 Y_m(\alpha_2 r) & r > a \end{cases}.$$

Continuity of $R(r)$ and dR/dr at the core–cladding interface leads to two equations as before. But now we have four unknowns C_1, E_1, E_2, and β. C_1 can be determined from the power carried by the mode and this leaves us with three unknowns and two equations of continuity. Therefore, we can not write an eigenvalue equation for β as was done in Section 2.4.1 for guided modes. In fact, β can take arbitrary values in the range $0 < \beta < k_0 n_2$. A connection with ray optics can be made by defining

$$\beta = k_1 \sin\phi_i = k_0 n_1 \sin\phi_i, \tag{2.80}$$

where ϕ_i is the angle of incidence, as shown in Fig. 2.23. The ray undergoes refraction as it goes from core to cladding if the angle of incidence $\phi_i < \phi_c$. When $\phi_i = \phi_c$, we have

$$\sin\phi_c = \frac{n_2}{n_1} \tag{2.81}$$

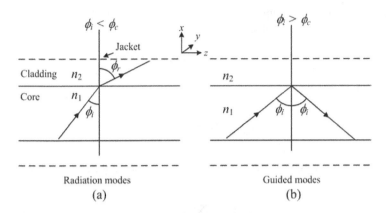

Figure 2.23 (a) Radiation modes corresponding to refraction with angle of incidence $\phi_i < \phi_c$. (b) Guided modes corresponding to total internal reflection with $\phi_i \geq \phi_c$.

and from Eq. (2.80), we obtain

$$\beta = \beta_c = k_0 n_2. \tag{2.82}$$

Therefore, the condition that $\beta < k_0 n_2$ for radiation modes corresponds to $\phi_i < \phi_c$ and the ray undergoes refraction, as shown in Fig. 2.23(a). Similarly, the condition that $k_0 n_2 < \beta < k_0 n_1$ for guided modes corresponds to $\phi_c < \phi < \pi/2$ and rays undergo total internal reflection for this range of angle of incidence, as shown in Fig. 2.23(b). The difference between guided modes and radiation modes is that the propagation constants of guided modes form a discrete set while those of radiation modes are continuous. Radiation modes do not propagate a longer distance since they are absorbed by the polymer jacket.

2.4.6 Excitation of Guided Modes

The total field in an optical fiber can be expressed as a superposition of fields due to guided modes and radiation modes. Radiation modes are attenuated strongly due to absorption by the polymer jacket. Therefore, the total field can be expressed as the superposition of fields due to guided modes given by Eq. (2.66):

$$\psi(r, \phi, z, t) = \sum_{m=-M}^{M} \sum_{n=1}^{N_m} A_{mn} R_{mn}(r) e^{-i(\omega t - \beta_{mn} z - im\phi)}, \tag{2.83}$$

where N_m is the number of solutions of the eigenvalue equation (2.57) for the given m, A_{mn} is the mode weight factor which is to be determined from launch conditions, and $R_{mn}(r)e^{im\phi}$ is the transverse field distribution given by Eq. (2.67). For convenience, Eq. (2.83) may be rewritten as

$$\psi(x, y, z, t) = \sum_{j=1}^{J} A_j \Phi_j(x, y) e^{-i(\omega t - \beta_j z)}, \tag{2.84}$$

where

$$A_j \equiv A_{mn},$$

$$\Phi_j(x, y) \equiv R_{mn}(r) e^{im\phi},$$

$$\beta_j \equiv \beta_{mn}, \tag{2.85}$$

and J is the total number of modes. Suppose the output of a laser is monochromatic and it is used as fiber input. The fiber input field can be written as

$$\psi(x, y, z = 0, t) = f(x, y)e^{-i\omega t}. \tag{2.86}$$

Using Eq. (2.86) in Eq. (2.84), we obtain

$$f(x, y) = \sum_{j=1}^{J} A_j \Phi_j(x, y). \tag{2.87}$$

To determine A_j, multiply Eq. (2.87) by $\Phi_k^*(x, y)$ and integrate over the cross-section to obtain

$$\int_{-\infty}^{+\infty} \int_{-\infty}^{+\infty} f(x, y)\Phi_k^*(x, y)dxdy = \sum_{j=1}^{J} A_j \int_{-\infty}^{+\infty} \int_{-\infty}^{+\infty} \Phi_j(x, y)\Phi_k^*(x, y)dxdy. \tag{2.88}$$

Using the orthogonality relation,

$$\int_{-\infty}^{+\infty} \int_{-\infty}^{+\infty} \Phi_j(x, y)\Phi_k^*(x, y)dxdy = \delta_{jk}, \tag{2.89}$$

where δ_{jk} is the Kronecker delta function defined as

$$\delta_{jk} = \begin{cases} 1, & \text{if } j = k \\ 0, & \text{otherwise} \end{cases}. \tag{2.90}$$

Eq. (2.88) reduces to

$$A_k = \int_{-\infty}^{+\infty} \int_{-\infty}^{+\infty} f(x, y)\Phi_k^*(x, y)dxdy. \tag{2.91}$$

Thus, for the given input field distribution $f(x, y)$, we can find the mode weight factors A_k using Eq. (2.91) and the total field distribution at any distance z is given by Eq. (2.84). Suppose the output of the laser has exactly the same transverse distribution as that of the fundamental mode of the step-index fiber, i.e., if $f(x, y) = \Phi_1(x, y) = R_{01}(r)$, from Eq. (2.91), we find $A_1 = 1$ and $A_m = 0$ for $m > 1$. Therefore, from Eq. (2.84), the field distribution at z is

$$\psi(x, y, z, t) = \Phi_1(x, y)e^{-i(\omega t - \beta_1 z)}. \tag{2.92}$$

Ideally speaking, the fundamental mode LP_{01} can be launched to the fiber, which propagates down the fiber without any change in shape and thereby intermodal dispersion can be avoided. However, in practice, the fiber imperfections and refractive index fluctuations due to temperature and stress can easily transfer power from the LP_{01} mode to higher-order modes. Therefore, the safest way to avoid intermodal dispersion is by ensuring that the fiber is single-moded at the operating wavelength.

Suppose a multi-mode fiber is excited with the Gaussian input

$$f(x, y) = \exp\left(-\frac{x^2 + y^2}{2R_0^2}\right).$$

The mode weight factors A_p can be calculated using Eq. (2.91) and are shown in Fig. 2.24. As can be seen, in this example, most of the power is carried by the LP_{01} mode ($p = 1$).

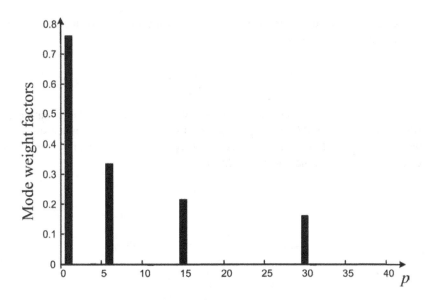

Figure 2.24 Mode weight factor versus mode index p. Core radius $= 31.25\,\mu m$, $\Delta = 0.01\,\mu m$, and $R_0 = 15\,\mu m$.

2.5 Pulse Propagation in Single-Mode Fibers

In the case of multi-mode fibers (MMFs), pulse broadening occurs because of the different times taken by the different rays (or modes) to propagate through the fiber. This broadening can be avoided by using single-mode fibers (SMFs). One may think that in the case of a SMF, there is only one path and hence pulses should not broaden, but this is not true. If a monochromatic light wave of infinite duration is launched to a SMF, it corresponds to a single ray path. However, such an optical signal does not convey any information. To transmit data over a fiber, the optical carrier has to be modulated. As a result, the optical signal propagating in the fiber consists of a range of frequency components. Since the propagation constant is frequency dependent (see Fig. 2.16), different frequency components undergo different amounts of delay (or phase shifts) and arrive at different times at the receiver, leading to pulse broadening even in a SMF. This is known as *intramodal dispersion*. The degree of pulse broadening in a SMF is much smaller than that in a MMF, but for high-rate transmission systems (>2.5 Gb/s) even the pulse spreading in a SMF could limit the maximum error-free transmission distance.

When the output of a CW (continuous wave) laser operating at frequency ω is incident on a single-mode fiber, the optical field distribution can be written as (Eq. (2.84) with $j = 1$)

$$\psi(x, y, z, t) = \Phi(x, y, \omega)A(\omega)e^{-i[\omega t - \beta(\omega)z]}. \tag{2.93}$$

The mode weight factor A and the transverse field distribution Φ could vary with frequency ω. So far we have assumed that the fiber is lossless. In the presence of fiber loss, the refractive index appearing in Eq. (2.29) should be complex and, as a result, the propagation constant β becomes complex,

$$\beta(\omega) = \beta_r(\omega) + i\alpha(\omega)/2, \tag{2.94}$$

where $\beta_r(\omega) = \text{Re}[\beta(\omega)]$ and $\alpha(\omega) = 2\text{Im}[\beta(\omega)]$. Using Eq. (2.94) in Eq. (2.93), we obtain

$$\psi(x, y, z, t) = \Phi(x, y, \omega)A(\omega)e^{-\alpha(\omega)z/2}e^{-i[\omega t - \beta_r(\omega)z]}. \tag{2.95}$$

If an optical fiber is excited with multiple frequency components, the total field distribution is the superposition of the fields due to each frequency component,

$$\psi(x, y, z, t) = \Phi(x, y)e^{-\alpha z/2} \sum_{n=1}^{N} A(\omega_n)e^{-i\omega_n t + i\beta_r(\omega_n)z}. \tag{2.96}$$

In Eq. (2.96), we have ignored the frequency dependency of the transverse field distribution Φ and also that of the loss coefficient α. This is valid if the frequency spread $\Delta\omega = |\omega_N - \omega_1|$ is much smaller than the mean frequency of the incident field. If the incident field envelope is a pulse, its frequency components are closely spaced and we can replace the summation in Eq. (2.96) with an integral

$$\psi(x, y, z, t) = \Phi(x, y)F(t, z) \tag{2.97}$$

where

$$F(t, z) = \frac{e^{-\alpha z/2}}{2\pi} \int_{-\infty}^{\infty} \tilde{A}(\omega)e^{-i[\omega t - \beta_r(\omega)z]} \, d\omega, \tag{2.98}$$

$$\tilde{A}(\omega) = 2\pi \lim_{\Delta\omega_n \to 0} \frac{A(\omega_n)}{\Delta\omega_n}. \tag{2.99}$$

From Eq. (2.98), we have

$$F(t, 0) = \frac{1}{2\pi} \int_{-\infty}^{+\infty} \tilde{A}(\omega)e^{-i\omega t} \, d\omega. \tag{2.100}$$

Eq. (2.100) represents the inverse Fourier transform of $\tilde{A}(\omega)$. Therefore, the Fourier transform $\tilde{A}(\omega)$ of the incident pulse $F(t, 0)$ is

$$\tilde{A}(\omega) = \int_{-\infty}^{+\infty} F(t, 0)e^{i\omega t} dt. \tag{2.101}$$

Thus, for the given incident pulse shape, we can calculate $\tilde{A}(\omega)$ using Eq. (2.101) and the optical field distribution at any z can be calculated using Eqs. (2.97) and (2.98). The impact of the fiber is characterized by $\beta(\omega)$. However, in practice, the dependence of the propagation constant on frequency for the commercially available fibers is not known. Besides, from the fiber-optic system design point of view, it is desirable to characterize the fiber using a few parameters. Therefore, we do the following approximation. The propagation constant at any frequency ω can be written in terms of the propagation constant and its derivative at some reference frequency (typically the carrier frequency) ω_0 using Taylor series,

$$\beta_r(\omega) = \beta_0 + \beta_1(\omega - \omega_0) + \frac{1}{2}\beta_2(\omega - \omega_0)^2 + \cdots, \tag{2.102}$$

where

$$\beta_0 = \beta_r(\omega_0), \tag{2.103}$$

$$\beta_1 = \frac{d\beta_r}{d\omega}\bigg|_{\omega=\omega_0} = \frac{1}{v_g}, \tag{2.104}$$

$$\beta_2 = \frac{d^2\beta_r}{d\omega^2}\bigg|_{\omega=\omega_0}. \tag{2.105}$$

β_1 is the inverse group velocity and β_2 is the second-order dispersion coefficient (see Section 1.10). If the signal bandwidth is much smaller than the carrier frequency ω_0, we can truncate the Taylor series after the second term on the right-hand side. To simplify Eq. (2.98), let us choose the variable $\Omega = \omega - \omega_0$. Using Eq. (2.102) in Eq. (2.98), we obtain

$$F(t,z) = \frac{1}{2\pi} \int_{-\infty}^{+\infty} \tilde{B}(\Omega) \exp\left[-\alpha z/2 - i(\omega_0 t - \beta_0 z) + i\beta_1 \Omega z + i\beta_2 \Omega^2 z/2\right] \exp\left(-i\Omega t\right) d\Omega$$

$$= \frac{\exp\left[-\alpha z/2 - i(\omega_0 t - \beta_0 z)\right]}{2\pi} \int_{-\infty}^{+\infty} \tilde{B}(\Omega) \exp\left(i\beta_1 \Omega z + i\beta_2 \Omega^2 z/2 - i\Omega t\right) d\Omega$$

$$= \frac{\exp\left[-i(\omega_0 t - \beta_0 z)\right]}{2\pi} \int_{-\infty}^{+\infty} \tilde{B}(\Omega) H_f(\Omega, z) \exp\left(-i\Omega t\right) d\Omega, \tag{2.106}$$

where

$$H_f(\Omega, z) = \exp\left(-\alpha z/2 + i\beta_1 \Omega z + i\beta_2 \Omega^2 z/2\right) \tag{2.107}$$

is called the *fiber transfer function* and

$$\tilde{B}(\Omega) \equiv \tilde{A}(\omega_0 + \Omega). \tag{2.108}$$

The linear phase shift $\beta_1 \Omega z$ corresponds to a delay in time domain. To see that, set $\beta_2 = 0$ in Eq. (2.107) and the fiber output at $z = L$,

$$F(t,L) = \frac{\exp\left[-\alpha z/2 - i(\omega_0 t - \beta_0 L)\right]}{2\pi} \int_{-\infty}^{+\infty} \tilde{B}(\Omega) \exp\left[-i\Omega(t - \beta_1 L)\right] d\Omega$$

$$= \exp\left[-\alpha z/2 - i(\omega_0 t - \beta_0 L)\right] B(t - \beta_1 L). \tag{2.109}$$

In a dispersion-free fiber ($\beta_2 = 0$), the pulse is simply delayed by $\beta_1 L$ at the fiber output without any change in pulse shape, as in the free-space propagation. Using Eqs. (2.98) and (2.106), the optical field distribution can be written as

$$\psi(x,y,z,t) = \underbrace{\Phi(x,y)}_{\text{transverse field}} \underbrace{\exp\left[-i(\omega_0 t - \beta_0 z)\right]}_{\text{carrier}} \underbrace{s(t,z)}_{\text{field envelope}}, \tag{2.110}$$

where

$$s(t,z) = \frac{1}{2\pi} \int_{-\infty}^{+\infty} \tilde{B}(\Omega) H_f(\Omega, z) \exp\left(-i\Omega t\right) d\Omega \tag{2.111}$$

and

$$\tilde{B}(\Omega) = \int_{-\infty}^{+\infty} s(t,0) \exp\left(i\Omega t\right) dt. \tag{2.112}$$

Eqs. (2.111) and (2.112) can be rewritten as

$$s(t,z) = \mathcal{F}^{-1}\left[\tilde{B}(\Omega) H_f(\Omega, z)\right], \tag{2.113}$$

$$\tilde{B}(\Omega) = \mathcal{F}[s(t,0)], \tag{2.114}$$

$$\tilde{B}(\Omega) \rightleftharpoons s(t,0), \tag{2.115}$$

where $\mathcal{F}$ and $\mathcal{F}^{-1}$ denote Fourier and inverse Fourier transforms, respectively, and $\rightleftharpoons$ indicates that they are Fourier transform pairs. In this section, we focus mainly on the field envelope $s(t,z)$. Let us assume that the transverse field distribution of the laser output is the same as that of the fiber, and therefore there is no change in the transverse field distribution along the fiber. Let the field envelope of the laser output be $s_i(t)$,

$$s_i(t) = s(t,0) \tag{2.116}$$

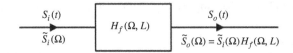

Figure 2.25 Optical fiber as a linear time-invariant system.

and

$$\tilde{B}(\Omega) = \mathcal{F}[s_i(t)] = \tilde{s}_i(\Omega). \tag{2.117}$$

The fiber can be imagined as a linear system with transfer function $H_f(\Omega, z)$ (see Fig. 2.25), The impact of the fiber nonlinearity is discussed in Chapter 10. Let the field envelope of the fiber output $s(t, L)$ be $s_o(t)$,

$$s(t, L) = s_o(t), \tag{2.118}$$

$$\mathcal{F}[s_o(t)] = \tilde{s}_o(\Omega) = H_f(\Omega, L)\tilde{s}_i(\Omega). \tag{2.119}$$

The optical signal propagation in a single-mode fiber can be summarized as follows.

Step 1: Input field envelope $s_i(t)$ is known. Take its Fourier transform to obtain $\tilde{s}_i(\Omega)$.
Step 2: Multiply $\tilde{s}_i(\Omega)$ by $H_f(\Omega, L)$ to get the output spectrum $\tilde{s}_o(\Omega)$.
Step 3: Take the inverse Fourier transform of $\tilde{s}_o(\Omega)$ to obtain the output field envelope $s_o(t)$.
Step 4: The total field distribution at the output is obtained by

$$\psi(x, y, L, t) = \Phi(x, y) \exp\left[-i(\omega_0 t - \beta_0 L)\right] s_o(t). \tag{2.120}$$

The advantage of this approach over that using Eq. (2.98) is that the fiber is characterized by three parameters β_0, β_1, and β_2 instead of $\beta(\omega)$. As the spectral width of the signal transmitted over the fiber increases, it may be necessary to include higher-order dispersion coefficients such as β_3 and β_4. β_1 and β_2 can be measured experimentally even if the fiber index profile is unknown. For example, by transmitting the output of a CW laser of angular frequency ω_0 over a fiber of length L, the time of flight ΔT_0 to traverse the distance L can be measured and $\beta_1(\omega_0)$ is $\Delta T_0/L$. Repeating the same experiment at $\omega_0 + \Delta\omega$, $\beta_1(\omega_0 + \Delta\omega)$ can be calculated. β_2 can be estimated as

$$\beta_2 \cong \frac{\beta_1(\omega_0 + \Delta\omega) - \beta_1(\omega_0)}{\Delta\omega}. \tag{2.121}$$

2.5.1 Power and the dBm Unit

The average power density of a plane wave is given by

$$P_z^{\text{av}} = \frac{|\tilde{E}_x|^2}{2\eta}, \tag{2.122}$$

where $\tilde{E}_x$ is the peak amplitude of the electric field intensity and η is the intrinsic impedance of the dielectric medium. A plane wave has infinite spatial extension in x- and y-directions and, therefore, the power carried by a plane wave is infinite. Under the LP-mode approximation, a fiber mode can be interpreted as a plane wave with finite spatial extension in the x- and y-directions and, therefore, power carried by a fiber mode can be obtained by integrating the absolute square of electric field intensity as done in Eq. (2.64),

$$P = \int_{-\infty}^{\infty} |\tilde{E}_x|^2 \frac{1}{2\eta} dx dy = \frac{|s(t, z)|^2}{2} \int_{-\infty}^{\infty} \frac{|\phi(x, y)|^2}{\eta} dx dy$$

$$= K|s(t, z)|^2. \tag{2.123}$$

Thus, we see that the power is proportional to the absolute square of the field envelope $s(t, z)$. Throughout this book, unless otherwise specified, we set $K = 1$ so that the absolute square of the electric field envelope is equal to the power.

Often it is convenient to use the logarithmic unit for power. The optical power in dBm units is expressed as

$$\text{power (dBm)} = 10 \log_{10} \left[\frac{\text{power (mW)}}{1 \text{ mW}} \right]. \tag{2.124}$$

In Eq. (2.124), 1 mW is chosen as a reference power and the letter "m" in dBm is a reminder of the 1 mW reference. For example, 1 mW of transmitter power corresponds to 0 dBm. If the transmitter power is increased to 2 mW, a factor of 2 in linear scale corresponds to 3 dB, and, therefore, in this case, the transmitter power is 3 dBm. Note that the optical power expressed in dBm units is not really a unit of power such as mW, but the ratio of the power in mW and 1 mW expressed in dB units. Typically, the loss and gain in a fiber-optic system are expressed in dB units. The advantage of using dBm units is that multiplications and divisions involving power and loss factors can be replaced by additions and subtractions as illustrated in Examples 2.8 and 2.9.

Inverting Eq. (2.124), we find

$$\text{power (mW)} = 10^{\text{power (dBm)}/10} \text{ mW}. \tag{2.125}$$

Example 2.4

The power transmitted in a fiber-optic system is 0.012 W. (a) Convert this into dBm units. (b) The received power is −5 dBm. Convert this into mW units.

Solution:
(a) From Eq. (2.124), the transmitted power in dBm units is

$$P_{\text{tr}}(\text{dBm}) = 10 \log_{10} \left[\frac{12 \text{ mW}}{1 \text{ mW}} \right] = 10.79 \text{ dBm}. \tag{2.126}$$

(b) The received power is

$$P_{\text{rec}}(\text{dBm}) = -5 \text{ dBm}. \tag{2.127}$$

Using Eq. (2.125), we find

$$P_{\text{rec}}(\text{mW}) = 10^{-5/10} \text{ mW} = 0.3162 \text{ mW}. \tag{2.128}$$

Example 2.5 Rectangular Pulse

The laser shown in Fig. 2.26 operates at 375 THz. It is turned on for 50 ps and then turned off. Sketch the field envelope at the screen if the medium is (a) free space, (b) fiber with $\beta_2 = 0$, (c) fiber with $\beta_2 = -21$ ps^2/km. Ignore fiber loss.

Solution:
Under steady-state conditions, the electric field intensity of a CW laser (ignoring the transverse field distribution) may be written as

$$F(t, 0) = f(t) = A \exp\left[-i2\pi f_0 t\right], \tag{2.129}$$

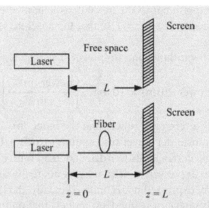

Figure 2.26 Pulse propagation in free space and optical fiber.

where $f_0 = 375\,\text{THz}$. When the laser is turned on for 50 ps and then turned off, a rectangular pulse is generated and, in this case, the electric field intensity is

$$F(t, 0) = f(t) = s_i(t) \exp\left[-i2\pi f_0 t\right], \tag{2.130}$$

where

$$s_i(t) = A \operatorname{rect}\left(\frac{t}{T_0}\right) \tag{2.131}$$

and $T_0 = 50\,\text{ps}$.

(a) In Section 1.6, the electric field intensity at the screen $(z = L)$ is found to be

$$F(t, L) = f(t - T_1) = s_o(t - T_1) \cos\left[2\pi f_0(t - T_1)\right], \tag{2.132}$$

where

$$s_o(t) = \operatorname{rect}\left(\frac{t - T_1}{T_0}\right), \tag{2.133}$$

$T_1 = L/c$, and c is the velocity of light in free space. The field is delayed by $T_1 = L/c$, which is the propagation delay as shown in Fig. 2.27.

(b) In the case of an optical fiber, let us first consider the case $\beta_2 = 0$.

Step 1:

$$s_i(t) = A \operatorname{rect}\left(\frac{t}{T_0}\right),$$

$$\tilde{s}_i(f) = A \int_{-T_0/2}^{T_0/2} \exp\left(i2\pi ft\right) dt = \frac{A \sin(\pi f T_0)}{\pi f}. \tag{2.134}$$

Step 2: The transfer function of a loss-free fiber in the absence of β_2 is

$$H_f(f, L) = \exp\left(i2\pi f \beta_1 L\right), \tag{2.135}$$

$$\tilde{s}_o(f) = \tilde{s}_i(f) H_f(f, L) = \frac{A \sin(\pi f T_0)}{\pi f} \exp\left(i2\pi f \beta_1 L\right). \tag{2.136}$$

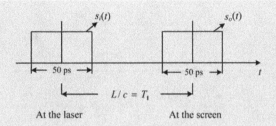

Figure 2.27   In free space, the pulse shape does not change.

Step 3:   The delay in time domain corresponds to a constant phase shift in frequency domain,

$$g(t - T_0) \rightleftharpoons \tilde{g}(f) \exp{(i2\pi f T_0)}. \tag{2.137}$$

Using Eqs. (2.134) and (2.136), the output field envelope may be written as

$$s_o(t) = s_i(t - \beta_1 L) = \text{rect} \left(\frac{t - \beta_1 L}{T_0} \right). \tag{2.138}$$

Fig. 2.28 shows the field envelope. As can be seen, there is no change in the pulse shape at $z = L$. It is simply delayed by $\beta_1 L$, similar to the case of free-space propagation.

(c) When $\beta_2 \neq 0$, Eq. (2.136) may be written as

$$\tilde{s}_o(f) = \frac{A \sin{(\pi f T_0)}}{\pi f} \exp{\left[i2\pi f \beta_1 L + i(2\pi f)^2 \beta_2 L/2 \right]} \tag{2.139}$$

$$s_o(t) = \mathcal{F}^{-1}[\tilde{s}_o(f)]. \tag{2.140}$$

It is not possible to do the inverse Fourier transform analytically. Fig. 2.29 shows the output field envelope $s_o(t)$ obtained using numerical techniques when $\beta_2 = -21 \, \text{ps}^2/\text{km}$ and $L = 80 \, \text{km}$. As can be seen, there is a significant pulse broadening after fiber propagation.

Step 4:   The total field distribution at the fiber output is

$$\psi(x, y, L, t) = \Phi(x, y) \exp{[-i(\omega_0 t - \beta_0 z)]} s_0(t). \tag{2.141}$$

Fig. 2.30 shows the total field distribution at the fiber input and output (transverse field distribution is not shown).

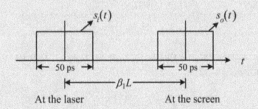

Figure 2.28   The field envelopes at the laser and at the screen. In optical fibers with $\beta_2 = 0$, the pulse shape does not change.

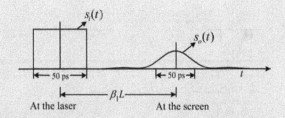

Figure 2.29   Optical field envelopes when $\beta_2 \neq 0$. $L = 80\,\mathrm{km}$, $\beta_2 = -21\,\mathrm{ps}^2/\mathrm{km}$.

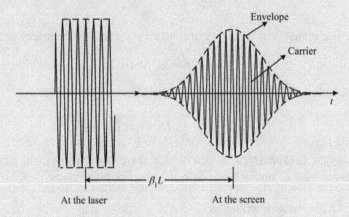

Figure 2.30   Total field distribution at the laser and at the screen when $\beta_2 \neq 0$.

Example 2.6 Gaussian Pulse

The input field envelope is

$$s_i(t) = A \exp\left(-t^2/2T_0^2\right), \tag{2.142}$$

where T_0 represents the half-width at $1/e$-intensity point and A is the peak amplitude. Find the output field envelope in a dispersive fiber. Ignore fiber loss and constant delay due to β_1.

Solution:
To relate T_0 to the full-width at half-maximum (FWHM), $T_{\mathrm{FWHM}}^{\mathrm{in}}$, let us first write an equation for power

$$P(t) = |s_i(t)|^2 = A^2 e^{-t^2/T_0^2}, \tag{2.143}$$

$$P_{\max} = P(0) = A^2. \tag{2.144}$$

Let t_h be the time at which the power is half of the peak power, as shown in Fig. 2.31. Since FWHM means the full-width at half-power point, we have

$$P(t_h) = A^2/2 = A^2 \exp\left(-t_h^2/T_0^2\right). \tag{2.145}$$

Taking logarithms on both sides, we obtain

$$t_h = T_0(\ln 2)^{1/2} \tag{2.146}$$

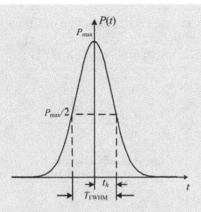

Figure 2.31   A Gaussian pulse.

and

$$T_{\text{FWHM}}^{\text{in}} = 2t_h = 2(\ln 2)^{1/2}T_0 \simeq 1.665T_0. \tag{2.147}$$

The transfer function of an optical fiber in the absence of fiber loss is given by Eq. (2.107) as

$$H_f(f,L) = \exp\left[i\beta_1(2\pi f)L + i\beta_2(2\pi f)^2 L/2\right]. \tag{2.148}$$

As mentioned before, the first term on the right-hand side introduces a constant delay and, hence, it can be ignored for the purpose of evaluating the output pulse shape. Using the following identity:

$$\exp\left(-\pi t^2\right) \rightleftharpoons \exp\left(-\pi f^2\right), \tag{2.149}$$

where $\rightleftharpoons$ indicates that they are Fourier transform pairs and using the scaling property

$$g(at) \rightleftharpoons \frac{1}{a}\tilde{g}(f/a), \quad \text{Re}(a) > 0, \tag{2.150}$$

the Fourier transform of $s_i(t)$ can be calculated. Taking

$$a = \frac{1}{\sqrt{2\pi}T_0}, \tag{2.151}$$

$$s_i(t) = A \exp\left[-\pi(at)^2\right] \rightleftharpoons \frac{A}{a}\exp\left[-\pi(f/a)^2\right] = \tilde{s}_i(f). \tag{2.152}$$

Therefore, we have

$$\tilde{s}_o(f) = \tilde{s}_i(f)H_f(f,L)$$

$$= \frac{A}{a}\exp\left[-\frac{\pi f^2}{a^2} + i\beta_2(2\pi f)^2 L/2\right]$$

$$= \frac{A}{a}\exp\left(-\pi f^2/b^2\right), \tag{2.153}$$

where

$$\frac{1}{b^2} = \frac{1}{a^2} - i2\pi\beta_2 L. \tag{2.154}$$

Using Eqs. (2.149) and (2.150), the inverse Fourier transform of $\tilde{s}_o(f)$ is

$$s_o(t) = \frac{Ab}{a} \exp\left[-\pi(tb)^2\right].$$

(2.155)

Using Eqs. (2.151) and (2.154), we have

$$\frac{a}{b} = \frac{[T_0^2 - i\beta_2 L]^{1/2}}{T_0},$$

(2.156)

$$\pi b^2 t^2 = \frac{t^2}{2(T_0^2 - i\beta_2 L)}.$$

(2.157)

Therefore, the output field envelope is

$$s_o(t) = \frac{AT_0}{(T_0^2 - i\beta_2 L)^{1/2}} \exp\left[-\frac{t^2}{2\left(T_0^2 - i\beta_2 L\right)}\right].$$

(2.158)

To find the pulse width at the output, let us first calculate the output power

$$P_o(t) = |s_o(t)|^2 = \frac{A^2 T_0^2}{\left|\left(T_0^2 - i\beta_2 L\right)^{1/2}\right|^2} \left|\exp\left[-\frac{t^2}{2\left(T_0^2 - i\beta_2 L\right)}\right]\right|^2.$$

(2.159)

Since

$$\frac{t^2}{2(T_0^2 - i\beta_2 L)} = \frac{t^2(T_0^2 + i\beta_2 L)}{2(T_0^4 + \beta_2^2 L^2)},$$

(2.160)

we obtain

$$P_o(t) = \frac{T_0 A^2}{T_1} \exp\left(-\frac{t^2}{T_1^2}\right)$$

(2.161)

where $T_1^2 = (T_0^4 + \beta_2^2 L^2)/T_0^2$. The FWHM at the output is given by

$$T_{\text{FWHM}}^{\text{out}} = 2(\ln 2)^{1/2} T_1 = 2(\ln 2)^{1/2} \frac{(T_0^4 + \beta_2^2 L^2)^{1/2}}{T_0}.$$

(2.162)

To determine the amount of pulse broadening, the ratio of output to input pulse widths is calculated as

$$\alpha = \frac{T_{\text{FWHM}}^{\text{out}}}{T_{\text{FWHM}}^{\text{in}}} = \frac{(T_0^4 + \beta_2^2 L^2)^{1/2}}{T_0^2}.$$

(2.163)

When

$$|\beta_2 L| = \sqrt{3} T_0^2$$

(2.164)

we find $\alpha = 2$, which means that the output pulse width is twice the input pulse width. Note that the amount of pulse broadening is independent of the sign of the dispersion coefficient β_2.

The frequency chirp or instantaneous frequency deviation is defined as

$$\delta\omega(t) = -\frac{d\phi}{dt},$$

(2.165)

where ϕ is the instantaneous phase of the field envelope and $\delta\omega(t)$ is the instantaneous frequency deviation from the carrier frequency. Note that the optical carrier is of the form $\exp(-i\omega_0 t)$. A negative sign is introduced above so that a positive value of $\delta\omega$ implies the frequency is up-shifted. At the fiber output, from Eq. (2.160), the instantaneous phase is

$$\phi(t) = -\frac{t^2 \beta_2 L}{2T_1^2 T_0^2} + \text{const.} \tag{2.166}$$

Substituting Eq. (2.166) in Eq. (2.165), we find

$$\delta\omega(t) = \left(\frac{\beta_2 L}{T_1^2 T_0^2}\right) t. \tag{2.167}$$

Fig. 2.32 illustrates the evolution of power and chirp along the fiber length. At the fiber output, the pulse becomes chirped and the sign of the chirp depends on the sign of the dispersion coefficient β_2. In Fig. 2.32 (b) and (c), we see that the trailing edge is down-shifted in frequency (or red-shifted) and the leading edge is up-shifted (or blue-shifted). This can be explained as follows. When the dispersion is anomalous ($\beta_2 < 0$), high-frequency components of the pulse travel faster than low-frequency components (see Section 1.10). Since these components arrive at different times, this leads to pulse broadening. Since the low-frequency components of the pulse (components whose frequency is lower than the carrier frequency) travel slower, they arrive later and, therefore, they are present near the trailing edge, which is another way of saying that the trailing edge is down-shifted in frequency.

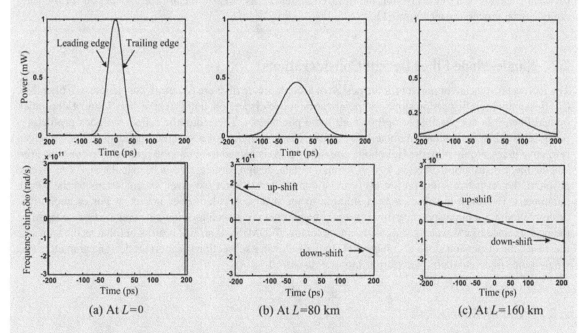

(a) At L=0   (b) At L=80 km   (c) At L=160 km

Figure 2.32   Evolution of unchirped Gaussian pulse in optical fiber. $\beta_2 = -21$ ps^2/km, $T_0 = 30$ ps.

2.6   Comparison between Multi-Mode and Single-Mode Fibers

MMFs have several advantages over SMFs. The core radius of a multi-mode fiber (25–35 μm) is much larger than that of a single-mode fiber (4–9 μm). Therefore, it is easier to launch optical power into a MMF and also to splice two MMFs. The large core of a MMF facilitates simple fiber-to-fiber or fiber-to-transceiver alignment and, hence, is best suited to local area network (LAN) applications [14]. The relative index difference Δ of a MMF is larger than that of a SMF. Therefore, the numerical aperture of a MMF is large, which implies more light can be launched to the fiber from an inexpensive optical source that has a large angular spread, such as a LED. To have a reasonable power coupling efficiency, SMFs are excited with laser diodes. Inexpensive short-haul fiber-optic links can be designed using LEDs and multi-mode fibers. However, multi-mode fibers are not used for long-haul and/or high-bit-rate applications because of intermodal dispersion. Although the dispersion can be reduced to some extent using graded-index multi-mode fibers, the pulse broadening increases linearly with distance (Eq. (2.24)) and becomes unacceptably large for a fiber-optic link that is hundreds of kilometers long. Typically, the transmission reach of a MMF fiber-optic link at a bit rate of 1 Gb/s is limited to a few kilometers. Intermodal dispersion would be absent if there was only one mode. Therefore, single-mode fibers are used for long-haul (1000 km–30,000 km) and high-bit-rate (10 Gb/s–100 Gb/s) applications.

From the information theory point of view, the channel capacity of a multi-mode fiber is larger than that of a single-mode fiber. This is because, in principle, each mode of a MMF can carry as much information as a SMF. When different modes of a MMF carry independent sets of data it is known as *mode-division multiplexing*, which has attracted significant attention recently [15–19]. In an ideal MMF with M guided modes, there is no power coupling among modes and the channel capacity can be enhanced by the factor M. However, due to refractive index fluctuations along the fiber, there is an exchange of power among modes, leading to cross-talk between channels of a mode division multiplexed system. This cross-talk can be compensated for by using digital signal processing techniques [15].

2.7   Single-Mode Fiber Design Considerations

The parameters that are important for the design of a single-mode fiber are (i) cutoff wavelength, (ii) fiber loss, (iii) dispersion, (iv) dispersion slope, (v) polarization mode dispersion, and (vi) spot size. Using a step-index optical fiber, it is not possible to optimize all these parameters. Therefore, the refractive index profile $n(r)$ is chosen so that the design parameters listed above are optimum for a specific application. For the given refractive index profile $n(r)$, the Helmholtz equation (2.28) is solved to obtain the propagation constant $\beta(\omega)$ and the mode distribution function $\Phi(x, y)$. From this data, design parameters can be calculated. As an inverse problem, the refractive index profile $n(r)$ can be constructed to meet the given specifications on the design parameters. However, in some cases, a solution to the inverse problem does not exist. For example, it is desirable to have a large spot size (to reduce nonlinear effects) as well as a low dispersion slope to improve the performance of a wavelength-division multiplexing (WDM) system. But it turns out that as the spot size increases, the dispersion slope also increases. In the following subsections, important design parameters of a single-mode fiber and their interrelationships are discussed.

2.7.1   Cutoff Wavelength

For high-bit-rate and long-haul applications, it is essential that the fiber is single-moded. The single mode condition for a step-index fiber is given by Eqs. (2.70) and (2.59),

$$V = \frac{2\pi a}{\lambda}(n_1^2 - n_2^2)^{1/2} \leq 2.4048. \qquad (2.168)$$

For example, if $\lambda = 1.55\,\mu m$, $a = 4\,\mu m$, and $(n_1^2 - n_2^2)^{1/2} = 0.1$, $V = 1.62$. Therefore, this fiber is single-moded at this wavelength. However, if $\lambda = 0.7\,\mu m$ corresponding to optical communication in the visible spectrum, V becomes 3.59 and the fiber is not single-moded at this wavelength. For the given fiber parameters, the cutoff wavelength is defined as

$$\lambda_c = \frac{2\pi a (n_1^2 - n_2^2)^{1/2}}{2.4048}$$

$$= \frac{2\pi a\ NA}{2.4048}.$$ (2.169)

If the operating wavelength λ is less than λ_c, the fiber will not be single-moded. For fibers with arbitrary index profiles, the Helmholtz equation (2.28) should be solved numerically to find the propagation constants β_n as a function of frequency, from which the conditions for the cutoff of higher-order modes can be established.

Example 2.7

The cutoff wavelength for a step-index fiber is $1.1\,\mu m$. The core index $n_1 = 1.45$ and $\Delta = 0.005$. Find the core radius. Is this fiber single-moded at $1.55\,\mu m$?

Solution:
From Eq. (2.8), we find
$$n_2 = n_1(1 - \Delta) = 1.4428.$$ (2.170)

Using Eq. (2.169), we find
$$a = \frac{2.4048\lambda_c}{2\pi(n_1^2 - n_2^2)^{1/2}} = 2.907\,\mu m.$$ (2.171)

Since the operating wavelength $\lambda = 1.55\,\mu m > \lambda_c$, it is single-moded at this wavelength.

2.7.2 Fiber Loss

Before the advent of optical amplifiers, the maximum transmission distance of a fiber-optic system was determined by the fiber loss, as the optical receivers need a certain amount of optical power to detect the transmitted signal reliably. Now the optical amplifiers are widely used and yet the maximum reach is affected by the fiber loss. This is because the optical amplifiers add noise whose power spectral density is proportional to the amplifier gain, which in turn is proportional to the fiber loss (see Chapter 6). In other words, the amount of noise in a long-haul communication system is directly related to fiber loss. In addition, if the fiber loss is small, the amplifier spacing can be increased, which reduces the system cost. So, it is important to design a fiber with the lowest possible loss.

Let us consider a CW input to the fiber. The optical field distribution is given by Eq. (2.95),

$$\psi(x, y, z, t) = \underbrace{\Phi(x, y, \omega)}_{\text{transverse distribution}}\ \underbrace{A(\omega)\exp{(-\alpha z/2)}}_{\text{field envelope}}\underbrace{\exp{[-i(\omega t - \beta_r(\omega)z)]}}_{\text{optical carrier}}.$$ (2.172)

The optical power is given by Eq. (2.123),

$$P(z) = |A(\omega)\exp{(-\alpha z/2)}|^2.$$ (2.173)

At the fiber input,

$$P_{\text{in}} = P(0) = |A(\omega)|^2. \tag{2.174}$$

At the fiber output $z = L$,

$$P_{\text{out}} = P(L) = |A(\omega)|^2 \exp(-\alpha L) = P_{\text{in}}(t) \exp(-\alpha L). \tag{2.175}$$

The optical power loss in dB units due to propagation in a fiber of length L is defined as

$$\text{loss(dB)} = -10 \log_{10} \frac{P_{\text{out}}}{P_{\text{in}}} = -10(-\alpha L) \log_{10} e = 4.343 \alpha L. \tag{2.176}$$

Here, α is the attenuation coefficient in units of km^{-1}. The loss per unit length is

$$\alpha(\text{dB}/\text{km}) = 4.343\alpha. \tag{2.177}$$

Next, let us consider the origin of fiber loss. The light wave is attenuated as it propagates in fiber mainly due to (i) Rayleigh scattering and (ii) material absorption. In the following subsections, we discuss these mechanisms in detail.

2.7.2.1 Rayleigh Scattering

Consider a perfect crystal with uniformly spaced atoms or molecules. When a light wave is incident on this crystal, electrons in the atoms oscillate and emit light waves of the same frequency as the incident light wave under a linear approximation (see Chapter 10). In other words, each atom acts as a tiny receiving and transmitting antenna. The light emitted by an atom could be in all directions. However, for a perfect crystal with uniformly spaced atoms or molecules, it can be shown that the emitted light waves add up coherently in the direction of the incident light wave; in any other direction, we get no light as they add up destructively [20]. In other words, in a perfect crystal, there is no scattering of incident light. Next, consider a crystal with defects such as atoms missing or irregularly placed in the lattice structure. In this case, light waves emitted by atoms may not add up destructively over a range of directions, which leads to scattering.

Rayleigh scattering is the scattering of light by atoms or molecules of size much smaller than the wavelength of the light. It is an important mechanism arising from local microscopic fluctuations in density and compositional variations. The fluctuations in density correspond to irregularly spaced atoms or molecules in a lattice structure and as a result, incident light is scattered over a range of angles as shown in Fig. 2.33. If the angle of scattering θ is less than the critical angle, it will escape to the cladding and then be absorbed at the polymer jacket. A part of the optical field is back-reflected as well, due to Rayleigh scattering which propagates as a backward-propagating mode. These effects lead to loss of power in the forward-propagating

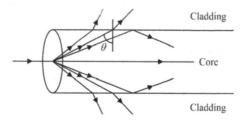

Figure 2.33 Rayleigh scattering in optical fibers.

direction. The loss coefficient due to Rayleigh scattering can be written as

$$\alpha_R \propto \frac{1}{\lambda^4}. \tag{2.178}$$

Because of the strong wavelength dependence of Rayleigh scattering, short wavelengths (blue) are scattered more than long wavelengths (red). The scattering at 400 nm is 9.2 times as great as that at 700 nm for equal incident intensity. Rayleigh scattering of sunlight in a clear atmosphere is the reason why the sky is blue. This also explains why the sun looks red in the morning/evening. The distance between the sun and an observer is large in the morning/evening and the light has to go through a thicker atmosphere, causing the lower wavelengths (violet, blue) to suffer higher losses (see Eq. (2.178)) and their intensities would be too low to detect.

The dominant contributions to the fiber loss come from Rayleigh scattering in the wavelength range of practical interest, 1550–1620 nm. One of the reasons why fiber-optic communication systems operate in the infrared region instead of the visible region (400–700 nm) is that the loss due to Rayleigh scattering is much smaller in the former region because of its λ^{-4} dependence. Fig. 2.34 shows the measured loss spectrum of a single-mode fiber with 9.4 μm core diameter and $\Delta = 0.0019$ [3]. As can be seen, the lowest fiber loss occurs at 1.55 μm wavelength. For the silica fiber, at $\lambda = 1.55$ μm, the loss due to Rayleigh scattering alone is $\alpha_R = 0.1559$ dB/km. Thus, for the fiber shown in Fig. 2.34, 77% of the total loss at 1.55 μm comes from Rayleigh scattering.

Conventional optical fibers are fabricated by doping the silica with GeO_2. The addition of a small amount of GeO_2 increases the refractive index and, therefore, enhancement of the core refractive index relative to the cladding index is achieved. However, the addition of GeO_2 increases the Rayleigh scattering. Therefore, efforts have been made to fabricate pure silica core fibers (PSCFs) in which the core is pure silica [21, 22]. The refractive index of cladding is reduced relative to the core index by adding a small amount of flourine. Since most of the light is confined to the core, the PSCF has lower Rayleigh scattering coefficient than the conventional silica–GeO_2 core fiber. The attenuation of 1570 nm for PSCFs is 0.154 dB/km [21], which is the lowest attenuation reported, whereas the lowest attenuation for the silica–GeO_2 core fiber is about 0.19 dB/km.

2.7.2.2 Material Absorption

Material absorption can be divided into two types: (a) intrinsic absorption and (b) extrinsic absorption.

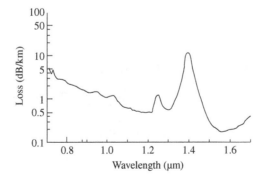

Figure 2.34 The measured loss spectrum of a single-mode fiber. Source: (After Ref. [3]. (c) IET. Reprinted with permission from [3]. Copyright (1979) IET.

Intrinsic Absorption

This loss is caused by the interaction of light with pure silica. An electron in the silica molecule absorbs light and it makes a transition from one electron state to another. This kind of resonance occurs in the ultraviolet region ($\lambda < 0.4\,\mu$m) for silica and the tail of the absorption band extends throughout the visible spectrum. A photon could interact with a molecule, causing a change in its vibrational state. This also leads to photon absorption or optical power loss. This kind of vibrational resonance occurs in the far infrared region $\lambda > 7\,\mu$m, and the tail of the vibrational resonances can be seen in Fig. 2.34 for $\lambda > 1.6\,\mu$m.

Extrinsic Absorption

This loss is caused by the interaction of light with impurities in silica. Metal impurities such as Cu, Fe, Cr, Ni and V lead to a strong signal attenuation. These impurities can be reduced to less than one part in 10^{10} by glass-refining techniques such as vapor-phase oxidation [23]. One of the major sources of extrinsic absorption is the water vapors present in silica fibers. The OH ion of the water vapor is bonded into the glass structure and has a fundamental vibrational resonance at $2.73\,\mu$m. Its overtones and combination tones with the fundamental silica vibrational resonances lead to strong absorption at 1.38, 1.24, 0.95, and $0.88\,\mu$m wavelengths. As shown in Fig. 2.34, the absorption at $1.31\,\mu$m is the strongest and its tail at $1.3\,\mu$m was the main hurdle for the development of fiber-optic communication systems at 1.3 microns. Efforts have been made to reduce the absorption at $1.31\,\mu$m to less than about 0.35 dB/km by reducing the water content in the glass [24, 25]. The majority of fiber-optic systems operate around the wavelength windows centered at $1.3\,\mu$m and $1.55\,\mu$m. This is because the window centered at $1.3\,\mu$m has the lowest dispersion for a standard SMF and the window at $1.55\,\mu$m has the lowest loss.

Example 2.8

A fiber of length 80 km has a loss coefficient of $0.046\,\text{km}^{-1}$. Find the total loss. If the power launched to this fiber is 3 dBm, find the output power in mW and dBm units.

Solution:

The loss per unit length (dB/km) $= 4.343 \times 0.046 = 0.2$ dB/km. Total loss $= 0.2 \times 80 = 16$ dB. From Eq. (2.124), we have

$$P(\text{dBm}) = 10 \log_{10} \frac{P(\text{mW})}{1\,\text{mW}}. \tag{2.179}$$

From Eq. (2.175), we find

$$P_{\text{out}}(\text{mW}) = P_{\text{in}}(\text{mW}) \exp(-\alpha L). \tag{2.180}$$

Dividing Eq. (2.180) by 1 mW and taking logarithms, we find

$$P_{\text{out}}(\text{dBm}) = 10 \log_{10} \left\{ \frac{P_{\text{in}}(\text{mW}) \exp(-\alpha L)}{1\,\text{mW}} \right\}$$

$$= P_{\text{in}}(\text{dBm}) + 10 \log_{10} \exp(-\alpha L)$$

$$= P_{\text{in}}(\text{dBm}) + 10 \log_{10} \frac{P_{\text{out}}}{P_{\text{in}}}. \tag{2.181}$$

Using Eq. (2.176), we find

$$P_{out}(dBm) = P_{in}(dBm) - loss(dB)$$

$$= 3\,dBm - 16\,dB$$

$$= -13\,dBm. \qquad (2.182)$$

Note that using dBm units, multiplication (Eq. (2.180)) is replaced by subtraction (Eq. (2.182)). Using Eq. (2.125), we find

$$P_{out}(mW) = 10^{-13/10}\,mW = 0.05\,mW. \qquad (2.183)$$

Example 2.9

Consider a fiber-optic system consisting of a fiber with loss F followed by an amplifier of gain G (see Fig. 2.35). The launch power is P_{in}. Calculate the output power of the amplifier in dBm units.

Solution:

$$\text{Fiber loss (dB)} = F(dB) = -10\log_{10}\frac{\text{Fiber-out}}{P_{in}}. \qquad (2.184)$$

$$\text{Fiber loss in linear units}, F = \frac{\text{Fiber-out}}{P_{in}}. \qquad (2.185)$$

$$\text{Fiber-out} = FP_{in}. \qquad (2.186)$$

Similarly, the amplifier output power is

$$P_{out} = GFP_{in}. \qquad (2.187)$$

From Eq. (2.124), we find

$$P(dBm) = 10\log_{10}\frac{P(mW)}{1\,mW}. \qquad (2.188)$$

Suppose that P_{out} and P_{in} in Eq. (2.187) are given in units of mW. Divide Eq. (2.187) by 1 mW and take logarithms on both sides to obtain

$$10\log_{10}\frac{P_{out}}{1\,mW} = 10\log_{10}\left[\frac{GFP_{in}}{1\,mW}\right], \qquad (2.189)$$

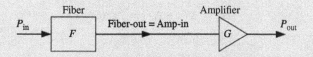

Figure 2.35 A fiber-optic link consisting of a fiber and an amplifier.

$$P_{\text{out}}(\text{dBm}) = 10 \log_{10} G + 10 \log_{10} F + 10 \log_{10} P_{\text{in}}$$

$$= G(\text{dB}) - F(\text{dB}) + P_{\text{in}}(\text{dBm}). \tag{2.190}$$

Note that any loss (such as fiber attenuation) in the system is subtracted from the input power in dBm units and any gain (such as amplifier gain) is added.

2.7.3 Fiber Dispersion

A medium is said to be dispersive if the group speed of light changes with the frequency of the optical wave. For example, a pulse p_1 with carrier frequency $f_1 = 193\,\text{THz}$ and inverse group speed $\beta_1(f_1) = 5\,\mu\text{s/km}$ is delayed by

$$\Delta T_1 = \beta_1(f_1)L = 50\,\mu\text{s} \tag{2.191}$$

after propagating through a 10-km fiber. Consider another pulse p_2 with a different carrier frequency $f_2 = f_1 + \Delta f$. If the fiber is not dispersive ($\beta_2 = 0$), the inverse group speed $\beta_2(f_2)$ is the same as $\beta_1(f_1)$ and, therefore, the pulse p_2 is delayed by the same amount

$$\Delta T_2 = \beta_1(f_2)L = 50\,\mu\text{s}. \tag{2.192}$$

In this case, the differential delay between the pulses is zero. Next, consider a dispersive fiber with $\beta_2 = 100\,\text{ps}^2/\text{km}$ at 193 THz. By definition,

$$\beta_2(\omega) = \frac{d^2\beta_0}{d\omega^2} = \frac{d\beta_1}{d\omega} = \frac{1}{2\pi}\frac{d\beta_1}{df}$$

$$\simeq \frac{1}{2\pi}\frac{\beta_1(f_1 + \Delta f) - \beta_1(f)}{\Delta f} \tag{2.193}$$

or

$$\beta_1(f_2) = \beta_1(f_1) + 2\pi\Delta f \beta_2. \tag{2.194}$$

Let $\Delta f = 1\,\text{THz}$. The pulse p_2 is delayed by

$$\Delta T_2 = \beta_1(f_2)L = \beta_1(f_1)L + 2\pi\Delta f \beta_2 L \tag{2.195}$$

$$= 50.00628\,\mu\text{s}. \tag{2.196}$$

The differential delay between the pulses is given by

$$\Delta T = \Delta T_2 - \Delta T_1 = 2\pi\Delta f \beta_2 L \tag{2.197}$$

$$= 6.28\,\text{ns}. \tag{2.198}$$

In other words, pulse p_2 arrives at the fiber output later than pulse p_1 by 6.28 ns. Instead of finding the derivative of β_1 with respect to frequency as in Eq. (2.193), we could define its derivative with respect to wavelength,

$$D = \frac{d\beta_1}{d\lambda}, \tag{2.199}$$

where D is called the *dispersion parameter*. Since

$$f = \frac{c}{\lambda},$$ (2.200)

$$df = -\frac{c}{\lambda^2} \, d\lambda.$$ (2.201)

Using Eq. (2.201) in Eq. (2.199) and making use of Eq. (2.193), we obtain

$$D = \frac{-2\pi c}{\lambda^2} \beta_2.$$ (2.202)

Substituting Eq. (2.202) in Eq. (2.197), the differential delay can be rewritten as

$$\Delta T = DL\Delta\lambda,$$ (2.203)

where $\Delta\lambda = -c\Delta f / f_1^2$. The above relation can be understood from the fact that β_1 is the delay per unit length and D is the delay per unit length per unit wavelength (Eq. (2.199)).

Fiber dispersion can be divided into two parts: (i) material dispersion and (ii) waveguide dispersion. Material dispersion is due to the frequency dependence of the refractive index of glass. Just like a prism spreads white light into a rainbow of colors (see Section 1.10), different frequency components travel at different speeds in glass, leading to pulse spreading. The second contribution to fiber dispersion comes from the waveguide effect and is known as waveguide dispersion. The dependence of the propagation constant on frequency can be varied by changing the refractive index profile. For example, if we change the refractive index profile from step index to parabolic index, the dispersion coefficient β_2 could vary significantly. In a hypothetical case in which the refractive index of core/cladding does not change with frequency, the fiber dispersion coefficient β_2 could be nonzero because of waveguide dispersion. The product of the dispersion parameters D and fiber length is called *accumulated dispersion*.

A curve fitting to an experimentally measured dispersion parameter of a standard single-mode fiber (SSMF) is given by

$$D(\lambda) = \frac{S_0}{4} \left[\lambda - \frac{\lambda_0^4}{\lambda^3} \right] \text{ ps/(nm} \cdot \text{km)},$$ (2.204)

where $\lambda_0 = 1317$ nm and $S_0 = 0.088$ ps/(nm$^2 \cdot$ km). From Eq. (2.204), at $\lambda = \lambda_0$, $D(\lambda_0) = 0$ and, therefore, λ_0 is called the *zero dispersion wavelength* and

$$\left. \frac{dD}{d\lambda} \right|_{\lambda=\lambda_0} = S_0.$$ (2.205)

S_0 is called the *dispersion slope* at λ_0. Fiber loss is the lowest at 1550 nm and, therefore, most of the optical communication systems operate in the wavelength range 1520–1620 nm. In this wavelength range, the dispersion parameter of a standard SMF is quite high, which leads to strong intersymbol interference (ISI) at the receiver. To avoid this problem, dispersion-shifted (DS) fibers were invented in the 1980s and 1990s [26, 27] which have $\lambda_0 = 1550$ nm. In the absence of fiber nonlinearity, the ideal characteristics of a fiber are low dispersion parameter $|D|$ and low loss, which can be achieved using DS fibers. However, it was soon realized that DS fibers are not suitable for WDM systems since nonlinear interactions between channels, such as four-wave mixing (FWM) and cross-phase modulation (XPM), are enhanced because of low dispersion (see Chapter 10). In the mid-1990s, nonzero dispersion-shifted fibers (NZ-DSFs) were invented, for which λ_0 is chosen to be out of the wavelength region 1530–1565 nm [28]. For NZ-DSFs, dispersion near 1550 nm is large enough to suppress the nonlinear effects and yet low enough to avoid strong ISI due to dispersion. Alternatively, the dispersion of the transmission fiber can be compensated for by using dispersion-compensating fibers (DCFs) (see Section 2.8) or using an equalizer in the electrical domain. This topic is discussed in detail in Chapter 11.

2.7.4   Dispersion Slope

The dispersion parameter depends on the wavelength. For a single-channel low-bit-rate optical communication system, the spectral width Δf (and therefore the wavelength spread $|\Delta \lambda|$) is quite small and the dispersion parameter $D(\lambda)$ can be considered as a constant over a small $\Delta \lambda$. However, for a high-bit-rate wide-band communication system, the dependence of D on the wavelength can not be ignored. For example, the spectral width of an optical non-return-to-zero (NRZ) signal at $B = 160$ Gb/s is roughly (see Chapter 4)

$$\Delta f \simeq 2B = 320 \text{ GHz.} \tag{2.206}$$

Since $c = f\lambda$,

$$\frac{\Delta \lambda}{\lambda} = -\frac{\Delta f}{f}. \tag{2.207}$$

At the carrier wavelength $\lambda_c = 1550$ nm, $\Delta \lambda = -2.56$ nm. The change in dispersion over such a large wavelength spread around 1550 nm can not be ignored. It is useful to define the dispersion slope as

$$S = \frac{dD}{d\lambda}. \tag{2.208}$$

Using Eq. (2.204), the dispersion slope for a standard SMF can be calculated as

$$S = \frac{S_0}{4} \left[1 + \frac{3\lambda_0^4}{\lambda^4} \right]. \tag{2.209}$$

If the dispersion parameter D_c and the dispersion slope S_c at the carrier wavelength λ_c are known, the dispersion parameter in the vicinity of the carrier wavelength can be obtained by the linear approximation [29]

$$D(\lambda) = D_c + S_c(\lambda - \lambda_0). \tag{2.210}$$

In Eq. (2.102), we have retained the Taylor expansion terms up to $\beta_2 \Omega^2 /2$. Under this approximation, β_2 or D is constant over the spectral width of the signal. To include the impact of the dispersion slope, we need to include a higher-order term in the Taylor expansion of Eq. (2.102),

$$\beta(\Omega) = \beta_0 + \beta_1 \Omega + \beta_2 \Omega^2 /2 + \beta_3 \Omega^3 /6, \tag{2.211}$$

where

$$\beta_3 = \left. \frac{d^3 \beta}{d\omega^3} \right|_{\omega=\omega_0} = \left. \frac{d\beta_2}{d\omega} \right|_{\omega=\omega_0}. \tag{2.212}$$

Substituting Eq. (2.211) in Eq. (2.98), we obtain

$$F(t, z) = \exp\left[-i \left(\omega_0 t - \beta_0 z \right) \right] \int_{-\infty}^{\infty} \tilde{B}(\Omega) H_f(\Omega, z) \exp\left(-i\Omega t \right) d\Omega, \tag{2.213}$$

where the fiber transfer function is modified as

$$H_f(\Omega, z) = \exp\left[-\alpha z + i\beta_1 \Omega z + i\beta_2 \Omega^2 z /2 + i\beta_3 \Omega^3 z /6 \right]. \tag{2.214}$$

The field envelope at the output is given by Eq. (2.113) as before,

$$s_o(t) = \mathcal{F}^{-1}\left[\tilde{s}_o(\Omega) \right] = \mathcal{F}^{-1}\left[H_f(\Omega, z) \tilde{s}_i(\Omega) \right]. \tag{2.215}$$

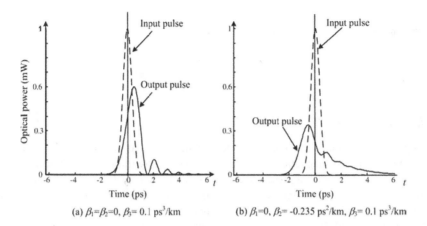

(a) $\beta_1 = \beta_2 = 0$, $\beta_3 = 0.1$ ps^3/km   (b) $\beta_1 = 0$, $\beta_2 = -0.235$ ps^2/km, $\beta_3 = 0.1$ ps^3/km

Figure 2.36 Impact of third-order dispersion on an ultra-short Gaussian pulse. FWHM $= 1.56$ ps, $L = 4.7$ km.

The dispersion slope, S and β_3 are related using Eqs. (2.208) and (2.202),

$$\beta_3 = \frac{d\beta_2}{d\omega} = \frac{d(-D\lambda^2/2\pi c)}{d\lambda}\frac{d\lambda}{d\omega} = S\left(\frac{\lambda^2}{2\pi c}\right)^2 + \frac{\lambda^3 D}{2\pi^2 c^2}. \tag{2.216}$$

Fig. 2.36(a) shows the field envelope at the output of a fiber in the presence of β_3 when $\beta_1 = \beta_2 = 0$. The effect of β_3 is to cause a pulse broadening asymmetrically. When $\beta_2 \neq 0$, pulses broaden symmetrically due to β_2 as well as asymmetrically due to β_3 (Fig. 2.36(b)). If the sign of β_3 is changed, the pulse distortion occurs at the other edge whereas if the sign of β_2 is changed, the amount of pulse broadening is not affected for an unchirped pulse (Eq. (2.162)).

Example 2.10

For a fiber-optic system, the channel wavelengths are in the range of 1530–1560 nm. Design a single-mode fiber such that the absolute accumulated dispersion should not exceed 1100 ps/nm over the wavelength range of interest. Assume that the dispersion varies linearly with wavelength. Fiber length = 80 km.

Solution:
The dependence of dispersion on wavelength is given by

$$D(\lambda) = (\lambda - \lambda_0)S. \tag{2.217}$$

If we choose the zero-dispersion wavelength λ_0 in the center of the band, the maximum allowable dispersion slope could be large. Therefore, we choose $\lambda_0 = 1545$ nm. Using Eq. (2.217), the absolute dispersion at the right edge of the band is

$$|D(1560\,\text{nm})| = |S|(1560 - 1545)\,\text{ps/nm/km}, \tag{2.218}$$

or

$$|S| = \frac{|D(1560\,\text{nm})|}{15}\,\text{ps/nm}^2/\text{km}. \tag{2.219}$$

Since

$$|D(1560\,\text{nm})L| \leq 1100\,\text{ps/nm}, \tag{2.220}$$

with $L = 80$ km and using Eqs. (2.219) and (2.220), we obtain

$$|S| \leq 0.917\,\text{ps/nm}^2/\text{km}. \tag{2.221}$$

2.7.5   Polarization Mode Dispersion

Weakly guiding approximation implies that the electromagnetic field propagates almost in the z-direction and, therefore, the electric and magnetic field components are nearly transverse (see Section 2.4). The scalar field ψ of Section 2.4 could either represent the electric field intensity E_x or E_y. Therefore, a single-mode fiber supports two independent propagation modes [30–32]. For one of the modes, the electric field component is along the $\hat{x}$-direction, the magnetic field component is along the $\hat{y}$-direction, and the propagation constant is β_x (x-polarization). For the other mode, the electric field component is along the $\hat{y}$-direction, the magnetic field component is along the $\hat{x}$-direction, and the propagation constant is β_y (y-polarization). If the fiber cross-section is perfectly circular, these two modes are degenerate, i.e., $\beta_x = \beta_y$. However, it is hard to fabricate a fiber whose cross-section is perfectly circular. Because of the asymmetry introduced during the fiber manufacturing process, and external factors such as bending or twisting, the propagation constants β_x and β_y differ. The inverse group speeds β_{1x} and β_{1y} corresponding to the x- and y-polarization components are also different. As a result, the x- and y-polarization components of the input signal arrive at the fiber output at different times, leading to pulse broadening if a direct detection receiver is used. This phenomenon is known as *polarization mode dispersion* (PMD) [33–37]. Owing to random fluctuations in the fiber refractive index along the fiber axis, there is an exchange of power between these polarization components that occurs randomly along the fiber. Therefore, the pulse broadening due to PMD is stochastic in nature.

Using Eq. (2.110), the x- and y-components of the electric field intensity can be written as

$$E_x(x, y, z, t) = s_x(t, z)\Phi(x, y)\exp\left[-i(\omega t - \beta_x z)\right],$$
$$E_y(x, y, z, t) = s_y(t, z)\Phi(x, y)\exp\left[-i(\omega t - \beta_y z)\right], \tag{2.222}$$

where s_x and s_y are electrical field envelopes, β_x and β_y are propagation constants for the x- and y-polarization components, respectively. The transverse field distributions are nearly the same for x- and y-polarization. As in Section 2.5, input and output electrical field envelopes are related by

$$\tilde{s}_{x,\text{out}}(\Omega) = \tilde{s}_{x,\text{in}}(\Omega)H_x(\Omega, L),$$
$$\tilde{s}_{y,\text{out}}(\Omega) = \tilde{s}_{y,\text{in}}(\Omega)H_y(\Omega, L), \tag{2.223}$$

where

$$H_a(\Omega, L) = \exp\left[-\alpha z/2 + i\left(\beta_{1a}\Omega z + \frac{\beta_2\Omega^2 z}{2}\right)\right], \quad a = x, y. \tag{2.224}$$

Eqs. (2.223) and (2.224) can be written in matrix form using Jones' vector notation (see Section 1.11):

$$\tilde{\mathbf{s}}_{\text{out}}(\Omega) = \begin{bmatrix} \tilde{s}_{x,\text{out}}(\Omega) \\ \tilde{s}_{y,\text{out}}(\Omega) \end{bmatrix}, \quad \tilde{\mathbf{s}}_{\text{in}}(\Omega) = \begin{bmatrix} \tilde{s}_{x,\text{in}}(\Omega) \\ \tilde{s}_{y,\text{in}}(\Omega) \end{bmatrix}, \tag{2.225}$$

$$\tilde{\mathbf{s}}_{\text{out}}(\Omega) = \mathbf{H}(\Omega, L)\tilde{\mathbf{s}}_{\text{in}}(\Omega), \tag{2.226}$$

$$\mathbf{H}(\Omega, L) = \begin{bmatrix} H_x(\Omega, L) & 0 \\ 0 & H_y(\Omega.L) \end{bmatrix}, \tag{2.227}$$

In the case of multi-mode fibers, modes propagate at different speeds and arrive at different times. Similarly, in a single-mode fiber, x-(y-)polarization component propagates at the speed of $1/\beta_{1x}$ ($1/\beta_{1y}$) and, therefore, the time delay ΔT between two polarization components at the output of the fiber of length L is

$$\Delta T = L|\beta_{1x} - \beta_{1y}|. \tag{2.228}$$

The above equation is valid when there is no coupling between x- and y-polarization components. However, for standard telecommunication fibers, there is a random coupling between these components due to perturbations such as stress and micro-bending. The fiber vector transfer function given by Eq. (2.227) does not take into account the random coupling between x- and y-polarization components. In general, the fiber vector transfer function can be written as [38]

$$\mathbf{H}(\Omega, L) = \begin{bmatrix} H_{xx}(\Omega, L) & H_{xy}(\Omega, L) \\ H_{yx}(\Omega, L) & H_{yy}(\Omega, L) \end{bmatrix}. \tag{2.229}$$

The transfer functions $H_{xy}(\Omega, L)$ and $H_{yx}(\Omega, L)$ represent the random coupling between x- and y-polarization components. Because of the random nature of the coupling, it is hard to characterize these functions. Nevertheless, these functions change over a time scale that is longer than the symbol period and, therefore, it is possible to estimate $\mathbf{H}(\Omega, L)$ and compensate for it using digital signal processing (see Chapter 11) in coherent communication systems.

2.7.6   Spot Size

The transverse extent of the field distribution of the fundamental mode plays an important role in determining splice loss between fibers, bending loss, fiber dispersion, and the threshold power required to have significant nonlinear effects (discussed in Chapter 10). The root mean square (r.m.s.) spot size or Petermann-1 spot size is defined as [39–40]

$$w_{p1} = \left[\frac{2 \int_0^\infty \Phi^2(r) r^3 dr}{\int_0^\infty \Phi^2(r) r dr} \right]^{1/2}, \tag{2.230}$$

where $\Phi(r)$ is the transverse field distribution of the fundamental mode, which is radially symmetric. When a fiber mode has a large transverse extent the spot size is large, leading to enhancement of bending losses. On the contrary, large spot size is desirable to reduce the effect of fiber nonlinearity on optical pulses and, thereby, transmission performance can be improved. This is because, for the given launch power, the power per unit cross-sectional area (= optical intensity) is larger when the spot size is smaller and the nonlinear change in refractive index is directly proportional to the optical intensity. Typically, as the spot size increases, the dispersion slope increases too. Therefore, the refractive index profile $n(r)$ of a fiber should be optimized so that (i) it has a single mode and has low loss at the desired wavelength range and (ii) the spot size is sufficiently large for the transmission performance to not be impaired by nonlinear effects and, yet, be small enough so that the dispersion slope and bending losses are not enhanced.

2.8   Dispersion-Compensating Fibers (DCFs)

For long-haul and/or high-bit-rate optical communication systems, the pulse broadening due to intramodal dispersion leads to intersymbol interference, which degrades transmission performance. The pulse broadening can be compensated using a DCF, as shown in Fig. 2.37. Using Eq. (2.107), the transfer functions of the

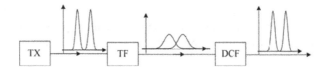

Figure 2.37 Fiber-optic system consisting of TF and DCF. TF = transmission fiber, DCF = dispersion-compensating fiber, TX = transmitter.

transmission fiber and DCF after ignoring β_1 and fiber loss can be written as

$$H_{TF}(f) = \exp\left[\frac{i(2\pi f)^2 \beta_2^{TF} L^{TF}}{2}\right],$$

$$H_{DCF}(f) = \exp\left[\frac{i(2\pi f)^2 \beta_2^{DCF} L^{DCF}}{2}\right], \qquad (2.231)$$

where the superscripts TF and DCF stand for transmission fiber and dispersion-compensating fiber, respectively. The total transfer function is

$$H_{tot}(f) = H_{TF}(f)H_{DCF}(f). \qquad (2.232)$$

To keep the output pulse width equal to the input pulse width, we require

$$H_{tot}(f) = 1, \qquad (2.233)$$

or

$$\beta_2^{TF} L^{TF} = -\beta_2^{DCF} L^{DCF}. \qquad (2.234)$$

When the loss is not ignored, the output of the DCF is attenuated by a factor $\exp\left[-\alpha_{TF}L_{TF} - \alpha_{DCF}L_{DCF}\right]$ without affecting the pulse broadening. When β_1 is included, the pulse is simply delayed by a factor $\beta_1^{TF}L^{TF} + \beta_1^{DCF}L^{DCF}$. Typically, SSMFs or NZDSFs are used as transmission fibers, which have anomalous dispersion. Therefore, the DCF should have normal dispersion. As mentioned before, the fiber dispersion coefficient can be altered by changing the amount of waveguide dispersion. The material dispersion of glass at 1550 nm is anomalous and the waveguide dispersion of the standard SMF is a small fraction of the total dispersion (at 1550 nm). If the sign of the waveguide dispersion is made opposite to that of the material dispersion by a proper choice of refractive index profile, the total fiber dispersion becomes normal. This is the underlying principle behind the design of a DCF. To design a DCF, the refractive index profile of a fiber is divided into several segments. Each segment is characterized by two or more parameters. For example, a segment could have a parabolic index profile or it could have a triangular index profile. By optimizing the parameters of these segments, the desired dispersion coefficient β_2 can be obtained. For WDM applications (see Chapter 9), it is desirable to compensate dispersion over a wide band. With proper design, the dispersion slope of the transmission fiber can also be compensated [41]. Such a fiber is called a dispersion-slope compensating fiber.

Example 2.11

A transmission fiber of length 80 km has a dispersion of -21 ps^2/km. The transmission fiber is followed by a DCF of dispersion 130 ps^2/km. (a) Find the length of the DCF such that the pulse width at the input of the transmission fiber is the same as that at the output of the DCF. (b) Suppose the power launched into the transmission fiber is 2 mW, losses of the transmission fiber and DCF are 0.2 dB/km and 0.5 dB/km, respectively. Calculate the power at the output of the DCF. Assume a splice loss of 0.5 dB between the transmission fiber and the DCF. (c) Find the gain of the amplifier such that the signal power at the output of the amplifier is the same as that at the input.

Solution:
(a) From Eq. (2.234), we have

$$\beta_2^{\mathrm{TF}} L^{\mathrm{TF}} = -\beta_2^{\mathrm{DCF}} L^{\mathrm{DCF}}, \tag{2.235}$$

$$\beta_2^{\mathrm{TF}} = -21\,\mathrm{ps}^2/\mathrm{km}, \quad L^{\mathrm{TF}} = 80\,\mathrm{km}, \tag{2.236}$$

$$\beta_2^{\mathrm{DCF}} = 130\,\mathrm{ps}^2/\mathrm{km}, \tag{2.237}$$

$$L^{\mathrm{DCF}} = \frac{-\beta_2^{\mathrm{TF}} L^{\mathrm{TF}}}{\beta_2^{\mathrm{DCF}}} = 12.9\,\mathrm{km}. \tag{2.238}$$

(b) The launch power in dBm units is given by Eq. (2.124) as

$$P_{\mathrm{in}}(\mathrm{dBm}) = 10\log_{10}\frac{2\,\mathrm{mW}}{1\,\mathrm{mW}} = 3\,\mathrm{dBm}. \tag{2.239}$$

Loss budget:

$$\text{loss in transmission fiber} = 0.2 \times 80\,\mathrm{dB} = 16\,\mathrm{dB},$$

$$\text{loss in DCF} = 0.5 \times 12.9\,\mathrm{dB} = 6.45\,\mathrm{dB},$$

$$\text{splice loss} = 0.5\,\mathrm{dB},$$

$$\text{total loss} = 16 + 6.45 + 0.5\,\mathrm{dB} = 22.95\,\mathrm{dB}.$$

The power at the output of the DCF is

$$P_{\mathrm{out,DCF}} = 3\,\mathrm{dBm} - 22.95\,\mathrm{dBm} = -19.95\,\mathrm{dBm}. \tag{2.240}$$

(c) To keep the signal power at the output of the amplifier the same as the input, the amplifier gain should be equal to the total loss in the system, i.e.,

$$\text{amplifier gain} = 22.95\,\mathrm{dBm}. \tag{2.241}$$

2.9   Additional Examples

Example 2.12

The numerical aperture of a multi-mode fiber is 0.2. Find the delay between the shortest and longest path. Fiber length = 2 km and core index = 1.45. Assume that the difference between the core index and the cladding index is small.

Solution:
The NA is given by Eq. (2.9) as

$$\mathrm{NA} = n_1 \sqrt{2\Delta}, \tag{2.242}$$

$$\Delta = \frac{(NA)^2}{2n_1^2},$$ (2.243)

$$NA = 0.2, n_1 = 1.45.$$

From Eq. (2.243), we find

$$\Delta = 0.0095.$$

When the difference between the core index and the cladding index is small, $n_1 \approx n_2$ and Eq. (2.17) can be approximated as

$$\Delta T \approx = \frac{n_1 L \Delta}{c},$$ (2.244)

$$n_1 = 1.45, L = 2\,\text{km}, \Delta = 0.0095, \text{ and } c = 3 \times 10^8\,\text{m/s}.$$

Substituting these values in Eq. (2.244), the delay between the shortest and longest paths is

$$\Delta T = 91.95\,\text{ns}.$$

Example 2.13

The propagation constant at the wavelength $\lambda_0 = 1550\,\text{nm}$ is 6×10^6 rad/m. Calculate the propagation constant at $\lambda_1 = 1551\,\text{nm}$. Assume $\beta_1 = 0.5 \times 10^{-8}$ s/m and $\beta_2 = -10\,\text{ps}^2/\text{km}$. Ignore $\beta_n, n > 2$.

Solution:
From Eq. (2.102), we find
$$\beta(\omega_1) = \beta_0 + \beta_1(\omega_1 - \omega_0) + \beta_2(\omega_1 - \omega_0)^2/2.$$ (2.245)

Using $c = f\lambda$, we have

$$\omega_0 = 2\pi f_0 = \frac{2\pi c}{\lambda_0} = 1.2161 \times 10^{15}\,\text{rad/s},$$

$$\omega_1 = 2\pi f_1 = \frac{2\pi c}{\lambda_1} = 1.2153 \times 10^{15}\,\text{rad/s},$$

$$\omega_1 - \omega_0 = -8.168 \times 10^{11}\,\text{rad/s},$$

$$\beta_0 = 6 \times 10^{16}\,\text{rad/m}.$$

Substituting these values in Eq. (2.245), we find

$$\beta(\omega_1) = 5.9959 \times 10^6\,\text{rad/s}.$$

Note that the change in propagation constant is very small.

Example 2.14

Consider a fiber-optic system as shown in Fig. 2.38. Fiber loss = 0.2 dB/km, length = 80 km, loss in optical filter = 0.5 dB, and amplifier gain = 15 dB. If the minimum power required at the receiver to have a good signal-to-noise ratio is −3 dBm, calculate the lower limit on the transmitter power in dBm and mW units.

Solution:

$$\text{Fiber loss,} \quad F_1(\text{dB}) = -0.2 \times 80 = -16 \,\text{dB}.$$

$$\text{Filter loss,} \quad F_2(\text{dB}) = -0.5 \,\text{dB}.$$

$$\text{Amplifier gain,} \quad G(\text{dB}) = 15 \,\text{dB}.$$

The minimum power required at the receiver is

$$P_{\text{out}}(\text{dBm}) = -3 \,\text{dBm},$$

$$P_{\text{out}}(\text{dBm}) = P_{\text{in}}(\text{dBm}) + F_1(\text{dB}) + F_2(\text{dB}) + G(\text{dB}).$$

Therefore, the lower limit on the transmitter power is

$$P_{\text{in}}(\text{dBm}) = -3 + 16 + 0.5 - 15 \,\text{dBm} = -1.5 \,\text{dBm}.$$

Using Eq. (2.125), the transmitter power in mW units is

$$P_{\text{in}} = 10^{0.1 P_{\text{in}}(\text{dBm})} = 0.7079 \,\text{mW}.$$

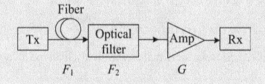

Figure 2.38 A fiber-optic system with loss and gain.

Example 2.15

The electric field envelope at the fiber input is

$$s_{\text{in}}(t) = A \cos (2\pi f_m t). \tag{2.246}$$

Show that the electric field envelope at the fiber output is

$$s_{\text{out}}(t) = A \cos (2\pi f_m t) \exp \left[i(2\pi f_m)^2 \beta_2 L/2 \right]. \tag{2.247}$$

Ignore fiber loss and β_1.

Solution:
Taking the Fourier transform of Eq. (2.246), we find

$$\tilde{s}_{\text{in}}(f) = \mathcal{F}[s_{\text{in}}(t)] = \frac{A}{2}[\delta(f - f_m) + \delta(f + f_m)], \tag{2.248}$$

where δ is the Dirac delta function. From Eq. (2.107), we have the fiber transfer function (after ignoring β_1 and loss)

$$H(f) = \exp[i\beta_2(2\pi f)^2 L/2]. \tag{2.249}$$

The output spectrum is

$$\tilde{s}_{\text{out}}(f) = \tilde{s}_{\text{in}}(f)H(f)$$

$$= \frac{A}{2}\left[\delta(f - f_m) + \delta(f + f_m)\right]\exp\left[i\beta_2(2\pi f)^2 L/2\right]. \tag{2.250}$$

Taking the inverse Fourier transform of Eq. (2.250), we obtain

$$s_o(t) = \frac{A}{2}\int_{-\infty}^{\infty}[\delta(f - f_m) + \delta(f + f_m)]\exp\left[-i2\pi f t + i(2\pi f)^2\frac{\beta_2 L}{2}\right]df. \tag{2.251}$$

Using the following relation:

$$\int_{-\infty}^{\infty}\delta(f - f_m)X(f)df = X(f_m), \tag{2.252}$$

Eq. (2.251) is simplified as

$$s_o(t) = \frac{A}{2}\left[\exp\left(-i2\pi f_m t\right) + \exp\left(i2\pi f_m t\right)\right]\exp\left[i(2\pi f_m)^2\beta_2 L/2\right]$$

$$= A\cos\left(2\pi f_m t\right)\exp\left[i(2\pi f_m)^2\beta_2 L/2\right]. \tag{2.253}$$

Comparing Eqs. (2.246) and (2.253), we find that if the field envelope is a sinusoid, it acquires only a phase shift.

Example 2.16

Consider a fiber-optic system as shown in Fig. 2.39. A Gaussian pulse is launched into the transmission fiber. Find the length of DCF so that the pulse width (FWHM) at the output of the DCF is twice the pulse width at the input of the TF. Assume $\beta_2^{\text{TF}} = -21\text{ ps}^2/\text{km}$, $\beta_2^{\text{DCF}} = 130\text{ ps}^2/\text{km}$, $L^{\text{TF}} = 80\text{ km}$. FWHM at the input of TF = 12.5 ps. Ignore loss and β_1.

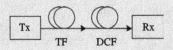

Figure 2.39   Fiber-optic system consisting of TF and DCF. TF = transmission fiber, DCF = dispersion-compensating fiber, Tx = transmitter, and Rx = receiver.

Solution:
The effective transfer function is given by Eq. (2.232) as

$$H_{\text{eff}}(f) = \exp\left\{ i\left[\left(\beta_2^{\text{TF}} L^{\text{TF}} + \beta_2^{\text{DCF}} L^{\text{DCF}} \right) \frac{(2\pi f)^2}{2} \right] \right\}$$

$$= \exp\left[i\beta_2^{\text{eff}} L^{\text{eff}} (2\pi f)^2 / 2 \right], \tag{2.254}$$

where

$$\beta_2^{\text{eff}} L^{\text{eff}} = \beta_2^{\text{TF}} L^{\text{TF}} + \beta_2^{\text{DCF}} L^{\text{DCF}}. \tag{2.255}$$

In Example 2.6, we found that when $|\beta_2 L| = \sqrt{3} T_0^2$, the output pulse width is twice the input pulse width. Replacing $\beta_2 L$ by $\beta_2^{\text{eff}} L^{\text{eff}}$, we find

$$|\beta_2^{\text{TF}} L^{\text{TF}} + \beta_2^{\text{DCF}} L^{\text{DCF}}| = \sqrt{3} T_0^2, \tag{2.256}$$

$$\beta_2^{\text{TF}} = -21 \, \text{ps}^2/\text{km}, \quad L^{\text{TF}} = 80 \, \text{km},$$

$$\beta_2^{\text{DCF}} = 130 \, \text{ps}^2/\text{km}, \quad T_{\text{FWHM}} = 12.5 \, \text{ps},$$

$$T_0 = T_{\text{FWHM}}/1.665 = 7.507 \, \text{ps}.$$

Eq. (2.256) may be written as

$$-21 \times 80 + 130 L^{\text{DCF}} = \pm\sqrt{3} \times 7.507^2.$$

Therefore, $L^{\text{DCF}} = 12.17 \, \text{km}$ or $13.67 \, \text{km}$. As the pulse propagates in the DCF, the pulse undergoes compression. At $L^{\text{DCF}} = 12.17 \, \text{km}$, the pulse width is twice the initial pulse width. If $\beta_2^{\text{TF}} L^{\text{TF}} = -\beta_2^{\text{DCF}} L^{\text{DCF}}$, the pulse width becomes equal to the initial pulse width. This corresponds to a propagation distance of 12.92 km in the DCF. After this, pulse broadening takes place and when $L^{\text{DCF}} = 13.67 \, \text{km}$, the output pulse width is twice the initial pulse width again.

Example 2.17

The zero dispersion wavelength of a transmission fiber (TF) is chosen as 1490 nm, so that the local dispersion in the desired wavelength range 1530–1560 nm is not zero (so as to avoid the enhancement of nonlinear effects). Find the accumulated dispersion of the DCF so that the net accumulated dispersion does not exceed 1100 ps/nm. Assume that the dispersion slopes of the TF and DCF are 0.08 ps/nm^2/km and 0 ps/nm^2/km, respectively. Total transmission distance = 800 km. Other parameters are the same as in Example 2.10.

Solution:
In the absence of DCF, the dispersion at 1560 nm is

$$D_{\text{TF}}(1560 \, \text{nm}) = 0.08(1560 - 1490) \, \text{ps/nm/km} = 5.6 \, \text{ps/nm/km}. \tag{2.257}$$

The accumulated dispersion at 1560 nm is

$$D_{\text{TF}} L_{\text{TF}} = 4480 \, \text{ps/nm}. \tag{2.258}$$

The net accumulated dispersion is

$$|D_{\text{TF}}L_{\text{TF}} + D_{\text{DCF}}L_{\text{DCF}}| \leq 1100 \, \text{ps/nm}. \qquad (2.259)$$

Therefore, the accumulated dispersion of the DCF should be

$$D_{\text{DCF}}L_{\text{DCF}} < -3380 \, \text{ps/nm}. \qquad (2.260)$$

Example 2.18

Let the input field envelope be

$$s_i(t) = A \exp\left[-\frac{t^2(1 + iC)}{2T_0^2}\right], \qquad (2.261)$$

where C is the chirp parameter. Show that the field envelope at the fiber output is

$$s_o(t) = \frac{AT_0}{T_1} \exp\left[-\frac{(1 + iC)t^2}{2T_1^2}\right],$$

where

$$T_1 = \left[(T_0^2 + \beta_2 LC) - i\beta_2 L\right]^{1/2}.$$

Plot the power and frequency deviation $\delta\omega$ at different fiber lengths for $\beta_2 C > 0$ and $\beta_2 C < 0$. Ignore fiber loss and β_1.

Solution:
Using Eqs. (2.150) and (2.152), we find

$$\tilde{s}_i(f) = \frac{A}{a} \exp\left[-\pi(f/a)^2\right], \qquad (2.262)$$

where

$$a^2 = \frac{1 + iC}{2\pi T_0^2}. \qquad (2.263)$$

The output field envelope is given by Eq. (2.155), with $b^2 = a^2/(1 - i2\pi\beta_2 La^2)$ as,

$$s_o(t) = \frac{AT_0}{T_1} \exp\left[-\frac{(1 + iC)t^2}{2T_1^2}\right], \qquad (2.264)$$

where

$$T_1 = [(T_0^2 + \beta_2 LC) - i\beta_2 L]^{1/2}. \qquad (2.265)$$

As in Example 2.6, the output pulse width can be calculated as

$$T_{\text{FWHM}}^{\text{out}} = T_{\text{FWHM}}^{\text{in}} \left[\left(1 + \frac{C\beta_2 L}{T_0^2}\right)^2 + \left(\frac{\beta_2 L}{T_0^2}\right)^2\right]^{1/2}. \qquad (2.266)$$

The field envelope may be written as

$$s(t,z) = A(t,z) \exp\left[i\phi(t,z)\right]. \tag{2.267}$$

The instantaneous frequency deviation from the carrier frequency is given by Eq. (2.165) as

$$\delta\omega(t,z) = -\frac{d\phi}{dt}. \tag{2.268}$$

At the fiber input, we have

$$\phi(t,0) = -\frac{t^2 C}{2T_0^2}. \tag{2.269}$$

So, the instantaneous frequency deviation from the carrier frequency is

$$\delta\omega(t,0) = \frac{Ct}{T_0^2}. \tag{2.270}$$

Fig. 2.40 shows the output pulse width as a function of propagation distance L. As can be seen from Eq. (2.266), the pulse broadening depends on the sign of $\beta_2 C$. When $\beta_2 C \geq 0$, the pulse width increases with distance monotonically. When $\beta_2 C < 0$, the first term within the square bracket of Eq. (2.266) becomes less than unity and, therefore, the output pulse width can be less than the input pulse width for certain distances. Fig. 2.40 shows that the pulse undergoes compression initially for $C = 4$ and $\beta_2 < 0$. The physical explanation for pulse compression is as follows. When $C > 0$, from Eq. (2.270), we see that the leading edge is down-shifted in frequency and the trailing edge is up-shifted at the fiber input. In an anomalous dispersion fiber ($\beta_2 < 0$), low-frequency (red) components travel slower than high-frequency (blue) components and, therefore, the frequency components at the leading edge travel slowly. In other words, they are delayed and move to the right (later time) as shown by the arrow in Fig. 2.41(a), and the frequency components at the leading edge move to the left (earlier time), leading to pulse compression as shown in Fig. 2.41(b). Since the frequency chirp imposed on the pulse at the input is of opposite sign to the frequency chirp developed via pulse propagation in an anomalous dispersion fiber, these two frequency chirps cancel at $L = 12.5$ km and the pulse becomes unchirped (see the bottom of Fig. 2.41(b)). At this distance, the pulse width is the shortest. Thereafter, pulse propagation is the same as discussed in Example 2.6, leading to pulse broadening.

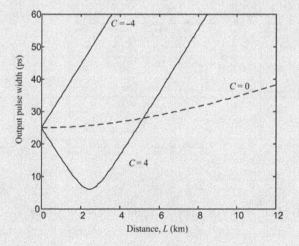

Figure 2.40   Output pulse width of a chirped Gaussian pulse. $\beta_2 = -21$ ps^2/km.

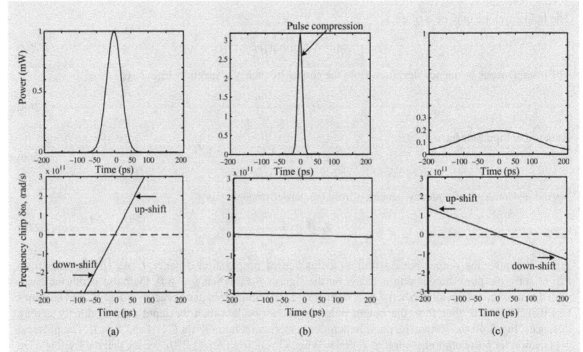

Figure 2.41 Power and frequency deviations of a chirped Gaussian pulse. $\beta_2 = -21$ ps^2/km, $C = +3$.

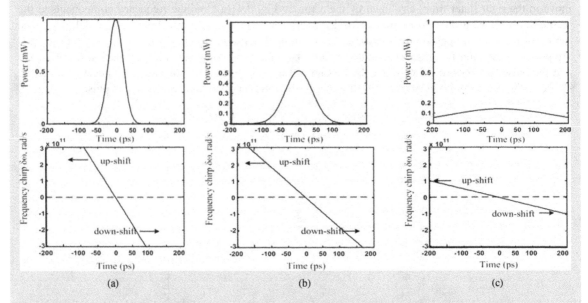

Figure 2.42 Power and frequency deviations of a chirped Gaussian pulse. $\beta_2 = -21$ ps^2/km, $C = -3$.

If $C < 0$, at the fiber input, the leading edge is up-shifted in frequency whereas the trailing edge is down-shifted. The frequency components corresponding to the leading edge travel faster than those corresponding to the trailing edge in an anomalous dispersion fiber. Therefore, as shown by the arrows in Fig. 2.42(a), the leading edge moves to the left (earlier time) and the trailing edge moves to the right, which leads to pulse broadening. The frequency chirp at the fiber input in this case has the same sign (as that due to dispersion as given by Eq. (2.167)) and, therefore, these two chirps add up, leading to enhanced broadening as seen in Fig. 2.40 ($C = -4$) compared with the case of an unchirped pulse.

Exercises

2.1 A step-index fiber has a cutoff wavelength = 900 nm, and NA = 0.22. (a) Calculate the core radius. (b) What could be the maximum allowable core radius to make this fiber single-moded at 500 nm?

(Ans: (a) 3.44 μm; (b) 2.29 μm.)

2.2 Consider a small fiber section of length ΔL as shown in Fig. 2.43. Let $F(\Delta L) = P(\Delta L)/P_{in}$. Next, consider a cascade of identical fiber sections as shown in Fig. 2.44. Let M be the total number of fiber sections. When $M \to \infty$ (or $\Delta L \to 0$), show that

$$F_{tot} = \frac{P_{out}}{P_{in}} = \exp(-\alpha L), \qquad (2.271)$$

where $\alpha = -dF/d(\Delta L)$.

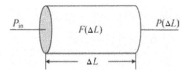

Figure 2.43 An infinitesimal fiber section.

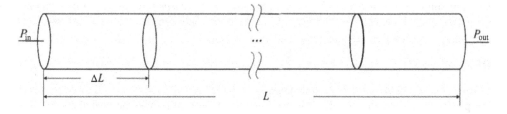

Figure 2.44 A fiber of length L with M sections of length ΔL.

Hint: $F_{tot} = [F(\Delta L)]^{L/\Delta L}$, expand $F(\Delta L)$ in a Taylor series with $F(0) = 1$ and use

$$e = \text{Lim}_{M \to \infty}(1 + 1/M)^M. \qquad (2.272)$$

2.3   A step-index multi-mode fiber has to be designed to support bit rates up to 10 Mb/s. The link length is 1.5 km. Calculate the upper limit on the relative index difference, Δ. Assume the core refractive index is approximately equal to the cladding refractive index.

(Ans: $\Delta \leq 0.02$.)

2.4   In a step-index multi-mode fiber, the critical angle for the core–cladding interface is 85°. The core refractive index is 1.46 and the core diameter is 100 μm. Find (a) the V-number and (b) the approximate number of guided modes M at the wavelength 1.3 μm.

(Ans: $V = 25.79$, $M = 332$.)

2.5   The maximum achievable bit rate–distance product in a step-index multi-mode fiber is 16 Mb/s·km. The core refractive index n_1 ($\approx$ cladding index n_2) = 1.45. Calculate (a) the numerical aperture and (b) the critical angle for the core–cladding interface.

(Ans: (a) 0.2795; (b) 1.3768 rad.)

2.6   A step-index multi-mode fiber has an acceptable angle of 0.2077 rad and the critical angle for the core–cladding interface is 1.4266 rad. Calculate the speed of light in the fiber core. It may be assumed that ray-optics theory is valid.

(Ans: 2.076×10^8 m/s.)

2.7   A single-mode fiber has to be designed to operate at 1.55 μm with a cutoff wavelength less than 1.5 μm. The core and cladding refractive indices are 1.45 and 1.445, respectively. Calculate the maximum allowable core radius.

(Ans: 4.77 μm.)

2.8   The power launched into a fiber is 5 mW and the power at the fiber output is 0.3 mW. Calculate the fiber loss in dBm units.

(Ans: 12.22 dBm.)

2.9   A 40-km single-mode fiber has a dispersion parameter $D = 10$ ps/nm·km. An optical signal of bandwidth 10 GHz is launched into the fiber. Find the delay between the highest- and lowest-frequency components. The carrier wavelength = 1.55 μm.

(Ans: 32.033 ps.)

2.10   A single-mode fiber has a zero-dispersion wavelength at 1550 nm. The dispersion slope = 0.06 ps/nm^2/km. Find the absolute dispersion $|D|$ at 1600 nm. Assume that the dispersion varies linearly with wavelength.

(Ans: 3 ps/nm/km.)

2.11   A Gaussian pulse is transmitted in a long single-mode fiber with dispersion coefficient $\beta_2 = -10$ ps^2/km. The input and output pulse widths (FWHM) are 8 ps and 32 ps, respectively. Calculate the fiber length.

(Ans: 8.9 km.)

2.12 The outputs of CW lasers of frequencies 193.5 THz and 193.6 THz are transmitted over a fiber of length 2 km. It is found that the lower-frequency optical wave arrived later than the higher-frequency component by 6 ps. Calculate the dispersion coefficient β_2.

(Ans: -4.77 ps^2/km.)

2.13 A fiber-optic link consists of a TF followed by a DCF of length 5 km. The transmission fiber has a dispersion parameter $D = 10$ ps/nm/km, loss = 0.25 dB/km, and length = 50 km. (a) Find the dispersion parameter of the DCF such that the pulse width (FWHM) at the output of the DCF is the same as the pulse width at the input of the TF. (b) The power launched into the TF is 2 mW and the power at the output of the DCF is -12 dBm. Find the loss coefficient of the DCF in dB/km.

(Ans: (a) -100 ps/nm/km; (b) 0.5 dB/km.)

Further Reading

K. Okamoto, *Fundamentals of Optical Waveguide*, 2nd edn. Academic Press, New York, 2006.
T. Okoshi, *Optical Fibers*. Academic Press, San Diego, CA, 1982.
L.B. Jeunhomme, *Single-mode Fiber Optics*, 2nd edn. Marcel Dekker, New York, 1990.
E.G. Neumann, *Single-mode Fibers*, Springer-Verlag, New York, 1988.
A.W. Snyder and J.D. Love, *Optical Waveguide Theory*. Chapman & Hall, London, 1983.
J.M. Senior, *Optical Fiber Communications*, 2nd edn. Prentice-Hall, London, 1992.
H. Kolimbris, *Fiber Optic Communications*. Prentice-Hall, Englewood Cliffs, NJ, 2004.
R.P. Khare, *Fiber Optics and Optoelectronics*. Oxford University Press, New York, 2004.

References

[1] C.K. Kao and G.A. Hockham, *Proc. IEE,* vol. **113**, p. 1151, 1966.
[2] R.D. Maurer, D.B. Keck, and P. Schultz, Optical waveguide fibers. US patent no. 3,711,262, 1973.
[3] T. Miya, Y. Terunuma, T. Husaka, and T. Miyoshita, *Electron. Lett.*, vol. **15**, p. 106, 1979.
[4] J. Gower, *Optical Communication Systems*, 2nd edn. Prentice-Hall, London, 1993.
[5] G.P. Agrawal, *Fiber-Optic Communication Systems*, 4th edn. John Wiley & Sons, New York, 2010.
[6] D. Marcuse, *Theory of Dielectric Optical Waveguides*, 2nd edn. Academic Press, New York, 1991.
[7] J.A. Buck, *Fundamentals of Optical Fibers*, 2nd edn. John Wiley & Sons, Hoboken, NJ, 2004.
[8] A. Ghatak and K. Thyagarajan, *Optical Electronics*. Cambridge University Press, Cambridge, 1991.
[9] D. Gloge, "Weakly guiding fibers," *Appl. Opt.*, vol. **10**, p. 2252, 1971.
[10] D. Marcuse, D. Gloge, and E.A.J. Marcatili, "Guiding properties of fibers." In S.E. Miller and A.G. Chynoweth (eds), *Optical Fiber Telecommunications*. Academic Press, New York, 1979.
[11] A.W. Snyder, *IEEE Trans. Microwave Theory Tech.*, vol. **MTT-17**, p. 1130, 1969.
[12] G. Keiser, *Optical Fiber Communications*, 4th edn. McGraw-Hill, New York, 2011.
[13] R.M. Gagliardi and S. Karp, *Optical Communications*, 2nd edn. John Wiley & Sons, New York, 1995.
[14] *Communication News*, March 2006. Nelson Publishing, www.comnews.com.
[15] R. Ryf *et al.*, *J. Lightwave Technol.*, vol. **30**, p. 521, 2012.
[16] X. Chen *et al.*, *Opt. Exp.*, vol. **20**, p. 14302, 2012.
[17] B. Neng *et al.*, *Opt. Exp.*, vol. **20**, p. 2668, 2012.
[18] N. Hanzawa *et al.*, Optical Fiber Conference, paper OWA4, 2012.
[19] H.S. Chen, H.P.A. van den Boom, and A.M.J. Koonen, *IEEE Photon. Technol. Lett.*, vol. **23**, 2011.
[20] R.P. Feynmann, R.B. Leighton, and M. Sands, *The Feyman Lectures on physics*, vol. **1**, chapter 32. Addison-Wesley, New York, 1963.

[21] H. Kanamori *et al.*, *J. Lightwave Technol.*, vol. **LT-4**, p. 1144, 1986.

[22] Y. Chigusa *et al.*, *J. Lightwave Technol.*, vol. **23**, p. 3541, 2005.

[23] H. Osanai *et al.*, *Electron. Lett.*, vol. **12**, p. 549, 1976.

[24] G.E. Berkey *et al.*, US patent no. EP1181254 B1, 2006.

[25] F.W. Dabby, US patent no. 20130025326 A1, 2013.

[26] A. Ohashi *et al.*, US patent no. 4,755,022, 1988.

[27] T. Kato *et al.*, US patent no. 5,721,800, 1998.

[28] Y. Liu and M.A. Newhouse, US patent no. EP1158323 A1, 1996.

[29] Corning product sheet information, www.corning.com/opticalfiber.

[30] I.P. Kaminov, *IEEE J. Quantum Electron.*, vol. **17**, p. 15, 1981.

[31] S.C. Rashleigh, *J. Lightwave Technol.*, vol. **LT-1**, p. 312, 1983.

[32] X.-H. Zheng, W.M. Henry, and A.W. Snyder, *J. Lightwave Technol.*, vol. **6**, p. 312, 1988.

[33] C.D. Poole, J.H. Winters, and J.A. Nagel, *Opt. Lett.*, vol. **16**, p. 372, 1991.

[34] G.J. Foschini and C.D. Poole, *J. Lightwave Technol.*, vol. **9**, p. 1439, 1991.

[35] J. Zhou and M.J. O'Mahony, *IEEE Photon. Technol. Lett.*, vol. **6**, p. 1265, 1994.

[36] P.K.A. Wai and C.R. Menyuk, *J. Lightwave Technol.*, vol. **14**, p. 148, 1996.

[37] A. Galtarossa and C.R. Menyuk (eds), *Polarization Mode Dispersion*. Springer-Verlag, New York, 2005.

[38] C.D. Poole and R.E. Wagner, *Electron. Lett.*, vol. **22**, p. 1029, 1986.

[39] K. Petermann, *Electron. Lett.*, vol. **19**, p. 712, 1983.

[40] C. Pask, *Electron. Lett.*, vol. **20**, p. 144, 1984.

[41] S. Bickham *et al.*, US patent no. 06671445, 2003.

3

Lasers

3.1 Introduction

LASER is an acronym for light amplification by stimulated emission of radiation. In 1917, Einstein postulated that an atom in the excited level is stimulated to emit radiation that has the same frequency and phase as the radiation [1]. This is known as stimulated emission, which remained a theoretical curiosity until Schewlow and Townes in the USA [2], and Basov and Prokhorov in the USSR, proposed that stimulated emission can be used in the construction of lasers. Townes, Gordon, and Zeiger built the first ammonia MASER (microwave amplification by stimulated emission of radiation) at Columbia University in 1953. This device used stimulated emission in a stream of energized ammonia molecules to produce amplification of microwaves at a frequency of about 24 GHz. In 1960, Maiman demonstrated the ruby laser, which is considered to be the first successful optical laser [3]. In 1962, the first semiconductor laser diode was demonstrated by a group led by Hall [4]. The first diode lasers were homojunction lasers. The efficiency of light generation can be significantly enhanced using double heterojunction lasers, as demonstrated in 1970 by Alferov and his collaborators [5].

Fiber-optic communications would not have progressed without lasers. Lasers have not only revolutionized fiber-optic communications, but also found diverse applications in laser printers, barcode readers, optical data recording, combustion ignition, laser surgery, industrial machining, CD players, and DVD technology.

In this chapter, we first discuss the basic concepts such as absorption, stimulated emission, and spontaneous emission. Next, we analyze the conditions for laser oscillations. After reviewing the elementary semiconductor physics, the operating principles of the semiconductor laser are discussed.

3.2 Basic Concepts

Consider two levels of an atomic system as shown in Fig. 3.1. Let the energy of the ground state be E_1 and that of the excited state be E_2. Let N_1 and N_2 be the atomic densities in the ground state and excited state, respectively. If radiation at an angular frequency

$$\omega = \frac{E_2 - E_1}{\hbar}, \tag{3.1}$$

where $\hbar = h/(2\pi)$, h = Planck's constant = 6.626×10^{-34} J $\cdot$ s, is incident on the atomic system, it can interact in three distinct ways [1].

(a) An atom in the ground state of energy E_1 absorbs the incident radiation and goes to the excited state of energy E_2, as shown in Fig. 3.2. In other words, an electron in the atom jumps from an inner orbit to an

Fiber Optic Communications: Fundamentals and Applications, First Edition. Shiva Kumar and M. Jamal Deen.
© 2014 John Wiley & Sons, Ltd. Published 2014 by John Wiley & Sons, Ltd.

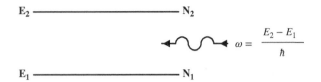

Figure 3.1   Two-level atomic system interacting with electromagnetic radiation.

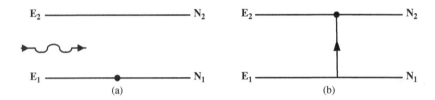

Figure 3.2   Two-level atomic system absorbing a photon (a) Before absorption and (b) After absorption.

outer orbit. To make this transition, atoms require energy corresponding to the difference in energy levels and this is provided by the incident electromagnetic radiation. The rate of absorption depends on the population density in the level E_1 and also on the energy spectral density per unit volume of the radiation. Einstein postulated that the number of atoms undergoing absorption per unit time per unit volume from level 1 to level 2 is given by

$$R_{\text{abs}} \equiv -\left(\frac{dN_1}{dt}\right)_{\text{abs}} = B_{12} u_s(\omega) N_1, \tag{3.2}$$

where $u_s(\omega)$ is the electromagnetic energy spectral density per unit volume, B_{12} is a constant, and R_{abs} is the rate of absorption. The negative sign in Eq. (3.2) indicates that the population density in level 1 decreases due to absorption. For example, consider an atomic system of volume $1\,\text{m}^3$. If 10^{15} atoms make an upward transition per second after absorbing the incident radiation in a volume of $1\,\text{m}^3$, the absorption rate is $R_{\text{abs}} = 10^{15}\,\text{s}^{-1}\,\text{m}^{-3}$. This also means that 10^{15} photons are absorbed per second per cubic meter.

(b) An atom which is in the excited state of energy E_2 is stimulated to emit radiation at frequency $\omega = (E_2 - E_1)/\hbar$ if the radiation at that frequency is already present. After emitting the radiation, it goes to state of energy E_1, as shown in Fig. 3.3. This process is called *stimulated emission*. Einstein postulated that the rate of emission is proportional to the energy spectral density of radiation at frequency ω and the population density at the excited state E_2,

$$R_{\text{stim}} \equiv -\left(\frac{dN_2}{dt}\right)_{\text{stim}} = B_{21} u_s(\omega) N_2, \tag{3.3}$$

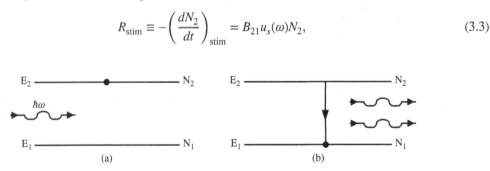

Figure 3.3   Two-level atomic system emitting a photon due to stimulated emission (a) Before stimulated emission and (b) After stimulated emission.

where R_{stim} is the stimulated emission rate that is equal to the number of atoms undergoing stimulated emission per unit time per unit volume, and B_{21} is a constant.

(c) An atom in the excited state of energy E_2 can also make a spontaneous emission and return to the ground state whether or not the radiation at frequency ω is present, as illustrated in Fig. 3.4. This occurs randomly and is called *spontaneous emission*. The rate of spontaneous emission is independent of the intensity of incident radiation. Therefore,

$$R_{spont} \boxtimes -\left(\frac{dN_2}{dt}\right)_{spont} = A_{21}N_2, \tag{3.4}$$

where A_{21} is a constant and R_{spont} is the spontaneous emission rate.

At thermal equilibrium between the atomic system and the radiation field, the number of upward transitions must be equal to the number of downward transitions. If not, for example if downward transition occurs more frequently than upward transition, this would result in an increase in radiation with time, which implies it is not in equilibrium. At thermal equilibrium, we have

$$R_{up} = R_{down}, \tag{3.5}$$

$$R_{abs} = R_{stim} + R_{spont}, \tag{3.6}$$

$$B_{12}u_s(\omega)N_1 = B_{21}u_s(\omega)N_2 + A_{21}N_2, \tag{3.7}$$

$$u_s(\omega) = \frac{A_{21}}{(N_1/N_2)B_{12} - B_{21}}. \tag{3.8}$$

According to Boltzmann's law, the ratio of populations of level 1 and level 2 at equilibrium is

$$\frac{N_2}{N_1} = \exp\left(-\Delta E/k_B T\right), \tag{3.9}$$

where $\Delta E = E_2 - E_1$ is the energy difference, $k_B = 1.38 \times 10^{-23}$ J/K is Boltzmann's constant, and T is the absolute temperature in Kelvin. Since the energy difference $\Delta E = \hbar\omega$, Eq. (3.9) can be written as

$$N_2 = N_1 \exp\left(-\hbar\omega/k_B T\right). \tag{3.10}$$

Eq. (3.9) is valid for any systems having different energy levels. For example, the number of air molecules decreases as we go to higher altitudes. If N_1 and N_2 are the number of molecules near the ground and at height h, respectively, at thermal equilibrium, their ratio is

$$\frac{N_2}{N_1} = \exp\left(-\frac{\Delta E}{k_B T}\right) = \exp\left(-\frac{mgh}{k_B T}\right), \tag{3.11}$$

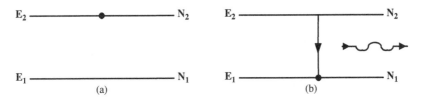

Figure 3.4 Two-level atomic system emitting a photon due to spontaneous emission (a) before spontaneous emission and (b) after spontaneous emission.

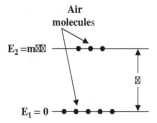

Figure 3.5   Air molecules at ground and at height h.

where $\Delta E = mgh$ is the energy required to lift a molecule of mass m to the height h, and g is the gravitational constant (see Fig. 3.5).

Using Eq. (3.10) in Eq. (3.8), we obtain

$$u_s(\omega) = \frac{A_{21}/B_{21}}{(B_{12}/B_{21}) \exp{(\hbar\omega/k_B T)} - 1}. \tag{3.12}$$

According to Planck's law, the energy spectral density per unit volume at thermal equilibrium is given by

$$u_s(\omega) = \frac{\hbar\omega^3 n_0^3}{\pi^2 c^3} \frac{1}{\exp{(\hbar\omega/k_B T)} - 1}, \tag{3.13}$$

where n_0 is the refractive index and c is the velocity of light in vacuum. Consider a hollow container heated to a temperature T by a furnace, as shown in Fig. 3.6. Under thermal equilibrium, if you make a very small hole and observe the spectrum of radiation, it would look like the curves shown in Fig. 3.7. For all these curves, the energy of low-frequency and high-frequency components of the electromagnetic waves approaches zero and the peak of the energy spectral density increases with temperature. A similar experiment was carried out by Rubens and Kurlbaum [6], and Planck developed a theoretical description for the enclosed radiation [7]. From his derivation it follows that the energy spectral density at thermal equilibrium is given by Eq. (3.13) and it is shown in Fig. 3.7, which is in good agreement with the measured data of Rubens and Kurlbaum [6]. Planck assumed that energy exchange between radiation and matter takes place as a discrete packet or quantum of

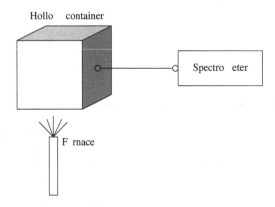

Figure 3.6   The radiation trapped in a hollow container under thermal equilibrium.

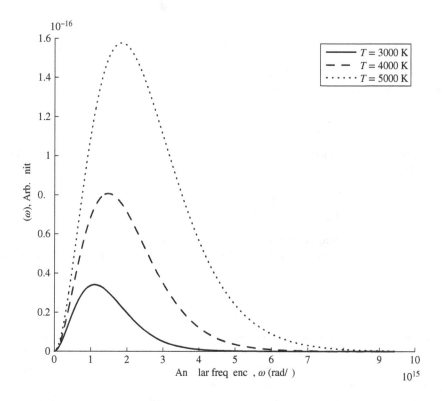

Figure 3.7   The energy spectral density as a function of angular frequency under thermal equilibrium.

energy, which is proportional to frequency. Planck's model and his assumptions to derive the radiation formula were milestones in the development of quantum mechanics.

The interesting fact is that the energy spectral density at thermal equilibrium depends only on the absolute temperature and not on the shape of the container or the material type. The radiation is continuously interacting with the walls of the container but when thermal equilibrium is reached, the intensity of radiation does not depend on the parameters characterizing the container, but only on the temperature.

Similarly, in the case of the atomic system of Fig. 3.1, radiation interacts with the atoms causing upward and downward transitions. At thermal equilibrium, the energy spectral density should depend only on temperature. Therefore, the energy spectral density per unit volume given by Eqs. (3.12) and (3.13) should be equal. Comparing Eqs. (3.12) and (3.13), we find

$$B_{12} = B_{21} \boxtimes B, \tag{3.14}$$

$$A_{21} = \frac{B\hbar\omega^3 n_0^3}{\pi^2 c^3} \boxtimes A. \tag{3.15}$$

The coefficients A and B are called *Einstein coefficients.* At thermal equilibrium, the ratio of spontaneous emission rate and stimulated emission rate is given by

$$\frac{R_{\text{spont}}}{R_{\text{stim}}} = \frac{A_{21}N_2}{B_{21}N_2 u_s(\omega)} = \exp\left(\frac{\omega}{\omega_c}\right) - 1, \tag{3.16}$$

where

$$\omega_c = \frac{k_B T}{\hbar}. \tag{3.17}$$

Typically for an optical source, $\omega > \omega_c$. Therefore, spontaneous emission dominates stimulated emission.
  As an example, consider an optical source at temperature $T = 300\,\mathrm{K}$:

$$\omega_c = \frac{k_B T}{\hbar} = \frac{1.38 \times 10^{-23} \times 300}{1.054 \times 10^{-34}} = 3.92 \times 10^{13}\,\mathrm{rad/s}. \tag{3.18}$$

For $\omega > 3.92 \times 10^{13}\,\mathrm{rad/s}$, radiation would mostly be due to spontaneous emission. If the operating wavelength of the optical source is 700 nm, $\omega = 2.69 \times 10^{15}\,\mathrm{rad/s}$ and

$$\frac{R_{\mathrm{spont}}}{R_{\mathrm{stim}}} = \exp\left(\frac{2.69 \times 10^{15}}{3.92 \times 10^{13}}\right) - 1 \simeq 6.34 \times 10^{29}. \tag{3.19}$$

The above equation indicates that, on average, one out of 6.34×10^{29} emissions is a stimulated emission. Thus, at optical frequencies, the emission is mostly due to spontaneous emission and hence the light from the usual light sources is not coherent. From Eqs. (3.2) and (3.3), we see that

$$\frac{R_{\mathrm{stim}}}{R_{\mathrm{abs}}} = \frac{N_2}{N_1}. \tag{3.20}$$

Therefore, the stimulated emission rate exceeds the absorption rate only when $N_2 > N_1$. This condition is called *population inversion*. For systems in thermal equilibrium, from Eq. (3.10), we find that N_2 is always less than N_1 and population inversion can never be achieved. Therefore, all lasers should operate away from thermal equilibrium. To achieve population inversion, atoms should be pumped to the excited state by means of an external energy source known as a *pump*. A flash pump could act as an optical pump and atoms are excited into higher-energy states through absorption of the pump energy. Alternatively, an electrical pump can be used to achieve population inversion as discussed in Section 3.8.
  The photons generated by stimulated emission have the same frequency, phase, direction, and polarization as the incident light. In contrast, the spontaneous emission occurs randomly in all directions and both polarizations, and often acts as noise. In lasers, we like to maximize the stimulated emission by achieving population inversion.
  The Einstein coefficient A is related to the spontaneous emission lifetime associated with state 2 to state 1 transition. Let us consider a system in which stimulated emission is negligible and atoms in the excited state spontaneously emit photons and return to the ground state. Considering only spontaneous emission, the decay rate of the excited level is given by Eq. (3.4),

$$\frac{dN_2}{dt} = -A_{21} N_2. \tag{3.21}$$

The solution of Eq. (3.21) is

$$N_2(t) = N_2(0) \exp\left(-t/t_{sp}\right), \tag{3.22}$$

where

$$t_{sp} = \frac{1}{A_{21}}. \tag{3.23}$$

At $t = t_{sp}$, $N_2(t) = N_2(0)e^{-1}$. Thus, the population density of level 2 reduces by e over a time t_{sp} which is known as the *spontaneous lifetime* associated with 2   1 transition.
  So far we have assumed that the energy levels are sharp, but in reality these levels consist of several sublevels or bands. The spectrum of the electromagnetic waves due to spontaneous or stimulated emission

from collection of atoms is not perfectly monochromatic; this is because the emission can take place due to transition from any of the sublevels. The interaction of electromagnetic waves over a range of frequencies with the two-band system is described by a lineshape function $l(\omega)$. Out of N_2 (or N_1) atoms per unit volume, $N_2 l(\omega)d\omega$ (or $N_1 l(\omega)d\omega$) atoms interact with electromagnetic waves in the frequency interval $[\omega, \omega + d\omega]$. The lineshape function is normalized such that

$$\int N_2 l(\omega)d\omega = N_2. \tag{3.24}$$

In other words, N_2 represents the total number of atoms per unit volume in band 2 and $N_2 l(\omega)d\omega$ represents the fraction of atoms per unit volume that could interact with radiation ranging from ω to $\omega + d\omega$. Therefore, the total stimulation rate per unit volume is

$$R_{\text{stim}} = \int B u_s(\omega) l(\omega) N_2 d\omega. \tag{3.25}$$

Similarly, the absorption rate per unit volume is modified as

$$R_{\text{abs}} = \int B u_s(\omega) l(\omega) N_1 d\omega. \tag{3.26}$$

Let us consider a special case in which the radiation is a monochromatic wave of frequency ω_0. The energy spectral density per unit volume of a monochromatic wave is an impulse function given by

$$u_s(\omega) = u\delta(\omega - \omega_0)/(2\pi). \tag{3.27}$$

Since the energy spectral density is energy per unit frequency interval, its integration over frequency is energy. Therefore,

$$\int u_s(\omega)d\omega = u \int \delta(\omega - \omega_0)d\omega/(2\pi) = u. \tag{3.28}$$

Here, u is energy per unit volume or *energy density*. Substituting Eq. (3.27) in Eqs. (3.25) and (3.26), we find

$$R_{\text{stim}} = Bul(\omega_0)N_2, \tag{3.29}$$

$$R_{\text{abs}} = Bul(\omega_0)N_1. \tag{3.30}$$

By letting $Bl(\omega_0)$ B, Eqs. (3.29) and (3.30) are the same as Eqs. (3.3) and (3.2), respectively, with the exception that u_s is replaced by energy density u.

xample 3.1

In an atomic sytem, the spontaneous lifetime associated with 2   1 transition is 2 ns and the energy difference between the levels is 2.4×10^{-19} J. Calculate the Einstein A and B coefficients. Assume that the velocity of light in the medium is 1.25×10^8 m/s.

Solution:
From Eq. (3.23), we have

$$A_{21} = \frac{1}{t_{sp}} = 5 \times 10^8 \text{ s}^{-1}.$$

The energy difference ΔE is

$$\Delta E = \hbar\omega,$$

$$\omega = \Delta E/\hbar = 2.28 \times 10^{15} \text{ rad/s}.$$

From Eq. (3.15), we have

$$B = \frac{A\pi^2 v^3}{\hbar\omega^3},$$

where $v = c/n_0$ is the velocity of light in the medium. With $v = 1.25 \times 10^8$ m/s, we obtain

$$B = \frac{5 \times 10^8 \times \pi^2 \times (1.25 \times 10^8)^3}{1.054 \times 10^{-34} \times (2.28 \times 10^{15})^3}$$

$$= 7.71 \times 10^{21} \text{ m}^3/\text{J} \cdot \text{s}^2.$$

xample 3.2

The energy levels of an atomic system are separated by 1.26×10^{-19} J. The population density in the ground state is 10^{19} cm^{-3}. Calculate (a) the wavelength of light emitted, (b) the ratio of spontaneous emission rate to stimulated emission rate, (c) the ratio of stimulated emission rate to absorption rate, and (d) the population density of the excited level. Assume that the system is in thermal equilibrium at 300 K.

Solution:
(a) The energy difference

$$\Delta E = \hbar\omega = \hbar 2\pi f$$

$$= hf,$$

where $h = \hbar 2\pi = 6.626 \times 10^{-34}$ J $\cdot$ s and

$$f = \frac{\Delta E}{h} = \frac{1.26 \times 10^{-19}}{6.626 \times 10^{-34}} = 191 \text{ THz}.$$

The wavelength of the light emitted is given by

$$\lambda = \frac{c}{f} = \frac{3 \times 10^8}{191 \times 10^{12}} = 1.56 \text{ }\mu\text{m}.$$

(b) The ratio of spontaneous emission to stimulated emission rate is given by Eq. (3.16),

$$\frac{R_{\text{spont}}}{R_{\text{stim}}} = e^{\hbar\omega/k_BT} - 1 = e^{\Delta E/k_BT} - 1,$$

$$k_B = 1.38 \times 10^{-23} \text{ J/K},$$

$$T = 300 \text{ K},$$

$$\frac{R_{\text{spont}}}{R_{\text{stim}}} = \exp\left(\frac{1.26 \times 10^{-19}}{1.38 \times 10^{-23} \times 300}\right) - 1$$

$$= 1.88 \times 10^{13}.$$

(c) From Eq. (3.20), we have

$$\frac{R_{\text{stim}}}{R_{\text{abs}}} = \frac{N_2}{N_1}.$$

According to Boltzmann's law,

$$N_2 = N_1 e^{-\hbar\omega/k_B T},$$

$$\frac{R_{\text{stim}}}{R_{\text{abs}}} = e^{-\hbar\omega/k_B T} = \exp\left(\frac{-1.26 \times 10^{-19}}{1.38 \times 10^{-23} \times 300}\right)$$

$$= 5.29 \times 10^{-14}.$$

(d) The population density of the excited level is

$$N_2 = N_1 e^{-\hbar\omega/k_B T}$$

$$= 5.29 \times 10^5 \, \text{cm}^{-3}.$$

3.3   Conditions for Laser Oscillations

Consider a lossless gain medium as shown in Fig. 3.8, in which the incident light wave is amplified by stimulated emission. The optical intensity at z can be phenomenologically described as

$$I(z) = I(0) \exp(gz), \tag{3.31}$$

where g is the gain coefficient associated with stimulated emission. For the atomic system with two levels, an expression for g can be obtained in terms of the population densities N_1, N_2 and the Einstein coefficient B (see Section 3.6 for details). By differentiating $I(z)$ with respect to z, Eq. (3.31) can be rewritten in differential form as

$$\frac{dI}{dz} = gI(0) \exp(gz)$$

$$= gI. \tag{3.32}$$

The optical field is attenuated in the gain medium due to scattering and other possible loss mechanisms similar to attenuation in optical fibers. The effect of loss is modeled as

$$I(z) = I(0) \exp(-\alpha_{\text{int}} z), \tag{3.33}$$

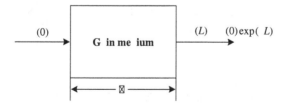

Figure 3.8   Light amplification in a gain medium.

where α_{int} is the coefficient of internal loss due to scattering and other loss mechanisms in the gain medium. The gain and attenuation occur simultaneously in the gain medium. So, we have

$$\boxtimes(z) = \boxtimes(0)\exp{(g_{\text{net}}z)}, \qquad (3.34)$$

where $g_{\text{net}} = g - \alpha_{\text{int}}$ is the coefficient of the net gain.

A laser is an oscillator operating at optical frequencies. Just like an electronic oscillator, the optical oscillator (laser) has three main components: (i) amplifier, (ii) feedback, and (iii) power supply, as shown in Fig. 3.9. The atomic system we have discussed before can act as a gain medium and the light is amplified by stimulated emission. The feedback is provided by placing the gain medium between two mirrors, as shown in Fig. 3.10. The optical or electrical pumps required to achieve population inversion are the power supply.

Consider the optical wave propagating in the Fabry–Perot (FP) cavity shown in Fig. 3.10. Let $\boxtimes(0)$ be the optical intensity at A. After passing through the gain medium, the intensity is $\boxtimes(0)\exp{(g_{\text{net}}L)}$, where L is the length of the cavity. The light wave is reflected by the mirror at B, whose reflectivity is R_2. This means that the reflected intensity at B is $R_2\boxtimes(0)\exp{(g_{\text{net}}L)}$. The reflected field passes through the gain medium again and is reflected by the mirror at A with reflectivity R_1. The optical intensity after a round trip is (see Fig. 3.11)

$$\boxtimes(0)R_1 R_2 \exp{[2(g - \alpha_{\text{int}})L]}. \qquad (3.35)$$

The condition for laser oscillation is that the optical intensity after one round trip should be the same as the incident intensity $\boxtimes(0)$. Otherwise, after several round trips, the optical intensity in the cavity would be too

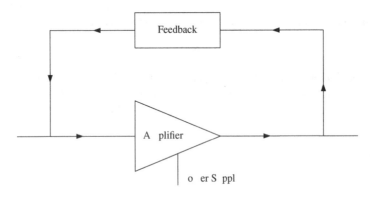

Figure 3.9   The structure of an optical oscillator (laser) or electronic oscillator.

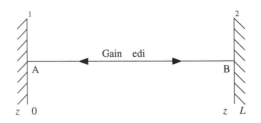

Figure 3.10   The Fabry–Perot cavity formed by mirrors.

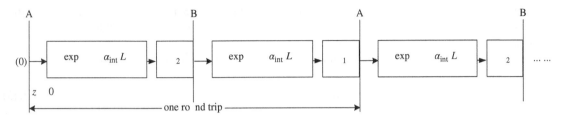

Figure 3.11 Illustration of multiple reflections in a FP cavity.

low or too high. For a stable laser operation, we need

$$\psi(0)R_1 R_2 \exp\left[2(g - \alpha_{\text{int}})L\right] = \psi(0). \tag{3.36}$$

Simplifying Eq. (3.36), we find

$$g = \alpha_{\text{int}} + \frac{1}{2L} \ln\left(\frac{1}{R_1 R_2}\right). \tag{3.37}$$

In Eq. (3.37), the second term represents the loss due to mirrors,

$$\alpha_{\text{mir}} = \frac{1}{2L} \ln\left(\frac{1}{R_1 R_2}\right). \tag{3.38}$$

Using Eq. (3.38) in Eq. (3.37), we find

$$g = \alpha_{\text{int}} + \alpha_{\text{mir}} = \alpha_{\text{cav}}, \tag{3.39}$$

where α_{cav} is the total cavity loss coefficient. Therefore, to have a stable laser operation, one of the essential conditions is that the total cavity loss should be equal to the gain. Suppose you are on a swing. Because of the frictional loss, the oscillations will be dampened and it will stop swinging unless you pump yourself or someone pushes you. To have sustained oscillations, the frictional loss should be balanced by the gain due to "pumping." In the case of a laser, the gain is provided by optical/electrical pumps. A monochromatic wave propagating in the cavity is described by a plane wave,

$$\psi = \psi_0 \exp\left[-i(\omega t - kz)\right]. \tag{3.40}$$

The phase change due to propagation from A to B is kL. And the phase change due to a round trip is $2kL$. The second condition for laser oscillation is that the phase change due to a round trip should be an integral multiple of 2π,

$$2kL = \frac{4\pi n}{\lambda_0}L = 2m\pi, \quad m = 0, \pm 1, \pm 2, \ldots \tag{3.41}$$

Otherwise, the optical field ψ at A would be different after each round trip. Here, λ_0 is the wavelength in free space and n is the refractive index of the medium. If the condition given by Eq. (3.41) is not satisfied, the superposition of the field components after N round trips,

$$\psi_N = \psi_0 \exp\left(-i\omega t\right) \sum_{n=0}^{N} \exp\left(i2knL\right), \tag{3.42}$$

approaches zero as $N \to \infty$. This is because sometimes the field component after a round trip may be positive and sometimes it may be negative, and the net sum goes to zero if m is not an integer. When m is an integer, the optical fields after each round trip add up coherently.

From Eq. (3.41), we see that only a discrete set of frequencies or wavelengths are supported by the cavity. They are given by

$$\lambda_m = \frac{2nL}{m}, \qquad m = 1, 2, \dots , \tag{3.43}$$

or

$$f_m = \frac{mc}{2nL}. \tag{3.44}$$

These frequencies correspond to the *longitudinal modes* of the cavity, and can be changed by varying the cavity length L. The laser frequency f must match one of the frequencies of the set $f_m, m = 1, 2, \dots$ The spacing Δf between longitudinal modes is constant,

$$\Delta f = f_m - f_{m-1} = \frac{c}{2nL}. \tag{3.45}$$

The longitudinal spacing Δf is known as the *free spectral range* (FSR). In a two-level atomic system, the gain would occur only for the frequency $\omega = (E_2 - E_1)/\hbar$. However, in practical systems, these levels are not sharp; each level is a broad collection of sublevels and, therefore, the gain would occur over a range of frequencies. Fig. 3.12 shows the loss and gain profiles of a FP laser. Many longitudinal modes of the FP cavity experience gain simultaneously. The mode for which the gain is equal to the loss (shown as the lasing mode) becomes the dominant mode. In theory, other modes should not reach the threshold since their gain is less than the loss of the cavity. In practice, the difference in gain between many modes of the cavity is extremely small, and one or two neighboring modes on each side of the main mode (lasing mode) carry a significant fraction of power. Such a laser is called a multi-longitudinal-mode laser. Fig. 3.13 shows the output of a multi-longitudinal-mode laser. If a multi-longitudinal-mode laser is used in fiber-optic communication systems, each mode of the laser propagates at a slightly different group velocity in the fiber because of dispersion, which leads to intersymbol interference at the receiver. Therefore, for high-bit-rate applications, it is desirable to have a single-longitudinal-mode (SLM) laser. A distributed Bragg grating is used to obtain a single longitudinal mode, as discussed in Section 3.8.5.

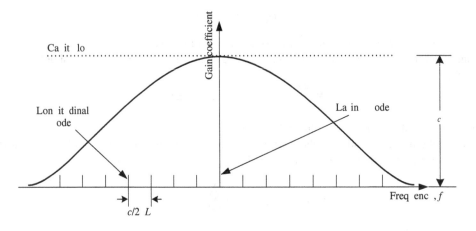

Figure 3.12 Loss and gain profiles of a Fabry–Perot laser.

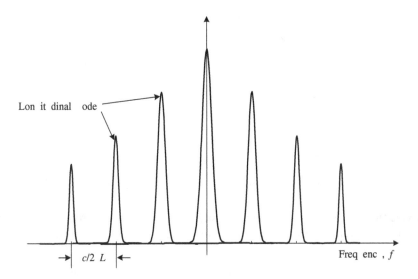

Figure 3.13   The output spectrum of a Fabry–Perot laser.

Eq. (3.31) provides the evolution of the optical intensity as a function of the propagation distance. Sometimes, it is desirable to find the evolution of the optical intensity as a function of time. To obtain the time rate of change of the optical intensity, we first develop an expression relating optical intensity I and energy density u. The optical intensity is power P per area S, which is perpendicular to the direction of propagation,

$$I = \frac{P}{S}. \tag{3.46}$$

The power is energy ΔE per unit time,

$$P = \frac{\Delta E}{\Delta t}, \tag{3.47}$$

where Δt is a suitably chosen time interval. Combining Eqs. (3.46) and (3.47), we find

$$I = \frac{\Delta E}{S\Delta t}. \tag{3.48}$$

Fig. 3.14 shows the optical intensity at z and $z + \Delta z$. The number of photons crossing the area S at $z + \Delta z$ over a time interval Δt is the same as the number of photons present in the volume $S\Delta z$ if

$$\Delta z = v\Delta t, \tag{3.49}$$

where v is the speed of light in the medium. For example, if Δt is chosen as 1 ns, Δz is 0.2 m assuming $v = 2 \times 10^8$ m/s. Using Eq. (3.49) in Eq. (3.47), Eq. (3.46) becomes

$$I = \frac{\Delta E v}{S\Delta z} = uv, \tag{3.50}$$

where u is the energy density or energy per unit volume. Since $I \propto u$, Eq. (3.32) can be written as

$$\frac{du}{dz} = gu. \tag{3.51}$$

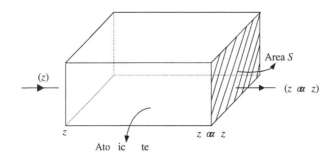

Figure 3.14 Optical intensity incident on the atomic system of volume $S\Delta z$.

Eq. (3.51) provides the rate of change of energy density as a function of the propagation distance in the gain medium. This can be converted to the time rate of change of energy density by using $dz = vdt$,

$$\frac{du}{vdt} = gu, \tag{3.52}$$

$$\frac{du}{dt} = Gu, \tag{3.53}$$

where

$$G = vg. \tag{3.54}$$

If we include the cavity loss, Eq. (3.53) should be modified as

$$\frac{du}{dt} = (G - v\alpha_{cav})u. \tag{3.55}$$

Note that the cavity loss has a contribution from internal loss and mirror loss. The mirror loss is lumped, whereas the internal loss is distributed. Therefore, Eq. (3.55) becomes inaccurate for time intervals less than the transit time $2L/v$.

xample 3.3

A Fabry–Perot laser has the following parameters: internal loss coefficient 50 dB/cm, $R_1 = R_2 = 0.3$, and distance between mirrors = 500 μm. Calculate the longitudinal mode spacing and the minimum gain required for laser oscillation. Assume that the refractive index $n = 3.5$.

Solution:
The longitudinal mode spacing Δf is given by Eq. (3.45),

$$\Delta f = \frac{c}{2nL} = \frac{3 \times 10^8}{2 \times 3.5 \times 500 \times 10^{-6}} = 85.71 \,\text{GHz}.$$

The minimum gain required is

$$g = \alpha_{int} + \alpha_{mir},$$

$$\alpha_{mir} = \frac{1}{2L} \ln \frac{1}{R_1 R_2}.$$

The internal loss is given in dB/cm. To convert this into cm^{-1}, consider a length of 1 cm. The loss over a length of 1 cm is 50 dB,

$$P_{out} = P_{in} \exp(-\alpha_{int}.1\,cm).$$

$$10\log_{10}\frac{P_{out}}{P_{in}} = -50\,dB,$$

$$10\log_{10}e^{-\alpha_{int}.1\,cm} = -\alpha_{int} \times 1\,cm \times 10\log_{10}e = -50\,dB,$$

$$\alpha_{int}\,(cm^{-1}) = \frac{50}{4.3429}\,cm^{-1} = 11.51\,cm^{-1}.$$

The distance between mirrors, $L = 0.05$ cm:

$$R_1 = R_2 = 0.3,$$

$$\alpha_{mir}\,(cm^{-1}) = \frac{1}{2 \times 0.05}\ln\frac{1}{0.3^2}$$

$$= 24.07\,cm^{-1},$$

and

$$g = \alpha_{int} + \alpha_{mir}$$

$$= (11.51 + 24.07)\,cm^{-1}$$

$$= 35.58\,cm^{-1}.$$

xample 3.4

In a gain medium, under steady-state conditions, the mean power is 20 mW. The area perpendicular to the direction of light propagation is $100\,\mu m^2$. The refraction index of the gain medium is 3.2. Calculate the energy density.

Solution:
The optical intensity $\boxtimes$ is power per unit area perpendicular to the direction of light propagation,

$$\boxtimes = \frac{P}{A} = \frac{20 \times 10^{-3}}{100 \times 10^{-12}} = 2 \times 10^8\,W/m^2.$$

The relation between optical intensity and energy density is given by Eq. (3.50),

$$I = uv$$

$$= \frac{uc}{n}.$$

The energy density is

$$u = \frac{nI}{c} = \frac{3.2 \times 2 \times 10^8}{3 \times 10^8} = 2.13\,J/m^3.$$

3.4 Laser xamples

3.4.1 Ruby Laser

The ruby laser was the first laser to be operated and it was demonstrated by Maiman [3] at the Hughes Research Laboratory in early 1960. The ruby laser is a three-level laser system, as shown in Fig. 3.15. The three-level system consists of a ground level with energy E_1, a metastable level with energy E_2, and a third level with energy E_3. Under thermal equilibrium, the number of chromium ions of the ruby crystal in level 2 is less than that in level 1 and, therefore, laser action can not be achieved. The chromium ions absorb the light of a flash lamp and make transitions to the level with energy E_3. The upper pumping level (E_3) is actually a broad collection of levels and, therefore, an optical source such as a flash lamp with wide spectrum can be used as the optical pump. The chromium ions drop down to a level with energy E_2 by rapid non-radiative transition. The chromium ions spend an unusually long time at the metastable state E_2. As a result, the population density of level 2 becomes more than that of the ground state, and the population inversion is achieved. The chromium ions make transitions from level 2 to level 1 by emitting photons of wavelength 694 nm corresponding to red light.

3.4.2 Semiconductor Lasers

The energy levels of electrons and holes are similar to the two-level atomic system we have discussed before. If an electron is recombined with a hole, the energy difference between them is released as radiation. To achieve population inversion, we need to have more electrons in the conduction band than in the valence band. This is done by means of an electrical pumping scheme. We will discuss these issues in detail in Section 3.8.

3.5 Wave–Particle Duality

We know that light acts as a wave in free space, and propagation is governed by the wave equation. However, when light interacts with matter, it may act like a particle. This is known as *wave–particle duality*. Although Maxwell's equations explain the effects such as interference and diffraction resulting from the wave nature of light, they fail to explain the effects associated with light–matter interaction such as the photoelectric effect. Light can be imagined to consist of particles known as photons of energy

$$E = \hbar\omega, \tag{3.56}$$

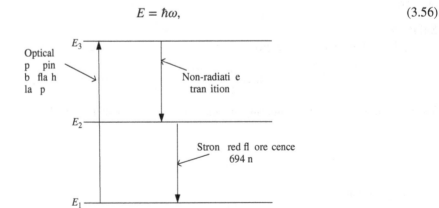

Figure 3.15   Ruby laser–a three-level laser system.

where ω is the angular frequency. In the wave picture, the optical field of a plane wave can be written as

$$\psi = A \exp\left[-i(\omega t - k_x x - k_y y - k_z z)\right]. \tag{3.57}$$

It has four degrees of freedom, ω, k_x, k_y, and k_z. If we imagine light as a particle, it has four degrees of freedom too. They are energy E and momenta p_x, p_y, and p_z in the x-, y- and z-directions, respectively. Energy (particle picture) and frequency (wave picture) are related by Eq. (3.56). Similarly, the wave vector components are related to the momenta by

$$p_x = \hbar k_x, \quad p_y = \hbar k_y, \quad p_z = \hbar k_z, \tag{3.58}$$

or

$$\mathbf{p} = \hbar \mathbf{k}, \tag{3.59}$$

where

$$\mathbf{p} = p_x \mathbf{x} + p_y \mathbf{y} + p_z \mathbf{z}, \tag{3.60}$$

$$\mathbf{k} = k_x \mathbf{x} + k_y \mathbf{y} + k_z \mathbf{z}. \tag{3.61}$$

From Eq. (3.59), we see that the photon carries a momentum in the direction of propagation. The magnitude of the momentum is

$$p = |\mathbf{p}| = \hbar |\mathbf{k}| = \hbar k. \tag{3.62}$$

If $\mathbf{k} = k_x \mathbf{x} + k_y \mathbf{y} + k_z \mathbf{z}$,

$$p = \hbar \sqrt{k_x^2 + k_y^2 + k_z^2}. \tag{3.63}$$

Light is a wave in free space, but it sometimes acts like a particle when it interacts with matter. In the early 1920s, De Broglie proposed that every particle (atom, electron, photon, etc.) has a wave nature associated with it. If a particle has energy E and momentum $\mathbf{p}$, the angular frequency associated with its wave part is $E/\hbar$ and the wave vector is $\mathbf{k} = \mathbf{p}/\hbar$ or the wavenumber is

$$k = \frac{2\pi}{\lambda} = \frac{p}{\hbar}, \tag{3.64}$$

or

$$\lambda = \frac{2\pi\hbar}{p}. \tag{3.65}$$

This wavelength is called the *De Broglie wavelength*. The wave nature of particles has been confirmed by several experiments. The first electron-diffraction experiment was done by Davisson and Germer in 1927 [8]. In this experiment, the incident beam was obtained by the acceleration of electrons in an electrical potential and diffraction of the electron beam by a single crystal is studied. This experiment showed that electrons behave as waves, exhibiting the features of diffraction and interference. From the electron interference pattern, it is possible to deduce the experimental value of the electron wavelength which is in good agreement with De Broglie's formula, Eq. (3.65). However, it turns out that these are not real waves, but probabilistic waves. If electrons were matter waves, we would expect that the intensity of the interference pattern should reduce as the intensity of the incident beam decreases, but the interference pattern should not become discontinuous; electron diffraction experiments contradict the above property of a matter wave. If the intensity of the incident electron beam in these experiments is reduced to a very low value, we would observe a single impact either on the central spot or on one of the diffraction rings, which shows the particle nature of electrons. The simplest interpretation we could give of wave–particle duality is a statistical interpretation: the intensity of the wave at each point on the diffraction pattern gives the probability of occurence of an impact at that point [9].

The energy and momentum of a free non-relativistic particle are related by

$$E = \frac{p^2}{2m_0} = \frac{\hbar^2 k^2}{2m_0} = \frac{\hbar^2}{2m_0}\left(\frac{2\pi}{\lambda}\right)^2, \tag{3.66}$$

where m_0 is the mass of a particle. Thus, an electron with high energy has a large momentum or in other words, has a short wavelength (see Eq. (3.65)). Typically, the highest achievable resolution of an optical microscope is of the order of the wavelength of the light used ($\sim 0.4 - 2\,\mu$m) and, therefore, it is hard to study the structure of a nanoparticle using optical microscopes. If we increase the energy of an electron, the corresponding De Broglie wavelength can be made quite small. This property is the principle behind electron microscopy.

The energy can be measured in several units, such as Joules (J) and electron volts (eV). Voltage is defined as the potential energy E per unit charge,

$$V = \frac{E}{q}. \tag{3.67}$$

The energy required to carry an electron of charge 1.602×10^{-19} C over a potential barrier of 1 V is 1 eV. With $q = 1.602 \times 10^{-19}$ C and V = 1 volt, from Eq. (3.67) we have

$$E = qV$$
$$= 1.602 \times 10^{-19} \times 1\,\text{J}$$
$$= 1\,\text{eV}. \tag{3.68}$$

Note that the electron volt is not a unit of voltage, but of energy.

xample 3.5

The energy difference between the two states of an ammonia maser is 10^{-4} eV. Calculate the frequency of the electromagnetic wave emitted by stimulated emission.

Solution:

$$E = 10^{-4}\,\text{eV} = 10^{-4} \times 1.602 \times 10^{-19}\,\text{J},$$
$$f = \frac{E}{2\pi\hbar} = \frac{10^{-4} \times 1.602 \times 10^{-19}}{2\pi \times 1.054 \times 10^{-34}} = 24\,\text{GHz}.$$

3.6 Laser Rate quations

In this section, we consider the gain rate and loss rate of photons and population densities in states 1 and 2 due to stimulated emission, spontaneous emission, and various loss mechanisms. This is similar to the population growth rate of a country. The population of a country increases due to new births and immigration, while it decreases due to death and migration to other countries. Suppose $N(t)$ is the population at t, the net rate of population growth may be modeled as

$$\frac{dN}{dt} = R_{\text{born}} + R_{\text{immigration}} + R_{\text{death}} + R_{\text{migration}}. \tag{3.69}$$

To model lasers, we follow a similar approach. Let us consider the atomic system with two levels. The population density of the excited state decreases due to stimulated emission, spontaneous emission, and non-radiative transition, while it increases due to absorption and external pumping. The net growth rate of the population density of state 2 is

$$\frac{dN_2}{dt} = R_{\text{pump}} + R_{\text{abs}} + R_{\text{stim}} + R_{\text{spont}} + R_{nr}. \tag{3.70}$$

Here, R_{abs}, R_{stim}, and R_{spont} are given by Eqs. (3.2), (3.3), and (3.4), respectively. R_{pump} refers to the pumping rate, which is the rate at which the population density of state 2 grows due to an external pump. A specific pumping scheme is discussed in Section 3.8.

An atom in state 2 could drop down to state 1 by releasing the energy difference as translational, vibrational, or rotational energies of the atom or nearby atoms/molecules. This is known as non-radiative transition, since no photon is emitted as the atom makes transition from state 2 to state 1 and R_{nr} represents the rate of non-radiative transition from state 2 to state 1. It is given by

$$R_{nr} = CN_2, \tag{3.71}$$

where C is a constant similar to the Einstein coefficient A.

Using Eqs. (3.2), (3.3), (3.4), and (3.71) in Eq. (3.70), we find

$$\frac{dN_2}{dt} = R_{\text{pump}} + BuN_1 - BuN_2 - (A + C)N_2. \tag{3.72}$$

The population density of the ground state increases due to stimulated emission, spontaneous emission, and non-radiative transition, while it decreases due to absorption. The rate of change of the population density of the ground state is

$$\frac{dN_1}{dt} = R_{\text{stim}} + R_{\text{spont}} + R_{nr} + R_{\text{abs}}$$

$$= BuN_2 + (A + C)N_2 - BuN_1. \tag{3.73}$$

Next, let us consider the growth rate of photons. Let N_{ph} be the photon density. When *an* atom makes a transition from the excited state to the ground state due to stimulated emission, it emits *a* photon. If there are R_{stim} transitions per unit time per unit volume, the growth rate of photon density is also R_{stim}. The photon density in a laser cavity increases due to stimulated emission and spontaneous emission, while it decreases due to absorption and loss in the cavity. The growth rate of photon density is given by

$$\frac{dN_{ph}}{dt} = R_{\text{stim}} + R_{\text{spont}} + R_{\text{abs}} + R_{\text{loss}}. \tag{3.74}$$

Here, R_{loss} refers to the loss rate of photons due to internal loss and mirror loss in the cavity. Since the energy of a photon is $\hbar\omega$, the mean number of photons present in the electromagnetic radiation of energy E is

$$n_{ph} = \frac{E}{\hbar\omega}. \tag{3.75}$$

The photon density N_{ph} is the mean number of photons per unit volume and the energy density u is the energy per unit volume. Therefore, they are related by

$$N_{ph} = \frac{n_{ph}}{V} = \frac{E}{\hbar\omega V} = \frac{u}{\hbar\omega}. \tag{3.76}$$

In Section 3.3, we developed an expression for the time rate of change of the energy density u in the presence of stimulated emission and loss as

$$\frac{du}{dt} = (G - v\alpha_{\text{cav}})u. \tag{3.77}$$

Since $u \propto N_{ph}$, the time rate of change of photon density is

$$\frac{dN_{ph}}{dt} = GN_{ph} - \frac{N_{ph}}{\tau_{ph}}, \tag{3.78}$$

where

$$\tau_{ph} = \frac{1}{v\alpha_{cav}} \tag{3.79}$$

is the *photon lifetime*. In the absence of gain ($G = 0$), Eq. (3.78) can be solved to yield

$$N_{ph}(t) = N_{ph}(0) \exp(-t/\tau_{ph}). \tag{3.80}$$

At $t = t_{ph}$, $N_{ph}(t) = N_{ph}(0)e^{-1}$. Thus, the photon density reduces by e over a time t_{ph}. In Eq. (3.78), G represents the net gain coefficient due to stimulated emission and absorption and, therefore, the first term on the right-hand side of Eq. (3.78) can be identified as

$$R_{stim} + R_{abs} = GN_{ph}, \tag{3.81}$$

or

$$BuN_2 - BuN_1 = GN_{ph}. \tag{3.82}$$

Since $u = N_{ph}\hbar\omega$, from Eq. (3.82) we find

$$G = B(N_2 - N_1)\hbar\omega. \tag{3.83}$$

In Eq. (3.78), the second term represents the loss, rate due to scattering, mirror loss, and other possible loss mechanisms,

$$R_{loss} = -\frac{N_{ph}}{\tau_{ph}}. \tag{3.84}$$

Eq. (3.78) does not include the photon gain rate due to spontaneous emission. Using Eqs. (3.81), (3.84), and (3.4) in Eq. (3.74), we find

$$\frac{dN_{ph}}{dt} = GN_{ph} + AN_2 - \frac{N_{ph}}{\tau_{ph}}. \tag{3.85}$$

Note that when $N_2 > N_1$, population inversion is achieved, $G > 0$ (see Eq. (3.83)) and amplification of photons takes place. In other words, the energy of the atomic system is transferred to the electromagnetic wave. When $N_2 < N_1$, the electromagnetic wave is attenuated and the energy of the wave is transferred to the atomic system.

Using Eq. (3.82), Eqs. (3.72) and (3.73) can be rewritten as

$$\frac{dN_2}{dt} = R_{pump} - GN_{ph} - \frac{N_2}{\tau_{21}}, \tag{3.86}$$

$$\frac{dN_1}{dt} = GN_{ph} + \frac{N_2}{\tau_{21}}, \tag{3.87}$$

where

$$\tau_{21} = \frac{1}{A + C} \tag{3.88}$$

is the lifetime associated with spontaneous emission and non-radiative decay from the excited state to the ground state. Eqs. (3.86) and (3.87) can be simplified further under the assumption that the population density of the ground state is negligibly small compared with the population density of the excited state [10, 11]. Assuming $N_1 \approx 0$, Eqs. (3.86) and (3.85) become

$$\frac{dN_2}{dt} = R_{\text{pump}} - GN_{ph} - \frac{N_2}{\tau_{21}}, \qquad (3.89)$$

$$\frac{dN_{ph}}{dt} = GN_{ph} + AN_2 - \frac{N_{ph}}{\tau_{ph}}, \qquad (3.90)$$

where

$$G = BN_2 \hbar \omega. \qquad (3.91)$$

The equations describing the population densities of electrons and photons in a semiconductor laser are similar to Eqs. (3.89) and (3.90). In Section 3.8, we will solve Eqs. (3.89) and (3.90) for a specific pumping scheme.

3.7 Review of Semiconductor Physics

In conductors, such as metals, electrons move around freely in the lattice and are available for conduction. These electrons can be drifted by applying an electric field across its terminals. In contrast, the insulators hardly have free electrons in the lattice and, therefore, they do not conduct. Semiconductor materials have conduction properties that are typically intermediate between that of a metal and an insulator. For example, silicon has four electrons in its outermost shell, by which it makes covalent bonds with its neighboring atoms, as shown in Fig. 3.16. These electrons are somewhat loosely bound. An electron can gain energy by external means, such as thermal energy, to break the covalent bond and, thereby, contribute to the *conduction band*.

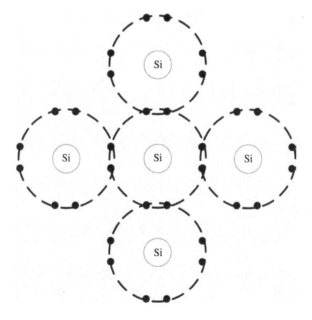

Figure 3.16 Covalent band structure of silicon.

Loosely speaking, if an electron is confined to the outermost shell of the atom, it is said to be in the *valence band* and if it is moving freely in the lattice, it is said to be in the conduction band. Strictly speaking, atoms of solid-state materials have such a strong interaction that they cannot be treated as individual entities. Valence electrons are not attached to individual atoms, instead, they belong to the system of atoms as a whole.

The conduction band and valence band are separated by an energy gap, or *band gap* E_g, as shown in Fig. 3.17. For Si, the band gap is 1.1 eV. At low temperature, the chance that an electron occupies the conduction band is approximately proportional to $\exp(-E_g/k_B T)$. Materials with a filled valence band and a large band gap (>3 eV) are insulators. Those for which the gap is small or non-existent are conductors. Semiconductors have band gaps that lie roughly in the range of 0.1 to 3 eV.

At very low temperature, the conduction band is nearly empty and, therefore, the valence band is nearly full, as shown in Fig. 3.18. As the temperature increases, electrons in the valence band gain energy to cross the band gap and get into the conduction band. This leads to a concentration of free electrons in the conduction band, which leaves behind equal numbers of vacancies or *holes* in the valence band. A hole refers to the absence of an electron and it acts as if it is a positive charge. Consider a semiconductor material connected to the terminals of a battery, as shown in Fig. 3.19. The electron in the leftmost region is attracted to the positive terminal of the battery and it leaves behind a hole (Fig. 3.19(a)). An electron from the neighboring atom jumps to fill the hole, thereby creating a hole as shown in Fig. 3.19(b). This process continues, and holes move to the right constituting a hole current. In addition, electrons moving freely in the lattice are also attracted to the positive terminal constituting the electron current. Free electrons can move far more easily around the lattice than holes. This is because the free electrons have already broken the covalent bond, whereas for a hole to travel through the structure, an electron must have sufficient energy to break the covalent bond each time a hole jumps to a new position.

When an electron comes out of the outermost shell of an atom after picking up thermal energy, it does not really become a free particle. This is because the electron is in periodic Coulomb potential due to atoms in the lattice, as shown in Fig. 3.20. Consider an electron in the vicinity of atom 1. There is a chance that it will

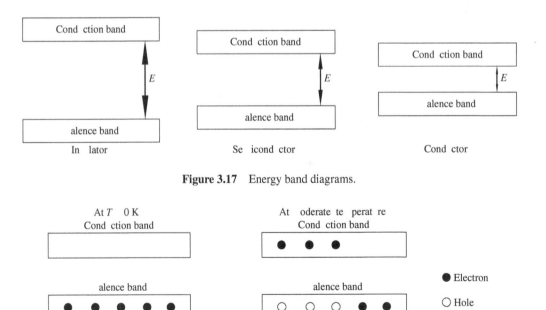

Figure 3.17   Energy band diagrams.

Figure 3.18   Temperature dependence of electron density in the conduction band.

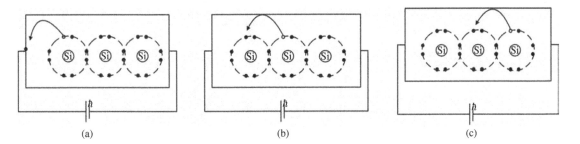

Figure 3.19 Electron and hole current in Si.

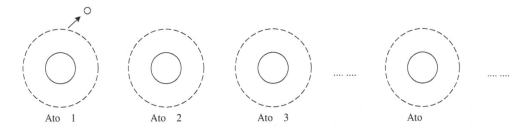

Figure 3.20 Electron in a periodic Coloumb potential.

be attracted toward the nucleus of atom 2. When it is in the vicinity of atom 2, there is a chance that it will be attracted toward atom 3 or atom 1. An alectron can hop on and off from atom to atom as if it were a free classical particle, but with the following difference. A free non-relativistic particle can acquire any amount of energy and the energy states are continuous. However, for an electron in periodic potential, there is a range of energy states that are forbidden and this is called the energy band gap. The existence of a band gap can only be explained by quantum mechanics. At very low temperature, electrons have energy states corresponding to valence bands. As the temperature increases, electrons occupy energy states corresponding to conduction bands, but are not allowed to occupy any energy states that are within the band gap, as shown in Fig. 3.21. For a free electron, the energy increases quadratically with k as given by Eq. (3.66),

$$E = \frac{\hbar^2 k^2}{2m_0},\tag{3.92}$$

where m_0 is the rest mass of an electron. Differentiating Eq. (3.92) twice, we find

$$m_0 = \frac{\hbar^2}{d^2 E/dk^2}.\tag{3.93}$$

The dotted line in Fig. 3.21 shows the plot of energy as a function of the wavenumber, k, for the free electron. For an electron in a pure semiconductor crystal, the plot of energy vs. wavenumber is shown as a solid line. We define the effective mass of an electron in the periodic potential as

$$m_{\text{eff}}(k) = \frac{\hbar^2}{d^2 E/dk^2}\tag{3.94}$$

in the allowed range of energy states, in analogy with the case of free electrons. The effective mass can be larger or smaller than the rest mass, depending on the nature of the periodic potential. For example, for GaAs,

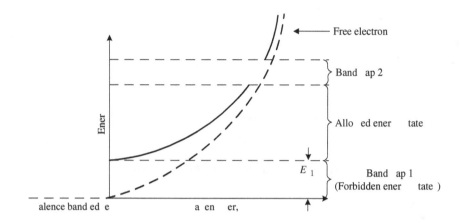

Figure 3.21 Plot of energy vs. wavenumber. The dotted line and solid line correspond to the energy of a free electron and an electron in a pure semiconductor crystal, respectively.

$m_{\text{eff}} = 0.07 m_0$ in the conduction band. The significance of the effective mass can be explained as follows. Suppose an electron in the pure semiconductor crystal is subjected to an external electric field intensity, ψ, then the equation of motion is given by Newton's law,

$$\frac{d(m_{\text{eff}} v)}{dt} = \text{force} = q\psi, \tag{3.95}$$

where v and q are the velocity and charge of an electron, respectively. Note that an electron in a pure semiconductor crystal behaves as if it is a free particle with effective mass m_{eff}. An electron with $m_{\text{eff}} < m_0$ experiences more acceleration than a free electron subjected to the same force $q\psi$.

The chance that an electron occupies an energy state E in thermal equilibrium is described by the Fermi–Dirac function [12]

$$f(E) = \frac{1}{\exp\left[(E - E_F)/k_B T\right] + 1}, \tag{3.96}$$

where E_F is called the Fermi level. From Eq. (3.96), we see that when $E = E_F$, $f(E) = 0.5$. Fig. 3.22 shows the Fermi–Dirac function as a function of energy. Note that the Fermi function $f(E)$ is not normalized and, therefore, it is not a probability density function. For an intrinsic semiconductor, the Fermi level is in the middle of the energy gap, as shown in Fig. 3.23.

For example, at the conduction band bottom, $E_c = E_v + E_g$ and $E_F = E_v + E_g/2$, at temperature $T = 300$ K, $k_B T \approx 0.025$ eV, and with $E_g = 1$ eV,

$$f(E_c) = \frac{1}{\exp\left[\frac{E_g}{2k_B T}\right] + 1} \approx \exp\left[-\frac{E_g}{2k_B T}\right] = \exp(-20). \tag{3.97}$$

Thus, at room temperature, the chance that an electron occupies the conduction band is very small and, therefore, the electrical conductivity of the intrinsic semiconductor is quite low. But the conductivity can be increased by adding impurity atoms. The basic semiconductor without doping is called an *intrinsic* semiconductor. Doping consists of adding impurities to the crystalline structure of the semiconductor. For example, a small amount of group V elements such as arsenic can be added to silicon. Arsenic has five electrons in the outermost shell; four electrons form a covalent bond with neighboring silicon atoms, as shown in Fig. 3.24, but there is one electron left over that can not take part in bonding. This fifth electron is very loosely attached

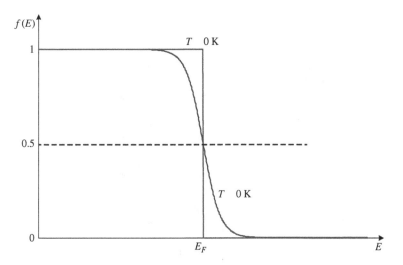

Figure 3.22   Fermi–Dirac function.

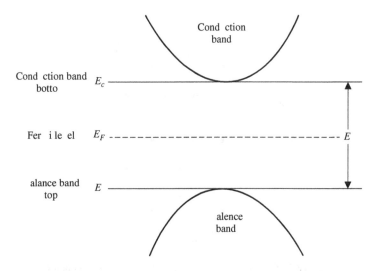

Figure 3.23   Energy-band diagram of an intrinsic semiconductor.

and it is free to move through the crystal when an electric field is applied. Thus the number of free electrons in the crystal is enhanced by doping with arsenic. Group V elements such as arsenic added to group IV elements are called *donors*, since they contribute free electrons. The resultant semiconductor material is known as an *n-type semiconductor*.

When a small amount of group III elements such as gallium are added to silicon, three valence electrons of gallium form a covalent bond with the neighboring three silicon atoms, while the fourth silicon atom shown in Fig. 3.25 is deprived of an electron to complete a total of eight electrons. The missing electron is a hole that can be filled by an electron that is in the neighborhood. Thus, the number of holes is increased by doping with group III elements, which are known as *acceptors*, and the resultant semiconductor is known as a *p-type semiconductor*.

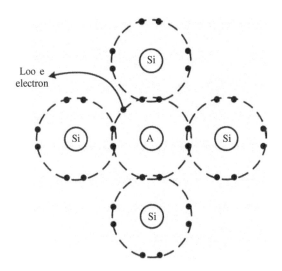

Figure 3.24   Molecular structure of an n-type semiconductor.

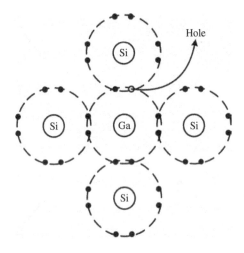

Figure 3.25   Molecular structure of a p-type semiconductor.

3.7.1   The PN Junctions

A PN junction or diode is formed by bringing p-type and n-type materials in contact, as shown in Fig. 3.26. The n-side is electron-rich while the p-side has only a few electrons. So, the electrons diffuse from the n-side to the p-side. Similarly, holes diffuse from the p-side to the n-side. As the electrons and holes cross the junction, they combine. When an electron combines with a hole, it means that it becomes part of the covalent bond. As a result, a region close to the junction is depleted of electrons and holes and hence this region is called a *depletion region*. Because of the addition of electrons on the p-side, the depletion region on the p-side consists of negative acceptor ions and similarly, the depletion region on the n-side consists of positive donor ions due to migration of holes. Note that the p-type material (in the absence of the n-type material on the right) is electrically neutral even though it has holes. This is because the charge of holes and that of the lattice cancel

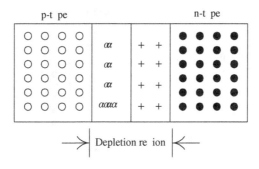

Figure 3.26   A PN junction or diode.

each other on average (if this was not the case, you would get a shock when you touched the p-type material). Similarly, n-type material is also charge neutral in the absence of the p-type material. However, when these materials are brought together to form a PN junction, we have negatively charged acceptor ions on the p-side and positively charged donor ions on the n-side, which acts as a battery. This is known as *contact potential*. This potential V_0 is about 0.6 to 0.7 V for silicon.

Next, let us consider the case when the diode is connected to the terminals of a voltage source. The diode is said to be *reverse-biased* when the positive terminal of the source is connected to the n-side and the negative terminal of the source is connected to the p-side, as shown in Fig. 3.27. Now, electrons in the rightmost region of the n-side are attracted to the positive terminal of the battery, and electrons closer to the PN junction move to the right, which enhances the width of the depletion region on the n-side. A similar effect happens on the p-side, which leads to the widening of the depletion region, as shown in Fig. 3.27.

The diode is said to be *forward-biased* when the positive terminal of the source is connected to the p-side and the negative terminal of the source is connected to the n-side, as shown in Fig. 3.28. The electrons on the n-side are attracted to the positive terminal of the voltage source and the holes on the p-side are attracted

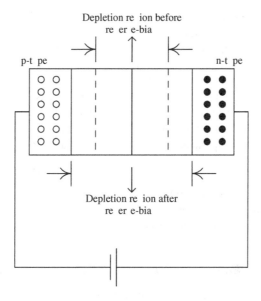

Figure 3.27   The PN junction under reverse-bias.

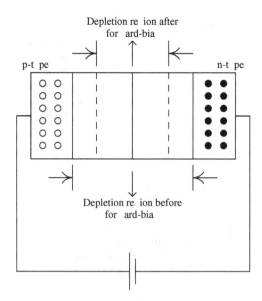

Figure 3.28 The PN junction under forward-bias.

to the negative terminal of the source. As a result, there is a current across the PN junction. The external voltage counteracts the contact potential and, therefore, the width of the depletion region decreases as shown in Fig. 3.28. We assume that the external voltage V is less than the contact potential. An external resistor must be inserted in series with the voltage source to protect the PN junction from an excess current flow.

3.7.2   Spontaneous and Stimulated Emission at the PN Junction

The conduction band and valence band are similar to the excited state and ground state of the atomic system discussed in Section 3.2, respectively. In the case of the atomic system, an atom in the ground state absorbs a photon and makes a transition to the excited state. Similarly, in a semiconductor, an electron in the valence band could jump to the conduction band by absorbing a photon if its energy exceeds the band-gap energy. As the electron moves to the conduction band, it leaves behind a hole in the valence band. In other words, a photon is annihilated to create an electron–hole pair. An electron in the conduction band is stimulated to emit a photon if a photon of the same kind is already present, and it jumps to the valence band. In other words, an electron combines with a hole, releasing the difference in energy as a photon. An electron in the conduction band could jump to the valence band spontaneously, whether or not a photon is present. This occurs randomly, leading to spontaneous emission.

Now let us consider the forward-biased PN junction. As electrons and holes cross the junction, they combine and release the difference in energy as photons. The spontaneously generated photons act as a seed for stimulated emission. As electrons are lost due to electron–hole recombination, the external voltage source injects electrons. Thus, the voltage source acts as an electrical pump to achieve population inversion.

3.7.3   Direct and Indirect Band-Gap Semiconductors

Fig. 3.29 shows a plot of energy as a function of wavenumber. Let E_g be the minimum energy required to excite an electron to the conduction band. If an electron absorbs the energy $E_1 > E_g$, the excess energy appears

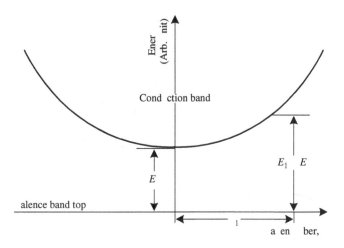

Figure 3.29 $E-k$ diagram assuming parabolic conduction band.

in the form of kinetic energy. If we assume that the energy depends on the wavenumber quadratically in the conduction band as in the case of a free particle, the energy of an electron in the conduction band is given by [13]

$$E_1 = E_g + \frac{\hbar^2 k_1^2}{2m_{\text{eff},1}},$$ (3.98)

where $m_{\text{eff},1}$ is the effective mass of an electron in the conduction band and $\hbar k_1$ is the momentum. If the bottom of the conduction band is aligned with the top of the valence band as shown in Fig. 3.30, such a material is called a *direct band-gap* material. For example, GaAs and InP are direct band-gap materials. For *indirect band-gap* materials, the conduction band minimum and valence band maximum occur at different values of momentum, as shown in Fig. 3.31. Silicon and germanium are indirect band-gap semiconductors.

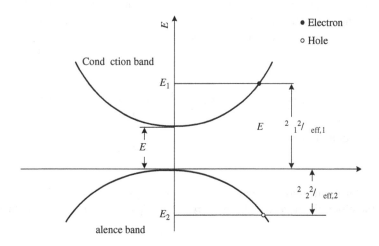

Figure 3.30 Simplified $E-k$ diagram for a direct band-gap material.

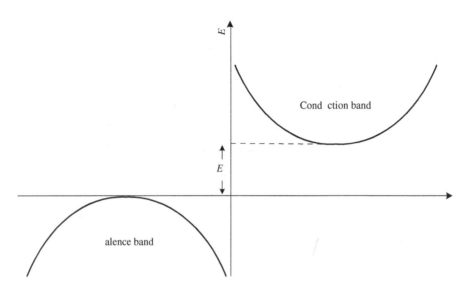

Figure 3.31   Simplified $E-k$ diagram for an indirect band-gap material.

Let us first consider a direct band-gap semiconductor. Let the energy of the top of the valence band be the reference with $E = 0$. The energy of an electron in the conduction band with effective mass $m_{\text{eff},1}$ is

$$E_1 = E_g + \frac{\hbar^2 k_1^2}{2m_{\text{eff},1}}. \tag{3.99}$$

The energy of an electron in the valence band, or equivalently that of a hole with effective mass $m_{\text{eff},2}$, is

$$E_2 = -\frac{\hbar^2 k_2^2}{2m_{\text{eff},2}}. \tag{3.100}$$

When an electron makes a transition from energy state E_1 to E_2, a photon of energy $\hbar\omega$ is emitted. In other words, an electron in the conduction band recombines with a hole in the valence band, releasing the energy difference as a photon. The conservation of energy yields

$$E_1 = E_2 + \hbar\omega,$$

$$\hbar\omega = E_1 - E_2 = E_g + \frac{1}{2}\left[\frac{\hbar^2 k_1^2}{m_{\text{eff},1}} + \frac{\hbar^2 k_2^2}{m_{\text{eff},2}}\right]. \tag{3.101}$$

Similarly, the conservation of momentum yields

$$\hbar k_1 = \hbar k_2 + \hbar k_{ph}, \tag{3.102}$$

where $\hbar k_{ph}$ is the photon momentum. At optical frequencies, $\hbar k_{ph} \ll \hbar k_j, j = 1, 2$. From Eq. (3.102), it follows that $k_1 \cong k_2$ and, therefore, the transition from energy state E_1 to E_2 in Fig. 3.30 is almost vertical. From Eq. (3.101), we find

$$\hbar\omega = E_g + \frac{\hbar^2 k_1^2}{2m_r}, \tag{3.103}$$

where

$$m_r = \frac{m_{\text{eff},1} m_{\text{eff},2}}{m_{\text{eff},1} + m_{\text{eff},2}} \qquad (3.104)$$

is the reduced effective electron mass [13]. For indirect band-gap materials, the momenta of electrons in the conduction band and in the valence band are different. Typically, the difference in momenta is much larger than the photon momentum and, therefore, the momentum can not be conserved in the electron–photon interaction unless the photon emission is mediated through a phonon. A phonon refers to the quantized lattice vibration or sound wave. If the momentum of the phonon is equal to the difference $\hbar(k_1 - k_2)$, the chance of photon emission as an electron jumps from conduction band to valence band increases. In other words, in the absence of phonon mediation, the event that an electron makes a transition from conduction band to valence band emitting a photon is less likely to happen. Therefore, indirect band-gap materials such as silicon and germanium are not used for making lasers, while direct band-gap materials such as GaAs and InP (and their mixtures) are used for the construction of lasers. However, silicon can be used in photo-detectors. As we will discuss in Chapter 5, an electron in the valence band jumps to the conduction band by absorbing a photon. We may expect that such an event is less likely to happen in silicon because of the momentum mismatch in the electron–photon interaction. But the crystal lattice vibrations (crystal momentum) provide the necessary momentum so that the momentum is conserved during the photon-absorption process. In contrast, during the photon-emission process, phonon mediation is harder to come by since the free electrons in the conduction band are not bound to atoms, and, therefore, they do not vibrate within the crystal structure [10].

xample 3.6

In a direct band-gap material, an electron in the conduction band having a crystal momentum of 7.84×10^{-26} Kg· m/s makes a transition to the valence band emitting an electromagnetic wave of wavelength $0.8\,\mu\text{m}$. Calculate the band-gap energy. Assume that the effective mass of an electron in the conduction band is $0.07m$ and that in the valence band is $0.5m$, where m is the electron rest mass. Assume parabolic conduction and valence band.

Solution:
The reduced mass m_r is related to the effective masses by Eq. (3.104),

$$m_r = \frac{m_{\text{eff},1} m_{\text{eff},2}}{m_{\text{eff},1} + m_{\text{eff},2}}.$$

$m_{\text{eff},1} = 0.07m$, $m_{\text{eff},2} = 0.5m$, electron mass $m = 9.109 \times 10^{-31}$ kg,

$$m_r = \frac{0.07 \times 0.5}{0.07 + 0.5} \times 9.109 \times 10^{-31} = 5.59 \times 10^{-32} \text{ kg}.$$

The electron momentum is

$$\hbar k_1 = 7.84 \times 10^{-26} \text{ kg} \cdot \text{m/s}.$$

The photon energy is

$$\hbar\omega = \hbar 2\pi \frac{c}{\lambda} = \frac{1.054 \times 10^{-34} \times 2\pi \times 3 \times 10^8}{0.8 \times 10^{-6}}$$

$$= 2.48 \times 10^{-19} \text{ J}.$$

From Eq. (3.103), we have

$$E_g = \hbar\omega - \frac{\hbar^2 k_1^2}{2m_r} = 2.48 \times 10^{-19} - \frac{(7.84 \times 10^{-26})^2}{2 \times 5.59 \times 10^{-32}} \text{ J}$$

$$= 1.93 \times 10^{-19} \text{ J}.$$

3.8   Semiconductor Laser Diode

The light emission in laser diodes is mostly by stimulated emission, whereas that in LEDs is mostly by spontaneous emission. Laser diodes can emit light at high powers (~ 100 mW) and also it is coherent. Because of the coherent nature of laser output, it is highly directional. The narrower angular spread of the output beam compared with a LED allows higher coupling efficiency for light coupling to single-mode fibers. An important advantage of the semiconductor laser is its narrow spectral width, which makes it a suitable optical source for WDM optical transmission systems (see Chapter 9). A semiconductor laser in its simplest form is a forward-biased PN junction. Electrons in the conduction band and holes in the valence band are separated by the band gap and they form a two-band system similar to the atomic system discussed in Sections 3.2 and 3.6. As electrons and holes recombine at the junction, the energy difference is released as photons, as discussed in Section 3.7.2. To obtain oscillation, optical feedback is required, which is achieved by cleaving the ends of the laser cavity. Cleaving provides flat and partially reflecting surfaces. Sometimes one reflector is partially reflecting and used as laser output port and the other has a reflectivity close to unity. By coating the side opposite the output with a dielectric layer, the reflection coefficient could be close to unity.

3.8.1   Heterojunction Lasers

The PN junction shown in Fig. 3.32 is called a *homojunction*. The problem with the homojunction is that when it is forward-biased, electron–hole recombination occurs over a wide region ($1 - 10$ μm). Therefore, high carrier densities can not be realized.

A *heterojunction* is an interface between two adjoining semiconductors with different band-gap energies. In Fig. 3.33, a thin layer is sandwiched between p-type and n-type layers. The band gap of this layer is smaller than that of the p-type and n-type layers, as shown in Fig. 3.34(b). This leads to two heterojunctions and such devices are called double heterostructures. The thin layer, known as the *active region*, may or may not be doped depending on the specific design. For example, the middle layer could be p-type GaAs and the surrounding layers p-type AlGaAs and n-type AlGaAs as shown in Fig. 3.33.

Double-heterojunction lasers have the following advantages: the band-gap difference between the active region and the surrounding layers results in potential energy barriers for electrons in the conduction band and

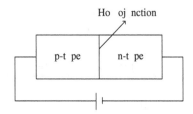

Figure 3.32   A homojunction.

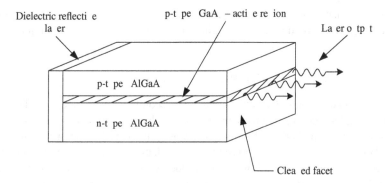

Figure 3.33   A double-heterojunction Fabry–Perot laser diode. The cleaved end functions as a partially reflecting mirror.

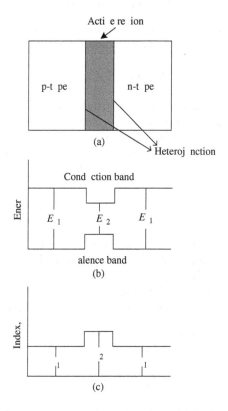

Figure 3.34   Double heterostructure: (a) heterojunctions; (b) band gap; (c) refractive index.

for holes in the valence band. Electrons and holes trapped in the active region could escape to the surrounding layers only if they have sufficient energy to cross the barriers. As a result, both electrons and holes are mostly confined to the active region. Because of the smaller band gap, the active region has a slightly higher refractive index. As discussed in Chapter 2, this acts as an optical waveguide and light is confined to the middle layer of the higher refractive index due to total internal reflection. Therefore, not only electrons and holes are confined to the active region, but also photons, which increases the interaction among them, and the efficiency of light generation in a double heterostructure is much higher than in the devices using homojunctions.

3.8.2   Radiative and Non-Radiative Recombination

When a PN junction is forward-biased, electrons and holes recombine to produce light. This is called radiative recombination. In a semiconductor, electrons and holes can also recombine non-radiatively. In this case, the energy difference is released as lattice vibrations or given to another electron or hole to increase its kinetic energy [14–16]. This type of recombination is called non-radiative recombination. In a practical light source, we like to maximize the radiative recombination by reducing the energy loss due to non-radiative recombination. Therefore, it is useful to define the internal quantum efficiency of a light source as

$$\eta_{int} = \frac{R_{rr}}{R_{tot}} = \frac{R_{rr}}{R_{rr} + R_{nr}}, \tag{3.105}$$

where R_{rr} is the radiative recombination rate, R_{nr} is the non-radiative recombination rate, and R_{tot} is the total recombination rate. Radiative recombination occurs in two different ways: (i) spontaneous emission and (ii) stimulated emission,

$$R_{rr} = R_{spont} + R_{stim}. \tag{3.106}$$

For direct band-gap materials, the radiative recombination rate could be larger than the non-radiative rate since the conservation of energy as well as momentum can be achieved when an electron makes a transition from the conduction band to the valence band emitting a photon. In contrast, for indirect band-gap materials, such as Si and Ge, the electron–hole recombination is mostly non–radiative and, therefore, the internal quantum efficiency is quite small. Typically, n_{int} is of the order of 10^{-5} for Si and Ge.

3.8.3   Laser Rate Equations

In Section 3.6 we developed the rate equations for an atomic system with two levels. In the atomic system, the interaction takes place among the photons, the atoms in the excited level, and in the ground level. Similarly, in the semiconductor laser diode, the interaction is between the electrons in the conduction band, holes in the valence band, and photons. Therefore, Eqs. (3.89) and (3.90) may be used to describe the time rate of change of electrons and photons in a cavity with N_2 being replaced by the electron density N_e,

$$\frac{dN_e}{dt} = R_{pump} + R_{stim} + R_{sp} + R_{nr}$$

$$= R_{pump} - GN_{ph} - \frac{N_e}{\tau_e}, \tag{3.107}$$

$$\frac{dN_{ph}}{dt} = R_{stim} + R_{sp} + R_{loss}$$

$$= GN_{ph} + R_{sp} - \frac{N_{ph}}{\tau_{ph}}. \tag{3.108}$$

Here, τ_e ⊠ τ_{21} represents the lifetime of electrons associated with spontaneous emission and non-radiative transition. In Section 3.3, we found that $G = gv$. This result was derived under the assumption that the light is a plane wave. But in a double-heterojunction laser, the active region has a slightly higher refractive index than the surrounding layers and, therefore, it acts as a waveguide. The tails of an optical mode extend well into the surrounding regions, but they do not contribute to the photon density in the active region. Since the electron–hole recombination by photon emission depends on the photon density in the active region, we introduce a confinement factor Γ,

$$G = \Gamma g v, \tag{3.109}$$

where Γ is the ratio of optical power in the active region to total optical power carried by the mode.

Let us consider the growth of photons due to stimulated emission alone. Eqs. (3.107) and (3.108) become

$$\frac{dN_e}{dt} = -GN_{ph}, \tag{3.110}$$

$$\frac{dN_{ph}}{dt} = GN_{ph}. \tag{3.111}$$

Adding Eqs. (3.110) and (3.111), we find

$$\frac{d(N_e + N_{ph})}{dt} = 0 \tag{3.112}$$

or

$$N_e + N_{ph} = \text{Const.} \tag{3.113}$$

This implies that the total number of electrons and photons is conserved under these conditions. In other words, if you lose 10 electrons per unit volume per unit time by recombination, you gain 10 photons per unit volume per unit time.

Now, let us find an expression for R_{pump}. The electrons and holes are consumed by stimulated emission. Therefore, the external power supply should inject electrons continuously. The current is

$$I = \frac{n_e q}{T}, \tag{3.114}$$

where n_e is the number of electrons, q is the electron charge $= 1.602 \times 10^{-19}$ C, and T is the time interval. The number of electrons crossing the active region per unit time is

$$\frac{n_e}{T} = \frac{I}{q}. \tag{3.115}$$

The above equation gives the electron pumping rate. We divide it by the volume of the active region to obtain the electron pumping rate per unit volume,

$$R_{pump} = \frac{n_e}{TV} = \frac{I}{qdwL}, \tag{3.116}$$

where d, w, and L are thickness, width, and length of the active layer, respectively, as shown in Fig. 3.35(b). Using Eq. (3.116) in Eqs. (3.107) and (3.108), we find

$$\frac{dN_e}{dt} = \frac{I}{qV} - GN_{ph} - \frac{N_e}{\tau_e}, \tag{3.117}$$

$$\frac{dN_{ph}}{dt} = GN_{ph} + R_{sp} - \frac{N_{ph}}{\tau_{ph}}. \tag{3.118}$$

In the case of an atomic system, we have derived an expression for G (see Eq. (3.91)). But in the case of a semiconductor laser, it is hard to find an exact analytical expression for the gain coefficient g. Instead, we use

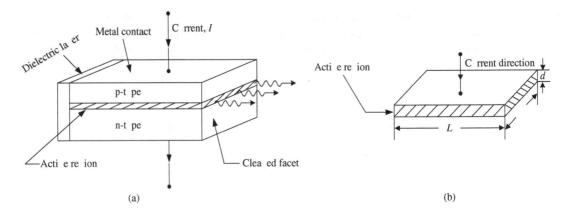

Figure 3.35   (a) Forward-biased heterojunction laser. (b) Active region.

the following approximation [16]:

$$g = \sigma_g(N_e - N_{e0}),$$ (3.119)

where σ_g and N_{e0} are parameters that depend on the specific design. σ_g is called the gain cross-section and N_{e0} is the value of the carrier density at which the gain coefficient becomes zero. Using Eq. (3.109), we find

$$G = \Gamma g v = G_0(N_e - N_{e0})$$ (3.120)

where

$$G_0 = \Gamma \sigma_g v.$$ (3.121)

3.8.4   Steady-State Solutions of Rate Equations

Eqs. (3.117) and (3.118) describe the evolution of electron density and photon density in the active region, respectively. In general, they have to be solved numerically on a computer. However, the steady-state solution can be found analytically under some approximations. First, we ignore the spontaneous emission rate since it is much smaller than the stimulated emission rate for a laser. Second, we use Eq. (3.120) for the gain, which is an approximation to the calculated/measured gain. Now, Eqs. (3.117) and (3.118) become

$$\frac{dN_{ph}}{dt} = GN_{ph} - \frac{N_{ph}}{\tau_{ph}},$$ (3.122)

$$\frac{dN_e}{dt} = -GN_{ph} - \frac{N_e}{\tau_e} + \frac{I}{qV}.$$ (3.123)

We assume that the current I is constant. Under steady-state conditions, the loss of photons due to cavity loss is balanced by the gain of photons due to stimulated emission. As a result, the photon density does not change as a function of time. Similarly, the loss of electrons due to radiative and non-radiative transitions is balanced by electron injection from the battery. So, the electron density does not change with time too. Therefore, under steady-state conditions, the time derivatives in Eqs. (3.122) and (3.123) can be set to zero,

$$\frac{dN_{ph}}{dt} = \frac{dN_e}{dt} = 0.$$ (3.124)

From Eq. (3.122), we have

$$G\tau_{ph} = 1. \tag{3.125}$$

Using Eq. (3.120), we obtain

$$G_0(N_e - N_{e0})\tau_{ph} = 1, \tag{3.126}$$

$$N_e = N_{e0} + \frac{1}{G_0\tau_{ph}}. \tag{3.127}$$

From Eq. (3.125), it follows that

$$\Gamma g = \alpha_{cav}, \tag{3.128}$$

which is a restatement of the fact that gain should be equal to loss. If the current I is very small, there will not be enough electrons in the conduction band to achieve population inversion. In this case, the gain coefficient will be much smaller than the loss coefficient and photons will not build up. For a certain current I, the gain coefficient Γg becomes equal to the loss coefficient α_{cav} and this current is known as the *threshold current*, I_{th}. If $I > I_{th}$, stimulated emission could become the dominant effect and the photon density could be significant. Under steady-state conditions, there are two possibilities.

Case (i) $I = I_{th-}$. Stimulated emission is negligible and $N_{ph} \cong 0$,

$$\frac{dN_e}{dt} = -\frac{N_e}{\tau_e} + \frac{I}{qV} = 0. \tag{3.129}$$

Let $N_e = N_{e,th}$. From Eq. (3.129), we have

$$I_{th} = \frac{N_{e,th}\, qV}{\tau_e}. \tag{3.130}$$

From Eq. (3.127), we have

$$N_{e,th} = N_{e0} + \frac{1}{G_0\tau_{ph}}. \tag{3.131}$$

Case (ii) $I > I_{th}$. When the current exceeds the threshold current, we may expect the electron density N_e to be larger than $N_{e,th}$. However, the electron density will be clamped to $N_{e,th}$ when $I > I_{th}$. This can be explained as follows. The threshold current is the minimum current required to achieve population inversion. When $I > I_{th}$, the excess electrons in the conduction band recombine with holes and, therefore, the photon density increases while the electron density would maintain its value at threshold. Fig. 3.36(a) and 3.36(b) shows the numerical solution of the laser rate equations for $I = 50$ mA and 100 mA, respectively. The threshold current in this example is 9.9 mA. In Fig. 3.36(a) and 3.36(b), after $t > 5$ ns, we may consider it as steady state since N_{ph} and N_e do not change with time. Comparing Fig. 3.36(a) and 3.36(b), we find that the steady-state electron density is the same in both cases, although the bias currents are different. In fact, it is equal to $N_{e,th}$ as given by Eq. (3.131). Using Eqs. (3.124) and (3.125) in Eq. (3.123), we obtain

$$\frac{N_{ph}}{\tau_{ph}} = \frac{I}{qV} - \frac{N_{e,th}}{\tau_e}. \tag{3.132}$$

Using Eq. (3.130), Eq. (3.132) can be written as

$$N_{ph} = \frac{(I - I_{th})\tau_{ph}}{qV}. \tag{3.133}$$

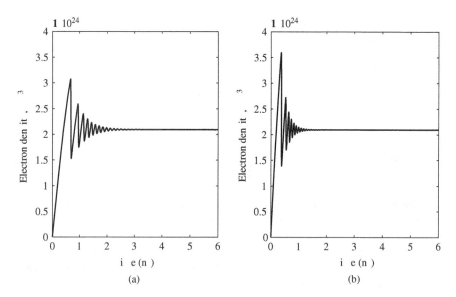

Figure 3.36 Numerical solution of the rate equation using typical parameters of an InGaAsP laser diode: (a) $I = 50\,\text{mA}$; (b) $I = 100\,\text{mA}$.

The next step is to develop an expression for the optical power generated as a function of the current. Since the energy of a photon is equal to $\hbar\omega$, the mean photon density of N_{ph} corresponds to the energy density,

$$u = N_{ph}\hbar\omega. \tag{3.134}$$

The relation between energy density and optical intensity is given by Eq. (3.50),

$$\boxtimes = uv = N_{ph}\hbar\omega v. \tag{3.135}$$

Since optical intensity is power per unit area perpendicular to photon flow, the mean optical power generated can be written as

$$P_{\text{gen}} = \boxtimes A = N_{ph}\hbar\omega v A, \tag{3.136}$$

where A is the effective cross-section of the mode. Using Eq. (3.133) in Eq. (3.136), we finally obtain

$$P_{\text{gen}} = \frac{(I - I_{th})\tau_{ph}\hbar\omega v A}{qwdL}. \tag{3.137}$$

Note that the above equation is valid only when $I > I_{th}$. If $I \boxtimes I_{th}$, $P_{\text{gen}} = 0$ under our approximations.

xample 3.7

A 1300-nm InGaAs semiconductor laser has the following parameters:

Active area width $w = 3\,\mu\text{m}$
Active area thickness $d = 0.3\,\mu\text{m}$

Length $L = 500\,\mu m$
Electron lifetime $= 1\,ns$ (associated with spontaneous and non-radiative recombination)
Threshold electron density $= 0.8 \times 10^{24}\,m^{-3}$
Internal cavity loss $= 46\,cm^{-1}$
Refractive index $= 3.5$
Reflectivity $R_1 = R_2 = 0.65$

Under steady-state conditions, calculate (a) the photon lifetime, (b) the threshold current, and (c) the current required to generate a mean photon density of $8.5 \times 10^{21}\,m^{-3}$.

Solution:
(a) The photon lifetime is given by Eq. (3.79),

$$\tau_p = \frac{1}{v(\alpha_{int} + \alpha_{mir})},$$ (3.138)

$$\alpha_{int} = 46\,cm^{-1} = 46 \times 10^2\,m^{-1}.$$ (3.139)

The mirror loss is, by Eq. (3.38),

$$\alpha_{mir} = \frac{1}{2L} \ln \left(\frac{1}{R_1 R_2} \right)$$

$$= \frac{1}{2 \times 500 \times 10^{-6}} \ln \left(\frac{1}{0.65^2} \right)$$

$$= 8.61 \times 10^2\,m^{-1},$$ (3.140)

$$v = \frac{c}{n} = \frac{3 \times 10^8}{3.5} = 8.57 \times 10^7\,m/s.$$ (3.141)

Using Eqs. (3.139), (3.140), and (3.141) in Eq. (3.138), we find

$$\tau_p = \frac{1}{8.57 \times 10^7 (46 \times 10^2 + 8.61 \times 10^2)} = 2.13\,ps.$$

(b) The threshold current I_{tn} is related to the threshold electron density by Eq. (3.130),

$$I_{th} = \frac{N_{e,th} q V}{\tau_e},$$

where V is the active volume,

$$V = wdL$$

$$= 0.3 \times 10^{-6} \times 3 \times 10^{-6} \times 500 \times 10^{-6}\,m^3$$

$$= 4.5 \times 10^{-16}\,m^3.$$

The electron lifetime $\tau_e = 1 \times 10^{-9}\,s$ and $N_{e,th} = 0.8 \times 10^{24}\,m^{-3}$,

$$I_{th} = \frac{0.8 \times 10^{24} \times 1.602 \times 10^{-19} \times 4.5 \times 10^{-16}}{1 \times 10^{-9}}\,A$$

$$= 52.7\,mA.$$

(c) The mean photon density and the current are related by Eq. (3.133),

$$N_{ph} = \frac{(I - I_{th})\tau_{ph}}{qV}$$

or

$$
\begin{aligned}
I &= I_{th} + \frac{N_{ph}qV}{\tau_{ph}} \\
&= 52.7 \times 10^{-3} + \frac{8.5 \times 10^{21} \times 1.602 \times 10^{-19} \times 4.5 \times 10^{-16}}{2.13 \times 10^{-12}} \\
&= 340.4 \text{ mA.}
\end{aligned}
$$

3.8.5 Distributed-Feedback Lasers

In Section 3.3, we saw that a Fabry–Perot laser supports many longitudinal modes. For many applications, it is desirable to have a single-longitudinal-mode laser. In the case of Fabry–Perot lasers, the cleaved facets act as mirrors. The mirrors can be replaced by periodically corrugated reflectors or Bragg gratings, as shown in Fig. 3.37(b). This type of laser is known as a *distributed Bragg reflector* (DBR) laser [17]. Bragg gratings are formed by periodically changing the refractive index. If Λ is the period of refractive index variations, the Bragg grating acts as a reflector with reflection maxima occurring at frequencies

$$f_m^{\text{Bragg}} = \frac{mc}{2n\Lambda}, \quad m = 1, 2, \dots, \tag{3.142}$$

where n is the effective mode index. The above condition is known as the *Bragg condition*. The longitudinal modes of the cavity which do not satisfy the Bragg condition do not survive, since the cavity loss (= internal loss + Bragg reflector loss) increases substantially for those longitudinal modes. The longitudinal modes of the cavity are given by Eq. (3.44),

$$f_l = \frac{lc}{2nL}, \quad l = 0, 1, 2, \dots \tag{3.143}$$

As an example, if $L = 300\,\mu\text{m}$ and $n = 3.3$, the frequency separation between longitudinal modes $= 0.15$ THz. If the main mode frequency $= 190$ THz, the frequency of two neighboring modes is 189.85 THz and 190.15 THz. The reflection is the strongest for first-order gratings ($m = 1$). If we choose the grating period such that $f_m^{\text{Bragg}} = 190$ THz for $m = 1$, from Eq. (3.142), we find $\Lambda = 0.24\,\mu\text{m}$. The neighboring

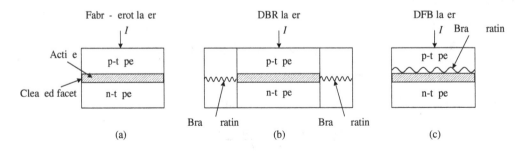

Figure 3.37 Different laser configurations: (a) FP laser, (b) DBR laser, (c) DFB laser.

modes do not satisfy the Bragg condition given by Eq. (3.142) and, therefore, they suffer huge losses. One drawback of the DBR is that the corrugated region is part of the cavity and it is somewhat lossy, which lowers the efficiency of the device. Instead, the corrugated region can be fabricated above the active region as shown in Fig. 3.37(c). Such a laser is known as a *distributed-feedback* (DFB) laser [18, 19]. The grating placed above the waveguide changes the effective index periodically and is equivalent to the waveguide with periodic index variation in the core region. This grating provides coupling between forward- and backward-propagating waves and maximum coupling occurs for frequencies satisfying the Bragg condition, Eq. (3.142). The advantage of a DFB laser is that the corrugated region is not part of the cavity and, therefore, cavity loss does not increase because of the grating. DFB lasers are widely used in applications such as CD players, transmitters in fiber-optic communications, and computer memory readers.

3.9   Additional   xamples

xample 3.8

For an atomic system under thermal equilibrium conditions, the ratio of spontaneous emission rate to stimulated emission rate is 2×10^{14}. Find the wavelength of the light emitted. Assume that the temperature is 30°C.

Solution:
From Eq. (3.16), we have

$$\frac{R_{spont}}{R_{stim}} = \exp\left(\frac{\hbar\omega}{k_B T}\right) - 1,$$

$$2 \times 10^{14} = \exp\left(\frac{\hbar\omega}{k_B T}\right) - 1,$$

$$2 \times 10^{14} \approx \exp\left(\frac{\hbar\omega}{k_B T}\right),$$

$$\frac{\hbar\omega}{k_B T} = \ln(2 \times 10^{14}).$$

With $T = 30°\,\mathrm{C} = 303\,\mathrm{K}$, $k_B = 1.38 \times 10^{-23}\,\mathrm{J/K}$, and $\hbar = 1.054 \times 10^{-34}\,\mathrm{J \cdot s}$,

$$\omega = 1.3 \times 10^{15}\ \mathrm{rad/s}.$$

The wavelength is

$$\lambda = \frac{c}{f} = \frac{2\pi c}{\omega} = \frac{2\pi \times 3 \times 10^8}{1.3 \times 10^{15}} = 1.44\ \mu\mathrm{m}.$$

xample 3.9

A laser diode operating at 1.3 μm has a cavity length of 300 μm and the refractive index n of the active region is 3.5. (a) What is the frequency separation between modes? (b) What is the wavelength separation between modes?

Solution:

(a) The frequency separation Δf is given by Eq. (3.45),

$$\Delta f = f_{n+1} - f_n = \frac{c}{2nL} = \frac{3 \times 10^8}{2 \times 3.5 \times 300 \times 10^{-6}} = 142.8 \text{ GHz}.$$

(b) Since

$$f = \frac{c}{\lambda},$$

we have

$$df = \frac{-c}{\lambda^2} d\lambda$$

or

$$|\Delta \lambda| = \frac{\lambda^2}{c} \Delta f = \frac{(1.3 \times 10^{-6})^2 \times (1.428 \times 10^{11})}{3 \times 10^8} \text{m} = 0.8 \,\mu\text{m}.$$

xample 3.10

In a direct band-gap material, an electron in the valence band having a crystal momentum of $9 \times 10^{-26} \text{ kg} \cdot \text{m/s}$ makes a transition to the conduction band absorbing a light wave of frequency 3.94×10^{14} Hz. The band gap is 1.18 eV and the effective mass of an electron in the conduction band is $0.07m$, where m is the electron rest mass. Calculate the effective mass of the electron in the valence band.

Solution:

From Eq. (3.103), we have

$$hf = E_g + \frac{\hbar^2 k_1^2}{2m_r}, \tag{3.144}$$

where

$$\frac{1}{m_r} = \frac{1}{m_{\text{eff},1}} + \frac{1}{m_{\text{eff},2}} \tag{3.145}$$

and

$$E_g = 1.18 \,\text{eV} = 1.18 \times 1.602 \times 10^{-19} \,\text{J} = 1.89 \times 10^{-19} \,\text{J}.$$

For a direct band-gap material, $k_1 \approx k_2$. The crystal momentum $\hbar k_1 = 9 \times 10^{-26} \text{ kg} \cdot \text{m/s}$ and the energy of the photon is

$$hf = 6.626 \times 10^{-34} \times 3.94 \times 10^{14} \,\text{J} = 2.61 \times 10^{-19} \,\text{J}.$$

From Eq. (3.144), we have

$$\frac{\hbar^2 k_1^2}{2m_r} = hf - E_g = (2.61 \times 10^{-19} - 1.89 \times 10^{-19}) \,\text{J} = 7.2 \times 10^{-20} \,\text{J},$$

$$m_r = \frac{(9 \times 10^{-26})^2}{14.4 \times 10^{-20}} = 5.62 \times 10^{-32} \,\text{kg}.$$

Since the electron rest mass, $m = 9.109 \times 10^{-31}$ kg, $m_{\text{eff},1} = 0.07m = 6.37 \times 10^{-32}$ kg, using Eq. (3.145), we obtain

$$\frac{1}{m_{\text{eff},2}} = \frac{1}{m_r} - \frac{1}{m_{\text{eff},1}} = \frac{1}{5.62 \times 10^{-32}} - \frac{1}{6.37 \times 10^{-32}} \text{ kg}^{-1}$$

$$m_{\text{eff},2} = 4.78 \times 10^{-31} \text{ kg}.$$

xample 3.11

A laser diode has a 320-μm cavity length, the internal loss coefficient is 10 cm^{-1}. The mirror reflectivities are 0.35 at each end. The refractive index of the active region is 3.3 under steady-state conditions. Calculate (a) the optical gain coefficient Γg required to balance the cavity loss and (b) the threshold electron density N_e. Assume that the gain can be modeled as $G = G_0(N_e - N_{e0})$, $G_0 = 1.73 \times 10^{-12}$ m^3/s and $N_{e0} = 3.47 \times 10^{23}$ m^{-3}.

Solution:
The total cavity loss coefficient is given by Eq. (3.39),

$$\alpha_{\text{cav}} = \alpha_{\text{int}} + \alpha_{\text{mir}},$$

where

$$\alpha_{\text{mir}} = \frac{1}{2L} \ln \left[\frac{1}{R_1 R_2} \right] = \frac{1}{2 \times 320 \times 10^{-6}} \ln \left[\frac{1}{0.35^2} \right] = 3.28 \times 10^3 \text{ m}^{-1}.$$

The internal loss coefficient is $\alpha_{\text{int}} = 10^3$ m^{-1},

$$\alpha_{\text{cav}} = 10^3 + 3.28 \times 10^3 \text{ m}^{-1} = 4.28 \times 10^3 \text{ m}^{-1}.$$

The optical gain coefficient Γg to balance the cavity loss is

$$\Gamma g = \alpha_{\text{cav}} = 4.28 \times 10^3 \text{ m}^{-1}.$$

(b) The threshold electron density is given by Eq. (3.131),

$$N_{e,th} = N_{e,0} + \frac{1}{G_0 \tau_{ph}},$$

where

$$\tau_{ph} = \frac{1}{v \alpha_{\text{cav}}},$$

$$v = \frac{c}{n} = \frac{3 \times 10^8}{3.3} = 9.09 \times 10^7 \text{ m/s},$$

$$\tau_{ph} = \frac{1}{9.09 \times 10^7 \times 4.28 \times 10^3} = 2.57 \text{ ps},$$

$$N_{e,th} = 3.47 \times 10^{23} + \frac{1}{1.73 \times 10^{-12} \times 2.57 \times 10^{-12}} \text{ m}^{-3}$$

$$= 5.71 \times 10^{23} \text{ m}^{-3}.$$

xercises

3.1 Explain the three processes by which a ligh twave interacts with an atom.

3.2 The operating wavelength of an optical source is 400 nm. Calculate the ratio of spontaneous to stimulated emission rate under thermal equilibrium. Assume $T = 293$ K. Is the optical source coherent? Provide an explanation.

(Ans: 4.12×10^{33}. No.)

3.3 In an atomic system, the spontaneous lifetime associated with 2   1 transition is 3 ns and the Einstein coefficient B is 6×10^{21} m^3/J $\cdot$ s^2. Calculate the energy difference between the levels 1 and 2. Assume that the speed of light is 1.5×10^8 m/s.

(Ans: 2.73×10^{-19} J.)

3.4 In an atomic system under thermal equilibrium conditions, the population density of the ground level is 2×10^{26} m^{-3} and the energy difference between the levels is 1.5 eV. Calculate the population density of the excited level. Assume that the temperature is 30° C.

(Ans: 22 m^{-3}.)

3.5 Under thermal equilibrium conditions, the ratio of spontaneous emission rate to stimulated emission rate is 2.33×10^{17}, the population density of the ground state is 1.5×10^{26} m^{-3}, and the temperature is 300 K. Calculate (a) the energy difference between the levels and (b) the population density of the excited level.

(Ans: (a) 1.65×10^{-19} J; (b) $6.47 \times 10^{\,8}$m^{-3}.)

3.6 An electron has a momentum of 4.16×10^{-26} kg $\cdot$ m/s. Calculate (a) the De Brogile wavelength and (b) the wavenumber.

(Ans: (a) 1.59×10^{-8}m; (b) 3.94×10^8 m^{-1}.)

3.7 A Fabry–Perot laser diode has a cavity length of 250 µm, the internal loss coefficient is 45 cm^{-1}, and the photon lifetime is 1.18 ps. Calculate the mirror reflectivity. Assume that the reflectivities are equal and the velocity of light in the active region is 9.09×10^7 m/s.

(Ans: $R_1 = R_2 = 0.299$.)

3.8 If one end of the laser cavity of Exercise 3.7 is coated with a dielectric reflector so that its reflectivity is 0.95, calculate the photon lifetime. Other parameters are the same as in Exercise 3.7.

(Ans: 1.56 ps.)

3.9 Show that the peak wavelength of the light emitted is related to the band-gap energy by

$$\lambda(\mu m) = \frac{1.24}{E_g \,(\text{eV})}.$$

Here, (µm) indicates that the wavelength is in units of µm.

3.10 The wavelength separation between longitudinal modes of a 1300-nm Fabry–Perot laser is $0.8\,\mu m$. Calculate the cavity length. Assume that the refractive index $n = 3.5$.

(Ans: $300\,\mu m$.)

3.11 The two cleaved facets of a 350-nm long semiconductor laser act as mirrors of reflectivity,

$$R_1 = R_2 = \left(\frac{n-1}{n+1}\right)^2, \tag{3.146}$$

where n is the refractive index of the gain medium. If the internal loss coefficient is $15\,cm^{-1}$, calculate the gain coefficient required to offset the loss.

(Ans: $50.75\,cm^{-1}$.)

3.12 Explain the difference between direct and indirect band-gap materials.

3.13 In a direct band-gap material, an electron in the conduction band makes transition to the valence band, emitting a light wave of frequency 75×10^{14} Hz. The band-gap energy is 1.8 eV. Calculate the crystal momentum of the electron. Assume that the effective mass of an electron in the conduction band and the valence band is $0.07m$ and $0.5m$, respectively, where m is the rest mass of an electron $= 9.109 \times 10^{-31}$ kg.

(Ans: 7.23×10^{-25} kg $\cdot$ m/s.)

3.14 The threshold electron density in a 800-nm Fabry–Perot laser diode is $4.2 \times 10^{23}\,m^{-3}$, the electron lifetime τ_e is 1.5 ns, and the volume of the active region is $5 \times 10^{-16}\,m^{-3}$. Calculate (a) the photon lifetime and (b) the threshold current. Assume that the gain can be modeled as $G = G_0(N_e - N_{e0})$, with $G_0 = 2 \times 10^{-12}\,m^3/s$ and $N_{e0} = 3.2 \times 10^{23}\,m^{-3}$.

(Ans: (a) 5 ps; (b) 22.4 mA.)

3.15 A semiconductor laser diode has the following parameters:

Active area width $w = 4\,\mu m$
Active area thickness $d = 0.5\,\mu m$
Length $L = 400\,\mu m$
Electron lifetime $= 1.5\,ns$
Internal cavity loss $= 10\,cm^{-1}$
Reflective index $= 3.3$
Reflectivity $R_1 = 0.32$
Reflectivity $R_2 = 0.92$
$G_0 = 1.5 \times 10^{-12}\,m^3/s$
$N_0 = 3.4 \times 10^{23}\,m^{-3}$

The bias current is 65 mA. Under steady-state conditions, calculate (a) the photon lifetime, (b) the threshold electron density, (c) the threshold current, (d) the photon density, and (e) the optical intensity.

(Ans: (a) 4.35 ps, (b) $4.93 \times 10^{23}\,m^{-3}$, (c) 42.1 mA, (d) $7.75 \times 10^{23}\,m^{-3}$, (e) 1.75×10^{10} W/m^2.)

3.16 Explain how the Bragg gratings can be used to make a single-longitudinal-mode laser.

3.17 A 1550-nm DFB laser has a cavity length of 300.07 μm. Find the grating period to have the strongest reflection (first-order grating) at 1550 nm. Assume $n = 3.2$.

Further Reading

A.E. Siegman, *Lasers*. University Science Books, Sausalito, CA, 1986.
P.W. Milonni and J.H. Eberly, *Laser Physics*. John Wiley & Sons, Englewood Cliffs, NJ, 2010.
G.P. Agrawal and N.K. Datta, *Semiconductor Lasers*, 2nd edn. Van Nostrand Reinhold, New York, 1993.
A.K. Ghatak and K. Thyagaratan, *Optical Electronics*. Cambridge University Press, Cambridge, 1990.
B.E.A. Saleh and M.C. Teich, *Fundamentals of Photonics*, 2nd edn. John Wiley & Sons, Englewood Cliffs, NJ, 2007.
J.M. Senior, *Optical Fiber Communications*, 2nd edn. Prentice-Hall, London, 1992.
G. Keiser, *Optical Fiber Communications*, 4th edn. McGraw-Hill, New York, 2011.
G. Lachs, *Fiber Optic Communications*, McGraw-Hill, New York, 1998.

References

[1] A. Einstein, *Phys. Z.*, vol. **18**, p. 121, 1917.
[2] A.L. Schewlow and C.H. Townes, *Phys. Rev.*, vol. **112**, p. 1940, 1958.
[3] T.H. Maiman, *Nature*, vol. **187**, p. 493, 1960.
[4] R.N. Hall, G.E. Fenner, T.J. Soltys, and R.O. Carlson, *Phys. Rev. Lett.*, vol. **9**, p. 366, 1962.
[5] Z. Alferov, *IEEE J. Spec. Top. Quant. Electr.*, vol. **6**, p. 832, 2000.
[6] H. Rubens and F. Kurlbaum, *Sbsr. Press. Akad. Wiss* p. 929, 1900.
[7] M. Planck, *Verh. Dt. Phys. Ges.*, vol. **2**, p. 237, 1900.
[8] C. Davisson and L.H. Germer, *Phys. Rev.*, vol. **30**, p. 705, 1927.
[9] A. Messiah, *Quantum Mechanics*. Dover Publications, New York, 1999.
[10] G. Lachs, *Fiber Optic Communications*. McGraw-Hill, New York, 1998.
[11] A.E. Siegman, *Lasers*. University Science Books, Sansalito, CA, 1986.
[12] C. Kittel, *Introduction to Solid State Physics*, 5th edn. John Wiley & Sons, Hoboken, NJ, 1976.
[13] A.F.J. Levi, *Applied Quantum Mechanics*, 2nd edn. Cambridge University Press, Cambridge, 2006.
[14] S.L. Chuang, *Physics of Optoelectronic Devices*, 2nd edn. John Wiley & Sons, Hoboken, NJ, 2008.
[15] G.P. Agrawal, *Fiber Optic Communication*, 4th edn. John Wiley & Sons, Hoboken, NJ, 2010.
[16] G.P. Agrawal and N.K. Datta, *Semiconductor Lasers*, 2nd edn. Van Nostrand Reinhold, New York, 1993.
[17] W.T. Tsang and S. Wang, *Appl. Phys. Lett.*, vol. **28**, p. 596, 1976.
[18] D.R. Scifres, R.D. Burham, and W. Streifer, *Appl. Phys. Lett.*, vol. **25**, p. 203, 1974.
[19] H. Ghafouri-shiraz, *Distributed Feedback Laser Diodes and Optical Tunable Filters*. John Wiley & Sons, Hoboken, NJ, 2004.

4

Optical Modulators and Modulation Schemes

4.1 Introduction

To convey a message, the amplitude, frequency, and phase of an optical carrier are switched in accordance with the message data. For example, bits '1' and '0' can be transmitted by turning a laser diode on and off, respectively. Typically, the message signal is in the form of binary data in an electrical domain, and optical modulators are used to convert the data into an optical domain. Sections 4.2 to 4.5 review the various line coders, pulse shapes, and digital modulation schemes. Sections 4.6 and 4.7 deal with different types of optical modulators and generation of modulated signals using optical modulators. The benefit of adding a controlled amount of ISI is discussed in Section 4.8. Section 4.9 deals with multi-level signaling, which enables higher transmission data rates without having to increase the bandwidth.

4.2 Line Coder

Digital data can be represented by electrical waveforms in a number of ways. This process is called *line coding*. In the binary case, bit '1' is sent by transmitting a pulse $p(t)$ and bit '0' is sent by transmitting no pulse. This line code is known as *unipolar* or *on–off*, as shown in Fig. 4.1(a). If a bit '1' and bit '0' are represented by $p(t)$ and $-p(t)$, respectively, such a line code is *polar*. This is shown in Fig. 4.1(b). If a bit '0' is represented by no pulse and a bit '1' is represented by $p(t)$ and $-p(t)$, such a line code is known as *bipolar*. In a bipolar line code, bit '1' is encoded by $p(t)$ if the previous bit '1' is encoded by $-p(t)$ and bit '1' is encoded by $-p(t)$ if the previous bit '1' is encoded by $p(t)$, as shown in Fig. 4.1(c). In other words, pulses representing consecutive bit '1's (no matter how many '0's are between the '1's) alternate in sign. Hence, this line code is also called *alternate mark inversion* (AMI).

4.3 Pulse Shaping

The message signal can be the internet data, voice data after analog-to-digital conversion (ADC), or any other form of digital data in an electrical domain. The widely used pulse shapes ($p(t)$) are *non-return-to-zero* (NRZ) and *return-to-zero* (RZ). In the case of NRZ, the signal does not return to a zero level if there are two consecutive '1's in a bit stream, whereas in the case of RZ, the signal returns to zero at the end of each bit

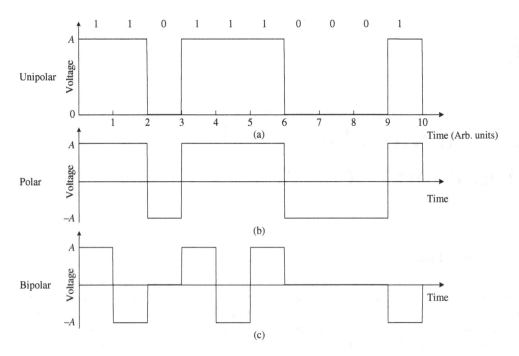

Figure 4.1   Various line codes: (a) unipolar, (b) polar, and (c) bipolar.

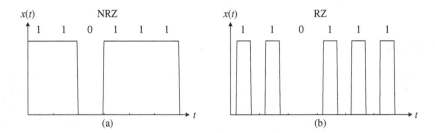

Figure 4.2   Pulse shapes: (a) NRZ, and (b) RZ.

slot, as shown in Fig. 4.2. The advantage of NRZ is that fewer transitions between '0' and '1' are required compared with RZ, since the signal amplitude remains the same if consecutive bits are '1' or '0'. Therefore, the bandwidth of a NRZ signal is less than that of a RZ signal. The wider spectral width of a RZ signal can also be understood from the fact that the pulse width of RZ pulse is shorter than that of a NRZ pulse. The message signal may be written as

$$m(t) = A_0 \sum_{n=-\infty}^{\infty} a_n p(t - nT_b),$$ (4.1)

where a_n is the binary data in the bit slot, $p(t)$ represents the pulse shape, and A_0 is a real constant.

An important parameter that characterizes a RZ signal is the *duty cycle*. This is defined as the time for which the light is turned on in a bit interval divided by the bit interval, i.e., the fraction of time over which the light is on "duty" within a bit interval. For example, for a 10-Gb/s system, the bit interval T_b is 100 ps and if the duration of the signal pulse is 30 ps, the duty cycle is 30%. The duty cycle of a NRZ signal can be considered to be 100%. In the above definition, we have assumed rectangular pulses. For pulses of arbitrary shape, the duty cycle x can be defined as the ratio of the FWHM of a pulse to the bit interval T_b,

$$x = \frac{\text{FWHM}}{T_b}. \tag{4.2}$$

When rectangular pulses are used, a RZ pulse in a bit interval $[-T_b/2, T_b/2]$ may be written as

$$p(t) = 1 \quad \text{for } |t| < xT_b/2$$

$$= 0 \text{ otherwise.} \tag{4.3}$$

4.4 Power Spectral Density

In this section, we find an expression for the power spectral density (PSD) of various line coders. Let the message signal be of the form

$$m(t) = A_0 \lim_{L \to \infty} \sum_{l=-L}^{L} a_l p(t - lT_b). \tag{4.4}$$

Noting that

$$\mathcal{F}\{p(t - lT_b)\} = \tilde{p}(f)e^{i2\pi f l T_b}, \tag{4.5}$$

the Fourier transform of $m(t)$ is

$$\tilde{m}(f) = A_0 \tilde{p}(f) \lim_{L \to \infty} \sum_{l=-L}^{L} a_l e^{i2\pi f l T_b}. \tag{4.6}$$

The PSD is defined as

$$\rho_m(f) = \lim_{T \to \infty} \frac{< |\tilde{m}(f)|^2 >}{T}, \tag{4.7}$$

where $T = (2L + 1)T_b$ and $< \ >$ denotes the ensemble average. From Eq. (4.6), we have

$$|\tilde{m}(f)|^2 = \tilde{m}(f)\tilde{m}^*(f)$$

$$= A_0^2 \tilde{p}(f) \lim_{L \to \infty} \sum_{l=-L}^{L} a_l e^{i2\pi f l T_b} \ \tilde{p}^*(f) \sum_{k=-L}^{L} a_k^* e^{-i2\pi f k T_b}$$

$$= A_0^2 |\tilde{p}(f)|^2 \lim_{L \to \infty} \sum_{l=-L}^{L} \sum_{k=-L}^{L} a_l a_k^* e^{i2\pi f(l-k)T_b}. \tag{4.8}$$

Using Eq. (4.8) in Eq. (4.7), we obtain

$$\rho_m(f) = A_0^2 |\tilde{p}(f)|^2 \lim_{L \to \infty} \frac{1}{(2L + 1)T_b} \sum_{l=-L}^{L} \sum_{k=-L}^{L} < a_l a_k^* > e^{i2\pi f(l-k)T_b}. \tag{4.9}$$

Let us first consider the case of a polar signal in which a_l is a random variable that takes values ±1 with equal probability. When $k \neq l$,

$$< a_l a_k^* >= 0. \tag{4.10}$$

This can be explained as follows. When $a_l a_k^* = 1$, it corresponds to $a_k = 1$ and $a_l = 1$, or $a_k = -1$ and $a_l = -1$; when $a_l a_k^* = -1$, it corresponds to $a_k = -1$ and $a_l = 1$, or $a_k = 1$ and $a_l = -1$. The chance that $a_l a_k^* = 1$ is the same as that of having $a_l a_k^* = -1$. Therefore, the ensemble average of $a_l a_k^*$ is zero when $k \neq l$. When $k = l$,

$$< a_l a_k^* > = < |a_k|^2 > = 1. \tag{4.11}$$

The terms in Eq. (4.9) can be divided into two groups; terms with $k = l$ and terms with $k \neq l$,

$$\rho_m(f) = A_0^2 |\tilde{p}(f)|^2 \lim_{L \to \infty} \frac{1}{(2L+1)T_b} \left[\sum_{k=-L}^{L} < |a_k|^2 > + \sum_{l=-L}^{L} \sum_{k=-L, k \neq l}^{L} < a_l a_k^* > e^{i2\pi f(l-k)T_b} \right]. \tag{4.12}$$

Using Eqs. (4.10) and (4.11) in Eq. (4.12), we find

$$\rho_m(f) = A_0^2 |\tilde{p}(f)|^2 \lim_{L \to \infty} \frac{2L+1}{(2L+1)T_b} = \frac{A_0^2 |\tilde{p}(f)|^2}{T_b}. \tag{4.13}$$

4.4.1   Polar Signals

Consider a polar signal with RZ pulses. The pulse shape function $p(t)$ is

$$p(t) = \text{rect}\left(\frac{t}{xT_b}\right), \tag{4.14}$$

$$\tilde{p}(f) = xT_b \text{sinc}\,(xT_b f), \tag{4.15}$$

where $\text{sinc}\,(y) = \sin(\pi y)/(\pi y)$. Using Eq. (4.15) in Eq. (4.13), we find

$$\rho_m^{RZ}(f) = A_0^2 x^2 T_b \text{sinc}^2(xT_b f). \tag{4.16}$$

When $x = 1$, this corresponds to a NRZ pulse and the PSD is

$$\rho_m^{NRZ}(f) = A_0^2 T_b \text{sinc}^2(T_b f). \tag{4.17}$$

Fig. 4.3 shows the PSD of a polar signal with NRZ and RZ pulses. As can be seen, the signal bandwidth of RZ pulses with 50% duty cycle is twice that of NRZ. For a polar signal with NRZ, the effective signal bandwidth (up to the first null) is B Hz. This is twice the theoretical bandwidth required to transmit B pulses per second [1]. Therefore, the polar signal is not the most bandwidth-efficient modulation format.

4.4.2   Unipolar Signals

The PSD of a unipolar signal is given by (see Example 4.5)

$$\rho_m(f) = \frac{A_0^2 |\tilde{p}(f)|^2}{4T_b} \left[1 + \frac{1}{T_b} \sum_{l=-\infty}^{\infty} \delta(f - \frac{l}{T_b}) \right]. \tag{4.18}$$

For RZ pulses, $\tilde{p}(f)$ is given by Eq. (4.15). Substituting Eq. (4.15) in Eq. (4.18), we find

$$\rho_m^{RZ}(f) = \frac{A_0^2 x^2 T_b}{4} \text{sinc}^2(xT_b f) \left[1 + \frac{1}{T_b} \sum_{l=-\infty}^{\infty} \delta(f - \frac{l}{T_b}) \right]. \tag{4.19}$$

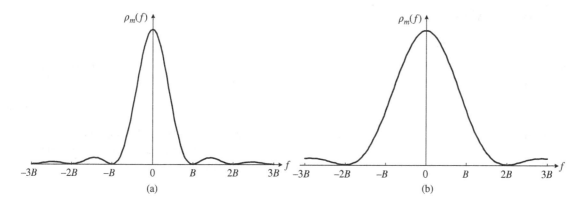

Figure 4.3 Spectrum of a polar signal: (a) NRZ, and (b) RZ with 50% duty cycle. B = bit rate.

When $x = 1$ (NRZ), the sinc function has nulls at $f = l/T_b$, $l \neq 0$, which coincide with the locations of the delta functions. Therefore, the PSD of a unipolar NRZ signal is

$$\rho_m^{\text{NRZ}}(f) = \frac{A_0^2 T_b}{4} \text{sinc}^2(T_b f) \left[1 + \frac{\delta(f)}{T_b} \right]. \tag{4.20}$$

Fig. 4.4 shows the PSD of unipolar NRZ and RZ signals. The PSD has continuous and discrete components corresponding to the first and second terms on the right-hand side of Eq. (4.19). For a RZ signal, the discrete components are located at $f = l/T_b$. However, the PSD of a unipolar NRZ signal has only a d.c. component ($f = 0$). The origin of the discrete components can be understood as follows. The unipolar NRZ signal can be imagined as a polar signal with constant bias. The PSD of this constant bias is the discrete component at $f = 0$. In the case of unipolar RZ, it can be imagined as a polar signal added to a periodic pulse train. Since the Fourier series expansion of the periodic pulse train leads to frequency components at the clock frequency $1/T_b$, and its harmonics, the PSD has discrete components at these frequencies.

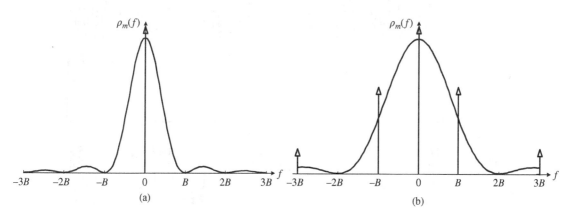

Figure 4.4 Spectrum of a unipolar signal: (a) NRZ, and (b) RZ with 50% duty cycle. B = bit rate. The arrows indicate delta functions.

4.5 Digital Modulation Schemes

4.5.1 Amplitude-Shift Keying

A laser is an optical carrier whose amplitude and/or phase can be varied in accordance with a message signal by means of an optical modulator. Let the laser output be (Fig. 4.5(a))

$$c(t) = A \cos (2\pi f_c t + \theta). \tag{4.21}$$

In Eq. (4.21), the amplitude A, frequency f_c, and phase factor θ are constants. When the amplitude A is varied in accordance with a message signal $m(t)$ while keeping f_c and θ constant, the resulting scheme is known as *amplitude modulation*. Suppose the amplitude is proportional to the message signal $m(t)$,

$$A(t) = k_a m(t), \tag{4.22}$$

where k_a is amplitude sensitivity. Now, the carrier is said to be amplitude modulated. The modulated signal can be written as

$$s(t) = k_a m(t) \cos (2\pi f_c t + \theta). \tag{4.23}$$

When the message signal $m(t)$ is a digital signal, such as shown in Fig. 4.5(b), the modulation scheme is known as *amplitude-shift keying* (ASK) or *on–off keying* (OOK). In general, to transmit bit '1', a sinusoid of certain amplitude A_1 is sent and to transmit bit '0', a sinusoid of amplitude A_2 is sent.

4.5.2 Phase-Shift Keying

When the phase θ of the carrier is varied in accordance with the message signal $m(t)$ while keeping the amplitude A and frequency f_c constant, the resulting scheme is known as *phase modulation*. Suppose the phase is proportional to the message signal,

$$\theta(t) = k_p m(t), \tag{4.24}$$

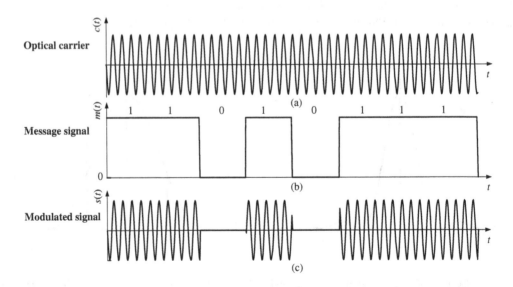

Figure 4.5 Modulation of the optical carrier by digital data: (a) carrier, (b) data, and (c) modulated signal.

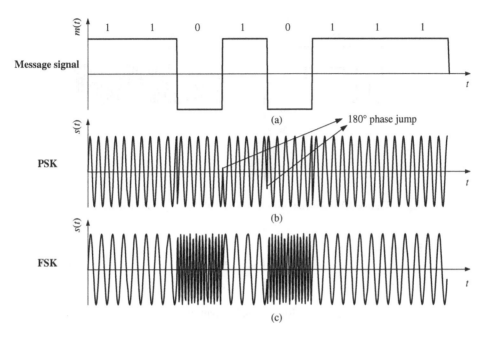

Figure 4.6 Phase and frequency modulation of an optical carrier. (a) Message signal, (b) Phase-shift keying, and (c) Frequency-shift keying.

where k_p is called the phase sensitivity. Now, the optical carrier is said to be phase modulated. The modulated signal can be written as

$$s(t) = A \cos \left[2\pi f_c t + k_p m(t) \right]. \tag{4.25}$$

For example,

$$m(t) = \begin{cases} -V & \text{for bit '1'} \\ V & \text{for bit '0'} \end{cases}, \tag{4.26}$$

where $V = \pi/(2k_p)$. From Eq. (4.25), it follows that

$$s(t) = \begin{cases} A \sin \left(2\pi f_c t \right) & \text{for bit '1'} \\ -A \sin \left(2\pi f_c t \right) & \text{for bit '0'} \end{cases}. \tag{4.27}$$

When the message $m(t)$ is a digital signal, such as shown in Fig. 4.6(a), the modulation scheme is known as *phase-shift keying* (PSK) or binary phase-shift keying (BPSK). Fig. 4.6(b) shows the modulated signal when the modulation scheme is PSK. Note that there is a 180° phase jump at the bit boundaries if the digital data in the consecutive bit intervals are different. In general, PSK can be described as a scheme in which a bit '1' is transmitted by sending a sinusoid of phase θ_1 and a bit '0' is transmitted by sending a sinusoid of phase θ_2. Fig. 4.7 shows the schematic of PSK generation.

4.5.3   Frequency-Shift Keying

FSK can be described as a scheme in which a bit '1' is transmitted by sending a sinusoid of frequency f_1 and a bit '0' is transmitted by sending a sinusoid of frequency f_2, as shown in Fig. 4.6(c). Let the message signal

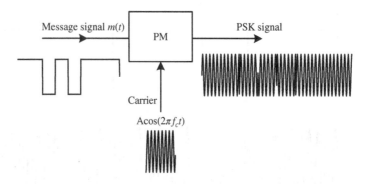

Figure 4.7   Generation of a PSK signal. PM = phase modulator.

be binary data of the form

$$m(t) = \begin{cases} m_1 & \text{for bit '1'} \\ m_2 & \text{for bit '0'} \end{cases}. \tag{4.28}$$

The transmitted signal within a bit interval $[0, T_b]$ can be written as

$$s(t) = A \cos [\phi(t)], \tag{4.29}$$

where

$$\phi(t) = \begin{cases} 2\pi f_1 t & \text{for bit '1'} \\ 2\pi f_2 t & \text{for bit '0'} \end{cases}, \tag{4.30}$$

$$f_i = f_c + k_f m_i, \quad i = 1, 2. \tag{4.31}$$

k_f is the frequency modulation index. Suppose the phase $\phi(t)$ in the bit interval $[0, T_b]$ is $2\pi f_1 t$ and the phase $\phi(t)$ in the next interval is $2\pi f_2 t$. At $t = T_b$, $\phi(T_b-) = 2\pi f_1 T_b$ and $\phi(T_b+) = 2\pi f_2 T_b$. This could cause phase discontinuity at the bit boundaries, which is undesirable in some applications. One possible way of avoiding phase discontinuities is to choose the frequencies such that the phase accumulated over a bit interval is an integral multiple of 2π,

$$f_c + k_f m_1 = \frac{n}{T_b}, \quad n \text{ is an integer,}$$

$$f_c + k_f m_2 = \frac{l}{T_b}, \quad l \text{ is an integer.} \tag{4.32}$$

Under these conditions, the phase would be continuous throughout and such a scheme is known as *continuous phase frequency-shift keying* (CPFSK). Note that the ASK signal has a constant frequency and the amplitude is varying, whereas FSK is a constant-amplitude signal but the instantaneous frequency is changing with time.

4.5.4   Differential Phase-Shift Keying

When a PSK signal is transmitted, it requires a complex receiver architecture to detect the phase. This is because the optical signal acquires a phase due to propagation, which fluctuates due to temperature, stress,

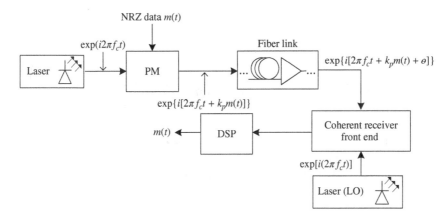

Figure 4.8 Block diagram of a fiber-optic system based on PSK.

and other factors. Therefore, to estimate the transmitted phase, we need a reference at the receiver which is provided by a laser known as a local oscillator (LO), whose phase is aligned with that of the received optical carrier (see Chapters 5 and 11). The phase alignment is achieved by phase estimation in digital domain using digital signal processing (DSP) techniques. This concept is illustrated in Fig. 4.8. The estimation of the absolute phase of the optical carrier is difficult, and also the linewidth of the LO should be quite small to avoid impairment due to laser phase noise. Instead of using the LO as a reference, the phase of the previous bit can be taken as a reference for the current bit. In this case, the local oscillator is replaced by the previous bit, which is obtained by delaying the current bit by the bit interval T_b. This scheme works relatively well under the assumption that the phase shift introduced due to propagation (θ in Fig. 4.8) is the same for both the current bit and the previous bit. In other words, the phase shift due to propagation should not fluctuate within a time period T_b. Under this condition, the previous bit could act as a reference to estimate the phase of the current bit. This is known as *differential phase detection*. The data at the transmitter should be encoded so that the phase of the current bit should be changed by 0 or π *relative* to the previous bit. In contrast, in the case of PSK, a phase of 0 or π is encoded relative to the *absolute* phase of the optical carrier corresponding to the current bit. Suppose the binary data to be transmitted is 010111. Let the message signal $m(t)$ in the interval $[0, T_b]$ be -1 V. This would serve as a reference for the next bit. To transmit the first bit '0' of the binary data 010111 in the interval $[T_b, 2T_b]$, we introduce no phase shift relative to the previous bit. To transmit the next bit '1' of the binary data, we introduce a phase shift of π relative to the previous bit, i.e., if the previous voltage level is -1 V, the current voltage level will be $+1$ V and vice versa. Continuing this process, we obtain the message signal shown in Table 4.1 and in Fig. 4.9. This type of coding is known as *differential coding*. In summary, the transmission of bit '1' is done by toggling from $+1$ V to -1 V or vice versa and the transmission of bit '0' is done by retaining the voltage of the previous bit. The above observation leads to a simple detection

Table 4.1 Differential coding.

Bit interval	$[T_b, 2T_b]$	$[2T_b, 3T_b]$	$[3T_b, 4T_b]$	$[4T_b, 5T_b]$	$[5T_b, 6T_b]$	$[6T_b, 7T_b]$
Binary data	0	1	0	1	1	1
Message signal $m(t)$	-1 V	$+1$ V	$+1$ V	-1 V	$+1$ V	-1 V

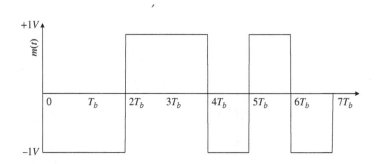

Figure 4.9 The DPSK signal corresponding to the binary data of Table 4.1.

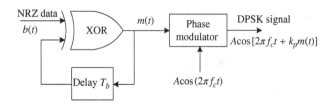

Figure 4.10 Generation of a DPSK signal.

scheme: take a product of the current bit (after demodulation) with the previous bit. If the product is negative, the transmitted data at the current bit is '1'. Otherwise, it is '0'.

The differential encoding can be realized using a XOR gate and a delay circuit as shown in Fig. 4.10. Let us assign a voltage level of $+1$ V and -1 V to the logic levels of 1 and 0, respectively. Let $m(t)$ be the encoded message signal. If the incoming binary data $b(t)$ to be transmitted is 0 and the previous message bit (of $m(t)$) is 0 (or -1 V), the current message bit will also be 0 since we introduce no phase shift relative to the previous message bit. If the binary data $b(t)$ is 1 and the previous message bit is 0 (1), the current bit will be 1 (0) since we need to introduce a π phase shift (or inversion of the amplitude) with respect to the previous message bit. Thus, we have

$$m(t) = b(t) \oplus m(t - T_b), \tag{4.33}$$

where $\oplus$ denotes the exclusive OR (XOR) operation. The truth table is shown in Table 4.2 and the waveforms for the input data $b(t)$ and the message signal $m(t)$ are shown in Fig. 4.11.

Table 4.2 Truth table.

$b(t)$		$m(t - T_b)$		$m(t)$	
Logic level	Voltage level (V)	Logic level	Voltage level (V)	Logic level	Voltage level (V)
0	-1	0	-1	0	-1
0	-1	1	1	1	1
1	1	0	-1	1	1
1	1	1	1	0	-1

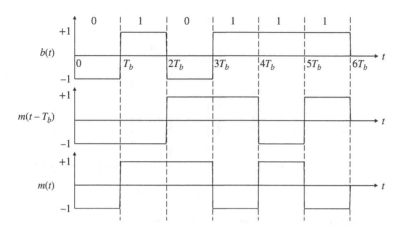

Figure 4.11   Input waveform $b(t)$ and message waveform $m(t)$.

Example 4.1

Consider the input data sequence

$$\{b_n\} = 01101110.$$

To proceed with the deferential encoding, add an extra bit to the encoder output. This extra bit may be chosen arbitrarily to be 0. Determine the encoded message sequence m_n.

Solution:
Let the first bit of the sequence m_n be 0 (or -1 V), which serves as the reference. To transmit the first bit 0 of $\{b_n\}$, we introduce no phase shift relative to the reference voltage. Therefore, the voltage level of the second bit of $\{m_n\}$ is -1 V (or 0). To transmit the second bit 1 of $\{b_n\}$, we introduce a π phase shift relative to the reference voltage (the voltage corresponding to the second bit of $\{m_n\}$). Therefore, the voltage level corresponding to the third bit of $\{m_n\}$ is $+1$ V (or 1). Continuing this process, we find

$$\{m_n\} = 001001011.$$

4.6   Optical Modulators

The simplest optical modulator we could think of is the switch of a flash light. Suppose we turn on a flash light for 1 second and turn it off for 1 second. We generate digital data '1' and '0', respectively, as shown in Fig. 4.12. In this example, the bit rate of the optical data generated from the flash light is 1 bit/s.

Optical modulation techniques can be divided into two types: (i) direct modulation of lasers and (ii) external modulation of lasers.

4.6.1   Direct Modulation

The laser drive current can be modulated by a message signal, as shown in Fig. 4.13. For example, when the message signal is bit '0' (bit '1'), the laser is turned off (on) and, therefore, the information in the electrical domain is encoded onto the optical domain. Directly modulated lasers (DMLs) have some major drawbacks.

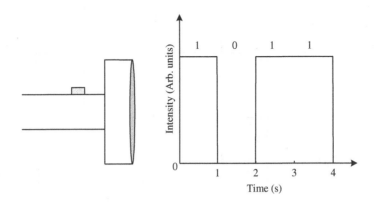

Figure 4.12 A simple optical modulator: the switch of a flash light.

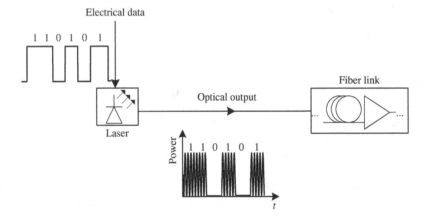

Figure 4.13 Direct modulation of a laser.

First, the instantaneous frequency of the laser output changes with time. This frequency chirping is caused by the refractive index changes of the active layer due to the carrier density modulation. The interaction of the positive chirp of the laser with the anomalous dispersion of the transmission fiber leads to pulse broadening (Example 2.18) and sets a limit on the maximum achievable transmission distance. However, the interaction of the laser chirp with the normal dispersion of the fiber leads to pulse compression initially (Example 2.18). In fact, the error-free transmission distance can be increased using positively chirped lasers and normal dispersion transmission fibers [2]. However, the pulses broaden eventually (even in normal dispersion fibers) and the laser chirp leads to transmission penalties for long-haul applications. Directly modulated lasers are usually used for transmission systems operating at low bit rates ($\leq$ 10 Gb/s) and for short-haul application (<100 km). The pulse distortion and frequency chirp prevent the use of directly modulated lasers for high-bit-rate applications.

4.6.2 External Modulators

Fig. 4.14 shows the schematic of a transmitter using external modulators. Widely used external modulators are: (i) the phase modulator, (ii) the Mach–Zehnder (MZ) interferometer modulator, and (iii) the electroabsorption (EA) modulator.

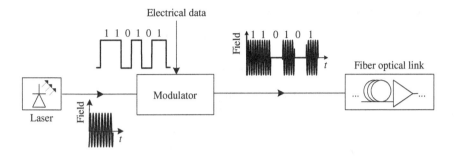

Figure 4.14   A transmitter using an external modulator.

4.6.2.1   Phase Modulators

The phase modulation of an optical carrier can be achieved in a number of ways. When an electric field is applied to an electro-optic crystal, the refractive index of the crystal changes and, therefore, the phase ($\propto$ refractive index) of an optical carrier propagating in the crystal also changes. The refractive index change is directly proportional to the applied electric field intensity [3], [4]. This effect is known as the *Pockels effect* or *linear electro-optic effect*.

Consider the light propagation in a LiNbO$_3$ crystal as shown in Fig. 4.15. Suppose E is the electric field intensity due to the applied voltage and an optical wave is propagating along the x-axis with its direction of polarization parallel to the z-axis. The dependence of the refractive index on the reflective field intensity is given by [3], [4]

$$n = n_0 - \frac{1}{2}n_0^3 r_{33} E_z \tag{4.34}$$

where n_0 is the refractive index in the absence of the applied electric field, and r_{33} is a coefficient describing the electro-optic effect. If V is the voltage applied across the crystal and d is the thickness of the crystal, the z-component of the electric field intensity is

$$E_z = V/d. \tag{4.35}$$

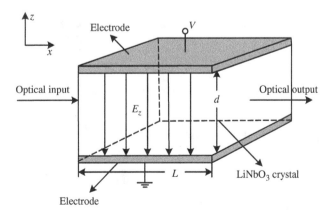

Figure 4.15   Phase modulation of an optical signal in a LiNbO$_3$ crystal.

Suppose the optical field of the incident optical wave is

$$\psi(t,0) \equiv \psi_{in}(t) = A_0 \exp(-i2\pi f_c t). \tag{4.36}$$

Using Eqs. (1.95), (4.34), and (4.35), the optical wave emerging from the LiNbO$_3$ crystal is

$$\psi_{in}(t,L) = A_0 \exp[-i(2\pi f_c t - \phi)], \tag{4.37}$$

where

$$\phi = \frac{2\pi}{\lambda_0} nL = \frac{2\pi L}{\lambda_0}\left(n_0 - \frac{n_0^3 r_{33} V}{2d}\right) = \phi_0 - \Delta\phi. \tag{4.38}$$

Here, ϕ_0 is the constant phase shift in the absence of the applied electric voltage, L is the length of the crystal, and

$$\Delta\phi = \frac{\pi L n_0^3 r_{33} V}{\lambda_0 d} \tag{4.39}$$

is the phase change. The required voltage to yield a phase change of π is known as the *half-wave voltage* or *switching voltage* V_π, and is given by

$$\Delta\phi = \pi = \frac{\pi L n_0^3 r_{33} V_\pi}{\lambda_0 d} \quad \text{or} \quad V_\pi = \frac{\lambda_0}{n_0^3 r_{33}}\frac{d}{L}. \tag{4.40}$$

Substituting Eq. (4.40) in Eq. (4.37), we obtain

$$\psi_{in}(t,L) = A_0 \exp\left[-i\left(2\pi f_c t - \phi_0 + \frac{\pi V}{V_\pi}\right)\right]. \tag{4.41}$$

Thus, we see that the phase change is directly proportional to the applied voltage. If $V(t)$ is a message signal, the phase of the optical carrier can be varied in accordance with the message signal. For the example, if $V(t) = V_\pi$, in a bit interval $0 < t < T_b$, the carrier phase is shifted by π. If $V(t) = 0$, no phase shift is introduced. Thus, the PSK or DPSK signal can easily be generated using a phase modulator.

Example 4.2

An electro-optic modulator operating at 1530 nm has the following parameters:

Thickness $d = 10\,\mu m$
Length $L = 5\,cm$
Index $n_0 = 2.2$
Pockel coefficient $r_{33} = 30\,pm/V$

Calculate the voltage required to introduce a phase shift of $\pi/2$.

Solution:
From Eq. (4.39), we have

$$V = \frac{\Delta\phi \lambda_0 d}{\pi L n_0^3 r_{33}}.$$

With $\Delta\phi = \pi/2$,

$$V = \frac{\pi/2 \times 1530 \times 10^{-9} \times 10 \times 10^{-6}}{\pi \times 5 \times 10^{-2} \times (2.2)^3 \times (30 \times 10^{-12})} \text{V}$$

$$= 0.47 \text{ V}.$$

4.6.2.2   Dual-Drive Mach–Zehnder Modulators (MZMs)

A MZM consists of two arms, as shown in Fig. 4.16. Voltages V_1 and V_2 are applied to the upper and lower arms, respectively. The exact analysis of the field propagation in this structure requires knowledge of the transverse field distributions. Instead, we follow an approximate approach in which the guided modes of the waveguide are replaced by plane waves propagating in the direction of the x-axis. We ignore the losses in the y-branches and arms. Let the electric field of the input optical beam be

$$\psi_{\text{in}} = A_0 \exp(-i2\pi f_c t), \tag{4.42}$$

where f_c is the frequency of the optical carrier. The first y-branch splits the input wave into two optical beams of equal power. Therefore, the electric field of the optical beam entering the upper (or lower) arm of the interferometer is

$$\frac{A_0}{\sqrt{2}} \exp(-i2\pi f_c t). \tag{4.43}$$

In Eq. (4.43), we have ignored a constant phase factor due to propagation in the y-branch. The factor $1/\sqrt{2}$ is introduced so that the total power is conserved. From the law of reciprocity, it follows that if the input and output of a y-branch are reversed, i.e., if the y-branch is used as a two-input/one-output device with inputs being $\psi_{\text{in}}/\sqrt{2}$, its output would be ψ_{in}, which is $1/\sqrt{2}$ times the addition of its inputs. This fact will be used later to find the output of the second y-branch.

The optical beams in the upper and lower arms undergo a phase shift of ϕ_1 and ϕ_2, respectively, in the presence of applied voltages. Using Eqs. (4.38) and (4.40), we have

$$\phi_j = \frac{2\pi L}{\lambda_0} \left(n_0 - \frac{1}{2} n_0^3 r_{33} V_j(t)/d \right), \quad j = 1, 2$$

$$= \phi_0 - \frac{V_j \pi}{V_\pi}. \tag{4.44}$$

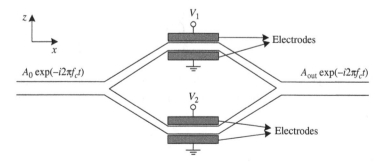

Figure 4.16   A dual-drive MZM.

The optical beams in the upper and lower arms are recombined via a second y-branch. The optical fields at the inputs of the second y-branch are

$$\psi_j = \frac{A_0}{\sqrt{2}} \exp\left[-i(2\pi f_c t - \phi_j)\right], \quad j = 1, 2. \tag{4.45}$$

The output of the second y-branch is

$$\psi_{\text{out}} = \frac{\psi_1 + \psi_2}{\sqrt{2}}. \tag{4.46}$$

Substituting Eqs. (4.45) and (4.44) in Eq. (4.46), we find that the output of the MZM is [5], [6]

$$\psi_{\text{out}} = \frac{A_0}{\sqrt{2}} \exp\left(-i\pi f_c t\right) \left[\frac{\exp\left(i\phi_1\right)}{\sqrt{2}} + \frac{\exp\left(i\phi_2\right)}{\sqrt{2}}\right] = A_{\text{out}} \exp\left(-i\pi f_c t\right), \tag{4.47}$$

where

$$A_{\text{out}} = A_0 \exp\left[i(\phi_1 + \phi_2)/2\right] \frac{\exp\left[i(\phi_1 - \phi_2)/2\right] + \exp\left[-i(\phi_1 - \phi_2)/2\right]}{2}$$

$$= A_0 \exp\left[i(\phi_1 + \phi_2)/2\right] \cos\left[(\phi_1 - \phi_2)/2\right]. \tag{4.48}$$

From Eq. (4.48), we see that the power is conserved when $\phi_1 = \phi_2$. When $\phi_1 = \phi_2 + \pi$, optical fields coming from two branches of the second y-branch do not excite a guided mode in the output waveguide; instead, radiation modes are excited which go out of the waveguide [3]. From Eq. (4.48), it appears that conservation of power is not satisfied when $\phi_1 \neq \phi_2$. But, if we take into account radiation modes, conservation of power is always satisfied. Using Eq. (4.44) in Eq. (4.48), we obtain

$$A_{\text{out}} = A_0 \exp\left(i\bar{\phi}\right) \cos\left\{\frac{[V_1(t) - V_2(t)]\,\pi}{2V_\pi}\right\}, \tag{4.49}$$

where

$$\bar{\phi} = \frac{\phi_1 + \phi_2}{2} = \phi_0 - \frac{[V_1(t) + V_2(t)]\pi}{2V_\pi}. \tag{4.50}$$

The instantaneous frequency shift or frequency chirp is given by (Eq. (2.165))

$$\omega_i = -\frac{d\bar{\phi}}{dt} = \frac{\pi}{2V_\pi}\left(\frac{dV_1}{dt} + \frac{dV_2}{dt}\right). \tag{4.51}$$

The output optical power is

$$P_{\text{out}} = |A_{\text{out}}|^2 = P_0 \cos^2\left\{\frac{[V_1(t) - V_2(t)]\,\pi}{2V_\pi}\right\}, \tag{4.52}$$

where $P_0 = A_0^2$. The frequency chirp combined with fiber dispersion could lead to pulse broadening (see Example 2.18) and performance degradations. Therefore, it is desirable to have zero chirp. From Eq. (4.51), we see that the chirp is zero if

$$\frac{dV_1(t)}{dt} = \frac{-dV_2(t)}{dt} \tag{4.53}$$

or

$$V_1(t) = -V_2(t) + V_{\text{bias}}, \tag{4.54}$$

where V_{bias} is a constant bias voltage. This driving condition is known as a *balanced driving* or *push–pull* operation. Let $V_1(t)$ be the message signal $m(t)$. Using Eq. (4.54) in Eqs. (4.48) and (4.52) and ignoring constant phase shift, we obtain

$$A_{out} = A_0 \cos\left\{\left[m(t) - \frac{V_{bias}}{2}\right]\frac{\pi}{V_\pi}\right\}, \tag{4.55}$$

$$P_{out} = P_0 \cos^2\left\{\left[m(t) - \frac{V_{bias}}{2}\right]\frac{\pi}{V_\pi}\right\}. \tag{4.56}$$

Let us consider two cases: (i) $V_{bias} = V_\pi/2$ and (ii) $V_{bias} = V_\pi$.
   Case (i) $V_{bias} = V_\pi/2$. Let

$$A_{out} = A_0 \cos\left[\frac{m(t)\pi}{V_\pi} - \frac{\pi}{4}\right]$$
$$= \frac{A_0}{\sqrt{2}}\left[\cos\left(\frac{m(t)\pi}{V_\pi}\right) + \sin\left(\frac{m(t)\pi}{V_\pi}\right)\right]. \tag{4.57}$$

When $m(t) \ll V_\pi/\pi$,

$$\cos\left[\frac{m(t)\pi}{V_\pi}\right] \cong 1 \quad \text{and} \quad \sin\left[\frac{m(t)\pi}{V_\pi}\right] \cong \frac{m(t)\pi}{V_\pi}. \tag{4.58}$$

Therefore, we have

$$\psi_{out} = \frac{A_0}{\sqrt{2}}\left[1 + \frac{m(t)\pi}{V_\pi}\right]\exp\left(-i2\pi f_c t\right). \tag{4.59}$$

The above equation corresponds to an amplitude modulated (AM) wave used in commercial AM broadcasting. Thus, in the small signal limit and with $V_{bias} = V_\pi/2$, the MZ modulator acts as an AM modulator.
   Case (ii) $V_{bias} = V_\pi$. Now, Eq. (4.55) becomes

$$A_{out} = A_0 \sin\left[\frac{m(t)\pi}{V_\pi}\right]. \tag{4.60}$$

When $m(t)\pi/V_\pi \ll 1$,

$$\psi_{out} = \left(\frac{A_0\pi}{V_\pi}\right)m(t)\exp\left(-i2\pi f_c t\right). \tag{4.61}$$

Now, the MZM acts as a product modulator which multiplies the message signal and optical carrier. The above equation also corresponds to a form of AM modulation known as double sideband with suppressed carrier (DSB-SC). The output power in the small-signal limit is

$$P_{out} \cong \frac{A_0^2\pi^2}{V_\pi^2}m^2(t). \tag{4.62}$$

D.c. Extinction Ratio

So far we have assumed equal power splitting between two arms of the interferometers. In practice, the power splitting may not be exactly 50 : 50 due to temperature or stress fluctuations. In general, the optical field distribution entering the arm j is

$$A_0 \alpha_j \exp{(-i2\pi f_c t)}, \quad j = 1, 2, \tag{4.63}$$

with $\alpha_1^2 + \alpha_2^2 = 1$. Now, Eq. (4.47) is modified as

$$\psi_{out} = \frac{A_0 \exp{(-i2\pi f_c t)}}{\alpha_1 + \alpha_2}[\alpha_1 \exp{(i\phi_1)} + \alpha_2 \exp{(i\phi_2)}] \tag{4.64}$$

and the output optical power is

$$P_{out} = \frac{P_0}{(\alpha_1 + \alpha_2)^2}[\alpha_1^2 + \alpha_2^2 + 2\alpha_1 \alpha_2 \cos{(\phi_1 - \phi_2)}]. \tag{4.65}$$

When $\phi_1 - \phi_2 = 0$, the interference is constructive and the output is maximum:

$$P_{out}^{max} = P_0. \tag{4.66}$$

When $\phi_1 - \phi_2 = \pi$, the interference is destructive and the output is minimum:

$$P_{out}^{min} = \frac{P_0(\alpha_1 - \alpha_2)^2}{(\alpha_1 + \alpha_2)^2}. \tag{4.67}$$

The d.c. extinction ratio is defined as the ratio of maximum to minimum power:

$$\delta = \frac{P_{out}^{max}}{P_{out}^{min}} = \left(\frac{\alpha_1 + \alpha_2}{\alpha_1 - \alpha_2}\right)^2. \tag{4.68}$$

In dB units, it may be expressed as

$$\delta(dB) = 10 \log_{10}\delta. \tag{4.69}$$

In the ideal case, $\alpha_1 = \alpha_2 = 1/\sqrt{2}$ and δ is infinite. For ASK, it is desirable to have zero power for bit '0'. However, because of the power-splitting imperfections, the minimum power is not zero and, as a result, the distance between constellation points becomes smaller, which leads to performance degradations (see Chapter 8).

Eq. (4.68) can be written in a different form [5], [6],

$$r = \frac{\sqrt{\delta} - 1}{\sqrt{\delta} + 1}, \tag{4.70}$$

where $r = \alpha_2/\alpha_1$. Eq. (4.54) provides the biasing condition to obtain zero chirp. However, Eq. (4.54) is obtained for the ideal case of infinite d.c. extinction ratio. For the case of finite extinction ratio, there will be residual chirp even when the zero-chirp biasing condition given by Eq. (4.54) is used, which degrades the performance [5].

4.6.2.3 Electroabsorption Modulator (EAM)

Electroabsorption refers to the dependence of the absorption coefficient of a semiconductor on the applied electric field. The band gap of a semiconductor decreases as the applied field increases. Consider an optical carrier with frequency $f_c < E_{g0}/h$, where E_{g0} is the band gap in the absence of the applied electric field. Let the driving voltage $V(t)$ vary from 0 to V_0 volts, as shown in Fig. 4.17. When $V(t) = 0$, the photon energy is less than the band gap and the optical carrier is not absorbed. When $V(t) = V_0$, the band gap decreases. Now the photon energy could be larger than the band gap and the optical carrier is absorbed, which generates electron–hole pairs. Thus, the information in the electrical domain is translated into the optical domain. Fig. 4.18 shows the typical dependence of the absorption coefficient on the wavelength. Suppose the carrier wavelength is λ_0. Let α_0 and α_1 be the absorption coefficients at $V(t) = 0$ and $V(t) = V_0$, respectively. If L is the length of the modulator, the optical power exiting the modulator is

$$P_{\text{out}} = \begin{cases} P_{\text{max}} = P_0 \exp\left(-\alpha_0 L\right) & \text{when } V(t) = 0 \\ P_{\text{min}} = P_0 \exp\left(-\alpha_1 L\right) & \text{when } V(t) = V_0 \end{cases} \tag{4.71}$$

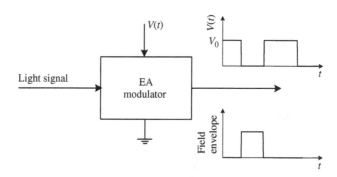

Figure 4.17 Amplitude modulation using an EAM.

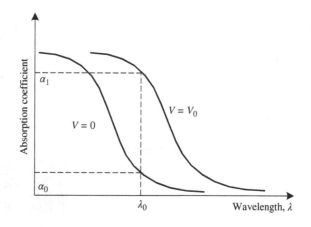

Figure 4.18 Typical dependence of absorption coefficient on wavelength.

where P_0 is the input power; the extinction ratio is

$$\delta = \frac{P_{\text{max}}}{P_{\text{min}}} = \frac{\exp(-\alpha_0 L)}{\exp(-\alpha_1 L)}.$$

(4.72)

To obtain the optimum performance, a high extinction ratio is desirable. To achieve this, typically InP-based semiconductors are used for 1300-nm or 1550-nm applications.

The absorption coefficient can be changed significantly by applying a relatively lower driving voltage. Therefore, the EAMs are very effective and the size could be quite small. The length of EAMs is typically 200 mm, whereas that of electro-optic modulators is a few centimeters. EAMs can easily be integrated with the laser diode, since both are based on similar semiconductor materials. The drawbacks of EAMs are as follows. (i) They have residual chirps similar to directly modulated lasers. The interaction of the chirp and fiber dispersion could lead to enhanced pulse broadening. (ii) The extinction ratio is typically ≤ 10 dB, which could lead to a power penalty [8].

4.7 Optical Realization of Modulation Schemes

4.7.1 Amplitude-Shift Keying

The optical ASK signal can be generated using a MZM, as shown in Fig. 4.19. The optical power of the MZM output may be written as (Eq. (4.56))

$$P_{\text{out}} = P_0 \cos^2\left[\frac{m(t)\pi}{V_\pi} - \frac{V_{\text{bias}}\pi}{2V_\pi}\right].$$

(4.73)

Let the message signal be a polar NRZ,

$$m(t) = \begin{cases} +V & \text{for bit '1'} \\ -V & \text{for bit '0'} \end{cases}.$$

(4.74)

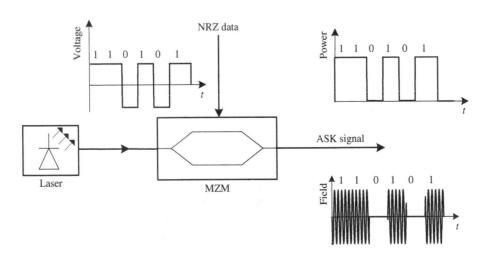

Figure 4.19 Generation of ASK signal using a MZM.

The desired Mach–Zehnder output power is

$$P_{\text{out}} = \begin{cases} P_0 & \text{for bit '1'} \\ 0 & \text{for bit '0'} \end{cases}. \tag{4.75}$$

For bit '1', substituting Eq. (4.74) in Eq. (4.73) and using Eq. (4.75), we obtain

$$P_0\cos^2\left[\frac{m(t)\pi}{V_\pi} - \frac{V_{\text{bias}}\pi}{2V_\pi}\right] = P_0,$$

$$\frac{V\pi}{V_\pi} - \frac{V_{\text{bias}}\pi}{2V_\pi} = j\pi, \quad j = 0, \pm 1, \pm 2, \ldots \tag{4.76}$$

Similarly, for bit '0',

$$P_0\cos^2\left[\frac{m(t)\pi}{V_\pi} - \frac{V_{\text{bias}}\pi}{2V_\pi}\right] = 0,$$

$$\frac{-V\pi}{V_\pi} - \frac{V_{\text{bias}}\pi}{2V_\pi} = \frac{l\pi}{2}, \quad l = \pm 1, \pm 3, \ldots \tag{4.77}$$

Subtracting Eq. (4.77) from Eq. (4.76), we find

$$V = \left(j - \frac{l}{2}\right)\frac{V_\pi}{2}. \tag{4.78}$$

Addition of Eqs. (4.76) and (4.77) leads to

$$V_{\text{bias}} = -\left(j + \frac{l}{2}\right)V_\pi. \tag{4.79}$$

If we choose $j = 0$ and $l = -1$, we find

$$V = \frac{V_\pi}{4}, \tag{4.80}$$

$$V_{\text{bias}} = \frac{V_\pi}{2}. \tag{4.81}$$

Thus, the polar NRZ in an electrical domain becomes a unipolar NRZ in an optical domain, as shown in Fig. 4.19. The process of modulation can be visualized using Fig. 4.20. When $V = V_\pi/4$ corresponding to bit '1', constructive interference occurs and the MZM power transmission is at its peak. When $V = -V_\pi/4$, destructive interference occurs and the MZM power output is zero. There are two approaches for the generation of RZ-ASK. The first approach is to apply the message signal, which is a polar RZ, as shown in Fig. 4.21(a). The output of the MZ modulator will be a RZ-ASK signal. However, the achievable bit rate is limited to 10 Gb/s using this approach [8]. The second approach is to introduce a RZ pulse carver in series with a MZ modulator, as shown in Fig. 4.21(b). The pulse carvers can be realized by driving the MZ modulator with a sinusoidal electrical signal [8]. The advantage of the second approach is that higher bit rates (≥ 40 Gb/s) can be realized.

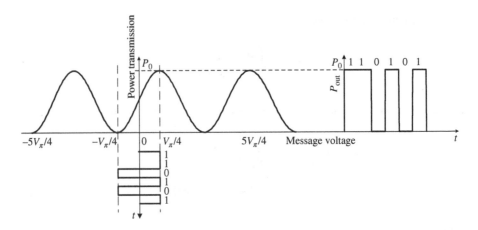

Figure 4.20 Power transmission function of the MZM. $V_{bias} = V_\pi/2$.

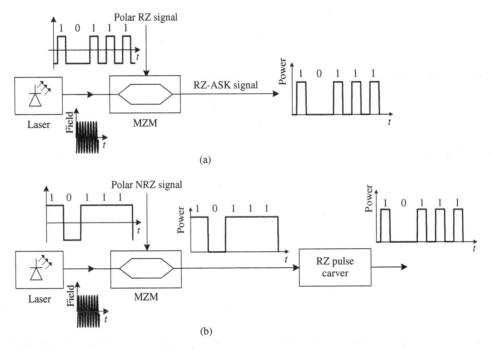

Figure 4.21 Generation of RZ-ASK. (a) Using RZ signal in electrical domain. (b) Using RZ pulse carver in optical domain.

4.7.2 Phase-Shift Keying

The optical field envelope of the MZM output can be written as (Eq. (4.55))

$$A_{out} = A_0 \cos\left[\frac{m(t)\pi}{V_\pi} - \frac{V_{bias}\pi}{2V_\pi}\right]. \tag{4.82}$$

Let the message signal be a polar NRZ given by Eq. (4.74). The desired field envelope of the Mach–Zehnder output is

$$A_{\text{out}} = \begin{cases} +A_0 & \text{for bit '1'} \\ -A_0 & \text{for bit '0'} \end{cases}. \tag{4.83}$$

For bit '1', substituting Eq. (4.74) in Eq. (4.82) and using Eq. (4.83), we obtain

$$A_0 \cos\left[\frac{V\pi}{V_\pi} - \frac{V_{\text{bias}}\pi}{2V_\pi}\right] = A_0,$$

$$\frac{V\pi}{V_\pi} - \frac{V_{\text{bias}}\pi}{2V_\pi} = 2j\pi, \quad j = 0, \pm 1, \pm 2, \ldots \tag{4.84}$$

Similarly, for bit '0', we have

$$A_0 \cos\left[\frac{-V\pi}{V_\pi} - \frac{V_{\text{bias}}\pi}{2V_\pi}\right] = -A_0,$$

$$-\frac{V\pi}{V_\pi} - \frac{V_{\text{bias}}\pi}{2V_\pi} = (2l + 1)\pi, \quad l = 0, \pm 1, \pm 2, \ldots \tag{4.85}$$

Simplifying Eqs. (4.84) and (4.85), we obtain

$$V = \frac{[2(j - l) - 1]V_\pi}{2},$$

$$V_{\text{bias}} = -[2(j + l) + 1]V_\pi. \tag{4.86}$$

If we choose $j = 0$ and $l = -1$, $V = V_\pi/2$ and $V_{\text{bias}} = V_\pi$. Fig. 4.22 shows a schematic of the PSK signal generation. Fig. 4.23 shows the MZM field transmission as a function of message signal $m(t)$. When the message signal $m(t) = +V_\pi/2$, the field transmission is maximum and when $m(t) = -V_\pi/2$, it is minimum. Note that the field envelope is negative (π phase) for bit '0' and positive (0 phase) for bit '1'. However, the power, which is the absolute square of the field, remains constant throughout. The PSK with NRZ rectangular

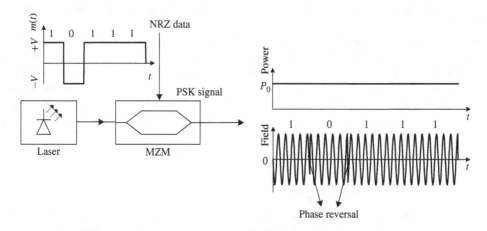

Figure 4.22   Generation of optical PSK signal.

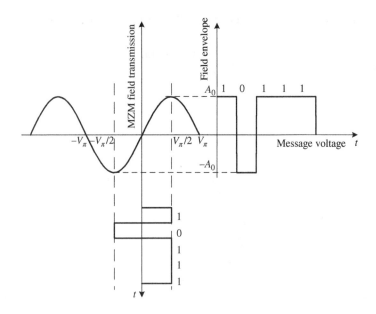

Figure 4.23 MZM field transmission. $V_{bias} = V_\pi$.

Table 4.3 Voltage levels of the message signal.

Binary data	1	0	1	1	1
$m(t)$	$V_\pi/2$	$-V_\pi/2$	$V_\pi/2$	$V_\pi/2$	$V_\pi/2$
Optical phase	0	π	0	0	0

pulses is a constant power signal and there is a reversal of phase at the bit boundaries when the data changes from '0' to '1' and vice versa. RZ-PSK can be generated using a RZ pulse carver in series with a MZM, similar to RZ-ASK.

PSK can also be generated using a phase modulator. When the message signal corresponds to bit '1', the phase modulator provides no phase change and when it corresponds to bit '0', the phase modulator changes the carrier phase by π (see Table 4.3). However, the performance characteristics of a phase-modulator-based PSK are worse than those of a MZM-based PSK [9]. This is because, in the case of the MZM, imperfections in driving conditions are translated into optical power variations, but the information-bearing optical phase is intact, whereas when a phase modulator is used, waveform imperfections distort the optical phase as well, which degrades the performance [9].

4.7.3 Differential Phase-Shift Keying

The DPSK signal generation is analogous to PSK generation except for a precoder which provides differential coding as discussed in Section 4.5.4. The differentially encoded signal $m(t)$ is used to drive a PM or a MZM as shown in Fig. 4.24.

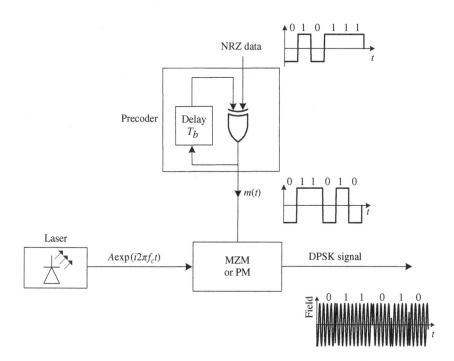

Figure 4.24   Optical DPSK signal generation.

4.7.4   Frequency-Shift Keying

In complex notation, the FSK signal in a bit interval $[0, T_b]$ can be written as (Eq. (4.29))

$$s(t) = A \exp \{ i[2\pi(f_c t + k_f m^{'}(t))] \}, \qquad (4.87)$$

where

$$m^{'}(t) = \begin{cases} m_1 t & \text{for bit '1'} \\ m_2 t & \text{for bit '0'} \end{cases}. \qquad (4.88)$$

Suppose the message signal $m(t)$ is a polar NRZ signal with $m_1 = +1$ V and $m_2 = -1$ V. The signal $m^{'}(t)$ can be obtained by integrating $m(t)$ within a bit interval $[0, T_b]$ and resetting it to zero at the end of the bit interval. The signal $m^{'}(t)$ is used to drive the phase modulator, as shown in Fig. 4.25.

4.8   Partial Response Signals*

If the symbols in the adjacent bit slots interfere, this leads to degradation in transmission performance. However, if we introduce a controlled amount of ISI, it is possible to correct for it at the receiver since the amount of ISI introduced is known. Suppose we add the kth bit and the $(k-1)$th bit, so that a known amount of ISI is introduced. This can be accomplished by a delay-and-add filter as shown in Fig. 4.26. The delay-and-add filter adds the signal $u_{in}(t)$ and the delayed version of $u_{in}(t)$,

$$u_{out}(t) = u_{in}(t) + u_{in}(t - T_b). \qquad (4.89)$$

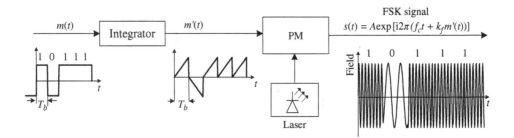

Figure 4.25 Optical FSK signal generation.

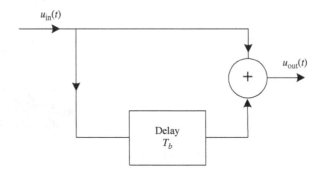

Figure 4.26 A duobinary encoder using delay-and-add filter.

If the voltage levels in the $(k-1)$th bit slot and the kth bit slot of $u_{\text{in}}(t)$ are both $+1$ V (or -1 V), this leads to a voltage level of $+2$ V (or -2 V) in the kth bit slot of $u_{\text{out}}(t)$. If the voltage level in the $(k-1)$th bit slot is of opposite polarity to that in the kth bit slot, this leads to a voltage level of 0 V in the kth bit slot of $u_{\text{out}}(t)$. Thus, a two-level voltage waveform is translated into a three-level waveform, as shown in Fig. 4.27. This scheme is called *duobinary encoding*. The above observation leads to the following decision rule: if the voltage at the sample instant is positive, the present and previous bits are both '1's. If the voltage at the sample instant is negative, the present and previous bits are both '0's. If the voltage is zero, the present bit is a complement of the previous bit. In this case, knowledge of the decision on the logic level of the previous bit is required to determine the current bit. The superposition of the $(k-1)$th bit and the kth bit leads to a known amount of ISI only at the sampling instant T_b. This scheme of introducing a known amount of ISI is called *correlative coding*. This is because there is a correlation between values of $u_{\text{out}}(t)$ between two successive bit intervals. This scheme is also referred to as a *partial response scheme* [10]. One of the disadvantages of the scheme shown in Fig. 4.26 is that the decision on the current bit at the receiver requires knowledge of the decision on the previous bit. This implies that if the decision on the previous bit is wrong, the decision on the current bit is also wrong, leading to error propagation. However, if we use the differential coding discussed in Section 4.5.4, error propagation can be prevented. Suppose the binary data to be transmitted at the kth bit slot is '1'. If differential coding is used, the current bit (kth bit slot) will have a phase shift of π relative to the previous bit, i.e., if the voltage level of the previous bit is $+1$ V, the voltage level of the current bit will be -1 V. Superposition of these two bits leads to a sample value of 0 V at kT_b. Similarly, if the binary data to be transmitted is 0, the voltage level of the current bit has a zero phase shift relative to the previous bit, i.e., the voltage level in the current and the previous bit slots is identical. In this case, the superposition of bits

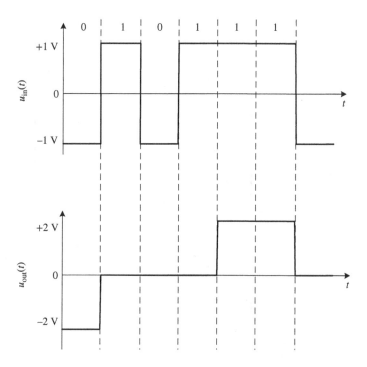

Figure 4.27 Duobinary encoding. The data in the interval $-T_b < t < 0$ of $u_{in}(t)$ is assumed to be '0' $(-1\,V)$.

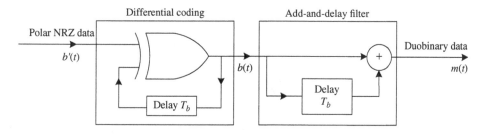

Figure 4.28 Duobinary encoder.

leads to a sample value of $\pm 2\,V$ at kT_b. This significantly simplifies the decision rule: if the absolute sample value is $\leq 1\,V$, '1' is transmitted. Otherwise, '0' is transmitted. Fig. 4.28 shows the realization of a duobinary encoder and Fig. 4.29 an example of duobinary encoding.

We can introduce a known amount of ISI such that a pulse in the 0th bit slot interferes only with a pulse in the first bit slot and does not interfere with pulses in other bit slots at sampling instants $t = nT_b$, where n is an integer. Such a pulse can be described by

$$p(nT_b) = \begin{cases} 1 & n = 0, 1 \\ 0 & \text{otherwise} \end{cases}. \tag{4.90}$$

An example of a pulse satisfying the requirement of Eq. (4.90) is a Nyquist pulse [1], [7],

$$p(t) = \frac{\sin(\pi Bt)}{\pi Bt(1 - Bt)}, \tag{4.91}$$

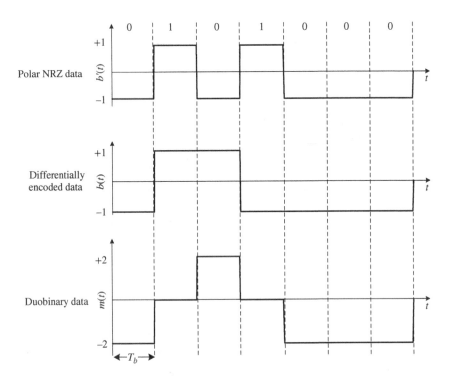

Figure 4.29 Duobinary encoding.

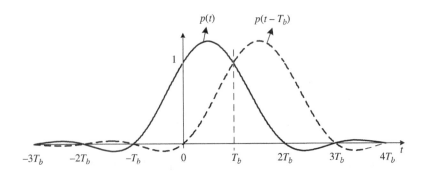

Figure 4.30 A duobinary pulse in the 0th bit slot interacts only with a pulse in the first slot at $t = T_b$. At any other sampling instants, $p(t) = 0$.

and is shown in Fig. 4.30. The pulse $p(t)$ is used to transmit a '1' and $-p(t)$ is used to transmit a '0'. As can be seen from Fig. 4.30, pulses $p(t)$ and $p(t - T_b)$ do not interfere at any sampling instants except at $t = T_b$. Although pulses interfere at other times, this does not lead to performance degradations since the decisions are made based on the sample values at $t = nT_b$. The duobinary signal may be written as

$$m(t) = \sum_{n=-\infty}^{\infty} b_n p(t - nT_b),$$

(4.92)

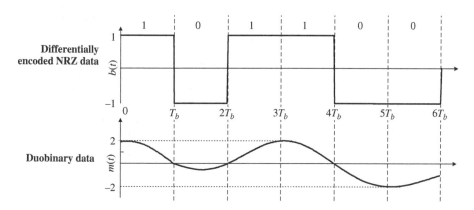

Figure 4.31 Input waveform $b(t)$ and duobinary waveform $m(t)$. The data in the interval $-T_b < t < 0$ is assumed to be '1'.

where b_n is the differentially encoded message data. Using Eq. (4.90), we find

$$m(t) = b_n + b_{n-1} \quad \text{at} \quad nT_b. \tag{4.93}$$

Superposition of pulses $p(t)$ and $p(t - T_b)$ leads to a sample value of 2 at T_b. If both the pulses are $-p(t)$ and $-p(t - T_b)$, the sample value would be -2 at T_b and if the pulses are $p(t)$ and $-p(t - T_b)$, the sample value would be zero, as illustrated in Fig. 4.31. The Fourier transform of the pulse described by Eq. (4.91) is (see Example 4.7)

$$\tilde{p}(f) = \frac{2}{B} \cos\left(\frac{\pi f}{B}\right) \text{rect}\left(\frac{f}{B}\right) \exp\left(-i\frac{\pi f}{B}\right). \tag{4.94}$$

From Fig. 4.32, we see that the bandwidth of the pulse is $B/2$ Hz. In contrast, the bandwidth of NRZ-OOK or RZ-OOK is $\geq B$ Hz. Eq. (4.94) can be rewritten as

$$\tilde{p}(f) = \frac{1}{B}[1 + \exp(i2\pi f T_b)]\text{rect}(f/B). \tag{4.95}$$

The factor $\exp(i2\pi f T_b)$ corresponds to a time delay of T_b and, therefore, the pulse $p(t)$ can be generated by cascading a delay-and-add filter and an ideal Nyquist filter with the transfer function

$$H_N(f) = \begin{cases} 1 & \text{for } |f| < B/2 \\ 0 & \text{otherwise} \end{cases}, \tag{4.96}$$

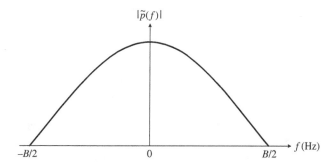

Figure 4.32 Spectrum of the duobinary pulse.

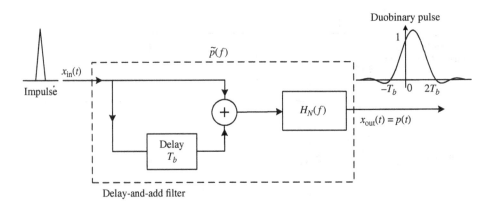

Figure 4.33 Duobinary pulse generation.

as shown in Fig. 4.33. Using Eq. (4.96), Eq. (4.95) can be written as

$$\tilde{p}(f) = \frac{1}{B}[1 + \exp{(i2\pi f T_b)}]H_N(f). \tag{4.97}$$

If an impulse is applied to the filter with the transfer function $\tilde{p}(f)$, the output is a duobinary pulse $p(t)$ (see Fig. 4.33). This is because the output $x_{out}(t)$ and input $x_{in}(t)$ of Fig. 4.33 are related by

$$\tilde{x}_{out}(f) = \tilde{x}_{in}(f)\tilde{p}(f). \tag{4.98}$$

For an impulse, $\tilde{x}_{in}(f) = 1$. Therefore, we have

$$x_{out}(t) = p(t). \tag{4.99}$$

Fig. 4.34 shows the duobinary encoding scheme using the pulse shown in Fig. 4.33. The impulse generator generates a positive impulse if the input is $+1$ V and it generates a negative impulse if the input is -1 V. The delay-and-add filter in conjunction with a Nyquist filter generates the corresponding duobinary pulses. Optical generation of partial response formats based on duobinary pulses is discussed in Ref. [8]. Fig. 4.35 shows the schematic of optical duobinary generation. The duobinary encoder shown in Fig. 4.35 could be realized using either a delay-and-add filter (Fig. 4.26) or a delay-and-add filter in conjunction with a Nyquist

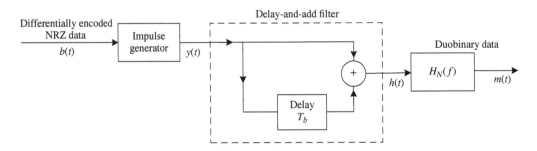

Figure 4.34 Duobinary encoder.

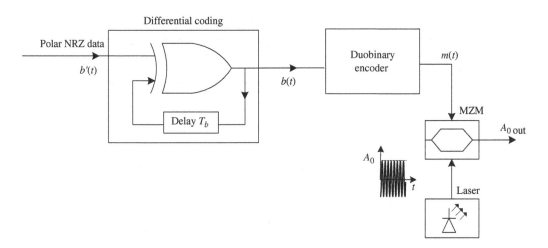

Figure 4.35 Optical duobinary signal generation.

filter (Fig. 4.34). After the duobinary encoder, the message signal $m(t)$ is used to drive the MZM. When $V_{\text{bias}} = V_\pi$, the MZM output is given by Eq. (4.60),

$$A_{\text{out}} = A_0 \sin\left[\frac{m(t)\pi}{V_\pi}\right], \tag{4.100}$$

$$A_{\text{out}} = \begin{cases} A_0 & \text{when } m(t) = V_\pi/2 \\ -A_0 & \text{when } m(t) = -V_\pi/2 \ . \\ 0 & \text{when } m(t) = 0 \end{cases} \tag{4.101}$$

Thus, a three-level electrical signal is converted to an optical signal with three levels in an optical field envelope, as shown in Fig. 4.36. The optical power of the MZM is

$$P_{\text{out}} = \begin{cases} |A_0|^2 & \text{when } m(t) = \pm V_\pi/2 \\ 0 & \text{when } m(t) = 0 \end{cases} . \tag{4.102}$$

Therefore, three voltage levels are translated into two power levels.

4.8.1 Alternate Mark Inversion

As discussed in Section 4.2, pulses representing two consecutive, '1's would have opposite signs no matter how many '0's are between these '1's. For example, the voltage levels of the AMI signal corresponding to a bit sequence $b_n = \{100110001\}$ are $\{A00 - AA000 - A\}$, where A is the amplitude. To generate an AMI signal, let us first consider the differential coding of the bit sequence b_n. The voltage levels of the differentially encoded data $b'(t)$ are given by $\{-1111 - 11111 - 1\}$, as shown in Fig. 4.37. From the figure, we observe that the occurrence of bit '1' of $b(t)$ leads to a voltage transition from -1 V to 1 V of $b'(t)$; the occurrence of the next bit '1' leads to a voltage transition from 1 V to -1 V, no matter how many '0' bits are between the '1's. If we delay the differentially encoded data by T_b and subtract it from the current bit, the resulting signal will have the desired properties of an AMI signal. The AMI signal is given by

$$m(t) = b'(t) - b'(t - T_b). \tag{4.103}$$

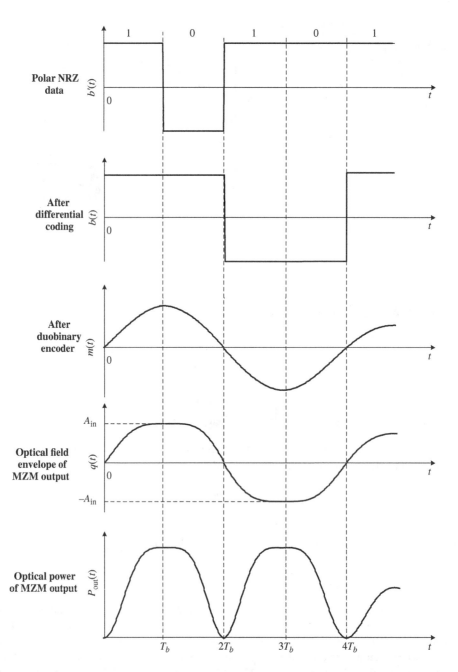

Figure 4.36   Electrical/optical waveforms at various stages of optical duobinary signal generation.

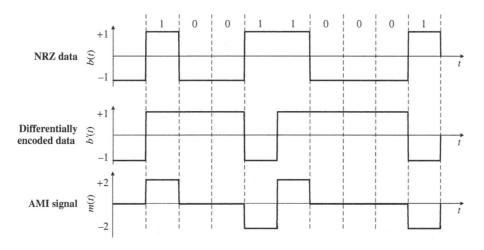

Figure 4.37 Waveforms of the input signal $b(t)$, the differentially encoded signal $b'(t)$, and the AMI signal $m(t)$.

If the voltage levels of the previous bit and the current bit of $b'(t)$ are -1 V and 1 V, respectively (corresponding to the bit '1' of b_n), delay and subtraction leads to $+2$ V. If the voltage levels of the previous bit and the current bit are $+1$ V and -1 V, respectively, the delay-and-subtract circuit gives -2 V. Since a bit '1' of b_n corresponds to a voltage change from -1 V to 1 V (or 1 V to -1 V) and the next '1' of b_n corresponds to a voltage change from 1 V to -1 V (or -1 V to 1 V), this ensures that alternate marks ('1's) are inverted. If the voltage levels of the adjacent bits are both $+1$ V (or -1 V) corresponding to '0' of the original bit sequence b_n, the delay-and-subtract circuit output is 0 V. Fig. 4.38 shows an optical realization of an AMI using MZMs. Biasing of a MZM is the same as in the duobinary case. Alternatively, an AMI can be generated using a delay-and-subtract operation in an optical domain [8]. This can be achieved using a MZ delay interferometer (DI), as shown in Fig. 4.39. A phase shift of π is introduced to one of the arms of the DI and a delay τ is also introduced. Therefore, the output optical field envelope can be written as (see Section 4.6.2.2)

$$u_{\text{out}}(t) = \frac{1}{2}[u_{\text{in}}(t) - u_{\text{in}}(t - \tau)]. \tag{4.104}$$

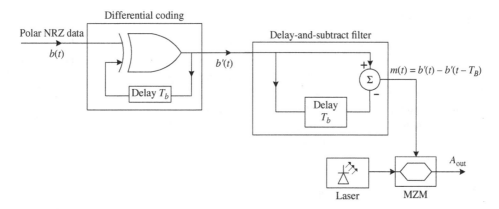

Figure 4.38 Generation of optical AMI signal.

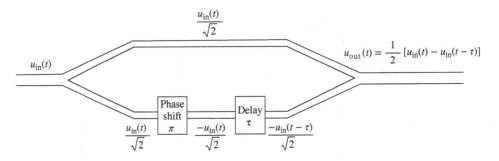

Figure 4.39 Mach–Zehner DI to perform delay-and-subtract operation.

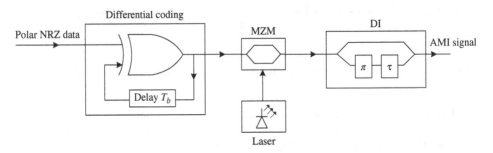

Figure 4.40 AMI signal generation using a Mach–Zehnder DI.

Thus, the DI acts as a delay-and-subtract circuit. When $\tau = T_b$, the AMI signals generated using the schemes shown in Figs. 4.38 and 4.40 are similar, and both techniques generate a NRZ-AMI signal in an optical domain. By varying τ, the RZ-AMI signals of different duty cycles can be generated [8]. Therefore, there is no need for an active RZ pulse-carving modulator. This is clearly an advantage of the optical realization of the delay-and-subtract operation. Although AMIs are used in non-optical communication systems to enable the use of a.c. coupling during transmission, this format is beneficial in optical communication systems as well. Because of the phase transitions, intrachannel four-wave mixing (IFWM)(see Chapter 10) can be suppressed using a RZ-AMI signal [11].

4.9   Multi-Level Signaling*

So far we have assumed that the message signal is binary data with two levels represented by the symbols '0' and '1'. Instead, the message signal could consist of multiple levels. For example, we could have four levels with voltages -3 V, -1 V, 1 V, and 3 V. These four levels can be represented by four symbols, '00', '01', '10', and '11'. We may do the following mapping: -3 V → '00', -1 V → '01', 1 V → '11', and 3 V → '10'. These four symbols could correspond not only to four amplitude levels (4-ASK), but also to four phase levels (4-PSK), or a combination of amplitude and phase levels (4-QAM). In this section, we discuss the following multi-level formats: (i) M-ASK, (ii) M-PSK, and (iii) quadrature amplitude modulation (QAM).

4.9.1   M-ASK

Here M stands for the number of symbols or levels. The simplest example we could think of is the switch of a flash light with two levels of brightness and an off button. In total, there are three power (or amplitude) levels corresponding to 3-ASK.

When $M = 4$, the four symbols can be represented by four amplitude levels $\pm 3A$, $\pm A$. The smallest separation between any two amplitude levels is $2A$, to ensure equal noise immunity. Fig. 4.41 shows the four symbols of 4-ASK. When $M = 8$, we need three binary digits or bits to represent eight symbols: '000', '001', ..., '111' and eight amplitude levels: $\pm 7A$, $\pm 5A$, $\pm 3A$, and $\pm A$. The M-ASK signal in an interval $0 \leq t \leq T_s$ may be written as

$$s_j(t) = m_j(t) \cos\left(2\pi f_c t\right), \tag{4.105}$$

where

$$m_j(t) = a_j p(t). \tag{4.106}$$

Here $p(t)$ represents the pulse shape in a symbol interval and a_j is a random variable that takes values $[-(M-1)A, -(M-3)A, \ldots, -3A, -A, A, 3A, \ldots, (M-1)A]$ with equal probability. Suppose the symbol interval is T_s, corresponding to a symbol rate of $B_s = 1/T_s$. M symbols convey information of $\log_2 M$ bits. For example, when $M = 8$, we have three bits of information encoded in a single symbol interval, i.e., if we were to use binary ASK (BASK), we would need three bit slots within a symbol interval to convey the same amount of information. Therefore, if we transmit B_s symbols per second, it is equivalent to transmitting $B_s \log_2 M$ bits/s,

$$B = B_s \log_2 M, \tag{4.107}$$

where B is the bit rate of an equivalent binary ASK signal. Equivalently, the data rate is enhanced by a factor of $\log_2 M$ compared with a binary ASK using the same symbol interval (= bit interval for BASK). For example, if $M = 4$, we have $\log_2 M = 2$ bits to represent all the four levels. Fig. 4.42(a) shows the waveform of a 4-ASK signal at a symbol rate of 10 GSym/s or 10 GBaud, with each symbol chosen out of the symbol set shown in Fig. 4.41. This is equivalent to transmitting a BASK signal at a bit rate of 20 Gb/s as shown in Fig. 4.42(b). Note that the symbol interval in Fig. 4.42(b) is half of that in Fig. 4.42(a). Typically, the bandwidth required to transmit a NRZ-BASK signal at a bit rate of B bits/s on a fiber channel is around $2B$ Hz. If we were to transmit the same amount of information by NRZ-MASK, the symbol interval T_s is $T_B \log_2 M$ where T_B is the bit interval and the required bandwidth to transmit NRZ-MASK would be $2B_s = 2B/\log_2 M$. Thus, the bandwidth reduces by a factor of $\log_2 M$. This reduction in bandwidth comes at the price of reduced power efficiency, i.e., the average transmitter power required to achieve the given performance increases as M^2 (see Example 4.3). This can be explained as follows: the symbol error rate is determined by the separation between

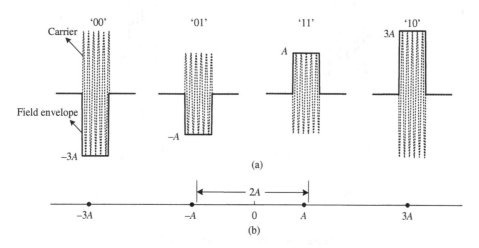

Figure 4.41 (a) Amplitude levels of 4-ASK. (b) Constellation diagram.

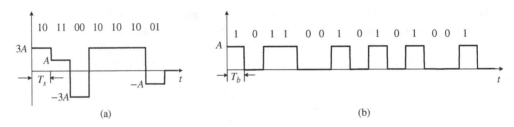

Figure 4.42   4-ASK and BASK signal for 20 Gb/s transmission: (a) 4-ASK with $T_s = 100$ ps, (b) BASK with $T_b = 50$ ps.

constellation points of Fig. 4.41(b). The larger the separation, the less is the chance of mistaking one symbol for another. If we fix the average power and the information rate of MASK to be the same as those of BASK, then the constellation points come closer and, therefore, the error rate increases. Equivalently, for a given error rate, the separation between constellation points for MASK should be the same as for BASK and then the spread of the amplitude levels would range from $-(M + 1)A$ to $(M + 1)A$, whereas the corresponding range is from $-A$ to A for BASK. Therefore, the average power of MASK ($M > 2$) increases relative to BASK. The trade-off between bandwidth and power efficiency is a common feature of all multi-level modulation formats.

4.9.2   M-PSK

When $M = 2$, we have a binary PSK or BPSK signal with two phase levels 0 or π, as shown in Fig. 4.43. See also Fig. 4.44. When $M = 4$, the signal is called quadriphase-shift keying (QPSK). The phase of the carrier takes on one of four values, 0, $\pi/2$, π, and $3\pi/2$, as shown in Fig. 4.45. See also Fig. 4.46. In general, the

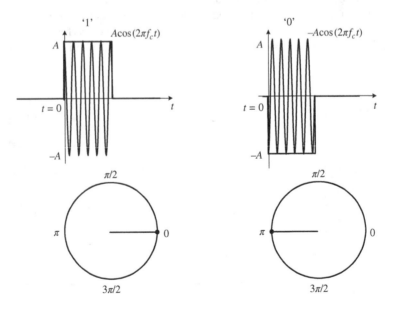

Figure 4.43   BPSK symbols '1' and '0'.

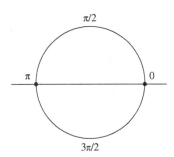

Figure 4.44 BPSK constellation diagram.

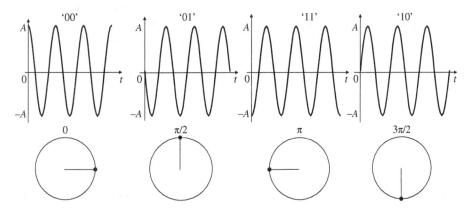

Figure 4.45 QPSK symbols.

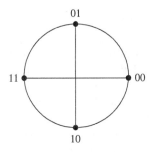

Figure 4.46 Constellation diagram of QPSK.

M-PSK signal in an interval $kT_s \leq t \leq (k+1)T_s$ may be written as

$$s_j(t) = Ap(t) \cos \left[2\pi f_c t + k_p m_j(t) + \theta_0 \right], \tag{4.108}$$

where

$$k_p m_j(t) = \frac{2\pi(j-1)}{M}, \quad j = 1, 2, \ldots, M \tag{4.109}$$

and θ_0 is a phase constant.

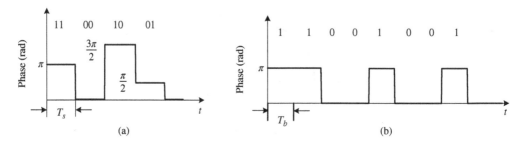

Figure 4.47   Time diagrams for (a) QPSK and (b) BPSK signals. $T_s = 100$ ps and $T_b = 50$ ps.

Fig. 4.47(a) shows the waveform of QPSK at a symbol rate of 10 GSym/s. This is equivalent to transmitting a BPSK signal at a bit rate of 20 Gb/s, as shown in Fig. 4.47(b). In a symbol interval $kT_s \leq t \leq (k+1)T_s$, any one of the messages $m_j(t)$ is sent. Using the formula $\cos(A + B) = \cos A \cos B - \sin A \sin B$, Eq. (4.108) can be rewritten as

$$s_j(t) = A[m_I(t) \cos(2\pi f_c t) + m_Q(t) \sin(2\pi f_c t)] \tag{4.110}$$

where

$$m_I(t) = p(t) \cos[k_p m_j(t) + \theta_0],$$
$$m_Q(t) = -p(t) \sin[k_p m_j(t) + \theta_0]. \tag{4.111}$$

$m_I(t)$ and $m_Q(t)$ can be imagined as two message signals modulating an in-phase carrier, $\cos(2\pi f_c t)$ and a quadrature carrier, $\sin(2\pi f_c t)$, respectively. See Table 4.4 and Fig. 4.48. $m_I(t)$ and $m_Q(t)$ are called the *in-phase* and *quadrature* components of the message signal, respectively. $m_I(t)\cos(2\pi f_c t + \theta_0)$ and $m_Q(t)\sin(2\pi f_c t + \theta_0)$ can be thought of as two amplitude-modulated waves on orthogonal carriers with a constraint that $[m_I^2(t) + m_Q^2(t)]/p^2(t) = 1$. For example, when $M = 4$, let us choose $\theta_0 = \pi/4$. If $k_p m(t) = 0$, $m_I(t) = p(t)/\sqrt{(2)}$ and $m_Q(t) = -p(t)/\sqrt{(2)}$. If $k_p m(t) = \pi/2$, $m_I(t) = -p(t)/\sqrt{(2)}$ and $m_Q(t) = -p(t)/\sqrt{(2)}$. For all the symbols of QPSK, it can be verified that $m_I(t) = \pm p(t)/\sqrt{(2)}$ and $m_Q(t) = \pm p(t)/\sqrt{(2)}$ (see Table 4.4). Thus, a QPSK signal can be generated using two polar NRZ data streams. The optical realization of QPSK can be achieved using a phase modulator, as shown in Fig. 4.49. However, a multi-level driving signal $m_j(t)$ is required, which degrades the system performance due to higher eye spreading when overlapping binary electrical signals with multi-level signals [12]. An alternative is to use the optical IQ modulator shown in Fig. 4.50, which is the optical analog of the scheme shown in Fig. 4.48 [12–14]. The output of the laser passes through the pulse carver and its output is split into two equal parts using a 3-dB coupler. The upper arm is known as the in-phase (I) arm and the lower arm is known as the quadrature (Q) arm. In the I-arm, the optical signal amplitude is modulated using a MZM. In the Q-arm, the optical signal is first phase-shifted by $-\pi/2$ and then its amplitude is modulated using a MZM. The d.c. bias of the MZMs is the same as used for a BPSK (see Section 4.7.2 and

Table 4.4   In-phase and quadrature amplitudes.

$k_p m(t)$	0	$\pi/2$	π	$3\pi/2$
$\dfrac{m_I(t)}{p(t)}$	$1/\sqrt{2}$	$-1/\sqrt{2}$	$-1/\sqrt{2}$	$1/\sqrt{2}$
$\dfrac{m_Q(t)}{p(t)}$	$-1/\sqrt{2}$	$-1/\sqrt{2}$	$1/\sqrt{2}$	$1/\sqrt{2}$

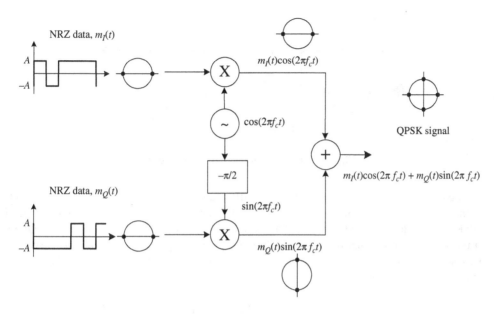

Figure 4.48   Generation of QPSK signal using two BPSK signals.

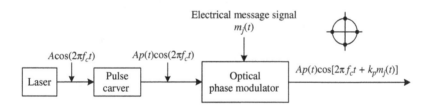

Figure 4.49   Optical QPSK generation using a phase modulator.

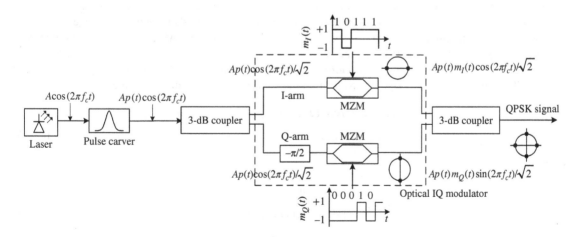

Figure 4.50   Optical QPSK generation using two MZMs.

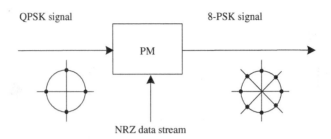

Figure 4.51   Generation of an 8-PSK signal.

Fig. 4.22). The optical signal is recombined using a 3-dB coupler to obtain the QPSK signal. The advantage of this scheme is that both $m_I(t)$ and $m_Q(t)$ are binary NRZ data streams. An 8-PSK signal can be obtained from a QPSK signal after passing through a phase modulator which changes the phase of the QPSK signal by $\pi/4$ or 0 depending on whether the electrical driving signal is $V_\pi/4$ or zero, respectively (see Fig. 4.51). This is because an 8-PSK signal is nothing but a phase-modulated QPSK signal. To see that, let us rewrite Eq. (4.108) in the complex notation

$$s_j^{\text{8-PSK}}(t) = Ap(t)\exp\left\{i\left[2\pi f_c t + (j-1)\frac{2\pi}{8}\right] + \theta_0\right\},\quad j = 1, 2, \ldots, 8$$

$$= Ap(t)\exp\left\{i\left[2\pi f_c t + \frac{(l-1)2\pi}{4} + \frac{k\pi}{4}\right] + \theta_0\right\},\quad \begin{array}{l} l = 1, 2, 3, 4 \\ k = 0, 1 \\ j = (2l-1) + k \end{array}$$

$$= Ap(t)s_l^{\text{QPSK}}(t)\exp\left(ik\pi/4\right).\tag{4.112}$$

4.9.3   Quadrature Amplitude Modulation

In M-ary ASK, the amplitude of the carrier is modulated in accordance with a message signal with the constraint that frequency f_c and θ of Eq. (4.21) are constant. This constraint leads to the straight-line constellation of Fig. 4.41. In M-ary PSK, the phase of the carrier is modulated in accordance with a message signal with the constraint that amplitude A and frequency f_c of Eq. (4.21) are constant, which leads to the circular constellation of Fig. 4.46. However, if we let the amplitude and phase of the carrier vary simultaneously, we get a modulation scheme known as an M-ary *quadrature amplitude modulation* (QAM) or M-ary *amplitude and phase-shift keying* (APSK). The signal waveform in the interval $0 \le t \le T_s$ may be expressed as

$$s_j(t) = A_j p(t)\cos\left(2\pi f_c t + \theta_j\right)\quad j = 1, 2, \ldots, M$$

$$= m_I(t)\cos\left(2\pi f_c t\right) + m_Q(t)\sin\left(2\pi f_c t\right),\tag{4.113}$$

where

$$m_I(t) = p(t)A_j\cos\left(\theta_j\right),$$

$$m_Q(t) = -p(t)A_j\sin\left(\theta_j\right).\tag{4.114}$$

The amplitude of the in-phase carrier $\cos\left(2\pi f_c t\right)$ is modulated by $m_I(t)$ and that of the quadrature carrier $\sin\left(2\pi f_c t\right)$ is modulated by $m_Q(t)$. Hence, this scheme is known as quadrature amplitude modulation. This is similar to QPSK, except that the amplitude $A_j\ (= \sqrt{m_I^2 + m_Q^2}/p(t))$ is constant in QPSK whereas it may be

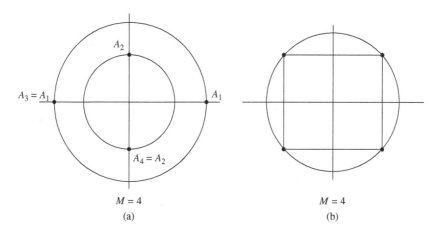

Figure 4.52   4-QAM constellations: (a) circular, (b) rectangular.

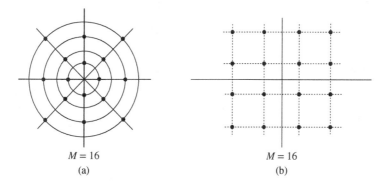

Figure 4.53   16-QAM constellations: (a) circular, (b) rectangular.

changing in QAM. For example, when $M = 4$, one possible way of realizing QAM is by choosing $A_1 = A_3$, $A_2 = A_4$, $\theta_1 = 0$, $\theta_2 = \pi/2$, $\theta_3 = \pi$, and $\theta_4 = 3\pi/2$. The corresponding constellation is shown in Fig. 4.52(a). The other possible way is to choose the four corners of a rectangle, $A_1 = A_2 = A_3 = A_4$, and $\theta_1 = 0$, $\theta_2 = \pi/2$, $\theta_3 = \pi$, and $\theta_4 = 3\pi/2$ (Fig. 4.52(b)). This scheme is the same as QPSK. Fig. 4.53 shows 16-QAM constellations.

In complex notation, assuming $p(t)$ to be real, the QAM signal waveform may be written as

$$s_j(t) = p(t)\tilde{A}_j \exp\left[i(2\pi f_c t + \theta_m)\right]$$
$$= \underbrace{p(t)\tilde{A}_j}_{\text{field envelope}}\ \underbrace{\exp(i2\pi f_c t)}_{\text{optical carrier}}, \tag{4.115}$$

where $\tilde{A}_j$ is the complex amplitude which is related to the real amplitude A_j by

$$\tilde{A}_j = A_j \exp(i\theta_m). \tag{4.116}$$

Thus, the simultaneous amplitude and phase modulation of the carrier are described by the complex variable $\tilde{A}_j$. Fig. 4.54 shows a possible realization of star 16-QAM [12, 13]. An 8-PSK signal can be generated using a QPSK modulator and a phase modulator (see Section 4.9.2). The 8-PSK signal passes through the MZM,

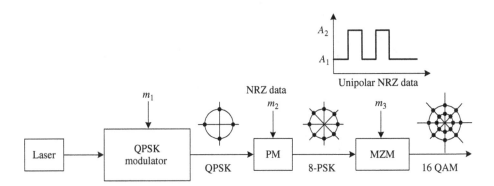

Figure 4.54   Schematic of 16-QAM generation using a QPSK modulator.

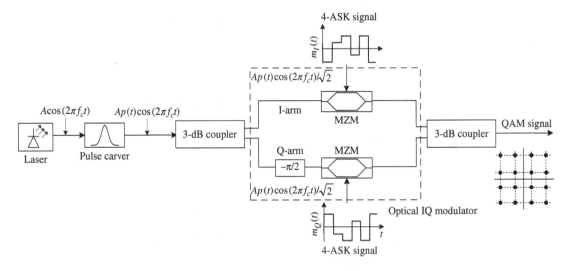

Figure 4.55   16-QAM generation using an optical IQ modulator.

which acts as an amplitude modulator. The bias conditions for a MZM to act as an amplitude modulator are discussed in Section 4.6.2.2. When the message signal $m_3 = A_1$ and A_2 $(> A_1)$ volts, we obtain the inner and outer circles of the constellation, respectively.

A QAM signal may be interpreted as a signal obtained by simultaneous amplitude modulation of in-phase and quadrature carriers (see Eq. (4.113)). This suggests that a QAM signal can be generated using an optical IQ modulator as shown in Fig. 4.55 [15]. To obtain a square 16-QAM signal, 4-ASK electrical signals are used to drive the optical IQ modulator, as discussed in Section 4.9.2.

Example 4.3   A MASK signal transmitted over the fiber channel

The driving voltage to the MZM is adjusted such that the optical field amplitudes are equally spaced. The MZM output in a symbol slot $0 < t < T_s$ is

$$A_{\text{out}} = \sqrt{P_0}(2m + 1), \tag{4.117}$$

where m is an integer, $m \in [-M/2, \ M/2 - 1]$, and M is even. Assume that the occurrence of any of these symbols is equally probable. Find the mean optical power.

Solution:
Since the M symbols could occur with equal probability, the mean power is

$$\bar{P}_{\text{out}} = \frac{P_0}{M} \sum_{m=-M/2}^{M/2-1} (2m+1)^2. \qquad (4.118)$$

Let $n = m + M/2 + 1$. Now, Eq. (4.118) becomes

$$\bar{P}_{\text{out}} = \frac{P_0}{M} \sum_{n=1}^{M} [2n - (M+1)]^2$$

$$= \frac{P_0}{M} \sum_{n=1}^{M} [4n^2 + (M+1)^2 - 4n(M+1)]. \qquad (4.119)$$

Using the following relations,

$$\sum_{n=1}^{M} n = \frac{M(M+1)}{2},$$

$$\sum_{n=1}^{M} n^2 = \frac{M(M+1)(2M+1)}{6}.$$

Eq. (4.119) is simplified as

$$\bar{P}_{\text{out}} = \frac{P_0}{M} \left[\frac{4M(M+1)(2M+1)}{6} + M(M+1)^2 - \frac{4M(M+1)^2}{2} \right]$$

$$= \frac{P_0(M^2-1)}{3}. \qquad (4.120)$$

Note that the mean power scales as M^2 for $M \gg 1$.

Example 4.4

Repeat Example 4.3 for an M-ary rectangular QAM signal.

Solution:
Let X and Y be the number of levels of in-phase and quadrature components with $M = XY$. The QAM signal in a symbol interval $0 < t < T_s$ may be written as (see Eq. (4.113))

$$A_{out}(t) = m_I(t) \cos(2\pi f_c t) + m_Q(t) \sin(2\pi f_c t), \qquad (4.121)$$

where

$$m_I(t) = \sqrt{P_0}(2x + 1) \ \ x \in [-X/2, \ \ X/2 - 1], \tag{4.122}$$

$$m_Q(t) = \sqrt{P_0}(2y + 1) \ \ y \in [-Y/2, \ \ Y/2 - 1]. \tag{4.123}$$

Here, x and y are integers. In complex notation, Eq. (4.121) can be written as

$$A_{\text{out}} = \sqrt{m_I^2 + m_Q^2} \ e^{i(2\pi f_c t + \theta)}, \tag{4.124}$$

where

$$\theta = \tan^{-1}\left[\frac{m_Q}{m_I}\right], \tag{4.125}$$

$$P_{\text{out}} = |A_{\text{out}}|^2 = m_I^2 + m_Q^2, \tag{4.126}$$

$$
\begin{aligned}
\bar{P}_{\text{out}} &= \frac{1}{XY} \sum_{x=-X/2}^{X/2-1} \sum_{y=-Y/2}^{Y/2-1} m_I^2 + m_Q^2 \\
&= \frac{P_0}{XY}\left[Y \sum_{x=-X/2}^{X/2-1} (2x+1)^2 + X \sum_{y=-Y/2}^{Y/2-1} (2y+1)^2 \right] \\
&= \frac{P_0[(X^2 - 1) + (Y^2 - 1)]}{3}.
\end{aligned}
\tag{4.127}
$$

4.10   Additional Examples

Example 4.5

Find the power spectral density of the unipolar signals.

Solution:
From Eq. (4.9), we have

$$\rho_m(f) = A_0^2 |\tilde{p}(f)|^2 \lim_{L \to \infty} \frac{1}{(2L+1)T_b} \sum_{l=-L}^{L} \sum_{k=-L}^{L} <a_l a_k> e^{i2\pi f(l-k)T_b}. \tag{4.128}$$

For unipolar signals, a_k takes values 1 or 0 with equal probability. Let us write

$$a_k = b_k + \frac{1}{2}, \tag{4.129}$$

where b_k is a random variable that takes values $\pm 1/2$ with equal probability similar to the random variable associated with the polar signal. Using Eq. (4.129), Eq. (4.128) can be expanded as

$$\rho_m(f) = A_0^2 |\tilde{p}(f)|^2 \lim_{L \to \infty} \frac{1}{(2L+1)T_b} \sum_{l=-L}^{L} \sum_{k=-L}^{L} \left[\langle b_k b_l \rangle + \frac{1}{2}\langle b_k \rangle + \frac{1}{2}\langle b_l \rangle + \frac{1}{4} \right] \exp\left(i2\pi f(l-k)T_b\right). \tag{4.130}$$

Since

$$\langle b_k b_l \rangle = \frac{1}{4} \quad \text{if} \quad k = l$$
$$= 0 \quad \text{otherwise} \tag{4.131}$$

and

$$\langle b_k \rangle = 0, \tag{4.132}$$

Eq. (4.130) reduces to

$$\rho_m(f) = A_0^2 |\tilde{p}(f)|^2 \left\{ \frac{1}{4T_b} + \lim_{L \to \infty} \frac{1}{4(2L+1)T_b} \sum_{l=-L}^{L} \sum_{k=-L}^{L} e^{i2\pi f(l-k)T_b} \right\}. \tag{4.133}$$

Using the following identities:

$$\lim_{L \to \infty} \sum_{l=-L}^{L} e^{i2\pi f l T_b} = \lim_{L \to \infty} \frac{1}{T_b} \sum_{l=-L}^{L} \delta(f - \frac{l}{T_b}),$$

$$\lim_{L \to \infty} \frac{1}{2L+1} \sum_{k=-L}^{L} e^{-i2\pi f k T_b} = 1, \tag{4.134}$$

Eq. (4.133) reduces to

$$\rho_m(f) = \frac{A_0^2 |\tilde{p}(f)|^2}{4T_b} \left\{ 1 + \frac{1}{T_b} \sum_{l=-\infty}^{\infty} \delta(f - \frac{l}{T_b}) \right\}. \tag{4.135}$$

Example 4.6

A raised-cosine pulse is defined as

$$p(t) = \frac{1}{2} \left[1 + \cos\left(\frac{\pi t}{T_B}\right) \right] \text{rect}\left(\frac{t}{2T_B}\right). \tag{4.136}$$

In a polar signaling scheme, raised-cosine pulses are used. Find the PSD.

Solution:
First let us calculate the Fourier transform of $p(t)$.

$$\mathcal{F}\left[\text{rect}\left(\frac{t}{2T_b}\right) \right] = 2T_b \text{sinc}\,(2fT_b), \tag{4.137}$$

$$\mathcal{F}\left[x(t) \cos\left(\frac{\pi t}{T_b}\right) \right] = \mathcal{F}\left[\frac{\exp\,(i2\pi f_0 t) + \exp\,(-i2\pi f_0 t)}{2} x(t) \right]$$

$$= \frac{\tilde{x}(f - f_0) + \tilde{x}(f + f_0)}{2}, \tag{4.138}$$

where $f_0 = 1/2T_b$. In Eq. (4.138), we have used the fact that a phase shift in time domain leads to a frequency shift in frequency domain. Using Eqs. (4.137) and (4.138), we find

$$\mathcal{F}\left[\cos\left(\frac{\pi t}{T_b}\right)\text{rect}\left(\frac{t}{2T_b}\right)\right] = T_b\text{sinc}\left[2T_b\left(f - \frac{1}{2T_b}\right)\right] + \text{sinc}\left[2T_b\left(f + \frac{1}{2T_b}\right)\right]$$

$$= -T_b\sin(2\pi T_b f)\left[\frac{1}{\pi(2T_b f - 1)} + \frac{1}{\pi(2T_b f + 1)}\right]$$

$$= \frac{-T_b}{\pi}\frac{\sin(2\pi T_b f)4T_b f}{4T_b^2 f^2 - 1}, \tag{4.139}$$

where we have used

$$\text{sinc}\left[2T_b\left(f \pm \frac{1}{2T_b}\right)\right] = \frac{\sin(2\pi T_b f \pm \pi)}{\pi(2T_b f \pm 1)}. \tag{4.140}$$

The Fourier transform of $p(t)$ is

$$\tilde{p}(f) = \frac{1}{2}\left[\frac{2T_b\sin(2\pi f T_b)}{2\pi f T_b} - \frac{T_b\sin(2\pi T_b f)4T_b f}{\pi(4T_b^2 f^2 - 1)}\right]$$

$$= \frac{T_b\text{sinc}(2fT_b)}{(1 - 4T_b^2 f^2)}. \tag{4.141}$$

The PSD is given by Eq. (4.13) as

$$\rho_m(f) = \frac{A_0^2|\tilde{p}(f)|^2}{T_b} = \frac{A_0^2 T_b\text{sinc}^2(2fT_b)}{(1 - 4T_b^2 f^2)^2}. \tag{4.142}$$

Example 4.7

Show that the Fourier transform of the duobinary pulse

$$p(t) = \frac{\sin(\pi Bt)}{\pi Bt(1 - Bt)} \tag{4.143}$$

is

$$\tilde{p}(f) = \frac{2}{B}\cos\left(\frac{\pi f}{B}\right)\text{rect}\left(\frac{f}{B}\right)\exp(i\pi f/B). \tag{4.144}$$

Solution:
Note that

$$\frac{1}{t(1 - Bt)} = \frac{1}{t} - \frac{1}{t - T_b}.$$

Eq. (4.143) can be written as

$$p(t) = \frac{\sin(\pi B t)}{\pi B t} - \frac{\sin(\pi B t)}{\pi B(t - T_b)}. \tag{4.145}$$

The first term on the right-hand side of Eq. (4.145) is in the form of a sinc function. To bring the second term into sinc form, consider

$$\sin[\pi B(t - T_b)] = \sin(\pi B t - \pi) = -\sin(\pi B t). \tag{4.146}$$

Using Eq. (4.146) in Eq. (4.145), we find

$$p(t) = \frac{\sin(\pi B t)}{\pi B t} - \frac{\sin(\pi B t)}{\pi B(t - T_b)} = \mathrm{sinc}(B t) + \mathrm{sinc}[B(t - T_b)]. \tag{4.147}$$

Using the following identities:

$$\mathcal{F}[\mathrm{sinc}(B t)] = \frac{1}{B}\mathrm{rect}(f/B),$$

$$\mathcal{F}[x(t - T_b)] = \tilde{x}(f)\exp(i2\pi f T_b),$$

the Fourier transform of Eq. (4.147) is

$$\tilde{p}(f) = \frac{1}{B}\mathrm{rect}(f/B)[1 + \exp(i2\pi f T_b)]$$

$$= \frac{2\exp(i\pi f T_b)\mathrm{rect}(f/B)}{B}\frac{[\exp(-i\pi f T_b) + \exp(i\pi f T_b)]}{2}$$

$$= \frac{2}{B}\exp(i\pi f/B)\mathrm{rect}(f/B)\cos(\pi f/B). \tag{4.148}$$

Exercises

4.1 Explain the differences between NRZ and RZ formats. Which of these formats has a wider spectrum?

4.2 Discuss the following modulation schemes: (i) ASK, (ii) PSK, and (iii) FSK.

4.3 The pulse shape of a RZ signal is described by

$$p(t) = \exp(-t^2/2T_0^2).$$

Find the PSD assuming (a) polar and (b) unipolar signaling.

4.4 Derive an expression for the PSD of a bipolar signal such as AMI. Assume rectangular pulses with 100% duty cycle.

4.5 Discuss the differences between binary PSK and DPSK. Does DPSK require a reference laser (local oscillator) at the receiver?

4.6 Explain the Pockels effect. Show that the phase change is proportional to the applied voltage in an electro-optic crystal.

4.7 An electro-optic modulator operating at 1550 nm has the following parameters:

Thickness $d = 8\mu m$
Index $n_0 = 2.2$
Pockel coefficient $r_{33} = 30\,pm/V$

It is desired that the half-wave voltage V_π is less than 2 V. Find the lower limit on the length L.

(Ans: 1.94 cm.)

4.8 Explain how an electroabsorption modulator can be used as an amplitude modulator.

4.9 The d.c. extinction ratio of a MZM is 13 dB. Calculate the power-splitting ratio $\alpha_1^2 : \alpha_2^2$. Assume that $\alpha_1^2 + \alpha_2^2 = 1$.

(Ans: 0.71 : 0.29.)

4.10 The input power to a dual-drive MZM is 0 dBm. The MZM is used to generate a NRZ-OOK signal. Find the drive voltage V and the bias voltage V_{bias}. If the d.c. extinction ratio is 10 dB, calculate the optical power levels corresponding to bit '1' and '0'. Assume $V_\pi = 4$ V.

(Ans: $V = 1$ V, $V_{bias} = 2$ V, power of bit '1' = 1 mW, power of bit '0' = 0.1 mW.)

4.11 Consider the input data sequence

$$\{b_n'\} = [110111011].$$

This data passes through the differential encoder and add-and-delay filter shown in Fig. 4.28. Determine the duobinary data voltage sequence m_n at instants nT_b $(=m(nT_b))$. To proceed with differential encoding, add an extra bit to the encoder output. State the decision rule.

(Ans: $\{m_n\} = [0\,0\,{-2}\,0\,0\,0\,2\,0\,0]$ V.)

4.12 Repeat Exercise 4.11 for the case of AMI generation as shown in Fig. 4.38. State the decision rule.

(Ans: $\{m_n\} = [2\,{-2}\,0\,2\,{-2}\,2\,0\,{-2}\,2]$.)

4.13 Explain how the correlative coding combined with differential coding simplifies the decision rule at the receiver.

4.14 The 4-ASK signal is transmitted over a fiber channel with a mean power of 0 dBm. Rectangular pulses with 50% duty cycle are used in each symbol slot. Sketch the waveform of the 4-ASK signal for a sequence $\{3, 1, 3, -1, -1, 1\}$ showing the peak powers of each symbol. Assume that the optical field amplitudes are equally spaced.

4.15 Repeat Exercise 4.14 for a square 16-QAM signal.

Further Reading

G.P. Agrawal, *Lightwave Technology*. John Wiley & Sons, New York, 2005.
S. Haykin and M. Moher, *Communication Systems*, 5th edn. John Wiley & Sons, New York, 2009.

D.P. Lathi and Z. Ding, *Modern Digital and Analog Communication Systems*, 4th edn. Oxford University Press, Oxford, 2009.

S. Haykin and M. Moher, *Introduction to Analog and Digital Communications*, 2nd edn. John Wiley & Sons, New York, 2007.

References

[1] D.P. Lathi and Z. Ding, *Modern Digital and Analog Communication Systems*, 4th edn. Oxford University Press, Oxford, 2009.

[2] I. Tomkos *et al.*, *IEEE J. Spec. Top. Quant. Elect.*, vol. **7**, p. 439, 2001.

[3] A. Yariv and P. Yeh, *Photonics*, 6th edn. Oxford University Press, Oxford, 2007, chapter 9.

[4] B.E.A. Saleh and M.C. Teich, *Fundamentals of Photonics*. John Wiley & Sons, Hoboken, NJ, 2007.

[5] S. Walklin and J. Conradi, *IEEE Photon. Technol. Lett.*, vol. **9**, p. 1400, 1998.

[6] J. Conradi, "Bandwidth-efficient modulation formats for digital fiber transmission systems." I. Kaminow and T. Li (eds.), *Optical Fiber Telecommunications IV B* Academic Press, San Diego, 2002.

[7] S. Haykin and M. Moher, *Communication Systems*, 5th edn. John Wiley & Sons, New York, 2009.

[8] P.J. Winzer and R.-J. Essiambre, *Proc. IEEE*, vol. **94**, p. 952, 2006.

[9] A.H. Gnauck and P.J. Winzer, *J. Lightwave Technol.*, vol. **23**, p. 115, 2005.

[10] P. Kabal and S. Pasupathy, *IEEE Trans. Commun.*, vol. **COM-23**, p. 921, 1975.

[11] A.V. Kanaev, G.G. Luther, V. Kovanis, S.R. Bickham, and J. Conradi, *J. Lightwave Technol.*, vol. **21**, p. 1486, 2003.

[12] M. Seimetz, M. Noelle, and E. Patzak, *J. Lightwave Technol.*, vol. **25**, p. 1515, 2007.

[13] M. Seimetz. In S. Kumar (ed.), *Impact of nonlinearities on Fiber-Optic communications*. Springer-Verlag, Berlin, 2011.

[14] K.P. Ho and H.-W. Cuei, *J Lightwave Technol.*, vol. **23**, p. 764, 2005.

[15] P.J. Winzer, A.H. Gnauck, C.R. Doerr, M. Magarini, and L.L. Buhl, *J. Lightwave Technol.*, vol. **28**, p. 547, 2010.

5

Optical Receivers

5.1 Introduction

In the past few decades, there have been tremendous advances in optoelectronic integrated circuits (OEICs), primarily because of their widespread use in optical communication systems. Among OEICs, some of the key drivers have been high performance, low cost, and small size of photoreceivers. And in photoreceivers and optical receivers, the photodetector and preamplifiers are critical components. The photodetector's function is to convert light (photons) or radiant energy into charge carriers, electrons and holes, which can then be processed, stored, or transmitted again [1]. Further, a monolithically integrated photoreceiver has several advantages–low parasitics, compact size, and low cost. To date, various designs and structures of photodetectors, transistors, and integrated circuits have been used to produce high-performance integrated photoreceivers. In the design of integrated photoreceivers, various devices and circuit parameters are involved. To obtain the best possible photoreceiver performance, the parameters of both the photodetector and the preamplifier should be optimized. Therefore, we concentrate on describing some important photodetector structures and optical receivers. An example of a typical optical detection system is shown in Fig. 5.1, [1–4]. In an optical communication system, the photodetector can be configured either as a direct or incoherent detector, or as a coherent detector.

In direct or incoherent detection, the "direct" detector converts the incident radiation into an electrical signal (sometimes called the photo-signal) that is proportional to the power of the incident light. There is no phase or frequency information and the photo-signal is then processed electronically using a low-noise preamplifier followed by signal processing circuits. The preamplifier should have very low noise and wide enough bandwidth to accurately reproduce the temporal characteristics of the input signal, which may be a 10 or 40 Gb/s pulse stream. Minimization of noise in an optical direct detection system is a critical issue. In particular, the various sources of noise from the background, the photodetector itself, biasing resistors, and other additional noise sources such as the signal processing circuits must be minimized if the optical detection system is to have an acceptable signal-to-noise ratio and low bit-error rates for a given input signal power.

A coherent detector, in contrast, is one in which the output electrical signal is related to the phase of the input as well as the input power. The coherent detector requires a local oscillator whose phase is "locked" onto the phase of the received signal or the phase difference between the two should be corrected dynamically using digital signal processing (DSP). More details on these two types of photodetection system will be presented later.

In this chapter, we will discuss various types of photodetector. We will describe photodetectors without internal gain, such as pn photodiodes, pin photodetectors (pin-PDs), Schottky barrier photodetectors, and

Fiber Optic Communications: Fundamentals and Applications, First Edition. Shiva Kumar and M. Jamal Deen.
© 2014 John Wiley & Sons, Ltd. Published 2014 by John Wiley & Sons, Ltd.

Figure 5.1   Simple schematic representation of a typical optical detector system.

metal–semiconductor–metal photodetectors (MSM-PDs). We will also describe photodetectors with internal gain, like avalanche photodetectors (APDs), photoconductive photodetectors, and phototransistors. Then, we will describe some advanced photodetectors, such as resonant cavity-enhanced photodetectors (RCE-PDs) and waveguide photodetectors (WG-PDs). We describe noise sources in photodetection systems as well as optical detection system architectures. Finally, it should be noted that some of the material in this chapter is common to that in chapter 8 of Ref. [1], which was written by one of the authors.

5.2   Photodetector Performance Characteristics

A photodetector is a device in which an electron–hole pair is generated by photon absorption. In the case of lasers, electrons and holes recombine (stimulated emission) and their energy difference appears in the form of light. In other words, an electron and a hole annihilate each other to create the photon. In the case of photodetectors, the reverse process takes place. A photon with energy $hf > E_g$, where E_g is the band-gap energy (see Fig. 5.2), is annihilated to create an electron–hole pair.

The photon energy (E_{ph}) decreases as the wavelength (λ) increases according to

$$E_{ph} = hf = \frac{hc}{\lambda}, \tag{5.1}$$

where h = Planck's constant (6.626×10^{-34} J · s), c = speed of light, f = frequency of light (Hz), and λ = wavelength of light (m). If the energy E_{ph} of the incident photon is greater than or equal to the band-gap energy E_g, an electron makes a transition from the valence band to the conduction band, absorbing the incident photon. Fig. 5.3 shows the dependence of the absorption coefficient on wavelength or photon energy. The wavelength λ_{co} at which the absorption coefficient α becomes zero is called the *cutoff wavelength*. If the incident wavelength λ is greater than λ_{co}, the photodiode will not absorb light. This is because, if $\lambda > \lambda_{co}$,

$$f < f_{co} = \frac{E_g}{h}. \tag{5.2}$$

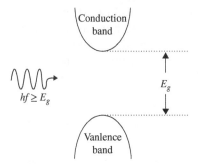

Figure 5.2   Photon absorption in a semiconductor.

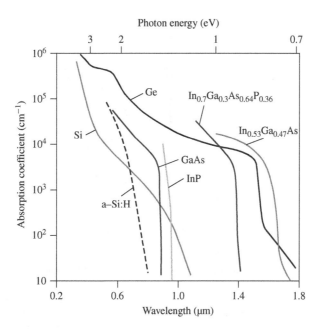

Figure 5.3   Absorption coefficient α versus wavelength (bottom x-axis) or photon energy (top x-axis) for seven common semiconductors.

Therefore, the energy of the photon ($\propto f$) will not be adequate to excite an electron into the conduction band if $\lambda > \lambda_{co}$, and such a photon will not be absorbed. Eq. (5.2) may be rewritten as

$$\lambda_{co} = \frac{hc}{E_g} \tag{5.3}$$

or

$$\lambda_{co} = \frac{1.2}{E_g(\text{eV})}(\mu m). \tag{5.4}$$

In a silicon photodiode, $\lambda_{co} \simeq 1.1\ \mu m$, so at 1.1 μm, the photon energy is just sufficient to transfer an electron across the silicon energy band gap, thus creating an electron–hole pair, as shown in Fig. 5.4 [5]. As this cutoff wavelength is approached, the probability of photon absorption decreases rapidly.

Table 5.1 shows some common semiconductors used as the active (absorption) materials in photodetectors and their corresponding cutoff wavelengths. This results in a spectral range of response of the photodetector, that is, the range of wavelengths over which the semiconductor material of the absorption layer of the photodetector is sensitive to input radiation. Also indicated in Table 5.1 are which semiconductors are direct band gap and which are indirect band gap.

In indirect band-gap semiconductors such as silicon or germanium, photon absorption requires the assistance of a phonon so that both momentum and energy are conserved (see Section 3.7.3). In this case, the absorption process can be sequential, with excited electron–hole pairs thermalizing within their respective energy bands by releasing some energy/momentum through phonons. Therefore, compared with absorption in a direct band gap where no phonons are involved, absorption in indirect band-gap semiconductors is less efficient. Below, we discuss briefly the features of different semiconductor absorption layer materials that have been used in commercial photodetectors.

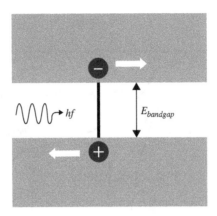

Figure 5.4   Absorption of photons with energies equal to or greater than the band gap.

Table 5.1   Some common semiconductor materials used in photodetectors with their E_g and λ_{co} values.

Semiconductor	Bandgap (eV) at 300 K	Cutoff wavelengths λ_{co} (μm)
Silicon	1.12 (indirect)	1.1
Germanium	0.66 (indirect)	1.85
GaAs	1.42 (direct)	0.87
GaSb	0.73 (direct)	1.7
AlAs	2.16 (direct)	0.57
InAs	0.36 (direct)	3.5
InP	1.35 (direct)	0.92
$In_{0.14}Ga_{0.86}As$	1.15 (direct)	1.08
$In_{0.47}Ga_{0.53}As$	0.75 (direct)	1.65

Silicon (Si)

- Indirect band-gap material with a small absorption coefficient.
- It has a high ratio of ionization coefficient of holes (β_i) to electrons (α_i). It is a good choice for avalanche photodetectors, especially for local area networks (LANs) or short-wavelength applications.
- It is not suitable for the long-haul communications that typically operate at 1.3 or 1.55 μm.

Germanium (Ge)

- Indirect band-gap material with a small absorption coefficient.
- High β_i/α_i ratio. It can be used for avalanche photodetectors for both local area networks and long-distance communications due to its long cutoff wavelength.

Gallium arsenide (GaAs)

- Direct band-gap material.
- Easy lattice matched to the InP substrate.

- It is not suitable for high-quality avalanche photodetectors since $\beta_i \approx \alpha_i$.
- It is not suited for long-distance applications due to its short cutoff wavelength.

Indium gallium arsenide (InGaAs)

- It can have a tunable band-gap energy depending on the ratio of Ga to In.
- It is a very good material for long-haul communications at 1.55 μm.
- It can be lattice matched to the InP substrate.

Indium gallium arsenide phosphide (InGaAsP)

- Suitable for both 1.3- and 1.55-μm applications.
- It can be lattice matched to the InP substrate.

In a semiconductor photodetector, there are two or three key processes depending on the type of photodetector.

(i) *Absorption and generation.* Here, the photons of appropriate energy (that is, the energy of the incoming photon should be at least equal to the active semiconductor material's band-gap energy) generate free electron–hole pairs (ehps) through the photoconductive (or internal photoemission) effect when they are absorbed in the photoresponsive (or active) region of the photodetector. Note that in the photoconductive effect, the photogenerated carriers remain in the semiconductor material and they result in an increase in its conductivity. This is in contrast to photoelectric emission, in which the photogenerated electrons escape from the material and are then free to move outside the material under an applied electric field. Photoelectric emission is used in photomultiplier tubes (PMTs).

(ii) *Transport.* The generated ehps drift under the influence of an applied electric field E. This results in a current that flows in the circuit.

(iii) *Amplification.* In some photodetectors, when the electric field is sufficiently large, the photogenerated carriers moving in the applied electric-field can gain sufficient energy to impact ionize. Upon impact ionization, additional carriers are generated, creating more ehps. In this way, one photogenerated ehp can result in many more ehps, leading to a photodetector with gain. In more detail, the gain of the photodetector is defined as the ratio of the number of collected ehps to the number of primary photogenerated pairs. Gain expresses the sensitivity of the photodetector at the operating wavelength. One popular photodetector with gain is the avalanche photodiode.

5.2.1   Quantum Efficiency

In a semiconductor photodetector, when a photon of energy $E_{ph} \geq E_g$ is absorbed, an ehp is formed. Then, a photocurrent is produced when the photon-generated ephs are separated in an applied electric field, with electrons moving to the n-region and holes to the p-region (Fig. 5.5). However, the photons of appropriate wavelength do not always generate ehps, nor are all ehps collected at the respective terminals. Therefore, quantum efficiency QE (or η) is defined as the probability that a photon incident on the photodetector generates an ehp (photocarrier) that contributes to the photodetector current and is given by

$$\eta = \frac{\text{number of photocarriers that contribute to the photocurrent}}{\text{number of incident photons}}. \tag{5.5}$$

Note that $0 < \eta \leq 1$, that is, the maximum value of η in a photodetector without gain is 1 or 100%, which means that each incident photon generates an ehp. The QE depends on the photon wavelength, type of semiconductor, and structure of the photodetector.

The mean number of photons, N_{ph}, in an optical wave of energy E and frequency f_0 is

$$N_{ph} = \frac{E}{hf_0}. \tag{5.6}$$

Therefore, the mean number of photons per unit time, or *photon rate* or *photon flux*, is given by

$$\frac{N_{ph}}{T} = \frac{E}{Thf_0} = \frac{P}{hf_0}. \tag{5.7}$$

If the incident optical power on the photodetector is P_I, the mean number of photons incident per unit time, or *photon incidence rate*, is

$$R_{\text{incident}} = \frac{P_I}{hf_0}. \tag{5.8}$$

Let the number of photocarriers generated be N_{PC}. Not all the photocarriers contribute to the photocurrent, as some of them recombine before reaching the terminals of the photodetector. Let ζ be the fraction of photocarriers that contribute to the photocurrent. The effective photocarrier generation rate may be written as

$$R_{\text{gen}} = \frac{\zeta N_{PC}}{T} = \frac{I_{PC}}{q}, \tag{5.9}$$

where q is the electron charge. Using Eqs. (5.8) and (5.9), Eq. (5.5) may be rewritten as

$$\eta = \frac{\text{photocarrier generation rate}}{\text{photon incidence rate}}$$

$$= \frac{I_{PC}/q}{P_I/hf_0} = \frac{I_{PC}}{P_I} \frac{hc}{q} \frac{1}{\lambda_0}. \tag{5.10}$$

From Eq. (5.10), it is noted that η is inversely proportional to wavelength λ_0. However, at short wavelengths, η decreases due to surface recombination because most of the light is absorbed very close to the surface. For example, if the absorption coefficient $\alpha = 10^5$ to 10^6 cm^{-1}, then most of the light is absorbed within the penetration distance $1/\alpha = 0.1$ to 0.01 μm. At these distances, close to the surface, the recombination lifetime is very short, so the majority of photogenerated carriers recombine before they can be collected at the terminals. This gives rise to the short-wavelength limit in the quantum efficiency of the photodetector. However, with careful surface treatment, it may be possible to extend the short-wavelength limit to lower values of wavelength λ.

An example of a simple pn-homojunction photodetector operating in the photoconductive mode (third quadrant of the $I-V$ characteristics) is shown in Figs. 5.5 and 5.6. In Fig. 5.6, the main absorption or photoactive region is the depletion region, where the electric field sweeps the photogenerated electrons to the n-side and holes to the p-side. This results in a photocurrent that is a drift current flowing in the reverse direction, that is, from the n-side (cathode) to the p-side (anode), and this is the main contribution to the total photocurrent.

In addition, if ehps are generated within one diffusion length of the depletion region boundaries, they can also contribute to the photocurrent. For example, the photogenerated minority carriers–holes on the n-side and electrons on the p-side–can reach the depletion boundary by diffusion before recombination happens. Once they reach the depletion region, they will be swept to the other side by the electric field. Thus, there is also a diffusion current flowing in the reverse direction and contributing to the photocurrent.

In contrast, in the bulk p- or n-regions, although the generation of ehps occurs by photon absorption, they do not contribute to the photocurrent. This is because there is negligible electric field to separate photogenerated

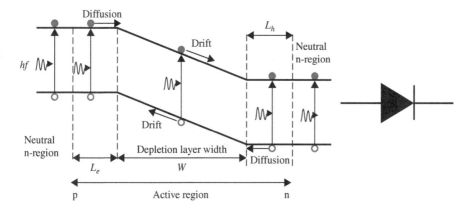

Figure 5.5   Photoexcitation and energy-band diagram of a pn photodiode and its symbol.

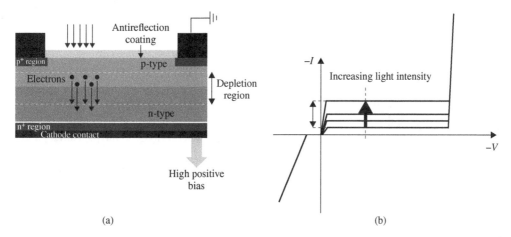

Figure 5.6   (a) Schematic representation of a simple photodiode with coating of reflectivity R_p. Note that only electrons are shown moving toward the n-type semiconductor from the depletion. An equivalent number of holes move in the opposite direction. (b) Typical reverse-bias characteristics where the photocurrent increases with light levels.

charges and hence they recombine randomly. For this pn-junction, if its cross-sectional area is A and the incident photons generate G electron–hole pairs per second per unit volume (ehp/s/cm^3), and if L_e and L_h are the respective minority diffusion lengths in the p- and n-regions, then the resulting photocurrent from the n- to the p-region is

$$I_{PC} = qAG(W + L_e + L_h),\qquad(5.11)$$

In practice, when computing η, we need to consider the details of the photodetector–the photoactive material through its absorption coefficient and geometry as well as its surface. A schematic representation of a pn photodiode with antireflection coating is shown in Fig. 5.6(a).

Let the optical power incident on one side of the pn-photodiode be P_I. If R_p is the power reflection coefficient at the air–semiconductor interface, the power transmitted at the interface is $(1 - R_p)P_I$. The transmitted power through the photodiode is

$$P_{\text{tr}} = (1 - R_p)P_I \exp{(-\alpha W)}, \tag{5.12}$$

where α is the absorption coefficient and W is the thickness of the depletion or active region of the photodetector. Therefore, the power absorbed in the photodiode is

$$P_{\text{abs}} = (1 - R_p)P_I - P_{\text{tr}} = (1 - R_p)P_I[1 - \exp{(-\alpha W)}]. \tag{5.13}$$

From Eq. (5.7), we find that the mean number of photons absorbed per unit time, or *photon absorption rate*, is $P_{\text{abs}}/(hf_0)$. If *a* photon is absorbed, *an* electron–hole pair is generated. Therefore, the number of electron-hole pairs generated per unit time is

$$\frac{N_{PC}}{T} = \frac{P_{\text{abs}}}{hf_0}. \tag{5.14}$$

Using Eqs. (5.14) and (5.13) in Eq. (5.10), we find

$$\eta = \frac{P_{\text{abs}}\zeta}{P_I}$$
$$= (1 - R_p)\zeta[1 - \exp{(-\alpha W)}]. \tag{5.15}$$

The quantum efficiency is equal to a product of:

1. the power transmission coefficient at the air–semiconductor interface, $1 - R_p$;
2. the photons absorbed in the active region of thickness W, given by the term $1 - \exp{(-\alpha W)}$; and
3. the fraction of photocarriers ζ that reach the device terminal and contribute to the measured photocurrent.

The third term is usually the most difficult to determine as it depends on a number of factors, such as carrier lifetimes, transit paths, surface properties, and the physical dimensions of the device.

Example 5.1

If the incident optical signal on a pn photodiode is at a wavelength of 550 nm, its absorption coefficient $\alpha = 10^4$ cm^{-1}, width of the active region $W = 3$ μm, and optical power 1 nW, calculate (a) the photon incidence rate, (b) the photon absorption rate, and, (c) the quantum efficiency. Assume $R_p = 0$ and $\zeta = 0.9$.

Solution:
(a) The energy of a photon is

$$E_{ph} = \frac{hc}{\lambda_0} = 3.6 \times 10^{-19} \text{ J}.$$

The photon incidence rate is given by Eq. (5.8),

$$R_{\text{incident}} = \frac{P_I}{E_{ph}} = \frac{1 \times 10^{-9} \text{ W}}{3.6 \times 10^{-19}} \text{photons/s}$$
$$= 2.77 \times 10^9 \text{ photons/s.}$$

(b) Using Eq. (5.13), with $R_p = 0$, the photon absorption rate is

$$R_{abs} = R_{incident}[1 - \exp(-\alpha W)]$$

$$= 2.77 \times 10^9 \times [1 - \exp(-10^4 \times 3 \times 10^{-4})] \text{ photons/s}$$

$$= 2.63 \times 10^9 \text{ photons/s.}$$

(c) The quantum efficiency is given by Eq. (5.15),

$$\eta = (1 - R_p)\zeta[1 - \exp(-\alpha W)]$$

$$= 0.9 \times (1 - \exp(-10^4 \times 3 \times 10^{-4}))$$

$$= 0.855.$$

5.2.2 Responsivity or Photoresponse

The responsivity or photoresponse (sometimes also called *sensitivity*) is a measure of the ability of the photodetector to convert optical power into an electrical current or voltage. It depends on the wavelength of the incident radiation, the type of photoresponsive (or active) material in the detector, and the structure and operating conditions of the photodetector. It is defined as

$$R = \frac{I_{PC}}{P_I}, \tag{5.16}$$

where I_{PC} is the photocurrent and P_I is the input optical power.

The photocurrent, in turn, depends on the absorption characteristics of the active (photoresponsive) material on the photodetector and the quantum efficiency. In a photodetector, the intrinsic quantum efficiency is the number of ehps generated per incident photon. In the ideal case, the quantum efficiency, which is a measure of the number of photogenerated ehps per incident photon, is 1 or 100%, that is, each photon of appropriate energy (equal to or greater than the energy band gap E_g of the active semiconductor material) generates one ehp. For a pn photodiode, using Eq. (5.10) in Eq. (5.16), we find

$$R = \frac{\eta q}{h f_0}. \tag{5.17}$$

If we insert the numerical values for q, c, and h and with $f_0 = c/\lambda_0$, Eq. (5.17) may be rewritten as

$$R(\text{A/W}) = \eta \frac{\lambda_0(\mu m)}{1.24}. \tag{5.18}$$

Note that the responsivity is proportional to both the quantum efficiency η and the free-space wavelength λ_0.

Fig. 5.7 shows schematically how the responsivity varies with wavelength. Notice that the responsivity curve falls at both longer and shorter wavelengths for all three photoresponsive materials. The long-wavelength drop is related to the energy band gap of the semiconductor. For example, for silicon, the energies of photons with wavelengths approaching 1.1 μm are close to its indirect band-gap energy, beyond which silicon is transparent. At the other extreme, at short wavelengths, as mentioned before, the quantum efficiency decreases rapidly due to surface recombination effects as most of the light is absorbed close to the surface.

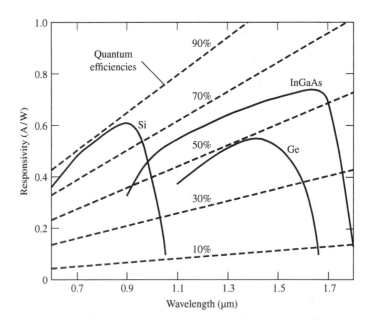

Figure 5.7 Schematic representation of responsivity vs. wavelength for three common absorption materials and various QEs from 10% to 90%.

Example 5.2

In a GaAs photodiode, if the quantum efficiency $\eta = 0.9$, band-gap energy $E_g = 1.42$ eV, and operating (free-space) wavelength $= 1.1$ μm, calculate (a) the responsivity R and (b) the cutoff wavelength λ_{co}.

Solution:
(a) From Eq. (5.18), we have

$$R = \eta \frac{\lambda_0(\mu m)}{1.24} \text{ A/W}$$

$$= \frac{0.9 \times 1.1}{1.24} \text{A/W} = 0.9 \text{ A/W}.$$

(b) The cutoff wavelength is given by Eq. (5.4),

$$\lambda_{co} = \frac{1.2}{E_g(eV)} \mu m$$

$$= \frac{1.2}{1.42} \mu m = 0.873 \; \mu m.$$

Example 5.3

Consider radiation of wavelength $\lambda = 700$ nm incident on a photodetector whose measured responsivity is 0.4 A/W. What is its quantum efficiency at this wavelength? If the wavelength is reduced to 500 nm, what is the new QE assuming that the responsivity is the same?

Solution:
Using Eq. (5.18), we get

$$0.4 \ \text{A/W} = \eta \frac{\lambda_0(\mu\text{m})}{1.24} \Rightarrow \eta = \frac{0.4 \times 1.24}{0.7} = 0.7086 (\approx 71\%).$$

For $\lambda_0 = 500$ nm, the new QE is

$$\eta = 0.7086 \times \frac{0.5}{0.7} = 0.506 (\approx 51\%).$$

5.2.3   Photodetector Design Rules

As shown in Eq. (5.15), to improve the quantum efficiency, we should minimize light reflections (R_p term) from the semiconductor surface or maximize the light transmitted into the semiconductor. For this, we can use an antireflection coating to achieve better light transmittance. If the light is incident from air (refractive index n_{air}) into the semiconductor (refractive index n_{sc}), then we should choose a material whose refractive index n_{AR} (refractive index of antireflection coating) is given by

$$n_{\text{AR}} = \sqrt{n_{\text{air}} n_{\text{sc}}}. \tag{5.19}$$

If we use a quarter-wavelength antireflection coating of a transparent material with a refractive index n_{AR}, then the thickness t_{AR} which causes minimum reflection of the incoming radiation is given by (see Section 6.6.3)

$$t_{\text{AR}} = \frac{\lambda}{4 n_{\text{AR}}}, \tag{5.20}$$

where λ is the free-space wavelength of the incident light onto the antireflection coating.

Example 5.4

If we use a silicon photodetector to detect red light at 680 nm, and refractive index of air (n_{air}) = 1, refractive index of silicon (n_{Si}) = 3.6, determine the refractive index and thickness of the antireflection coating.

Solution:
The required antireflection coating should have a refractive index $n_{\text{AR}} = \sqrt{n_{\text{air}} n_{\text{Si}}} = 1.9$ and its thickness should be $t_{\text{AR}} = \lambda/(4 n_{\text{AR}}) = 680$ nm$/(4 \times 1.9) \approx 90$ nm. At 680 nm, the refractive index of silicon nitride (Si_3N_4) ~ 2 and that of silicon dioxide (SiO_2) ~ 1.5. Therefore, Si_3N_4 would be a good (though not perfect)

choice for the antireflection coating. Note that if we do not use an antireflection coating, then the reflectivity of silicon at 680 nm is ~0.32. This means that ~68% of the incident light is transmitted into the silicon photoactive region. Note that the reflection coefficient R_p is computed using,

$$R_p = \left(\frac{n_{Si}(\lambda) - 1}{n_{Si}(\lambda) + 1} \right)^2. \tag{5.21}$$

In addition to the antireflection coating, we also need to design a suitable absorption layer thickness (W) or appropriately select the reverse-bias voltage of the photodiode to be large enough so that adequate light is absorbed. However, as will be shown later, if the absorption region is too thick, then the speed of a transit-time-limited photodetector will be degraded. A good rule of thumb is to have the absorption layer thickness satisfy the following inequality:

$$\frac{2}{\alpha} < W < \frac{1}{\alpha}. \tag{5.22}$$

where α is the absorption coefficient and its inverse ($1/\alpha$) is the penetration depth of the incident light.
For the above example of a silicon detector used for 680-nm light, the inverse of the absorption coefficient is $1/(2.21 \times 10^3) \cong 4.5$ µm. Therefore, according to the design rule in Eq. (5.22), the absorption layer thickness should be between 4.5 and 9 µm. However, in practice, using standard silicon semiconductor technology, this large a thickness may be difficult to achieve, so the QE would be degraded.

5.2.4   Dark Current

The dark current is the current generated in the photo-detector without an incident optical signal or when it is in the dark. This current originates from the generation of an ehp due to thermal radiation or stray light. Here, we briefly introduce typical theoretical current expressions for three mechanisms that contribute to the dark current in a photodetector under reverse bias. The main mechanisms of transport in a reverse-bias homojunction diode are diffusion of minority carriers J_{diff}, generation–recombination current (for example, due to trap-assisted and band-to-band tunneling) J_{GR}, or surface leakage J_S. These three mechanisms can be described by the following expression:

$$J_{diff} = \frac{qn_i^2}{A} \left(\frac{1}{N_A} \sqrt{\frac{D_n}{\tau_n}} + \frac{1}{N_D} \sqrt{\frac{D_p}{\tau_p}} \right) [\exp(qV/kT) - 1], \tag{5.23}$$

$$J_{GR} = \frac{qn_i W}{A\tau_{GR}} [\exp(qV/2kT) - 1], \tag{5.24}$$

and

$$J_S = \frac{B_s V T^{3/2}}{A} \exp(-E_g/2kT). \tag{5.25}$$

In these expressions, A is the area of the photodiode, W is the depletion width at reverse-bias voltage V, n_i is the intrinsic carrier concentration, N_A and N_D are acceptor and donor densities, respectively, E_g is the band-gap energy, D's and τ's are minority carrier diffusion constants and lifetimes, and B_s is a fitting parameter. The corresponding D's and diffusion lengths L can be calculated using the Einstein relation $D/\mu = kT/q$ where $\mu = $ mobility and $T = $ absolute temperature. Note that if dislocations are present, we also need to include a model for the leakage current originating from the dislocations.

5.2.5   Speed or Response Time

The speed of response or bandwidth of a pin photodetector, shown in Fig. 5.8, depends on the following factors.

1. The transit time τ_t of the photogenerated carriers through the depletion or active region, given by

$$\tau_t = \frac{W}{v}, \tag{5.26}$$

where v is the speed of the carrier. If the carriers are not traveling at their saturation velocity v_{sat}, then $v = \mu E$ where μ is the mobility of the carrier traveling in an electric field E. The electric field intensity is in turn computed from $E \sim V/W$, where V is the voltage across the depletion region W. Therefore, we can write τ_t as

$$\tau_t = \begin{cases} W/v_{\text{sat}}, & \text{for carriers traveling at their saturation velocity} \\ W^2/(\mu V), & \text{for carriers traveling below their saturation velocity.} \end{cases} \tag{5.27}$$

2. The slower (relative to drifting carriers) diffusion of carriers occurs outside the depletion region. To minimize this diffusion time effect, generally the depletion region is made as large as possible. For example, a pin photodiode (Fig. 5.8) may be used instead of a pn photodiode where the i-region is much larger than a typical reverse-biased depletion region. Also, because the doping concentration in the i-region is significantly lower than that in the p- or n-regions in a pin photodiode, then most of the depletion width is the i-region and the carrier transit time is drift dominated.

3. The RC time constant τ_{RC} is due to the resistance R (the sum of the diode's parasitic resistance R_S and the load resistance R_L) and the capacitance C of the diode. In this case, the RC time constant is given by

$$\tau_{\text{RC}} = RC. \tag{5.28}$$

Therefore, the total response time τ_{tot} can be written as the root-mean-square value

$$\tau_{\text{tot}} = \sqrt{\tau_{\text{RC}}^2 + \tau_t^2}. \tag{5.29}$$

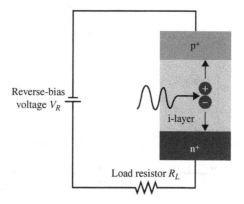

Figure 5.8   Schematic representation of a pin photodiode showing the photoactive region where electron–hole pairs are generated.

Note that a fast transit time implies a thin detector active region, while low capacitance and high responsivity require a thicker active region. Thus, there are trade-offs between fast transit times and low capacitance for high-speed response, high quantum efficiency, low dark current, and good coupling efficiency when used in a fiber system. For example, a fast transit time requires a thin detector photoactive region, while low capacitance and high responsivity (or quantum efficiency) require a thick active region. It is very generally favorable to design the absorption region to be larger than the penetration depth using expression (5.22).

Also, a smaller detector active area leads to lower dark current and smaller junction capacitance, but may be inefficient for detector coupling to the fiber when used in fiber-coupled systems. Therefore, building on the above examples, a silicon-based sensor that is optimized for 680-nm detection should be designed to have the thickness of the semiconductor within 4.5–9 μm.

5.2.6   Linearity

Typically, reverse-biased photodetectors are highly linear devices (Fig. 5.9). Detector linearity means that the output electrical current (photocurrent) of the photodiode is linearly proportional to the input optical power. Reverse-biased photodetectors remain linear over an extended range (six decades or more) of photocurrent before saturation occurs. Output saturation occurs at input optical power levels typically greater than 1 mW. Because fiber-optic communication systems operate at low optical power levels, detector saturation is generally not a problem.

5.3   Common Types of Photodetectors

As mentioned in Section 5.1, semiconductor photodetectors can be broadly classified into those without internal gain and those with internal gain. In the first category are pn photodiodes, pin photodetectors, Schottky barrier photodetectors, and MSM-PDs. In the second category are photoconductors, phototransistors, and APDs. These second types of photodetector are used to improve the overall sensitivity of the front-end photoreceiver.

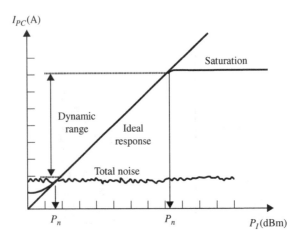

Figure 5.9   Response characteristics of a typical photodetector. Important features of the response characteristics are indicated in the figure.

5.3.1 pn Photodiode

A pn photodiode is basically a pn junction diode operating under reverse bias. It has already been described in Section 5.2.1 and Fig. 5.5. In a pn photodiode, the incident photons may be absorbed in both depletion and diffusion regions, and the number of ehps generated is proportional to the optical power (see Eqs. (5.11) and (5.16)). The ehps are separated in the depletion region and are induced to drift under the influence of the external applied electric field. In designing a pn photodiode, the depletion or absorption region should be wide enough to achieve high quantum efficiency, but at the same time, it should not be too wide because the drift time increases with the width, resulting in a decrease of the bandwidth of the photodiode. As mentioned before, this exemplifies the compromise between quantum efficiency and speed for almost all photodetectors.

5.3.2 pin Photodetector (pin-PD)

The pin photodetector is one of the popular types used in fiber-optic communications and it was previously introduced in Section 5.2.5. The performance of pin-PDs surpasses that of pn photodiodes because they can easily be tailored for optimum quantum efficiency and bandwidth (see the sensitivity–bandwidth trade-off discussed in Section 5.2.5). The basic pin-PD consists of three regions: heavily doped p^+ and n^+ layers and an intrinsic i-layer that is typically a much lower-doped semiconductor. The i-layer sandwiched between the p^+ and n^+ layers is shown in Fig. 5.10. In a pin-PD, the photon absorption takes place primarily in the intrinsic region that is depleted when reverse-bias voltage is applied to its terminals. The collection process for the generated carriers is therefore fast and efficient. Thus, the intrinsic bandwidth is very high, and the overall bandwidth that is limited by the extrinsic effects can be tens of gigahertz. Fabrication of pin photodetectors is relatively easy, with well-established semiconductor processes, and the fabricated devices are very reliable and of low noise. An example of a front-illuminated InP/InGaAs homojunction pin photodiode is shown in Fig. 5.10. Typically, pin-PDs are combined with erbium-doped fiber amplifiers (EDFAs) in order to increase the overall sensitivity of the receiver.

Similar to the pn photodiode, there is a compromise between quantum efficiency and bandwidth when designing a pin-PD. For an optimized structure, the quantum efficiency–bandwidth product (η-BW) is approximately constant. For example, the bandwidth can be increased from a few tens of gigahertz to more

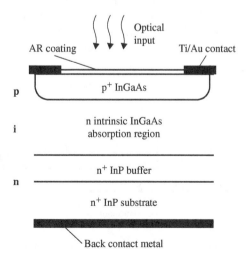

Figure 5.10   Schematic structure of a front-side-illuminated InGaAs–InP pin photodiode.

than 100 GHz [6] by using a matching network and decreasing the device size, but this reduces the quantum efficiency. If a side-entry or a waveguide-fed pin-PD is fabricated, then the η-BW product can be further improved [7]. The η-BW product can also be increased by inserting the pin-PD inside a resonant cavity as in RCE-PDs, where the quantum efficiency is enhanced even for a thin absorption layer due to reflections of light through the bottom and top mirrors of the cavity, resulting in multi-passes of light through the absorption layer. This is discussed in more detail later in section 5.3.8.1.

5.3.3   Schottky Barrier Photodetector

The Schottky barrier photodiode is made of metal−semiconductor−metal rectifying junctions rather than pn semiconductor junctions. Fig. 5.11 shows a simple schematic structure of a Schottky barrier photodetector. Schottky barrier photodiodes have narrow active layers compared with pin or pn PDs. An important advantage of Schottky barrier photodetectors is the very short carrier transit time, resulting in a very high bandwidth. However, this advantage due to the narrow active layer comes at the expense of poor quantum efficiency.

Schottky photodetectors using InGaAs/InP-based material for the 1.3−1.55 μm wavelength region are also useful photodetectors in the visible and ultraviolet wavelength range due to their large absorption coefficients at these shorter wavelengths. However, one of the technical challenges in manufacturing the Schottky diode is how to avoid surface traps and recombination centers that cause substantial loss of photogenerated carriers at the surface, and therefore a reduction in quantum efficiency. Also, because the quantum efficiency is very low in the 1.3−1.55 μm wavelength region, the Schottky barrier photodetectors are not widely used in optical fiber communication systems.

5.3.4   Metal−Semiconductor−Metal Photodetector

A MSM-PD uses an absorption layer of semiconductor material that is sensitive to the wavelength of interest. On top of this layer, metal electrodes are deposited as interdigitated fingers to form back-to-back Schottky diodes with a suitable antireflection coating between them. Each electrode forms a Schottky barrier contact with the semiconductor. It is connected to a large contact pad for subsequent interconnection to the external circuit. The MSM-PD is a variation of a Schottky barrier photodetector with both contacts made on the same side of the substrate. A top view of a MSM-PD is shown in Fig. 5.12, while a schematic of the cross-section of this photodetector is shown in Fig. 5.13. The rectified *I-V* characteristic of a Schottky diode is similar to that of a pn junction. However, a Schottky barrier occurs only for certain metal−semiconductor junctions

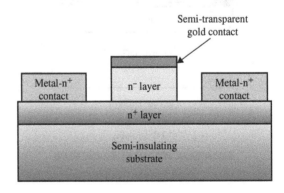

Figure 5.11   Schematic structure of Schottky barrier photodetector.

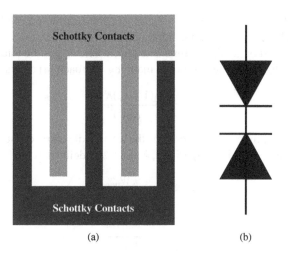

(a)   (b)

Figure 5.12   (a) Top view of a MSM-PD. and (b) Circuit schematic of a MSM-PD.

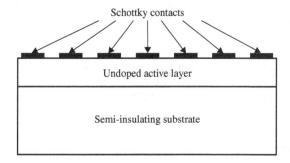

Figure 5.13   A schematic structure of the cross-section of a MSM-PD.

which can also be ohmic. The doping levels in the p- and n-semiconductors control the potential barrier of a pn junction. However, in a Schottky diode, the potential barrier (φ_b) is controlled by the work functions of the metal and semiconductor materials. A key difference between a pn diode and a Schottky diode is that pn junctions allow both electrons and holes to flow under forward bias, while a Schottky diode is a majority carrier device (only one type of carrier flows).

A MSM-PD is planar and requires only a single photolithography step for fabrication. The electrode deposition stage may generally be combined with other metallizations in the fabrication of an integrated receiver circuit. This makes the fabrication and monolithic integration with other electronic devices, such as a metal–semiconductor field-effect transistor (MESFET), relatively easy.

The MSM-PD also has the same compromise problem between quantum efficiency and bandwidth as with the other photodetector structures discussed above. While the quantum efficiency can be maintained by using a multi-finger interdigitated layout, the bandwidth is increased by reducing the effective absorption layer thickness. This can be achieved by artificially restricting the electric field within a certain layer near the surface by introducing, for example, a highly doped layer at a certain depth. The impulse response for a MSM-PD shows a tail response due to the photogenerated carrier distribution away from the surface. The slow tail response can be removed by tailoring the carrier lifetime in the material so that the lifetime is almost equal to the transit time between the electrodes.

Some MSM-PD technology can provide very thin fingers and narrow spacing between fingers. Therefore, the transit time can be made very small, but the limiting factor is now the capacitance C_d and the speed of the MSM-PD is mainly controlled by its $R_{tot}C_d$ time constant, where R_{tot} is the total resistance of the PD. The capacitance may be calculated using the conformal mapping approach [8] and is given by

$$C_d = \frac{\varepsilon_0(1 + \varepsilon_r)K(k)}{K(k')}, \tag{5.30}$$

where ε_0 is the absolute permittivity of vacuum, ε_r is the relative dielectric constant of semiconductor, and K is the elliptic integral of the first kind. In Eq. (5.27), k and k' are defined as

$$k = \tan^2\left(\frac{\pi W_{\text{finger}}}{4(W_{\text{finger}} + L_{\text{gap}})}\right) \tag{5.31}$$

and

$$k' = \sqrt{1 - k^2}, \tag{5.32}$$

where W_{finger} and L_{gap} are the width and spacing of the fingers, respectively. For $W_{\text{finger}} = L_{\text{gap}} = 0.5$ µm, and 40 fingers, each 10 µm long, the capacitance of the photodetector becomes 24 fF, which is a very small value compared with other conventional photodetectors. This value can further be reduced by using a series–parallel configuration of the fingers [8].

Regarding the geometry of a MSM-PD, it is noted that the signals are coupled to photodetectors through optical fibers which have circular cross-sections. Therefore, a rectangular structure does not help in utilizing the entire surface area and thus the responsivity is reduced. Circular structures can instead be used for improving the responsivity and the capacitance of a MSM-PD [5].

5.3.5   Photoconductive Detector

A photoconductive photodetector is made up of an absorptive semiconductor together with two electrical terminals. When it is illuminated, the electrical conductivity increases because the photogenerated carriers carry an electrical current. The internal gain mechanism arises from the space-charge neutrality requirement. The photogenerated carriers move toward their respective collecting terminals with different velocities. The carriers moving faster reach the terminal first, resulting in an excessive charge in the photoconductor. The excessive charge draws additional carriers into the conducting layer, until the slowest carrier is recombined or collected so that total charge neutrality in the photoconductor is satisfied. In this way, we can define the internal gain as the ratio of the transit time of the slow carrier to that of the fast carrier. One limitation is its bandwidth, which is inversely proportional to the transit time of the slow carrier. Therefore, photoconductors typically have bandwidths up to a few hundred megahertz. Higher bandwidths, in the gigahertz range, can be achieved by optimizing the distribution of photogenerated carriers along the device. Photoconductive devices are easy to fabricate and incorporate into OEICs. However, they are not often used in optical fiber communication systems due to their limited bandwidth and large leakage current.

5.3.6   Phototransistor

A phototransistor is similar to a bipolar junction transistor (BJT), but with normally only two terminals–the collector and the emitter–used as electrical contacts. The base and the base–collector junction are used as the absorption layer. The photogenerated holes in the absorption region accumulate in the base. This excessive

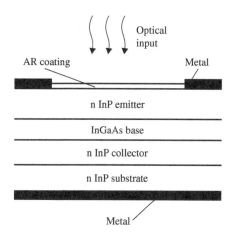

Figure 5.14   Schematic structure of an n–p–n InGaAs/InP phototransistor.

charge results in electrons being injected from the emitter. The current gain mechanism is the same as in normal electrical operation of a BJT. The schematic structure of an n–p–n InGaAs/InP phototransistor is shown in Fig. 5.14.

Heterojunction phototransistor (HPT) technology is promising, and can be built using different semiconductor structures. For example, resonant-cavity-enhanced HPTs can be developed to improve both quantum efficiency and responsivity. The HPT can also be integrated with a traveling wave device that is optimized for microwave performance [9].

5.3.7   *Avalanche Photodetectors*

APDs are the most important photodetectors with internal gain that have been widely used in fiber-optic communication systems. The APD is commonly used for detection of extremely low-intensity optical radiation due to its high sensitivity characteristics [10–14]. The APD can be made using Si, Ge, or III–V semiconductor materials. Its internal gain comes from the avalanche multiplication process through impact ionization events. The impact ionization phenomenon has been extensively investigated, both theoretically and experimentally [15–19], and a schematic representation is shown in Fig. 5.15.

Unlike the photodetector structures discussed above, an APD operates under sufficiently high reverse voltage to generate a high enough electric field in which highly energized photogenerated ehps can impact ionize. In more detail, under a high electric field, the high-energy conduction band electrons initially scatter with an electron in the valence band and knock it out into the conduction band, resulting in multiplication of the number of electrons in the conduction band and holes in the valence band. This results in a multiplication of the number of current-carrying charges in this avalanche process. This avalanche process could also happen to high-energy valence band holes that impact ionize. To cause impact ionization, the required minimum carrier energy is the ionization threshold energy that should be larger than the band-gap energy.

The process of ionization is exponentially dependent on the magnitude of the electric field. The ionization coefficients of electrons α and holes β are defined as the inverse of the mean distance between ionization collisions, but electrons and holes can lose energy in non-ionizing collision processes such as phonon scattering. One carrier undergoing the impact ionization process creates a pair of free carriers. All three carriers get accelerated, and then continue to undergo impact ionization events and generate more free carriers. This

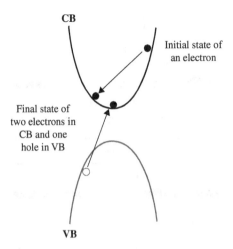

Figure 5.15 Schematic representation of impact ionization in a direct band-gap semiconductor. Note: CB = conduction band and VB = valence band.

process terminates when all the free carriers are swept out of the high-electric-field region. In the end, one initial electron or hole generates M extra e−h pairs, where M is called the multiplication gain of the photodetector.

A generalized theory regarding the impact ionization phenomena in semiconductor materials has been developed by Baraff [20] in terms of threshold ionization energy (E_i), optical phonon scattering energy (E_r), and carrier mean free path (λ) limited due to optical phonon scattering. However, Baraff's expression for the impact ionization could be evaluated only numerically. A simple expression for the ionization parameters of the charged carrier in a semiconductor as a function of an electric field and a lattice temperature has been developed by Okuto and Crowell [17, 19]. They expressed the ionization parameters in terms of a power-series expansion of the functions of an electric field (F), optical phonon energy, carrier mean free path due to optical phonon scattering, and threshold energy for ionization. By fitting Baraff's numerical results for α and β versus electric field at low field values and imposing energy conservation conditions, Okuto and Crowell obtained a semi-analytical expression for ionization coefficients. The expression is given by

$$\alpha; \beta = \frac{qF}{E_{ie;h}} \exp\left\{ 0.217\left(\frac{E_{ie;h}}{E_{re;h}}\right)^{1.14} - \sqrt{\left[0.217\left(\frac{E_{ie;h}}{E_{re;h}}\right)^{1.14}\right]^2 + \left[\frac{E_{ie;h}}{qF\lambda_{e;h}}\right]^2} \right\}, \tag{5.33}$$

where $E_{ie;h}$ denotes the threshold energy of electron and hole ionization, $\lambda_{e;h}$ and $E_{re;h}$ represent the mean free path and average energy loss of carriers per collision, respectively, due to optical phonons. The temperature dependence of α and β in the above relation comes from the temperature dependence of $E_{ie;h}$, $\lambda_{e;h}$, and $E_{re;h}$ and is given below (E_g is for the case in InP, as an example)

$$E_g(T) = 1.421 - \frac{3.63 \times 10^{-4}}{(T + 162)} T^2, \tag{5.34}$$

$$E_{ie}(T) = E_g(T) \cdot \frac{E_{ieo}}{E_g(300)},$$

$$E_{ih}(T) = E_g(T) \cdot \frac{E_{iho}}{E_g(300)},$$

$$\lambda_{e;h} = \lambda_{0e;h} \tanh\left(\frac{R_{r0e;h}}{2kT}\right),$$

$$E_{re;h} = E_{r0e;h} \tanh\left(\frac{R_{r0e;h}}{2kT}\right),$$

where $\lambda_{0e;h}$, $E_{r0e;h}$, and k are the carrier mean free path at $T = 0$ K, average energy loss per collision for an electron and a hole, respectively, and Boltzmann's constant. E_g is the semiconductor (e.g., InP) band gap, and E_{ie0} and E_{ih0} are the electron and hole ionization energies at 0 K. From Eq. (5.34), the breakdown conditions are expected to be temperature independent in the low-temperature limit.

The mathematical description of the avalanche multiplication has been well documented [21–24]. The theoretical research in these reviews is based on the assumption that the multiplication layer is thick, and the multiplication process is continuous. Therefore, statistically, the discrete nature of the multiplication process is averaged out. This is reflected, for example, in the fact that there is no "dead" space–a space where ionization is impossible because the carrier has not acquired enough energy to initiate the multiplication process. Under these assumptions, the multiplication occurs only in the multiplication region of thickness W (between 0 and W). For details, see Fig. 5.16.

The electron component $J_n(x)$ of the total current increases, and the hole component $J_p(x)$ of the total current decreases along the positive x-direction. The rate equation for $J_n(x)$ and $J_p(x)$ in this case can be written as

$$\frac{d}{dx}J_n(x) = \alpha(x) \cdot J_n(x) + \beta(x) \cdot J_p(x) + q \cdot G(x), \tag{5.35}$$

$$-\frac{d}{dx}J_p(x) = \alpha(x) \cdot J_n(x) + \beta(x) \cdot J_p(x) + q \cdot G(x), \tag{5.36}$$

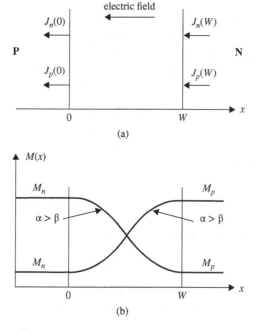

(a)

(b)

Figure 5.16 (a) Avalanche multiplication within a multiplication region of thickness W and (b) gain along the x-direction for the cases $\alpha > \beta$ and $\alpha < \beta$.

where $G(x)$ is the space charge generation rate due to optical or thermal generation. The total current density $J = J_n(x) + J_p(x)$ is a constant throughout the structure, to satisfy current continuity. Substituting this condition into Eqs. (5.35) and (5.36), and integrating both sides from 0 to W, the total current density can be obtained in two equivalent forms:

$$ J = \frac{J_p(W) + J_n(0) \exp\left[\phi(W)\right] + q \exp\left[\phi(W)\right] \int_0^W G(x) \exp\left[\phi(x)\right] dx}{1 - \int_0^W \beta \exp\left[\phi(W) - \phi(x)\right] dx}, \tag{5.37} $$

$$ J = \frac{J_p(W) \exp\left[-\phi(W)\right] + J_n(0) + q \int_0^W G(x) \exp\left[-\phi(x)\right] dx}{1 - \int_0^W \alpha \exp\left[-\phi(x)\right] dx}, \tag{5.38} $$

where $\exp\left(-\int_0^x (\alpha - \beta) dx\right) \equiv \exp\left(-\phi(x)\right)$. The multiplication gains with pure electron injection are $J_p(W) = 0$, M_n, and with pure hole injection are $J_n(W) = 0$, M_p, which can respectively be derived as

$$ M_n = \frac{J}{J_n(0)} = \frac{\exp\left[\phi(W)\right]}{1 - \int_0^W \beta \exp\left[\phi(W) - \phi(x)\right] dx} = \frac{1}{1 - \int_0^W \alpha \exp\left[-\phi(x)\right] dx}, \tag{5.39} $$

$$ M_p = \frac{J}{J_p(W)} = \frac{1}{1 - \int_0^W \beta \exp\left[\phi(W) - \phi(x)\right] dx} = \frac{\exp\left[-\phi(W)\right]}{1 - \int_0^W \alpha \exp\left[-\phi(x)\right] dx}. \tag{5.40} $$

Further, it can be shown that for both pure electron and hole injections, the breakdown condition when the multiplication gain is infinite is the same. This condition is expressed as

$$ 1 - \int_0^W \alpha \exp\left[-\phi(x)\right] dx = 1 - \int_0^W \beta \exp\left[\phi(W) - \phi(x)\right] dx = 0. \tag{5.41} $$

It can also be shown that if $\alpha > \beta$, then $M_n > M_p$, and vice versa. If the carriers with larger ionization coefficient are injected, then a higher gain should be obtained. This is also the condition for lower excess noise and higher bandwidth. Note that the formalism for photogain presented above was based on the assumption that the multiplication layer was thick. It did not take the discrete and statistical nature of the multiplication process into consideration. More sophisticated theories for ionization multiplication can be found in the research literature.

The multiplication gain described in Eqs. (5.40) and (5.41) represents only the average gain. The avalanche multiplication process is actually a stochastic process, and the statistical variation of the multiplication gain is responsible for the multiplication excess noise associated with the current. The total mean square noise spectral density can be described by

$$ \langle I_M \rangle^2 = 2q \left(2 \left[I_n(0) M_n^2 + I_p(W) M_p^2 + \int_0^W G(x) M^2(x) dx \right] + I \left[2 \int_0^W \left[\alpha M_n^2(x) dx - M_p^2 \right] \right] \right). \tag{5.42} $$

Here, I is the total current. Eq. (5.42) can be simplified in the case of pure electron or pure hole injection as

$$ \langle I_M \rangle^2 = 2q I_{p0} M^2 F. \tag{5.43} $$

Note that I_{p0} is the primary photocurrent, M is the multiplication, and F is the excess noise factor, which in turn is a function of the multiplication, device, and material parameters. If I_p is the total photocurrent, then the expression for M is

$$ M = \frac{I_p}{I_{p0}} \tag{5.44} $$

The expression for F is given by

$$F = M \cdot \left[1 - (1-k) \cdot \frac{(M-1)^2}{M^2} \right] = kM(1-k)(2-1/M). \tag{5.45}$$

Here, k is replaced by either k_{eff} or k'_{eff}:

$$k_{\text{eff}} = \frac{k_2 - k_1^2}{1 - k_2}, \tag{5.46}$$

$$k'_{\text{eff}} = \frac{k_{\text{eff}}}{k_1^2}, \tag{5.47}$$

and

$$k_1 = \frac{\int_0^W \beta(x) \cdot M(x) dx}{\int_0^W \alpha(x) \cdot M(x) dx}, \tag{5.48}$$

$$k_2 = \frac{\int_0^W \beta(x) \cdot M^2(x) dx}{\int_0^W \alpha(x) \cdot M^2(x) dx}. \tag{5.49}$$

For the case of a uniform electric field in the multiplication region, it turns out that $k_{\text{eff}} = \beta/\alpha$ and $k'_{\text{eff}} = \alpha/\beta$. From the computed results of excess noise factor F vs. gain M, it can be shown that, for lower excess noise, the carriers with higher ionization coefficient should be injected. Also, the ionization coefficients for electrons and holes should be significantly different for better performance of the APD. An APD provides gain without the need for an amplifier. However, its main limitation comes from its bandwidth. Because of the long avalanche build-up time, the inherent bandwidth of APDs is small. But APDs are very important because their internal gain is suitable for long-haul communication systems with a minimum number of repeaters, and also for dense wavelength-division multiplexing systems. Special structures can be used to improve the high-frequency performance. Unlike pin-PDs, even for moderate or high applied bias, the absorption layer may be at a low bias. This is because the multiplication layer should be under a high field for impact ionization. Therefore, the absorption and multiplication regions should be decoupled. The separate absorption and multiplication (SAM) structure with a bulk InP multiplication layer and an InGaAs absorption layer is shown in Fig. 5.17. Here, the objective is to make avalanche multiplication occur in a wider band-gap layer, such as InP, but for absorption to occur in a narrower band-gap layer, such as InGaAs.

To improve the performance of a SAM APD, a grading (G) layer is introduced in order to smooth out a band discontinuity between InP and InGaAs. This reduces a hole pile-up at the interface and, therefore, improves its frequency response at low biasing voltage [25–27]. Next, a charge (C) layer is used to control the electric field distribution between the absorption and multiplication layers. This is necessary because the electric field should be high enough to initiate impact ionization in the InP multiplication layer, and low enough to suppress ionization in the absorption layer, which could lead to a lower bandwidth. Depending on the particular design, different layers can be merged to perform the same tasks. For example, the charge layer can be merged with the multiplication layer, resulting in a very narrow multiplication layer. This leads to the SAGCM APD, which, despite its complexity, has an electric field profile that can be optimized for gain–bandwidth performance. It offers the most flexibility in terms of tuning the electric field profile for a particular application. Also, edge breakdown is suppressed because the charge sheet density is higher in the center of the device due to the mesa structure, which results in a higher electric field in the center of the device than in its periphery.

In order for this structure to operate properly as an APD, some practical conditions must be satisfied. For example, the electric field at the InGaAs/InP heterointerface should be smaller than 15 V/μm at operating

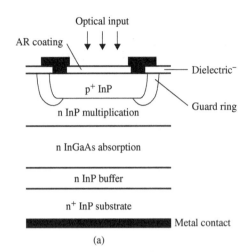

Figure 5.17   (a) Schematic structure of a planar SAM InP/InGaAs APD.

bias voltages to avoid significant tunneling currents at this heterointerface. The absorption layer should be completely depleted with minimum electric field of 10 V/μm to ensure that the photogenerated carriers are swept at their saturated velocities to the InP multiplication region at the operating bias voltages. Further, the absorption layer width should be large enough to obtain good quantum efficiency by ensuring adequate absorption, but not too large so as to increase unnecessarily the carrier transit time and reduce the bandwidth. The maximum electric field in the InP multiplication layer should be larger than 45 V/μm to achieve significant avalanche multiplication. And the doping concentration in the InP multiplication layer should be smaller than 2×10^{17} cm^{-3} to avoid the large tunneling current [25, 26].

In [11] and [21], a manufacturable bulk InP avalanche photodiode suitable for 10-Gb/s applications was introduced. It builds on the SAM structure with a bulk InP multiplication layer and an InGaAs absorption layer. Then, three InGaAsP grading layers between the field control and absorption layer are included to minimize hole trapping resulting from the valence band discontinuity that forms at an InGaAs–InP heterointerface. Also, edge breakdown is controlled by shaping the diffusion profile using a double-diffusion technique to create a wider multiplication region with a higher breakdown voltage around the device periphery. The back-illuminated geometry minimizes capacitance and dark current for a given optical coupling diameter. This separate absorption, grading, charge, and multiplication (SAGCM) APD structure is one in which the absorption, grading, charge, and absorption layers occur in sequence, as shown in Fig. 5.18, and it has excellent optoelectronic performance characteristics.

5.3.8   Advanced Photodetectors*

The performance of photoreceivers can be improved by using some advanced photodetector structures. As mentioned before, in the conventional surface-illuminated photodetectors, there is a trade-off between the quantum efficiency and the bandwidth as thinner absorption layer results in a higher bandwidth (transit-time limited) but a low quantum efficiency. Resonant cavity-enhanced (RCE) structures with a very thin active layer are a possible solution as the thin layer gives rise to a large bandwidth, and the multiple passes of light in the absorption layer in the resonant cavity increase the quantum efficiency [28–34]. A thin absorption region is placed in an asymmetric Fabry–Prot cavity. The top and bottom reflectors, which can be fabricated by distributed Bragg reflectors (DBRs), form the cavity. This is discussed in more detail in Section 5.3.8.1.

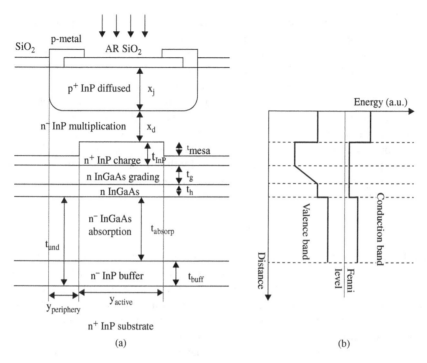

Figure 5.18 (a) Schematic structure of a mesa overgrown planar SAGCM InP/InGaAs APD. (b) Schematic representation of the band profile within the active region without an applied bias voltage.

Another means to improve the quantum efficiency as well as the transit-time limited bandwidth is to use edge-coupled structures. In this type of photodetector, the requirements for high efficiency and high bandwidth are decoupled by illuminating the photodetector from the side of the absorption layer. Therefore, the quantum efficiency is a function of the length of the absorption layer and not its thickness. So, a long thin absorption layer is good enough to get high efficiency and concurrently, the transit-time limited bandwidth is increased. Thus, the edge-coupled structure uses the attributes of waveguides [35–40] to improve both speed and quantum efficiency.

5.3.8.1   Resonant Cavity Enhancement

As mentioned earlier in this chapter, in a conventional surface-illuminated photodetector, there is a trade-off between the quantum efficiency and the bandwidth, two of the most important characteristics of a photodetector. The response time or speed of the photodetector is limited by two time constants.

(1) First, we have the transit time, defined as the time taken by photogenerated carriers to travel through the absorption region and to get collected by the electrical contacts. This time constant gives rise to a 3-dB bandwidth called the intrinsic bandwidth because it is linked to the intrinsic properties of the photodetector and the charge carrier.

(2) Second is the $R_{tot}C_d$ time constant, where C_d is the photodetector's capacitance and R_{tot} is the sum of its equivalent resistance and the load resistance. This bandwidth is called the extrinsic bandwidth. The 3-dB bandwidth depends on the thickness of the absorption layer of the photodetector, and can be increased by reducing this thickness. But a thin absorption layer which gives a large bandwidth (transit-time limited) results in a low quantum efficiency.

One possible solution to this trade-off is to use a RCE structure such as a Fabry–Perot cavity with a very thin active layer for a large bandwidth, and multiple passes of light in the absorption layer in the resonant cavity to increase the quantum efficiency. In this structure, a thin absorption region is placed in an asymmetric Fabry–Perot cavity (see Section 3.3). The top and bottom reflectors, which can be fabricated by DBRs, form the cavity. Below, we discuss some relevant details of the resonant cavity structure.

The Fabry–Perot cavity generally has two parallel mirrors comprised of quarter wavelength stacks (QWS) with a periodic modulation of the refractive index. That is, we use alternating materials that have different refraction indices in a multilayer form, which results in reflection at the interfaces. This multilayer structure is exploited to create optical mirrors, which reflect the light back to the optical absorption layer, thus effectively increasing the absorption width of the otherwise physically thin layer. Having mirrors on both sides of the absorption layer, the light gets "trapped" in the "cavity" between the mirrors until absorbed. The resonant frequency of a Fabry–Perot cavity is given by Eq. (3.43),

$$\frac{\lambda}{n} = 2L/\text{integer},$$
(5.50)

where λ is the light wavelength in vacuum, n is the refraction index of the material in the cavity, so that λ/n is the light wavelength in the material, and L is the distance between the mirrors. Owing to the electromagnetic wave property of the light, the "cavity" of mirrors can cause constructive or destructive summation of the waves, depending on the ratio of the distance between the mirrors to the wavelength of the light. Therefore, the spectrum of the enhancement becomes a quasi-oscillatory function of the light wavelength. The peak-to-valley ratio also depends on the energy loss between the mirrors, or more precisely, the photon absorption. For particular wavelengths, there is electromagnetic resonance in the cavity, which is the origin of the term "resonant cavity enhancement".

The classical arrangement of a RCE photodetector structure shown in Fig. 5.19(b) has a DBR as the front mirror on the side where the light is incident and a metal reflector as the back mirror on the other side for better reflection. Both mirrors can be electrically conductive, or one can place additional conductive sheets on the sides of the absorption layer with width W, and the photogenerated carriers are swept out by the bias of the photodetector traversing the structure once, thus, the transit time of the photodetector is virtually unaffected by the optical mirrors. A requirement is for the incident light to pass through the front mirror, and then to be reflected by the front mirror when traveling back into the cavity. Unfortunately, such a "diode-like" behavior of optical mirrors is not strong, although it can be achieved with proper selection of anti-reflection coating of the surface of the photodetector. Therefore, the front mirror is semitransparent, having similar reflectivities for incident and cavity light, $R_S \approx R_1 < 1$, and consequently, similar transmissivities $(1 - R_S) \approx (1 - R_1) < 1$ in both directions. Ideally, the back mirror should have a very high reflectivity, $R_2 \sim 1$. The materials (Si and Ge) in the cavity and in the absorption layer usually have similar refractive indices, weakly dependent on bias and charge or doping concentrations; so, we can neglect reflections inside the cavity. For other materials, this can be different; for example, in III–V semiconductors, organic semiconductors. The operation of the photodiode in a RCE was analyzed in Ref. [41], and the responsivity is given in Refs. [41, 42].

$$\frac{I_{ph}}{P_{\text{opt}}} = \left\{ \frac{q\eta}{hf_0} \times [1 - \exp(-\alpha W)] \times (1 - R_S) \right\} \times \text{RCE},$$

where
(5.51)

$$\text{RCE} = \frac{1 + \dfrac{R_2}{\exp(\alpha W)}}{1 - 2\dfrac{\sqrt{R_1 R_2}}{\exp(\alpha W)} \cos(2\beta L + \Psi_1 + \Psi_2) + \dfrac{R_1 R_2}{\exp(2\alpha W)}}$$

with $(q\eta/hf_0)$ the responsivity of the absorption material of infinite thickness, W the width of the absorption layer with absorption coefficient α, and $R_S \approx R_1$ the reflection from the photodiode surface. Thus, $(1 - R_S)$

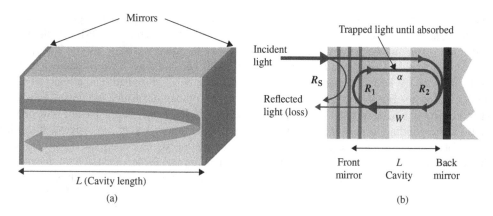

Figure 5.19 Fabry–Perot resonator: (a) concept and (b) arrangement in a photodetector.

is the transmissivity of incident light into the cavity, R_1 and R_2 are the magnitudes of the reflections of the front and back mirrors on both sides of the cavity, Ψ_1 and Ψ_2 are the phase shifts in the reflections introduced by the mirrors (since the reflection in real mirrors is at a certain depth in the mirror, but not exactly from the surface, especially for a DBR), L is the cavity length, and $\beta = 2n\pi/\lambda$ is the propagation constant of the light in the cavity with refraction index n. The terms in the curly bracket in equation Eq. (5.51) are the responsivity of the photodetector without a RCE, and the second line describes the resonant cavity enhancement. The RCE is a periodic function of L and λ or, more precisely, their ratio. Constructive resonance in the cavity occurs at the condition

$$(2\beta L + \Psi_1 + \Psi_2) = (4n\pi L/\lambda_o + \Psi_1 + \Psi_2)$$

$$= 2\pi \times \text{integer in steps of halves of the wavelength, } L$$

$$= \text{integer} \times \frac{1}{2}\left(\frac{\lambda_o}{n}\right). \tag{5.52}$$

With this condition for resonance at a wavelength λ_0, the RCE is maximum and is given by

$$\text{RCE}_{\text{max}} = \frac{1 + R_2 \exp(-\alpha W)}{[1 - \sqrt{R_1 R_2} \exp(-\alpha W)]^2}$$

$$\approx \frac{1 + R_2}{(1 - \sqrt{R_1 R_2})^2}, \quad \text{when } \alpha W \ll 1, \text{ since } \exp(-\alpha W) \approx 1$$

$$\approx \frac{2}{(1 - \sqrt{R_1})^2}, \text{when also } R_2 \approx 1. \tag{5.53}$$

Thus, the RC enhancement increases with the reflectivities of the mirrors. However, the enhancement is in a narrow optical "bandwidth" given by

$$\frac{1}{F} = \frac{\text{FWHM}}{\text{FSR}} = \frac{1 - \sqrt{R_1 R_2} \exp(-\alpha W)}{\pi\sqrt{\sqrt{R_1 R_2} \exp(-\alpha W)}}$$

$$\approx \frac{1 - \sqrt{R_1}}{\pi\sqrt{\sqrt{R_1}}} \sim \frac{1 - \sqrt{R_1}}{\pi}, \quad \text{when } \alpha W \ll 1, R_2 \approx 1, \text{ and } R_1 \to 1. \tag{5.54}$$

Here F is called the "finesse" of the resonator and F is the ratio of the "free spectral range" FSR $= \lambda_0^2/(2nL)$, which is the spacing between RC enhancement maxima and the spectral FWHM, which is the optical "bandwidth." The finesse in resonators corresponds to the quality factor $Q_{LC} = 2\Delta f_{3dB}/f_o$ in LC (inductor–capacitor) electrical resonators at frequency f_o, with $2\Delta f_{3dB}$ being the 3-dB bandwidth around f_o, which is the same as FWHM. The principal difference between an LC resonator and a Fabry–Perot cavity is that the LC resonator is a lumped-element resonator with only one resonance frequency f_o, whereas the Fabry–Perot cavity is a delay resonator with periodic resonance frequencies and spacing FSR between the wavelengths corresponding to these frequencies. The last expression in Eq. (5.51) suggests that a high R_1 is favorable for RC enhancement, e.g., $R_1 = 0.9$ would provide an RC enhancement ≈ 750. However, the narrowing of the bandwidth of RCE at high R_1 sets tight requirements for precision of distances in the Fabry–Perot cavity, so that the inaccuracy $\delta_L = (\Delta L/L) < (0.5/F)$ of the resonator length is a fraction of the RCE bandwidth.

Example 5.5

Given a high $R_1 = 0.9$ in the last approximate expression in Eq. (5.54), the RCE relative bandwidth is $1/F = (1 - \sqrt{0.9})/\pi = 1.6\%$. Using a conservative value for the refractive index of Si, $n = \sqrt{(\varepsilon_{Si}/\varepsilon_o)} = \sqrt{(11.9)} \approx 3.5$, and choosing $\lambda_o = 850\,\text{nm}$, then for integer $= 4$, the resonator length is $L = \text{integer} \times \lambda_o/(2n) = 4 \times 850\,\text{nm}/(2 \times 3.5) \approx 486\,\text{nm}$. Consequently, to be within the resonance, the resonator should be fabricated with inaccuracy $\Delta L < L \times 0.5/F = 486\,\text{nm} \times 0.5 \times 1.6\% \approx 4\,\text{nm}$.

Comments: The fabrication of a stack of heterogeneous materials for two mirrors and silicon in-between with accuracy of 4 nm is not simple. The materials might not be perfect, or the calculation might not be accurate, in order to guarantee 0.8% accuracy; e.g., the refractive index of Si is not 3.5, but 3.65 at $\lambda_o = 850\,\text{nm}$, which gives a much larger error of 4% in the calculation. Thus, we cannot really exploit RCE with large values. In real structures, RCE is in the range of 10, partially because of inaccuracy and additionally because $R_2 > 0.9$ for the back mirror is also difficult to achieve (metals have reflections of about this value and DBR requires more than four undulations of Si–SiO$_2$ for higher reflections from the Bragg mirror).

Another limitation for RCE is that the external quantum efficiency is not a monotonic function of reflections and absorption in the Fabry–Perot cavity. The terms in Eq. (5.51) that determine the maximum RCE external quantum efficiency $\eta_{\max}$ as a ratio of the material quantum efficiency at the resonant condition

$$(2\beta L + \Psi_1 + \Psi_2) = (4n\pi L/\lambda_o + \Psi_1 + \Psi_2) = 2\pi \times \text{integer} \tag{5.55}$$

are arranged in the following equation:

$$\eta_{\max}(4nL/\lambda_0 = 2 \times \text{integer}) = \frac{I_{ph}/P_{opt}}{hf_0/q\eta}$$

$$= \frac{[1 - \exp(-\alpha W)][1 + R_2 \exp(-\alpha W)]}{[1 - \sqrt{R_1 R_2}\exp(-\alpha W)]^2}(1 - R_S), \tag{5.56}$$

with $R_S \approx R_1$. Eq. (5.58) is for the minimum RCE external quantum efficiency $\eta_{\min}$ as a ratio of the material quantum efficiency at the antiresonant condition

$$(2\beta L + \Psi_1 + \Psi_2) = (4n\pi L/\lambda_o + \Psi_1 + \Psi_2) = \pi + 2\pi \times \text{integer}, \tag{5.57}$$

$$\eta_{min}(4nL/\lambda_0 = 1 + 2 \times \text{integer}) = \frac{I_{ph}/P_{opt}}{hf_0/q\eta}$$

$$= \frac{[1 - \exp(-\alpha W)][1 + R_2 \exp(-\alpha W)]}{[1 + \sqrt{R_1 R_2} \exp(-\alpha W)]^2}(1 - R_S), \qquad (5.58)$$

with $R_S \approx R_1$. As follows from Eqs. (5.56) and (5.58), the undulation of RC enhancement between peak and valley values, at resonance and antiresonance, is

$$\frac{\eta_{max}}{\eta_{min}} = \left[\frac{1 + \sqrt{R_1 R_2} \exp(-\alpha W)}{1 - \sqrt{R_1 R_2} \exp(-\alpha W)}\right]^2$$

$$\approx \left(\frac{1 + \sqrt{R_1 R_2}}{1 - \sqrt{R_1 R_2}}\right)^2, \quad \text{when } \alpha W \ll 1, \text{ since } \exp(-\alpha W) \approx 1;$$

$$\approx \left(\frac{1 + \sqrt{R_1}}{1 - \sqrt{R_1}}\right)^2, \quad \text{when } \alpha W \ll 1, \text{ and also } R_2 \approx 1;$$

$$\leq \frac{4}{(1 - \sqrt{R_1})^2} \sim \frac{1}{1 - \sqrt{R_1}} \approx \frac{F}{\pi}, \quad \text{when } \alpha W \ll 1, \quad R_2 \approx 1, \text{ and } R_1 \to 1, \qquad (5.59)$$

where the last line shows that the cavity finesse F and RCE undulation η_{max}/η_{min} between peak and valley values are related, recalling the approximate relations in Eq. (5.54).

The behavior of RCE is illustrated in Fig. 5.20 and compared with a non-resonant photodetector. The horizontal axis is reversed, considering that a particular photodetector has fixed width W of the absorption layer, and the absorption coefficient decreases for longer wavelengths, thus, the left-hand sides of the plots correspond to shorter wavelengths, while the right-hand sides correspond to longer wavelengths. The behavior of RCE is discussed further below. At shorter wavelengths in photodetectors with thick absorption layers, the product $\alpha \times W$ is larger than 1. In this case, the light is absorbed before reaching the back mirror, and the RCE structure behaves identically with the non-resonant photodetector–all lines overlap for $\alpha \times W > 3$ in Fig. 5.20 and the back mirror, if any, is irrelevant. Of course, a portion of the incident light is reflected by the front mirror (or the surface of the photodetector), and we desire $R_1 = R_S$ to be as low as possible.

At longer wavelengths in photodetectors with thin absorption layers, the product $\alpha \times W$ is smaller than 1, and RCE becomes relevant. If the back mirror is ideal ($R_2 = 1$, left-hand plot in Fig. 5.20), the resonance in the Fabry–Perot resonator would help to increase the quantum efficiency (non-monotonic thin lines) and even restore the ideal value $\eta_{max} = 1$ at condition $\exp(-\alpha W) = \sqrt{R_1}$. However, real mirrors have reflection $R_2 < 1$, and the decrease in back mirror reflection R_2 to 0.9 and 0.8 (still high) degrades the ability of RCE to restore the quantum efficiency, as seen in the middle and right-hand plots of Fig. 5.20, especially for high front mirror reflection R_1 (which is also the reflection from the photodetector surface, $R_S \approx R_1$). The condition for maximum quantum efficiency becomes a complicated expression:

$$\sqrt{R_1 R_2} = \frac{1 + R_2 [2 \exp(-\alpha W) - 1]}{2 - (1 - R_1) \exp(-\alpha W)}, \qquad (5.60)$$

but tends to $(R_1 \times R_2) \approx (1 - 2\alpha W)$ when $(\alpha W) \leq 0.1$ and $R_2 \geq 0.8$. In addition, if the resonator is not tuned at the wavelength, then RCE will suppress the quantum efficiency, as shown with symbols on dashed lines in Fig. 5.20. The suppression is less than 3 dB, which is not a dramatic decrease in responsivity, but we realize that the RCE is not favorable if the cavity is not precisely tuned at the wavelength of interest, e.g., in cases of

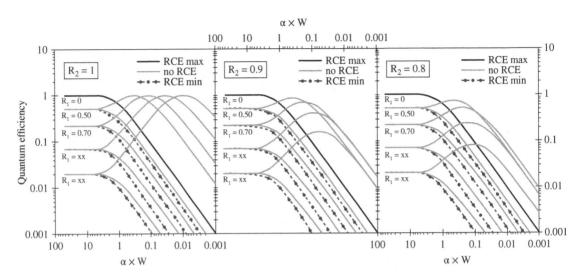

Figure 5.20 Resonant cavity enhancement (RCE max, non-monotonic thin lines) of photodetectors with Fabry–Perot resonator as a function of the product $\alpha \times W$ of the absorption coefficient α and the absorption layer width W compared with non-resonant photodetectors (thick lines, $R_2 = 0$) and suppression in RCE (RCE min, dashed lines with symbols) for several reflections of the front mirror ($R_1 = R_S$, as indicated by labels on the lines) and back mirror ($R_2 = 1$, 0.9 and 0.8, in the plots from left to right).

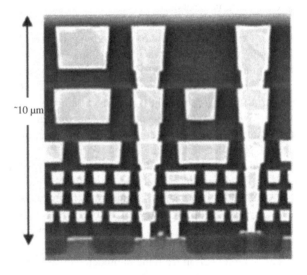

Figure 5.21 Cross-section of the oxide stack in a six-metal CMOS, around year 2000.

broad-spectrum photosensors for imager arrays, or by temperature and fabrication variations. Note that not all the curves in Fig. 5.20 exceed unity, which means that RCE can only remedy incomplete absorption in thin layers.

Submicrometer Si technologies usually have a thick stack of oxide layers, in the range of $3–10\,\mu m$ as shown in Fig. 5.21 from Ref. [43], since many metal layers need to be accommodated for electrical interconnections.

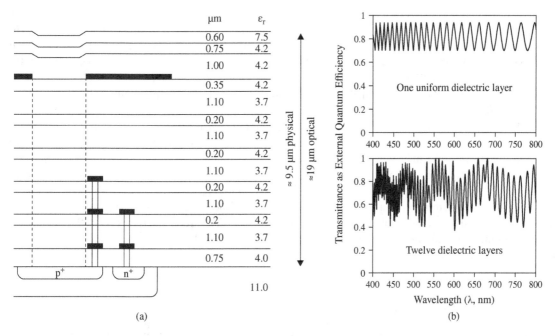

(a)   (b)

Figure 5.22 Layer content corresponding to Fig. 5.21 around year 2000. (a) Thicknesses (in microns) and relative permittivity of the layers. (b) Transmittance of the dielectric stack calculated at assumptions for uniform (upper plot) and alternated (bottom plot) refractive index.

Apart from the problem of having an unobstructed optical path in the web of metal wires, the oxide stack is not uniform but alternate silicon nitride and silicon dioxide layers in order to enable nanometer-scale lithography and etching for metal wires and vias for electrical interconnection, as shown in Fig. 5.22(a). The optical transmittance of the stack, being a multiplier in the external quantum efficiency, has been calculated in Ref. [43] with two assumptions, as shown in Fig. 5.22(b). One assumption is to consider a nearly uniform dielectric with an average refractive index, since the majority of layers have similar permittivity, thus $\sqrt{4.2} \approx \sqrt{3.7} \approx 2$. Using this assumption, the external quantum efficiency of the photodetector is undulating between two values, as shown in the upper plot. The bottom plot is the result from calculation with a transfer matrix method, showing additional modulation with slowly varying components due to thinner cavities in the dielectric stack. In either case, the spacing between minima and maxima can be estimated by FSR $= (\lambda^2/L_{opt})$, where $L_{opt} \approx (\Sigma nL) \approx 19.5$ μm, here, n is the refractive index. Depending on the wavelength λ, the values of FSR are between 9 nm and 32 nm. These values are relatively small for broadband light in natural scenes. Thus, the RCE effects do not compromise the use of CMOS technology for imager arrays. However, for special cases in spectrometry, it is clear that standard CMOS may not be the best choice.

5.4   Direct Detection Receivers

Fig. 5.23 shows the block diagram of a digital optical receiver. It has three parts: (i) front end, (ii) linear channel, and (iii) data recovery section (see Ref. [3] for more details). The front end consists of a reverse-biased photodiode. The photodiode converts the optical data into electrical data. The output of the photodiode passes through a preamplifier.

The linear channel consists of a high-gain amplifier (Amp) followed by a low-pass filter (LPF). The LPF is used to truncate the noise spectrum. Since the variance of receiver noise is proportional to the receiver bandwidth, it is desirable to keep the bandwidth of a LPF sufficiently low. However, the LPF truncates the signal spectrum too. The bandwidth of the LPF should be optimized so that the signal-to-noise ratio (SNR) is maximum. Typically, the SNR is maximum if the receiver bandwidth is of the order of the bit rate frequency. To have the best SNR, the receiver transfer function should be matched to the transmitted signal (see Chapter 8), which is hard to achieve in practice.

The data-recovery section consists of a decision circuit and a clock-recovery circuit. Suppose the bit pattern at the input end of a fiber-optic link is '1011', as shown in Fig. 5.24. Some of these bits may be corrupted due to the noise added by the fiber-optic link or by the receiver. Suppose that the peak current at the receiver is 100 mA and the threshold current is 30 mA. If the received current is more than 30 mA, at the sampling time (usually at the middle of the bit interval), the decision circuit interprets it as '1', otherwise the received bit is zero. In the figure, the third bit transmitted is '1', but because of the noise, the received current corresponding to that bit is less than the threshold current. Therefore, the decision circuit interprets it as '0', causing a bit error. The sampling time for the decision is provided by a clock which is extracted from the received signal using a clock-recovery circuit.

5.4.1   Optical Receiver ICs

To form a complete photoreceiver, the photodetector is integrated with a preamplifier. The most common preamplifiers that are combined with photodetectors are built with either high-electron mobility transistors (HEMTs) or heterojunction bipolar transistors (HBTs).

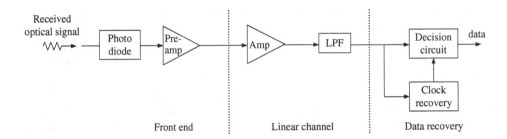

Figure 5.23   Block diagram of a direct detection receiver. LPF = low-pass filter.

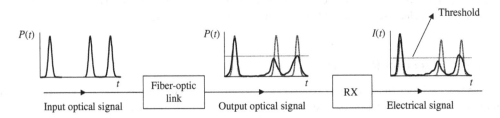

Figure 5.24   Bit patterns at the various stages of a fiber-optic system.

5.4.1.1 pin-HEMT

Monolithic pin-HEMT photoreceivers were demonstrated in many research publications, for example, in [44, 45]. In [44], an InAlAs/InGaAs pin-HEMT was reported, where the HEMT epitaxial layers are fabricated first and then the InGaAs pin-PD layers are grown. A schematic structure of this photoreceiver is shown in Fig. 5.25. This photoreceiver has a good response of 1.6 Gb/s non-return-to-zero (NRZ) signals using a standard common-emitter/common-collector amplifier. In [45], a transimpedance amplifier (TIA) configuration with a feedback resistance of 500 Ω had a high sensitivity of -17.3 dBm with a monolithically integrated pin-HEMT photoreceiver at 10 Gb/s for a BER of 10^{-9} using a $(2^{23} - 1)$ pseudorandom binary sequence NRZ lightwave signal. The receiver bandwidth is 7.4 GHz. When an erbium doped fiber amplifier (EDFA) was inserted before the pin-HEMT receiver, there was a significant improvement in its sensitivity (-30.6 dBm) [45]. Also, a high-impedance design may produce a high performance if it is followed by a suitable equalizer.

5.4.1.2 pin-HBT

HBTs have several advantages compared with compound semiconductor FETs. For example, HBTs have tremendous potential for high-speed circuits with very modest lithographic design rules. A monolithically integrated front-end pin-HBT photoreceiver can be fabricated by first growing pin-PD layers and then growing HBT epitaxial layers implemented with a single epitaxial growth technique and a self-aligned fabrication technology. The advantage of this type of fabrication is that both pin-PD and HBT designs can be individually controlled to obtain their optimum performance. Such a pin-HBT photoreceiver is demonstrated in Ref. [46], where the receiver circuit incorporating a cascode preamplifier and capacitors showed a bandwidth of 2.8 GHz and a sensitivity of -21 dBm. This technique has the disadvantages of increased parasitics and non-planarity.

The fabrication technique that uses shared-layer integration becomes attractive as shown in Fig. 5.26, where the subcollector, collector, and base layers are the same as the n-, i- and p-layers of pin-PD, respectively [47]. This structure has the advantage of minimizing the number of growth layers.

Note that the photodetector and HBT designs are not completely independent. For example, the design used in [49] for an integrated pin-PD and InP/InGaAs HBT photoreceiver is suitable for the HBT, but not for the photodetector. The photoreceiver consisting of a pin-PD and a transimpedance amplifier operated at 2.5 Gb/s, though the amplifier bandwidth is as high as 19 GHz as the performance is limited mainly by the characteristics of the photodetector.

For the heterojunction in the HBT, we can use InAlAs/InGaAs. In [50], an integrated InAlAs/InGaAs pin-HBT transimpedance photoreceiver using a three-stage amplifier configuration results in an improvement in the bandwidth ($\sim$7.1 GHz). In this preamplifier, the first stage is a TIA and the last two stages are emitter

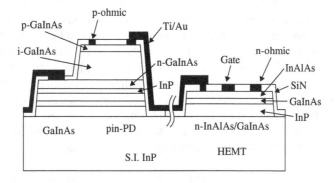

Figure 5.25 Schematic structure of a monolithic InAlAs/InGaAs pin-HEMT. Adapted from [44].

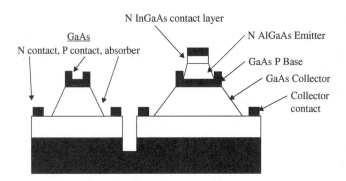

Figure 5.26   Shared-layer monolithic pin-HBT . Adapted from [59].

followers. With an optimum choice of collector thickness and use of a booster inductance, the InAlAs/InGaAs pin-HBT receiver achieved a bandwidth of 20 GHz [51].

Double heterojunction bipolar transistor (DHBT) configurations are superior to single heterojunction bipolar transistor (SHBT) in terms of radio-frequency (RF) performance and breakdown voltage. In [52], a monolithic pin-HBT receiver using DHBT of various collector thicknesses was fabricated. The collector layers are composed of undoped InGaAs, pn pair doped InGaAs, and n-InP. With this double-heterojunction technology, a receiver bandwidth of 26.7G Hz with a large transimpedance gain of 48.9 dBΩ was obtained. However, the SHBT structure with shared-layer integration has the advantage of fewer layers to fabricate, thus the cost would be less. Therefore, suitable device design and circuits can be used to obtain improved performance from SHBT systems. For example, using a (1 μm × 5 μm) HBT and a 9-μm diameter photodetector for a monolithic pin-SHBT photoreceiver [53], bandwidths as high as 46 GHz were achieved. Using a common base configuration as an input stage, further improvement in bandwidth was observed [54]. In [54], by a simple reduction of the photodetector's area and an adjustment of the feedback resistance, a bandwidth of 60 GHz was predicted.

5.4.1.3   MSM-HEMT

The integration of metal-semiconductor-metal (MSM) photo-detector and HEMT is usually based on stacked-layer technology. A schematic diagram of the layers for an integrated MSM-HEMT photoreceiver is shown in Fig. 5.27. As shown in Fig. 5.27(a), if the HEMT layers are grown over the MSM layers, the structure may be termed a HEMT/MSM structure. Another possibility is, therefore, a MSM/HEMT configuration. The overall photoreceiver performance depends on the performance of each individual component. In [55, 56], it has been shown that MSM/HEMT structures are superior to HEMT/MSM structures.

The intrinsic bandwidth of a MSM-PD is very high, and the main limitation of its high-frequency performance comes from the preamplifier. The transimpedance amplifier configuration is a good choice with respect to high bandwidth, wide dynamic range, and low noise at high frequencies. The feedback resistance is very important in a TIA. It can be either a fixed metallic resistance formed on the substrate, or a voltage-controlled variable resistance [55]. In [55], a two-stage variable TIA with a common-gate HEMT was used as the feedback resistance (see Fig. 5.28). By adjusting the d.c. voltage to the gate of this HEMT, transimpedance gains ranging from 55.8 dBΩ to 38.1 dBΩ with corresponding 3-dB cutoff frequencies from 6.3 GHz to 18.5 GHz were obtained.

Noise consideration is a critical factor for the performance of the photoreceivers. It is usually expressed in terms of the equivalent input current noise spectral density. Current noise spectral densities of 7.5, 8,

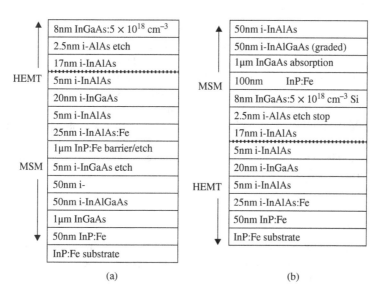

Figure 5.27   Schematic structure of (a) HEMT/MSM and (b) MSM/HEMT. Adapted from [54].

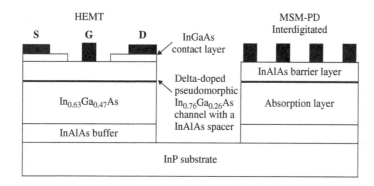

Figure 5.28   MSM and HEMT grown in parallel. Information for schematic taken from [57].

and $12\,\text{pA/Hz}^{1/2}$ were obtained for bandwidths of 6.3, 8, and 13.7 GHz, respectively in [55]. The noise performance of a receiver in the case of a digital signal is usually defined by its sensitivity, or the minimum detectable signal power in the presence of noise for a particular bit error rate (BER). As standard practice, a BER of 10^{-9} is used for characterization of most receivers. For example, the sensitivity of an InP-based MSM-HEMT photoreceiver described in [57] is measured to be $-10.7\,\text{dBm}$ at 10 Gb/s with 1 μm spacing of the MSM electrodes. The noise performance is frequently shown by eye diagrams. The more open the eye is, the better is the noise performance and, hence, the inter-symbol interference (ISI) is less.

Although the performance of InP-based structures is very good, the electronic device technology is more established for GaAs-based structures. This significantly reduces their cost. Thus, GaAs-based structures are more popular and attractive for commercial applications. Also, compared with pin-PDs, MSM-PDs have the advantage of high intrinsic bandwidth, ultra-low capacitance, and easy planar integrability with HEMTs and HBTs [59, 60].

5.4.1.4 APD-HEMT

In [12], an optical APD photoreceiver operating at 10 Gb/s was introduced. The optical front end consists of a SAM-APD and a TIA preamplifier constructed from HEMT devices using hybrid circuit techniques. The planar InGaAs SAM-APD is back-illuminated and has a reflecting P-contact to maximize the quantum efficiency. This photoreceiver had a maximum gain bandwidth of 75 GHz, a quantum efficiency of 80%, and a parasitic capacitance of 13 pF. The three-transistor preamplifier uses a feedback resistance of 330 Ω after the second stage, while the third resistance is used to perform impedance matching to 50 Ω. High sensitivity of −29.4 dBm at 10 Gb/s is achieved using a hybrid APD-HEMT photoreceiver [13]. A noise-matching network between APD and HEMT amplifier stages helps to increase the sensitivity. It is implemented using an InAlAs/InGaAs superlattice APD and HEMTs.

5.5 Receiver Noise

An optical receiver not only translates the data in the optical domain into an electrical domain, but it also adds noise. Two important noise sources are (i) shot noise and (ii) thermal noise.

5.5.1 *Shot Noise*

As mentioned in Section 5.2.1, if the power of an optical signal in an interval T is P, it corresponds to a photon rate of P/hf_0 where f_0 is the frequency of the light wave. In other words, the mean number of photons in this optical signal is $\bar{n} = PT/hf_0$. Owing to the quantum nature of photons, the actual number of photons in the interval T is random—sometimes more than $\bar{n}$ and sometimes less. For light from an ideal laser, the number of photons in an optical signal of power P in a time interval T obeys the Poisson probability distribution [61]

$$p(n) = \frac{\bar{n}^n \exp{(-\bar{n})}}{n!}, \quad n = 0, 1, 2, \ldots \tag{5.61}$$

Here, n is the number of photons. Fig. 5.29 shows the Poisson probability distribution. As can be seen, the curve becomes broader as the mean photon number $\bar{n}$ increases. An important property of the Poisson distribution is that the mean is the same as the variance,

$$< n > = \sigma^2 = < n^2 > - < n >^2 = \bar{n}. \tag{5.62}$$

Thus, if the mean number of photons is 16, the actual number of photons is approximately in the range 16 ± 4. Next, let us consider an ideal photodiode with quantum efficiency $\eta = 1$. If n photons are incident on this photodiode, the number of photocarriers generated is also n. Since the number of photons in an interval T is random (even though the optical power is fixed), the number of photocarriers generated in that interval is also random and it obeys the same Poisson probability distribution as the photon given by Eq. (5.61). For a non-ideal photodiode (with quantum efficiency $\eta < 1$), the probability of the event that a photon incident on the photodiode generates an electron–hole pair that contributes to the photocurrent is η, or the probability of this event not happening is $1 - \eta$. The randomly generated photocarriers lead to fluctuations in photocurrent, which is known as *shot noise*. In summary, the noise component in a photocurrent has two contributing factors. (i) Owing to the quantum nature of photons, the photon arrival times within an interval T are random, which leads to ehp generation at random times. (ii) Not all the photons incident are absorbed, and not all the photocarriers generated contribute to the photocurrent as there is a chance of recombination before reaching the terminals of the photodetector. When an optical signal of power P_I falls on the *pin* photodiode, the current can be written as

$$I(t) = I_{PC} + i_{\text{shot}}(t), \tag{5.63}$$

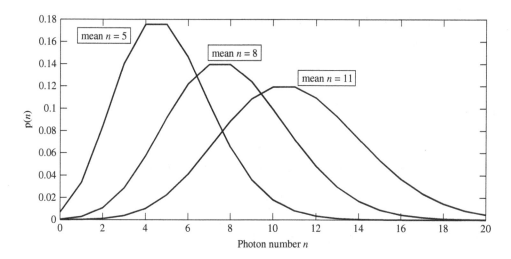

Figure 5.29 Poisson probability distribution for different mean photon numbers $\bar{n}$.

where $I_{PC} = RP_I$ is the deterministic part of the current and $i_{shot}(t)$ is the noise component of the current due to shot noise. The shot noise current $i_{shot}(t)$ is a random variable with zero mean, i.e., $< i_{shot}(t) >= 0$. The noise power dissipated due to $i_{shot}(t)$ in a 1-Ω resistor is $i^2_{shot}(t)$. The mean noise power is

$$N^1_{shot} =< i^2_{shot}(t) >= \sigma^2_{shot}, \qquad (5.64)$$

where σ_{shot} is the standard deviation and superscript 1 indicates that it is the noise power dissipated in a 1-Ω resistor. Thus, the variance is the same as the mean noise power dissipated in a 1-Ω resistor.

The shot noise is a white noise process, and its power spectral density is constant. Using Eq. (5.62), it can be shown that the PSD of the shot noise is [61]

$$\rho_{shot}(f) = qI_{PC}, \qquad (5.65)$$

where q is the electron charge.

For simplicity, let us assume that the receiver is an ideal low-pass filter with bandwidth B_e. The receiver transfer equation $\tilde{H}_e(f)$ is given by

$$\tilde{H}_e(f) = \begin{cases} 1, & \text{if } |f| < B_e \\ 0, & \text{otherwise.} \end{cases} \qquad (5.66)$$

The shot noise generated in the photodiode passes through the rest of the receiver circuit. The PSD at the receiver output is

$$\rho_{shot,out}(f) = \rho_{shot}(f)|\tilde{H}_e(f)|^2. \qquad (5.67)$$

The PSD refers to the mean power per unit frequency interval and, therefore, the mean noise power at the receiver output (dissipated in a 1-Ω resistor) can be obtained by integrating the PSD over frequency,

$$N^1_{shot} = \int_{-\infty}^{\infty} \rho_{shot,out}(f)df. \qquad (5.68)$$

Using Eqs. (5.65) and (5.66) in Eq. (5.68), we obtain

$$N_{\text{shot}}^1 = qI_{PC} \int_{-\infty}^{\infty} |\tilde{H}_e(f)|^2 df$$

$$= 2qI_{PC}B_e. \tag{5.69}$$

Using Eq. (5.64), we find the variance of shot noise as

$$\sigma_{\text{shot}}^2 = N_{\text{shot}}^1 = 2qI_{PC}B_e. \tag{5.70}$$

Eq. (5.70) is valid for arbitrary filter shapes if the effective bandwidth B_e is defined as

$$B_e = \frac{1}{2} \int_{-\infty}^{\infty} |\tilde{H}_e(f)|^2 df. \tag{5.71}$$

When the dark current I_d is not negligible, Eq. (5.70) is modified as

$$\sigma_{\text{shot}}^2 = 2q(I_{PC} + I_d)B_e. \tag{5.72}$$

For APD receivers, the variance of shot noise is given by (see Section 5.3.7) [62]

$$\sigma_{\text{shot}}^2 = 2qM^2 F(RP_I + I_d)B_e, \tag{5.73}$$

where M is the multiplication factor and F is the excess noise factor.

5.5.2 Thermal Noise

Electrons move randomly in a conductor. As the temperature increases, electrons move faster and therefore the electron current increases. However, the mean value of the current is zero since, on average, there are as many electrons moving in one direction as there are in the opposite direction. Because of the random motion of electrons, the resulting current is noisy and is called "thermal noise" or "Johnson noise".

In the presence of thermal noise, the current in the receiver circuit may be written as

$$I(t) = I_{PC} + i_{\text{thermal}}(t), \tag{5.74}$$

where I_{PC} is the mean photocurrent (deterministic), and $i_{\text{thermal}}(t)$ is the thermal noise current.

For low frequencies ($f \ll k_B T/h$), thermal noise can be regarded as white noise, i.e., its power spectral density is constant. It is given by

$$\rho_{\text{thermal}}(f) = 2k_B T/R_L, \tag{5.75}$$

where k_B is Boltzmann's constant, R_L is the load resistance, and T is the absolute temperature. If B_e is the effective bandwidth of the receiver, the noise variance can be calculated as before,

$$\sigma_{\text{thermal}}^2 = < i_{\text{thermal}}^2 > = 4k_B TB_e/R_L. \tag{5.76}$$

Eq. (5.76) does not include the noise sources in the amplifier circuit, such as that coming from resistors and active elements. Eq. (5.76) can be modified to account for the noise sources within the amplifier as [24, 63]

$$\sigma_{\text{thermal}}^2 = 4k_B TB_e F_n/R_L, \tag{5.77}$$

where F_n is the amplifier noise factor.

5.5.3   Signal-to-Noise Ratio, SNR

Let us first consider *pin* receivers. The mean signal power is

$$S = I_{PC}^2 R_L = (RP_I)^2 R_L. \tag{5.78}$$

Using Eq. (5.72), the mean noise power dissipated in the resistor R_L due to the shot noise current is

$$N_{\text{shot}} = <i_{\text{shot}}^2(t)> R_L = 2q(I_{PC} + I_d)B_e R_L. \tag{5.79}$$

Using Eq. (5.77), the mean noise power due to thermal noise is

$$N_{\text{thermal}} = <i_{\text{thermal}}^2 R_L > = 4k_B T B_e F_n. \tag{5.80}$$

The total mean noise power is

$$N = N_{\text{shot}} + N_{\text{thermal}} = (\sigma_{\text{shot}}^2 + \sigma_{\text{thermal}}^2)R_L. \tag{5.81}$$

The SNR is defined as

$$\text{SNR}_{\text{pin}} = \frac{\text{mean signal power}}{\text{mean noise power}}$$

$$= \frac{S}{N} = \frac{I_{PC}^2}{\sigma_{\text{shot}}^2 + \sigma_{\text{thermal}}^2}$$

$$= \frac{R^2 P_I^2}{[2q(RP_I + I_d) + 4k_B T F_n / R_L]B_e}. \tag{5.82}$$

For APD receivers, the signal power is

$$S = (MRP_I)^2 R_L. \tag{5.83}$$

Using Eq. (5.73), the SNR can be calculated as

$$\text{SNR}_{APD} = \frac{(MRP_I)^2}{[2qM^2 F(RP_I + I_d) + 4k_B T F_n / R_L]B_e}. \tag{5.84}$$

5.6   Coherent Receivers

In the 1990s, coherent detection was pursued mainly because the fiber-optic systems were then loss limited. Coherent receivers have higher sensitivity than direct detection receivers and, hence, coherent detection is more attractive for loss-limited systems (see Chapter 7). However, with the development and deployment of EDFAs in the mid-1990s, fiber loss was no longer a problem and coherent receivers were not pursued due to the technical hurdles associated with aligning the phase and polarization of the local oscillator (LO) with those of the received signal. Owing to the rapid advances in digital signal processing (DSP), coherent detection has drawn renewed interest recently [64–73]. The polarization and phase alignment can be performed using DSP as discussed in Chapter 11, which is easier than optical phase-locked loop (PLL) used in 1990s' coherent receivers. Unlike direct detection, coherent detection allows us to retrieve both amplitude and phase information. The advantages of the coherent receiver are the following. (i) In the case of direct detection, the detection process is nonlinear–photocurrent is proportional to the absolute square of the optical field. Since the phase information is lost during the detection, it is hard to compensate for dispersion and polarization mode dispersion (PMD) in an electrical domain. In the case of coherent detection, the detection process is linear–the complex optical field envelope is linearly translated into an electrical domain and, therefore, the inverse fiber transfer function can be realized using DSP to compensate for dispersion and PMD. (ii) With coherent detection, higher spectral efficiencies can be realized using multi-level modulation formats that make use of both amplitude and phase modulation.

5.6.1 Single-Branch Coherent Receiver

In this section, we assume that the polarization of the received signal is perfectly aligned with that of the local oscillator (LO), and we use scalar notation. The polarization effects are considered in Section 5.6.5. Let the transmitted signal be

$$q_T(t) = A_T s(t) \exp(-i\omega_c t), \tag{5.85}$$

where $s(t)$ is the complex field envelope, ω_c is the frequency of the optical carrier, and A_T^2 is the peak transmitter power. Let us assume a perfect optical channel that introduces neither distortion nor noise. However, the phase of the optical carrier changes due to propagation, and the field amplitude may be attenuated. Let the received signal be

$$q_r(t) = A_r s(t) \exp[-i(\omega_c t + \phi_c)] \tag{5.86}$$

and the local oscillator output be

$$q_{LO}(t) = A_{LO} \exp[-i(\omega_{LO} t + \phi_{LO})], \tag{5.87}$$

where $A_{LO}^2 = P_{LO}$ is the LO power, and ϕ_{LO} is the phase. These two signals are combined using a 3-dB coupler and pass through a photodetector (PD), as shown in Fig. 5.30[1]. The photocurrent is proportional to the absolute square of the incident optical field. Therefore, the photocurrent is

$$I(t) = R|[q_r(t) + q_{LO}(t)]/\sqrt{2}|^2$$
$$= \frac{R}{2}\{|A_r s(t)|^2 + |A_{LO}|^2 + 2A_r A_{LO} \mathrm{Re}\{s(t) \exp[-i(\omega_{IF} t + \phi_c - \phi_{LO})]\}\} \tag{5.88}$$

where

$$\omega_{IF} = \omega_c - \omega_{LO} \tag{5.89}$$

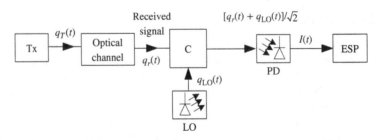

Figure 5.30 Block diagram of a single-branch coherent receiver. Tx = transmitter, LO = local oscillator, C = 3-dB coupler, PD = photo-detector, ESP = electrical signal processor.

---

[1] An unbalanced directional coupler can be used to combine the signals to maximize SNR [74]. In this section, a 3-dB coupler is used for simplicity.

is called the *intermediate frequency*. To obtain Eq. (5.88), we have used the formula

$$\text{Re}(X) = (X + X^*)/2. \tag{5.90}$$

When the LO power P_{LO} is much larger than the signal power A_r^2, the first term in Eq. (5.88) can be neglected. Since the LO output is cw, A_{LO}^2 is a constant and it leads to a d.c. component in the photocurrent which can be removed by capacitive coupling from the photodetector to the front end of the electrical amplifier. Therefore, the signal that goes to the front end can be written as

$$I_d(t) = RA_r A_{LO} \text{Re} \{s(t) \exp[-i(\omega_{IF}t + \phi_c - \phi_{LO})]\}. \tag{5.91}$$

With

$$P_r = A_r^2, \tag{5.92}$$

$$P_{LO} = A_{LO}^2, \tag{5.93}$$

Eq. (5.91) may be rewritten as

$$I_d(t) = R\sqrt{P_r P_{LO}} \text{Re} \{s(t) \exp[-i(\omega_{IF}t + \phi_c - \phi_{LO})]\}. \tag{5.94}$$

5.6.1.1 Homodyne Single-Branch Receiver

If $\omega_{IF} = 0$, such a receiver is known as a *homodyne receiver*. If $\omega_{IF} < 2\pi B_s$, where B_s is the symbol rate, it is sometimes referred to as *intradyne*. Otherwise, it is called a *heterodyne receiver*. For a homodyne receiver, the phase of the received carrier ϕ_c should be exactly the same as the phase of the local oscillator. This can be achieved using an optical phase-locked loop or it can be post-corrected using digital phase estimation techniques (see Chapter 11). When the phases are exactly aligned ($\phi_c = \phi_{LO}$), Eq. (5.94) can be written as

$$I_d(t) = RP_0 \text{Re}[s(t)], \tag{5.95}$$

where $P_0 = \sqrt{P_r P_{LO}}$. If the transmitted signal is real such as that corresponding to binary OOK or PSK, the real part of $s(t)$ has all the information required to retrieve the transmitted data. If the transmitted signal is complex, as in the case of amplitude and phase-modulated signals, in-phase and quadrature (IQ) receivers are required to estimate the transmitted information, which will be discussed in Sections 5.6.3 and 5.6.4. Note that in Eq. (5.95), the responsivity R is multiplied by $\sqrt{P_r P_{LO}}$. If we choose a very large local oscillator power, P_{LO}, the effective responsivity, RP_0, could be increased and, therefore, the sensitivity of the coherent receiver was significantly larger than that of the direct detection receiver. This was one of the reasons for pursuing coherent receivers in the 1980s. If the phase of the local oscillator is not fully aligned with the phase of the carrier, Eq. (5.94) can be written as

$$I_d(t) = RP_0 \text{Re}[s(t) \exp(-i\Delta\phi)], \tag{5.96}$$

where $\Delta\phi \equiv \phi_c - \phi_{LO}$ is the phase error. If $\Delta\phi$ is $\pi/2$ and $s(t)$ is 1 within a bit interval, $I_d(t) = 0$ and, therefore, the phase error leads to bit errors.

Example 5.6

A BPSK-NRZ signal is transmitted over a fiber of length 100 km. The peak power of the signal at the transmitter is 12 dBm. The fiber loss is 0.2 dB/km. Assuming that the receiver is a homodyne single-branch receiver, find the peak current if (a) LO power = 10 dBm, (b) LO power = −10 dBm. Assume $R = 0.9$ A/W.

Solution:
The power at the transmitter,

$$P_T(\text{dBm}) = 12 \text{ dBm}.$$

Fiber loss,

$$\text{loss(dB)} = 0.2 \text{ dB/km} \times 100 \text{ km} = 20 \text{ dB}.$$

The power at the receiver,

$$P_r(\text{dBm}) = P_T(\text{dBm}) - \text{loss(dB)} = (12 - 20) \text{ dBm} = -8 \text{ dBm}.$$

(a)

$$P_{\text{LO}}(\text{dBm}) = 10 \text{ dBm},$$

$$P_{\text{LO}} = 10^{0.1 P_{\text{LO}}(\text{dBm})} \text{ mW} = 10 \text{ mW},$$

$$P_r = 10^{0.1 P_r(\text{dBm})} \text{ mW} = 0.1585 \text{ mW}.$$

For a BPSK signal assuming rectangular NRZ pulses, $s(t)$ takes values ± 1. From Eq. (5.95), we have

$$\text{Peak current} = |I_d| = R\sqrt{P_r P_{\text{LO}}} = 0.9 \times \sqrt{10 \times 0.1585} \text{ mA}$$

$$= 1.1331 \text{ mA}.$$

If we use Eq. (5.88) after ignoring the d.c. term, we find the peak current as

$$|I| = RP_r/2 + R\sqrt{P_r P_{\text{LO}}} = 1.204 \text{ mA}. \tag{5.97}$$

Note that $|I| \approx |I_d|$. The difference $|I - I_d|$ is known as intermodulation cross-talk.

(b)

$$P_{\text{LO}}(\text{dBm}) = -10 \text{ dBm},$$

$$P_{\text{LO}} = 10^{0.1 \times -10} \text{ mW} = 0.1 \text{ mW},$$

$$|I_d| = 0.9\sqrt{0.1 \times 0.1585} = 0.1133 \text{ mA}.$$

The peak current $|I|$ after ignoring the d.c. term is

$$|I| = RP_r/2 + R\sqrt{P_r P_{\text{LO}}} = 0.1846 \text{ mA}. \tag{5.98}$$

In this case, the intermodulation cross-talk is comparable to I_d.

5.6.1.2 Heterodyne Single-Branch Receiver

When the frequency offset between the transmitter laser and LO is in the microwave range, the signal $I_d(t)$ given by Eq. (5.91) may be interpreted as the message $s(t)$ modulating the microwave carrier of frequency ω_{IF}.

Let us assume that $s(t)$ is real and Eq. (5.91) can be written as

$$I_d(t) = I_0 s(t) \cos(\omega_{\text{IF}} t + \Delta\phi), \tag{5.99}$$

where $I_0 = RP_0$. The corresponding signal spectrum is shown in Fig. 5.31(b). Suppose the bandwidth of the signal $s(t)$ is $\omega_B/2\pi$ (Fig. 5.31(a)). The bandwidth of $I_d(t)$ is $2\omega_B$, as shown in Fig. 5.31(b). The photocurrent $I_d(t)$ in a homodyne receiver is proportional to $s(t)$ and, therefore, the bandwidth of the homodyne receiver circuit is approximately $\omega_B/2\pi$, whereas the signal spectrum is centered around ω_{IF} in the case of a heterodyne receiver with bandwidth ω_B/π. Therefore, the bandwidth of the heterodyne receiver circuit should be approximately $(\omega_{IF} + \omega_B)/(2\pi)$. The large bandwidth requirement is one of the disadvantages of the heterodyne receiver. The signal $I_d(t)$ is multiplied by a microwave oscillator whose phase is aligned with that of $I_d(t)$, as shown in Fig. 5.32. The resulting signal is

$$I_1(t) = I_0 s(t) \cos^2\left(\omega_{IF} t + \Delta\phi\right) = \frac{I_0 s(t)}{2}\{1 + \cos\left[2(\omega_{IF} + \Delta\phi)\right]\}. \tag{5.100}$$

The first term on the right-hand side of Eq. (5.100) corresponds to the baseband and the second term corresponds to a signal with its spectrum centered around $2\omega_{IF}$, as shown in Fig. 5.33. If we introduce a LPF

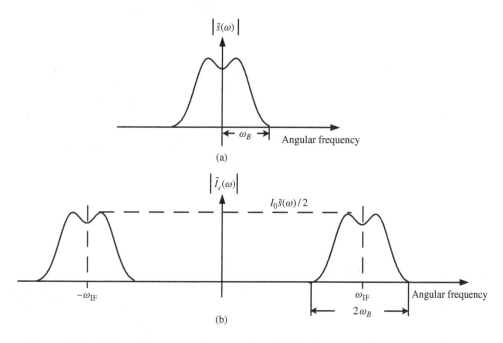

Figure 5.31   (a) Signal spectrum at the transmitter. (b) Signal spectrum after the photo-detector.

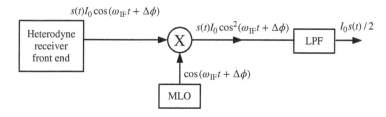

Figure 5.32   Block diagram of a single-branch heterodyne receiver. MLO = microwave local oscillator, LPF = low-pass filter.

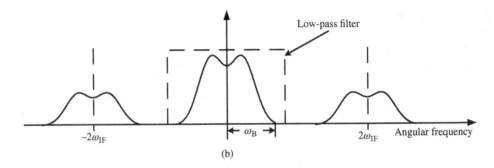

Figure 5.33   Signal spectrum after multiplication.

with its bandwidth roughly equal to the bandwidth of $s(t)$ $(= \omega_B/2\pi)$, shown in Fig. 5.32, we can remove the spectrum centered around $2\omega_{IF}$. After passing through the LPF, the final output is

$$I_2(t) = \frac{I_0 s(t)}{2},\tag{5.101}$$

which is proportional to the transmitted signal $s(t)$. Historically, homodyne receivers required optical phase-locked loops to align the phase of the LO with the received carrier. For heterodyne receivers, it is not essential to employ the optical PLL to align the phase of the optical carrier with the phase of the optical LO. This is because the phase of $I_d(t)$ can be arbitrary. However, the microwave LO phase should be aligned with $I_d(t)$ using an electrical PLL. The electrical PLL is easier to implement than an optical PLL. However, with the advent of high-speed digital signal processors, phase estimation can be done in the digital domain for homodyne/heterodyne systems and, therefore, analog optical/electrical PLL is no longer required (see Chapter 11).

5.6.2   Balanced Coherent Receiver

In the case of a single-branch receiver, the intermodulation cross-talk ($|A_r s(t)|^2$ of Eq. (5.88)) can lead to penalty unless the LO power is very large. This problem can be avoided by using the balanced receiver shown in Fig. 5.34. The outputs of the optical channel and local oscillator are connected to a directional coupler (DC). The directional coupler is a two-input/two-output device. The DC can be fabricated by bringing two optical waveguides or fibers very close, as shown in Fig. 5.35. If the light is launched to the DC through input port In 1 only, it gets coupled to the waveguide 2. The length L and spacing d can be chosen so that at the output, the power is equally distributed between output ports Out 1 and Out 2. In this case, the DC acts as a 3-dB power splitter. Let q_j^{in} and q_j^{out} be the optical fields at the input port In j and output port Out j, respectively. The output fields when $q_2^{in} = 0$ can be written as

$$|q_1^{out}| = |q_1^{in}|/\sqrt{2},\tag{5.102}$$

$$|q_2^{out}| = |q_1^{in}|/\sqrt{2}.\tag{5.103}$$

When $q_2^{in} \neq 0$, the DC can be designed such that the output fields can be given by [75]

$$q_1^{out} = (q_1^{in} - iq_2^{in})/\sqrt{2},\tag{5.104}$$

$$q_2^{out} = (-iq_1^{in} + q_2^{in})/\sqrt{2}.\tag{5.105}$$

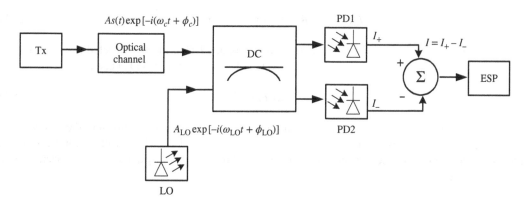

Figure 5.34 Block diagram of a balanced coherent receiver. Tx = transmitter, DC = directional coupler, PD = photodetector, ESP = electronic signal processing.

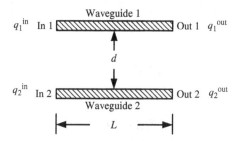

Figure 5.35 Directional coupler.

In other words, the transfer function of the DC is

$$\mathbf{T}_{DC} = \frac{1}{\sqrt{2}} \begin{bmatrix} 1 & -i \\ -i & 1 \end{bmatrix}. \tag{5.106}$$

In Fig. 5.34, the received signal field and local oscillator output are the input fields of the DC. The DC outputs with these inputs are

$$q_1^{\text{out}} = \{A_r s(t) \exp\left[-i(\omega_c t + \phi_c)\right] - iA_{\text{LO}} \exp\left[-i(\omega_{\text{LO}} t + \phi_{\text{LO}})\right]\}/\sqrt{2}, \tag{5.107}$$

$$q_2^{\text{out}} = \{-iA_r s(t) \exp\left[-i(\omega_c t + \phi_c)\right] + A_{\text{LO}} \exp\left[-i(\omega_{\text{LO}} t + \phi_{\text{LO}})\right]\}/\sqrt{2}. \tag{5.108}$$

The outputs of the DC are fed to two identical photo-detectors. The photocurrents are given by

$$I_+(t) = R|q_1^{\text{out}}|^2$$

$$= \frac{R}{2}\{A_r^2 |s(t)|^2 + A_{\text{LO}}^2 - 2A_r A_{\text{LO}} \text{Im}\{s(t) \exp\left[-i(\omega_{\text{IF}} t + \Delta\phi)\right]\}\}, \tag{5.109}$$

$$I_-(t) = R|q_2^{\text{out}}|^2$$

$$= \frac{R}{2}\{A_r^2 |s(t)|^2 + A_{\text{LO}}^2 + 2A_r A_{\text{LO}} \text{Im}\{s(t) \exp\left[-i(\omega_{\text{IF}} t + \Delta\phi)\right]\}\}. \tag{5.110}$$

$$I_d(t) = I_+(t) - I_-(t) = -2RA_rA_{LO}\text{Im}\{s(t)\exp[-i(\omega_{IF}t + \Delta\phi)]\}. \tag{5.111}$$

From Eq. (5.111), we see that the intermodulation cross-talk and DC terms are canceled because of the balanced detection. For homodyne receivers, $\omega_{IF} = 0$. When $\Delta\phi = \pi/2$, Eq. (5.111) becomes

$$I_d(t) = 2RP_0\text{Re}\{s(t)\}. \tag{5.112}$$

For binary modulation schemes, $s(t)$ is real and in this case the current is proportional to the received signal $s(t)$. For M-ary signals, $s(t)$ is complex and an IQ receiver is required, which is discussed in Sections 5.6.3 and 5.6.4. For a heterodyne receiver, the current $I_d(t)$ should be multiplied by a microwave carrier, as is done in Section 5.6.1.2.

Example 5.7

Repeat Example 5.6 assuming that the balanced receiver is used instead of the single-branch receiver. Comment on the intermodulation cross-talk in a single-branch receiver and the balanced receiver.

Solution:
From Eq. (5.112), we have

$$|I_d(t)| = 2RP_0.$$

(a)

$$P_{LO} = 10 \text{ mW},$$

$$P_0 = \sqrt{P_rP_{LO}} = 1.259 \text{ mW},$$

$$|I_d(t)| = 2 \times 0.9 \times 1.259 \text{ mW} = 2.2662 \text{ mW}.$$

(b)

$$P_{LO} = 0.1 \text{ mW},$$

$$P_0 = \sqrt{P_rP_{LO}} = 0.1258 \text{ mW},$$

$$|I_d(t)| = 2 \times 0.9 \times 0.1258 \text{ mA} = 0.2266 \text{ mA}.$$

Since $P_{LO} < P_r$, in the case of a single-branch receiver, the first term in Eq. (5.88) cannot be ignored. Therefore, a single-branch receiver would have a significant amount of cross-talk. In contrast, for a balanced receiver, intermodulation cross-talk is canceled out due to the balanced detection (see Eqs. (5.109)–(5.112)).

5.6.3 Single-Branch IQ Coherent Receiver

So far we have assumed that the message signal $s(t)$ is real. Now we consider a more generalized case, in which $s(t)$ is complex. The examples of complex signals are multi-level PSK and QAM. To recover the real part of $s(t)$, the LO phase should be aligned with that of the optical carrier. Similarly, to recover the imaginary part of $s(t)$, the LO phase should be shifted by $\pi/2(\exp(i\pi/2) = i)$ with respect to the optical carrier. Fig. 5.36 shows a block diagram of the single-branch IQ coherent receiver. The received signal and LO outputs are divided

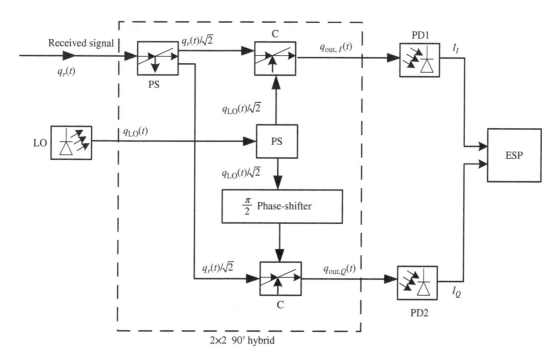

Figure 5.36 Block diagram of a single-branch IQ receiver. PS = power splitter, C = combiner, LO = local oscillator, PD = photodetector, ESP = electrical signal processor.

into two parts using power splitters and are mixed together as done in Section 5.6.1. Let us first consider the in-phase component. The optical input of PD1 is

$$q_{\text{out},I}(t) = \frac{1}{\sqrt{2}} \left[\frac{q_r(t)}{\sqrt{2}} + \frac{q_{\text{LO}}(t)}{\sqrt{2}} \right].$$
(5.113)

The corresponding photocurrent after ignoring the intermodulation cross-talk and DC terms is (see Section 5.6.1)

$$I_I(t) = R|q_{\text{out},I}|^2$$
(5.114)

$$\approx \frac{RA_r A_{\text{LO}}}{2} \text{Re} \left\{ s(t) \exp\left[-i(\omega_{\text{IF}} t + \Delta\phi) \right] \right\}$$

$$\approx \frac{RA_r A_{\text{LO}}}{2} |s(t)| \cos(\omega_{\text{IF}} t + \phi_s(t) + \Delta\phi),$$
(5.115)

where

$$s(t) = |s(t)| \exp\left[-i\phi_s(t) \right].$$
(5.116)

The other part of the LO is 90° phase-shifted and is mixed with the received signal. The optical input of PD2 is

$$q_{\text{out},Q}(t) = \frac{1}{\sqrt{2}} \left[\frac{q_r(t)}{\sqrt{2}} + i \frac{q_{\text{LO}}(t)}{\sqrt{2}} \right].$$
(5.117)

Note that $i = e^{i\pi/2}$ corresponding to a 90° phase shift as shown in Fig. 5.36. The corresponding photocurrent is

$$I_Q(t) = R|q_{\text{out},Q}|^2 \tag{5.118}$$

$$\approx \frac{RA_rA_{\text{LO}}}{2}\text{Im}\{s(t)\exp[-i(\omega_{\text{IF}}t + \Delta\phi)]\}$$

$$\approx \frac{RA_rA_{\text{LO}}}{2}|s(t)|\sin(\omega_{\text{IF}}t + \phi_s(t) + \Delta\phi). \tag{5.119}$$

To obtain Eq. (5.118), we have used the formula

$$2i\text{Im}\{X\} = (X - X^*). \tag{5.120}$$

For a homodyne receiver, $\omega_{\text{IF}} = 0$ and when the phase mismatch $\Delta\phi = 0$, we have

$$I_I(t) = \frac{RA_rA_{\text{LO}}}{2}\text{Re}[s(t)], \tag{5.121}$$

$$I_Q(t) = \frac{RA_rA_{\text{LO}}}{2}\text{Im}[s(t)]. \tag{5.122}$$

The electrical signal processing unit forms the complex current $I(t) = I_I(t) + iI_Q(t) = RA_rA_{\text{LO}}s(t)/2$. Thus, the transmitted complex signal could be retrieved. In Fig. 5.36, the components inside the rectangle constitute a 2×2 90° *optical hybrid*. It is a device with two inputs and two outputs, as shown in Fig. 5.37. The transfer matrix of an ideal 2×2 90° hybrid can be written as

$$\mathbf{T} = \frac{1}{2}\begin{bmatrix} 1 & 1 \\ 1 & i \end{bmatrix}. \tag{5.123}$$

Let the input of the 2×2 90° hybrid be

$$\mathbf{q}_{\text{in}} = \begin{bmatrix} q_r(t) \\ q_{\text{LO}}(t) \end{bmatrix}, \tag{5.124}$$

where $q_r(t)$ and $q_{\text{LO}}(t)$ are the complex fields of received signal and local oscillator, respectively, as shown in Fig. 5.37. Let the outputs of the 2×2 90° hybrid be

$$\mathbf{q}_{\text{out}} = \begin{bmatrix} q_{\text{out},I} \\ q_{\text{out},Q} \end{bmatrix}. \tag{5.125}$$

Now, Eqs. (5.113) and (5.117) can be rewritten as

$$\mathbf{q}_{\text{out}} = \mathbf{T}\mathbf{q}_{\text{in}}. \tag{5.126}$$

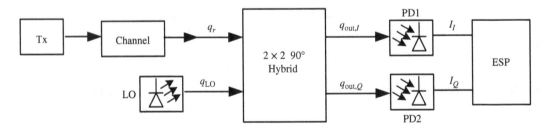

Figure 5.37   Single-branch coherent IQ receiver using a 2×2 90° hybrid.

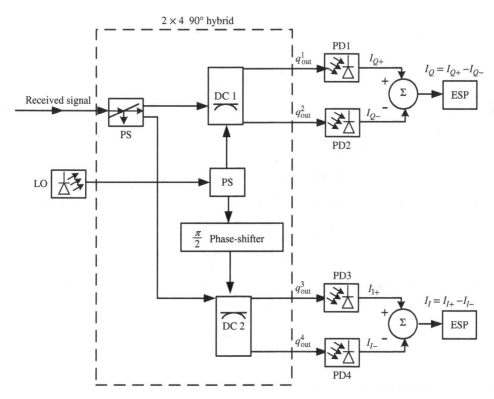

Figure 5.38 Block diagram of a balanced IQ receiver. DC = directional coupler, PS = power splitter, LO = local oscillator, PD = Photodetector.

5.6.4 *Balanced IQ Receiver*

It is straightforward to modify the block diagram of Fig. 5.36 to obtain the balanced IQ receiver, shown in Fig. 5.38. The inputs of DC1 are

$$q_1^{\text{in}} = \frac{q_r}{\sqrt{2}}, \tag{5.127}$$

$$q_2^{\text{in}} = \frac{q_{\text{LO}}}{\sqrt{2}}. \tag{5.128}$$

The transfer function of DC1 is given by Eq. (5.106),

$$\mathbf{T}_{DC} = \frac{1}{\sqrt{2}} \begin{bmatrix} 1 & -i \\ -i & 1 \end{bmatrix}. \tag{5.129}$$

Therefore, the output of DC1 is

$$\mathbf{q}^{\text{out}} = \mathbf{T}_{dc}\mathbf{q}^{\text{in}}, \tag{5.130}$$

$$q_1^{\text{out}} = \frac{1}{2}[q_r - iq_{\text{LO}}], \tag{5.131}$$

$$q_2^{\text{out}} = \frac{1}{2}[-iq_r + q_{\text{LO}}], \tag{5.132}$$

$$\mathbf{q}^{\text{out}} = \begin{bmatrix} q_1^{\text{out}} \\ q_2^{\text{out}} \end{bmatrix}. \tag{5.133}$$

The photocurrents I_{Q+} and I_{Q-} can be calculated as in Section 5.6.2. Proceeding as before, the quadrature current I_Q is given by

$$I_Q = I_{Q+} - I_{Q-} \tag{5.134}$$

$$= -RA_r A_{\text{LO}} \text{Im} \{ s(t) \exp [-i(\omega_{\text{IF}} t + \Delta\phi)] \}. \tag{5.135}$$

For the I-branch, the LO output is phase-shifted by 90°. The in-phase component of the current is

$$I_I = I_{I+} - I_{I-} \tag{5.136}$$

$$= RA_r A_{\text{LO}} \text{Re} \{ s(t) \exp [-i(\omega_{\text{IF}} t + \Delta\phi)] \}. \tag{5.137}$$

For homodyne receivers, $\omega_{\text{IF}} = 0$. When $\Delta\phi = 0$, from Eqs. (5.136) and (5.134), we find

$$I_I = RA_r A_{\text{LO}} \text{Re} \{ s(t) \}, \tag{5.138}$$

$$I_Q = -RA_r A_{\text{LO}} \text{Im} \{ s(t) \}, \tag{5.139}$$

$$I = I_I - iI_Q = RA_r A_{\text{LO}} s(t). \tag{5.140}$$

The components inside the rectangular line constitute a 2×4 90° hybrid. The transfer matrix of the 2×4 90° hybrid can be written as

$$\mathbf{T} = \begin{bmatrix} 1 & -i \\ -i & 1 \\ 1 & 1 \\ -i & i \end{bmatrix}. \tag{5.141}$$

Let the output of the 2×4 90° hybrid be

$$\mathbf{q}^{\text{out}} = \begin{bmatrix} q_{\text{out}}^1 \\ q_{\text{out}}^2 \\ q_{\text{out}}^3 \\ q_{\text{out}}^4 \end{bmatrix}. \tag{5.142}$$

Now, the input–output relationship of the 2×4 90° hybrid can be written as

$$\mathbf{q}^{\text{out}} = \mathbf{T}\mathbf{q}^{\text{in}}. \tag{5.143}$$

Example 5.8

Repeat Example 5.7(a) with the BPSK signal replaced by the QPSK signal. Let $s(t)$ be $1\angle 3\pi/4$. Find the in-phase and quadrature components of the current of a balanced IQ receiver.

Solution:

$$P_{\text{LO}} = 10 \text{ mW},$$

$$s(t) = e^{i3\pi/4} = \frac{-1}{\sqrt{2}} + i\frac{1}{\sqrt{2}},$$

$$P_r = 0.1585 \text{ mW},$$

$$P_0 = \sqrt{P_r P_{\text{LO}}} = 1.259 \text{ mW}.$$

From Eq. (5.138), we find the in-phase component of the current as

$$I_I = RP_0 \text{Re}\left[s(t)\right] = 0.9 \times 1.259 \times \frac{-1}{\sqrt{2}} \text{ mA} = -0.8012 \text{ mA}.$$

From Eq. (5.139), we find

$$I_Q = -RP_0 \text{Im}\left[s(t)\right] = -0.9 \times 1.259 \times \frac{1}{\sqrt{2}} \text{ mA} = -0.8012 \text{ mA}.$$

5.6.5   Polarization Effects

So far we have ignored the state of polarization of light. In this section, we consider the optical signal modulated in orthogonal polarizations. Let the transmitted signal be

$$\mathbf{q}_T = \begin{bmatrix} q_{T,x} \\ q_{T,y} \end{bmatrix}, \tag{5.144}$$

where

$$q_{T,x} = \frac{A_T}{\sqrt{2}} s_x(t) \exp\left(-i\omega_c t\right), \tag{5.145}$$

$$q_{T,y} = \frac{A_T}{\sqrt{2}} s_y(t) \exp\left(-i\omega_c t\right), \tag{5.146}$$

$s_x(t)$ and $s_y(t)$ are the data in x- and y-polarizations, respectively. In fiber, the propagation constants of x- and y-polarization components are slightly different due to a possible asymmetry in fiber geometry and, as a result, these components acquire different amounts of phase shift. Besides, because of the perturbations during the propagation, there is a power transfer between the x- and y-components. These effects can be taken into account by a channel matrix (see Chapter 2):

$$\mathbf{M} = \begin{bmatrix} M_{xx} & M_{xy} \\ M_{yx} & M_{yy} \end{bmatrix}. \tag{5.147}$$

The output of the fiber-optic link may be written as

$$\mathbf{q}_r = \begin{bmatrix} q_{r,x} \\ q_{r,y} \end{bmatrix} = \mathbf{M}\mathbf{q}_T, \tag{5.148}$$

$$q_{r,x} = [M_{xx}q_{T,x} + M_{xy}q_{T,y}]e^{i\phi_c}, \tag{5.149}$$

$$q_{r,y} = [M_{yx}q_{T,x} + M_{yy}q_{T,y}]e^{i\phi_c}, \tag{5.150}$$

where ϕ_c is the common phase of both polarizations. Owing to fluctuations in ambient properties, the matrix **M** changes with time. Typically, the rate of change of the matrix elements of **M** is much slower than the transmission data rate. Therefore, the matrix elements can be estimated using digital signal processing, which is discussed in Chapter 11.

Let the output of the local oscillator be

$$\mathbf{q}_{LO} = \begin{bmatrix} q_{LO,x} \\ q_{LO,y} \end{bmatrix}, \tag{5.151}$$

$$q_{LO,j} = \frac{A_{LO}}{\sqrt{2}} \exp\left[-i(\omega_{LO}t + \phi_{LO})\right], \quad j = x \text{ or } y. \tag{5.152}$$

Fig. 5.39 shows a schematic of the receiver. The x- and y-components of the received field and LO output are separated using polarization beam splitters PBS1 and PBS2, respectively. The x-components of the received field and the LO output are combined using a 2×4 90° optical hybrid and pass through the four photodetectors as discussed in Section 5.6.4. With $\omega_{IF} = \Delta\phi = 0$, the outputs $I_{I,x}$ and $I_{Q,x}$ are given by

$$I_{I,x} = \frac{RA_T A_{LO}}{2} \text{Re}\left[M_{xx}s_x + M_{xy}s_y\right], \tag{5.153}$$

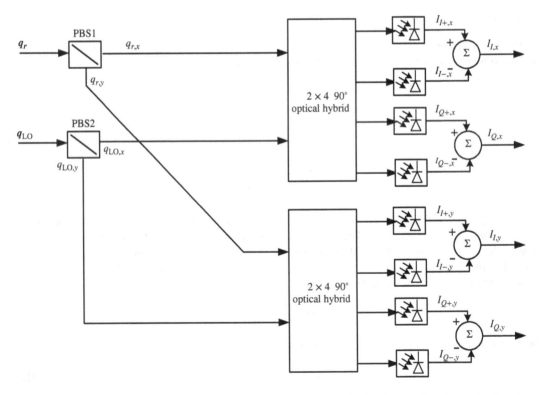

Figure 5.39 Block diagram of a dual polarization IQ receiver PBS = polarization beam splitter, LO = local oscillator.

$$I_{Q,x} = -\frac{RA_TA_{LO}}{2}\,\text{Im}\,[M_{xx}s_x + M_{xy}s_y]. \tag{5.154}$$

The complex photocurrent corresponding to x-polarization is

$$I_x = I_{I,x} - iI_{Q,x} \tag{5.155}$$

$$= \frac{RA_TA_{LO}}{2}[M_{xx}s_x + M_{xy}s_y]. \tag{5.156}$$

Similarly, the y-components of the received field and LO output pass through a second balanced IQ receiver. Its outputs are

$$I_{Q,x} = \frac{RA_TA_{LO}}{2}\,\text{Re}\,[M_{yx}s_x + M_{yy}s_y], \tag{5.157}$$

$$I_{Q,y} = -\frac{RA_TA_{LO}}{2}\,\text{Im}\,[M_{yx}s_x + M_{yy}s_y]. \tag{5.158}$$

The complex photocurrent corresponding to y-polarization is

$$I_y = I_{I,y} - iI_{Q,y} \tag{5.159}$$

$$= \frac{RA_TA_{LO}}{2}[M_{yx}s_x + M_{yy}s_y]. \tag{5.160}$$

Eqs. (5.156) and (5.159) can be rewritten as

$$\frac{RA_TA_{LO}}{2}\begin{bmatrix} M_{xx} & M_{xy} \\ M_{yx} & M_{yy} \end{bmatrix}\begin{bmatrix} s_x \\ s_y \end{bmatrix} = \begin{bmatrix} I_x \\ I_y \end{bmatrix}. \tag{5.161}$$

In the DSP unit of the coherent receiver, the matrix elements of $\mathbf{M}$ are adaptively estimated and its inverse is calculated (see Chapter 11). Multiplying Eq. (5.161) by $\mathbf{M}^{-1}$, we find

$$\begin{bmatrix} s_x \\ s_y \end{bmatrix} = \frac{2}{RA_TA_{LO}}\mathbf{M}^{-1}\begin{bmatrix} I_x \\ I_y \end{bmatrix}. \tag{5.162}$$

Thus, the transmitted data can be estimated using Eq. (5.162).

Example 5.9

Repeat Example 5.8 with a polarization modulated (PM) QPSK signal given by

$$s(t) = \begin{bmatrix} s_x \\ s_y \end{bmatrix} = \begin{bmatrix} 1\angle\pi/4 \\ 1\angle5\pi/4 \end{bmatrix}.$$

Assume $\mathbf{M}$ to be of the form

$$\mathbf{M} = \begin{bmatrix} 1 & 0 \\ 0 & 1 \end{bmatrix}e^{-\alpha L/2},$$

where α is the fiber-loss coefficient and L is the transmission distance. From Eq. (5.161), we have

$$I_x = \frac{RA_TA_{LO}}{2}e^{-\alpha L/2}s_x$$

$$= \frac{RA_rA_{LO}}{2}s_x$$

$$= \frac{RP_0}{2}s_x$$

$$= \frac{0.9 \times 1.259}{2}1\angle \pi/4 \text{ mA} = 0.566\angle \pi/4,$$

$$I_{I,x} = \text{Re}\,(I_x) = 0.4006 \text{ mA},$$

$$I_{Q,x} = -\text{Im}\,(I_x) = -0.4006 \text{ mA}.$$

Similarly,

$$I_y = \frac{RA_TA_{LO}}{2}e^{-\alpha L/2}s_y$$

$$= \frac{RA_rA_{LO}}{2}s_y$$

$$= \frac{RP_0}{2}s_y$$

$$= \frac{0.9 \times 1.259}{2}1\angle 5\pi/4 \text{ mA} = 0.566\angle 5\pi/4,$$

$$I_{I,y} = \text{Re}\,(I_y) = -0.4006 \text{ mA},$$

$$I_{Q,y} = -\text{Im}\,(I_y) = 0.4006 \text{ mA}.$$

Exercises

5.1 Compare qualitatively the features of a direct detection receiver with a coherent detection one.

5.2 Discuss the strengths and weaknesses of pn, pin, and Schottky barrier photodiodes. Which one would you choose for a 10-Gb/s fiber-optic receiver? State the reasons for your choice of structure.

5.3 Compare the characteristics of avalanche photodiodes and pin photodiodes. Which one would you choose for a 10-Gb/s fiber-optic receiver? If the bit rate is increased to more than 40 Gb/s, would your choice remain the same or would you switch to the other photodiode. Why?

5.4 Avalanche photodiodes can be made as in the following structures: normal pn junctions, SAM, SACM, and SAGCM. Compare and contrast the characteristics of each structure. Which structure is most suitable for a low-Gb/s fiber-optic receiver?

5.5 Discuss the key design features of a RCE photodiode.

5.6   Compare vertical illumination RCE, edge-coupled, and waveguide-coupled photodiode structures. Discuss how each structure addresses the bandwidth–quantum efficiency trade-off. Discuss also their ease of manufacturing.

5.7   The band-gap energy of diamond and silicon carbide is 5.5 eV and 3.0 eV, respectively. Determine their long-wavelength cutoff values.

5.8   A 500-nm light source delivers $10\,\mu W$ of power uniformly on a photodetector with a photosensitive area of $400\,\mu m^2$. Determine the light intensity in W/cm^2. Next, determine the photon flux per unit time, that is, the number of photons/cm^2/s. If the quantum efficiency of this photodiode is 90%, what is its responsivity?

5.9   A large-area photodiode has a radius of 200 μm and is used to detect light at 650 nm. When illuminated with light at $100\,\mu W/cm^2$, 60 nA of photocurrent flows in the detector. Determine the photodiode's responsivity and quantum efficiency.

5.10   Given the responsivity of a photodiode of 0.6 A/W for light at 900 nm, what is its quantum efficiency? For the same quantum efficiency, if the wavelength is decreased to 400 nm, what is the new value of the responsivity?

5.11   The wavelength of the laser diode in a music CD player is ~800 nm and the ideal responsivity of the photodiode is ~0.6 A/W. If the laser diode emits 2 mW, how much current flows in the photodiode?

5.12   The refractive index of silicon is 3.5. If light is incident from air onto a silicon photodiode, how much of it is lost through reflection. Describe quantitatively how you would reduce this reflection loss.

5.13   The noise equivalent power (NEP) of a photodiode is defined as the ratio of the root-mean-square (r.m.s.) noise current to its responsivity. If the current in a photodiode is 1 nA (gives shot noise), its shunt resistance is 0.5 GΩ (gives thermal noise), and its responsivity is 0.6 A/W, then for a bandwidth of 1 Hz, determine the total noise current (in A) and the NEP (in W). The electronic charge is 1.6×10^{-19} C and Boltzmann's constant is 1.38×10^{-23} J/K. Assume that the photodiode is operating at room temperature or 300 K.

5.14   A commercial silicon photodiode has a responsivity of 0.6 A/W at 850 nm and a NEP in a 1-Hz bandwidth of 40 fW. Determine the total r.m.s noise of the photodiode.

5.15   Plot the excess noise factor F versus multiplication M for various values of the ratio of ionization coefficients k (where $k = \alpha/\beta$) is varied from 0 to 1.0 in steps of 0.1. Comment on your results.

5.16   The impact ionization for electrons and holes is given by

$$(\alpha, \beta) = \frac{qF}{E_i} \exp\left\{ 0.217 \left(\frac{E_i}{E_r} \right)^{1.14} - \sqrt{\left[0.217 \left(\frac{E_i}{E_r} \right)^{1.14} \right]^2 + \left[\frac{E_i}{qF\lambda} \right]^2} \right\} \qquad (5.163)$$

with

$$\frac{E_r}{E_{ro}} = \frac{\lambda}{\lambda_0} = \tanh\left(E_{ro}/2kT \right), \qquad (5.164)$$

where F is the electric field, E_i is the threshold energy, λ and λ_0 are the mean free path, and E_r and E_{ro} are the optical phonon energy at temperatures T K and 0 K respectively. Given that $E_{ro} = 28.4$ meV and for electrons: $\lambda_e = 55$ Å, $\lambda_o = 100$ Å, $E_{i,e} = 0.682\,\text{eV}$, and for holes: $\lambda_h = 56$ Å, $\lambda_o = 112$ Å, and $E_{i,h} = 1.2\,\text{eV}$, plot the ionization coefficients versus $1/F^2$ for F between 10^5 V/cm and 4×10^5 V/cm.

5.17 Explain the different types of noise mechanism in a pin receiver.

5.18 Explain the differences between homodyne and heterodyne coherent receivers. Which of these receivers requires a larger bandwidth?

References

[1] M.J. Deen and P.K. Basu, *Silicon Photonics – Fundamentals and Devices*. John Wiley & Sons, New York, 2012.
[2] A. Bandyopadhyay and M.J. Deen. In H.S. Nalwa (ed.), *Photodetectors and Fiber Optics*. Academic Press, New York, pp. 307–368, 2001.
[3] G.P. Agrawal, *Fiber-Optic Communication System*, 4th edn. John Wiley & Sons, Hoboken, NJ, 2010.
[4] G. Keiser, *Optical Fiber Optic Communications*, 4th edn. McGraw-Hill, New York, p. 688, 2010.
[5] J. Burm *et al.*, *IEEE Photon. Technol. Lett.*, vol. **6**, p. 722, 1994.
[6] Y.-G. Wey *et al.*, *IEEE J. Lightwave Technol.*, vol. **13**, p. 1490, 1995.
[7] K.S. Giboney *et al.*, *IEEE Photon. Technol. Lett.*, vol. **7**, p. 412, 1995.
[8] Y.C. Lim and R.A. Moore, *IEEE Trans. Electron Dev.*, vol. **15**, p. 173, 1968.
[9] D.P. Prakash *et al.*, *IEEE Photon. Technol. Lett.*, vol. **9**, p. 800, 1997.
[10] Z. Bielecki, *IEEE Proc.-Optoelectron.*, vol. **147**, p. 234, 2000.
[11] M.A. Itzler *et al.*, Optical Fiber Communication Conference, vol. **4**, p. 126, 2000.
[12] A.H. Gnauck, C.A. Burrus, and D.T. Ekholm, *IEEE Photon. Technol. Lett.*, vol. **4**, p. 468, 1992.
[13] T.Y. Yun *et al.*, *IEEE Photon. Technol. Lett.*, vol. **8**, p. 1232, 1996.
[14] A. Torres-J and E.A. Gutiérrez-D, *IEEE Electron Dev. Lett.*, vol. **18**, p. 568, 1997.
[15] A.G. Chynoweth, *Semiconductors and Semimetals*, Vol. **4**. Academic Press, New York, 1968.
[16] G.E. Stillman and C.M. Wolfe, *Semiconductor and Semimetals*, Vol. **12**. Academic Press, New York, 1977.
[17] Y. Okuto and C.R. Crowell, *Phys. Rev. B*, vol. **6**, p. 3076, 1972.
[18] F. Capasso, *Semiconductors and Semimetals*, Vol. **22D**. Academic Press, New York, 1985.
[19] Y. Okuto and C.R. Crowell, *Phys. Rev. B*, vol. **10**, p. 4284, 1974.
[20] G.A. Baraff, *Phys. Rev.*, vol. **128**, p. 2507, 1962; vol. **133**, p. A26, 1964.
[21] C.L.F. Ma, M.J. Deen, and L. Tarof, *Adv. Imag. Electron Phys.*, vol. **99**, p. 65, 1998.
[22] M. Casalino *et al.*, Physica E: Low-dimensional Systems and Nanostructures, Proceedings of the E-MRS 2008 Symposium C: Frontiers in Silicon-Based Photonics, Vol. **41**, p. 1097, 2009.
[23] C. Li *et al.*, *J. Appl. Phys.*, vol. **92**, p. 1718, 2002.
[24] M. Ghioni *et al.*, *IEEE Trans. Electron Dev.*, vol. **43**, p. 1054, 1996.
[25] N.R. Das and M.J. Deen, *J. Vac. Sci. Technol. A*, vol. **20**, p. 1105, 2002.
[26] N.R. Das and M.J. Deen, *IEEE J. Quant. Electron.*, vol. **37**, p. 1574, 2001.
[27] N.R. Das and M.J. Deen, *IEEE J. Quant. Electron.*, vol. **37**, p. 69, 2001.
[28] G. Kinsey *et al.*, *IEEE Photon. Technol. Lett.*, vol. **10**, p. 1142, 1998.
[29] R.G. Decorby, A.J.P. Hnatiw, and G. Hillier, *IEEE Photon. Technol. Lett.*, vol. **11**, p. 1165, 1999.
[30] H. Nie *et al.*, *IEEE Photon. Technol. Lett.*, vol. **10**, p. 409, 1998.
[31] C. Lenox *et al.*, *IEEE Photon. Technol. Lett.*, vol. **11**, p. 1162, 1999.
[32] M.S. Ünlü and S. Strite, *J. Appl. Phys.*, vol. **78**, p. 607, 1995.
[33] H.H. Tung and C.P. Lee, *IEEE J. Quant. Electron.*, vol. **33**, p. 753, 1997.
[34] J.A. Jervase and Y. Zebda, *IEEE J. Quant. Electron.*, vol. **34**, p. 1129, 1998.
[35] A. Umbach, M. Leone, and G. Unterbrsch, *J. Appl. Phys.*, vol. **81**, p. 2511, 1997.
[36] L. Giraudet *et al.*, *IEEE Photon. Technol. Lett.*, vol. **11**, p. 111, 1999.
[37] K. Kato, *IEEE Trans. Microw. Theory Tech.*, vol. **47**, p. 1265, 1999.
[38] C.L. Ho *et al.*, *IEEE J. Quant. Electron.*, vol. **36**, p. 333, 2000.

[39] St. Kollakowski *et al.*, *IEEE Photon. Technol. Lett.*, vol. **9**, p. 496, 1997.

[40] G.S. Kinsey *et al.*, *IEEE Photon. Technol. Lett.*, vol. **12**, p. 416, 2000.

[41] K. Kishino *et al.*, *IEEE J. Quant. Electron.*, vol. **27**, p. 2025, 1991.

[42] C. Li *et al.*, *IEEE Photon. Technol. J.*, vol. **12**, p. 1373, 2000.

[43] Y. Ardeshirpour and M.J. Deen, CMOS photodetectors, unpublished results, 2005.

[44] G. Sasaki *et al.*, *Electron. Lett.*, vol. **24**, p. 1201, 1988.

[45] Y. Akatsu *et al.*, *IEEE Photon. Technol. Lett.*, vol. **5**, p. 163, 1993.

[46] S. Chandrasekhar *et al.*, *Electron. Lett.*, vol. **26**, p. 1880, 1990.

[47] K.D. Pedrotti *et al.*, IEEE GaAs IC Symposium, p. 205, 1991.

[48] K. Takahata *et al.*, *Electron. Lett.*, vol. **33**, p. 1576, 1997.

[49] E. Sano *et al.*, *J. Lightwave Technol.*, vol. **12**, p. 638, 1994.

[50] J. Cowles *et al.*, *IEEE Photon. Technol. Lett.*, vol. **6**, p. 963, 1994.

[51] K. Yang *et al.*, *J. Lightwave Technol.*, vol. **14**, p. 1831, 1996.

[52] E. Sano *et al.*, *IEEE Trans. Electron Dev.*, vol. **43**, p. 1826, 1996.

[53] D. Huber *et al.*, *Electron. Lett.*, vol. **35**, p. 40, 1999.

[54] D. Huber *et al.*, *J. Lightwave Technol.*, vol. **18**, p. 992, 2000.

[55] P. Fay *et al.*, *J. Lightwave Technol.*, vol. **15**, p. 1871, 1997.

[56] P. Fay, C. Caneau, and I. Adesida, IEEE Proceedings on Microwave and Optoelectronics Conference, SBMO/IEEE MTT-S IMOC'99, Vol. **2**, p. 537, 1999.

[57] P. Fay *et al.*, *IEEE Photon. Technol. Lett.*, vol. **9**, p. 991, 1997.

[58] U. Hodel *et al.*, Proceedings of the International Conference on Indium Phosphide and Related Compounds/2000, p. 466, 2000.

[59] P. Bhattacharya, *Semiconductor Optoelectronic Devices*. Prentice-Hall, Englewood Cliffs, NJ, 1994.

[60] N.R. Das, P.K. Basu, and M.J. Deen, *IEEE Trans. Electron Dev.*, vol. **47**, p. 2101, 2000.

[61] B.E.A. Saleh and M.C. Teich, *Fundamentals of Photonics*, 2nd edn. John Wiley & Sons, Hoboken, NJ, 2007.

[62] R.J. McIntyre, *IEEE Trans. Electron. Dev.*, vol. **19**, p. 703, 1972; vol. **13**, p. 164, 1966.

[63] J.M. Senior, *Optical Fiber Communications*, 2nd edn. Prentice-Hall, London, 1992.

[64] Y. Han and G. Li, *Opt. Expr.*, vol. **13**, p. 7527, 2005.

[65] K. Kikuchi and S. Tsukamoto, *J. Lightwave Technol.*, vol. **26**, p. 1817, 2008.

[66] B. Zhang *et al.*, *Opt. Expr.*, vol. **20**, p. 3225, 2012.

[67] S.J. Savory, *IEEE J. Select. Top. Quant. Electron.*, vol. **16**, p. 1164, 2010.

[68] C.R. Doerr *et al.*, *IEEE Photon. Technol. Lett.*, vol. **23**, p. 694, 2011.

[69] K. Kikuchi. In I.P. Kaminow, T. Li, and A.E. Willner (eds), *Optical Fiber Telecommunications, V*, Vol. B. Elsevier, Amsterdam, chapter 3, 2008.

[70] M. Birk *et al.*, *J. Lightwave Technol.*, vol. **29**, p. 417, 2011.

[71] Y. Painchaud *et al.*, *Opt. Expr.*, vol. **17**, p. 3659, 2009.

[72] D.-S. Ly-Gagnon *et al.*, *J. Lightwave Technol.*, vol. **29**, p. 12, 2006.

[73] E. Ip *et al.*, *Opt. Expr.*, vol. **16**, p. 758, 2008.

[74] S. Betti, G. DeMarchis, and E. Iannone, *Coherent Optical Communication Systems*. John Wiley & Sons, New York, 1995.

[75] K. Okamoto, *Fundamentals of Optical Waveguides*. Academic Press, San Diego, 2000.

6

Optical Amplifiers

6.1 Introduction

The optical amplifier may be considered as a laser without feedback, or one in which the feedback is suppressed. In the 1980s, optical amplifiers were not commercially available and long-haul fiber-optic communication systems used electrical amplifiers to compensate for the fiber loss. The optical signal was first converted to the electrical signal (O/E conversion) using a photodetector and then converted back to the optical domain (E/O conversion) after amplification in the electrical domain. However, this type of optoelectronic regenerator is expensive for multi-channel optical communication systems. With the advent of optical amplifiers, the optical signal can be amplified directly without having to do O/E and E/O conversion.

There are different physical mechanisms that can be used to amplify the optical signal. In semiconductor optical amplifiers (SOAs), an electrical pump (power supply) is used to achieve population inversion. In the presence of signal photons that have energy close to the band gap, electrons are stimulated to recombine with holes and, thereby, emit photons due to stimulated emission. Thus, the input signal photons are amplified. In EDFAs, an optical pump is used to achieve population inversion. In the presence of signal photons, the erbium ions in the excited state emit light by stimulated emission and make transitions to the ground state. In Raman amplifiers, an optical pump gives up its energy to create a signal photon of lower energy and the rest appears as molecular vibration (or optical photons). This is known as stimulated Raman scattering (SRS). If a signal photon of lower energy is already present, it is amplified by SRS.

In this chapter, we focus mainly on three types of optical amplifiers: (1) the semiconductor optical amplifier, (2) the erbium-doped fibers amplifier, and (3) the Raman amplifier. In each case, the physical principles, governing equations, noise amplifications, and practical applications are discussed. From Sections 6.2 to 6.5, we consider a generic amplifier and the system impact of noise is discussed. In Sections 6.6 to 6.8, we focus on specific amplifiers.

6.2 Optical Amplifier Model

In Sections 6.2 to 6.5, we consider a simple amplifier model in which the amplifier magnifies the input power by a factor of G and adds white noise, as shown in Fig. 6.1. Let the signal field envelope at the input and output of an amplifier be ψ_{in} and ψ_{out}, respectively. They are related by

$$\psi_{out} = \sqrt{G}\psi_{in} \qquad (6.1)$$

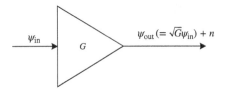

Figure 6.1   Simple amplifier model.

or

$$P_{\text{out}} = |\psi_{\text{out}}|^2 = G|\psi_{\text{in}}|^2 = GP_{\text{in}}, \tag{6.2}$$

where G is the amplifier power gain. The amplifier adds noise $n(t)$ and the total field envelope at the amplifier output is

$$\psi_{\text{tot}} = \psi_{\text{out}} + n(t) = \sqrt{G}\psi_{\text{in}} + n(t). \tag{6.3}$$

The gain is provided by the stimulated emission and the noise in the amplifier is mainly due to the spontaneous emission, which is discussed in the next section. We assume that the samples of $n(t)$ are identically distributed Gaussian complex random variables.

6.3   Amplified Spontaneous Emission in Two-Level Systems

The light wave generated due to stimulated emission has the same polarization, frequency, and phase as that of the incident wave, whereas the light wave generated by spontaneous emission has random phase and frequency, and it propagates in all directions. The spontaneously emitted photons are amplified by the amplifier and this is known as *amplified spontaneous emission* (ASE). ASE is the main source of noise in the amplifiers. The mean number of photons at the output of an amplifier of length L is given by (see Appendix A)

$$n_{ph}(L) = n_{ph}(0)G + n_{sp}(G - 1), \tag{6.4}$$

where $n_{ph}(0)$ is the mean number of photons at the amplifier input ($z = 0$), G is the amplifier gain, and n_{sp} is the *spontaneous emission factor* or *population-inversion factor*, given by

$$n_{sp} = \frac{N_2}{N_2 - N_1}, \tag{6.5}$$

where N_1 and N_2 are the population densities at state 1 and 2, respectively. Eq. (6.4) is of fundamental significance. The first and second terms on the right-hand side represent the photon gain due to stimulated emission and spontaneous emission, respectively. When there is full population inversion, $N_1 = 0$ and $n_{sp} = 1$. This corresponds to an ideal amplifier. For a realistic amplifier, $N_1 \neq 0$ and n_{sp} is greater than 1.

The mean number of photons $n_{sp}(G - 1)$ corresponds to the mean noise power P_{ASE} in the frequency range of f_0 to $f_0 + \Delta f$ in a single polarization (see Appendix A):

$$P_{\text{ASE},sp} = n_{sp}hf_0(G - 1)\Delta f. \tag{6.6}$$

Here, the subscript *sp* refers to single polarization. The noise power given by Eq. (6.6) is the noise power per mode. In a single-mode fiber, there are actually two modes corresponding to two polarizations. Therefore, the noise power in two polarizations is

$$P_{\text{ASE},dp} = 2n_{sp}(G - 1)hf_0\Delta f. \tag{6.7}$$

The noise power per unit frequency interval is the power spectral density (PSD), which is given by

$$\rho_{ASE,dp} = \frac{P_{ASE,dp}}{\Delta f} = 2n_{sp}(G-1)hf_0. \tag{6.8}$$

Note that the PSD given by Eq. (6.8) is single-sided, i.e., the frequency components are positive. The power spectral density is constant over the bandwidth $\Delta f \ll f_0$, and the ASE can be considered as a white noise process. The single-sided PSD per polarization is

$$\rho_{ASE,sp} = n_{sp}hf_0(G-1). \tag{6.9}$$

Example 6.1

An optical amplifier operating at 1550 nm has a one-sided ASE power spectral density of 5.73×10^{-17} W/Hz in both polarizations. Calculate the gain G. Assume $n_{sp} = 1.5$.

Solution:
From Eq. (6.8), we have

$$\rho_{ASE,dp} = 2n_{sp}(G-1)hf_0,$$

$$f_0 = \frac{c}{\lambda} = \frac{3 \times 10^8}{1550 \times 10^{-9}} = 193.55 \text{ THz},$$

$$G = \frac{\rho_{ASE,dp}}{2n_{sp}hf_0} + 1$$

$$= \frac{5.73 \times 10^{-17}}{2 \times 1.5 \times 6.626 \times 10^{-34} \times 193.55 \times 10^{12}} + 1$$

$$= 150.$$

6.4   Low-Pass Representation of ASE Noise

The complex ASE noise field in a single polarization may be written as

$$\phi_n(t) = n(t)\exp[-i(2\pi f_0 t)], \tag{6.10}$$

where $n(t)$ is the slowly time-varying field envelope of noise. Taking the Fourier transform of Eq. (6.10), we find

$$\tilde{n}(f) = \tilde{\phi}_n(f - f_0). \tag{6.11}$$

Note that $\phi_n(t)$ is a band-pass noise process, and $n(t)$ is its low-pass equivalent. Fig. 6.2(a) and 6.2(b) shows the absolute of the Fourier transform of $\phi_n(t)$ and $n(t)$, respectively. As can be seen, $\tilde{\phi}_n(f)$ occupies a spectral region $f_0 - B_o/2 \leq f \leq f_0 + B_o/2$ and $\tilde{n}(f)$ is band-limited to $B_0/2$. Let us first consider the ASE noise as a band-pass process with the single-sided PSD given by Eq. (6.9),

$$\rho_{ASE,sp} = n_{sp}hf_0(G-1). \tag{6.12}$$

The noise power in a bandwidth of B_o is

$$N = \rho_{ASE,sp}B_o. \tag{6.13}$$

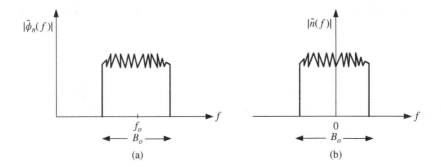

Figure 6.2 Fourier transform of (a) the complex noise field, (b) its envelope.

Fig. 6.3 shows the single-sided PSD of the band-pass signal $\phi_n(t)$.

Next let us consider the equivalent low-pass representation. Since $n(t)$ is the low-pass signal, its PSD, by definition, is

$$\rho_{\text{ASE},sp}^{\text{LP}} = \lim_{T \to \infty} \frac{< |\tilde{n}(f)|^2 >}{T}, \tag{6.14}$$

where T is a long time interval. Fig. 6.4 shows the low-pass representation of the ASE PSD. $\rho_{\text{ASE},sp}^{\text{LP}}$ can be determined by the condition that the noise power in the low-pass representation should be the same as that in the band-pass representation, as given by Eq. (6.13), i.e.,

$$N = \rho_{\text{ASE},sp}^{\text{LP}} B_o = \rho_{\text{ASE},sp} B_o, \tag{6.15}$$

$$\rho_{\text{ASE},sp}^{\text{LP}} = \rho_{\text{ASE},sp} = n_{sp} h f_0 (G - 1). \tag{6.16}$$

In other words, the double-sided PSD of the low-pass signal $n(t)$ is the same as the single-sided PSD, $\rho_{\text{ASE},sp}$ of $\phi_n(t)$. From now on, we omit the subscripts and superscripts and denote the PSD of the $n(t)$ as ρ_{ASE},

$$\rho_{\text{ASE}} \equiv \rho_{\text{ASE},sp}^{\text{LP}} = \rho_{\text{ASE},sp} = n_{sp} h f_0 (G - 1). \tag{6.17}$$

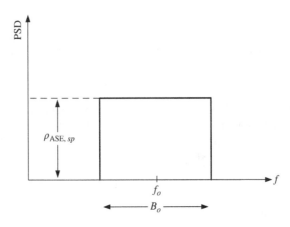

Figure 6.3 Band-pass representation of ASE PSD.

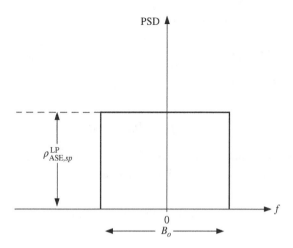

Figure 6.4   Low-pass representation of ASE PSD.

6.5   System Impact of ASE

Consider an amplifier with gain G. Let ρ_{ASE} be the power spectral density of ASE noise and ψ_{in} be the optical field envelope at the amplifier input. We assume that the input of the amplifier is CW, i.e., ψ_{in} is a constant, and consider the case of a single polarization. The impact of ASE for the case of dual polarizations is discussed in Section 6.5.4. The optical field envelope at the amplifier output can be written as

$$\psi_{\text{tot}} = \psi_{\text{out}} + n(t), \tag{6.18}$$

where $\psi_{\text{out}} = \sqrt{G}\psi_{\text{in}}$ is the output signal field envelope and $n(t)$ is the noise field envelope due to ASE. The amplifier output passes through an optical filter, as shown in Fig. 6.5. Let the transfer function of the optical filter be $\tilde{H}_{\text{opt}}(f)$. The noise output of the optical filter is

$$\tilde{n}_F(f) = \tilde{n}(f)\tilde{H}_{\text{opt}}(f), \tag{6.19}$$

where

$$\tilde{n}_F(f) = \mathcal{F}[n_F(t)],$$

$$\tilde{n}(f) = \mathcal{F}[n(t)].$$

The power spectral density of $n_F(t)$ is

$$\rho_{n_F}(f) = \rho_{\text{ASE}}|\tilde{H}_{\text{opt}}(f)|^2, \tag{6.20}$$

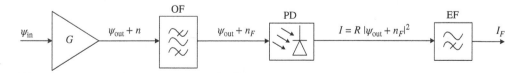

Figure 6.5   A fiber-optic system consisting of an amplifier, an optical band-pass filter, a photodetector, and an electrical low-pass filter. OF = optical filter, PD = photodetector, EF = electrical filter.

and is shown in Fig. 6.6. The mean noise power is

$$P_{ASE} = <|n_F(t)|^2> = \int_{-\infty}^{\infty} \rho_{n_F}(f)df = \rho_{ASE}\int_{-\infty}^{\infty}|\tilde{H}_{opt}(f)|^2 df$$

$$= \rho_{ASE}B_o, \tag{6.21}$$

where

$$B_o = \int_{-\infty}^{+\infty}|H_{opt}(f)|^2 df \tag{6.22}$$

is the *effective bandwidth* of the optical filter. Since the field envelope is a low-pass signal, we model the optical band-pass filter as the low-pass filter. An ideal band-pass filter is modeled as an ideal low-pass filter with the transfer function

$$\tilde{H}_{opt}(f) = 1 \quad \text{for} \quad |f| < f_o/2$$

$$= 0 \quad \text{otherwise.} \tag{6.23}$$

Here, f_o is the full bandwidth of the optical filter. Using Eq. (6.23), Eq. (6.22) becomes

$$B_o = \int_{-f_o/2}^{f_o/2} df = f_o. \tag{6.24}$$

The optical filter output passes through the photodetector and the photocurrent I is proportional to the incident power,

$$I = R|\psi_{out} + n_F(t)|^2$$

$$= R[|\psi_{out}|^2 + |n_F(t)|^2 + \psi_{out}n_F^*(t) + \psi_{out}^*n_F(t)]. \tag{6.25}$$

Let

$$I_0 = R|\psi_{out}|^2, \tag{6.26}$$

$$I_{s-sp} = R[\psi_{out}n_F^*(t) + \psi_{out}^*n_F(t)], \tag{6.27}$$

$$I_{sp-sp} = R|n_F(t)|^2. \tag{6.28}$$

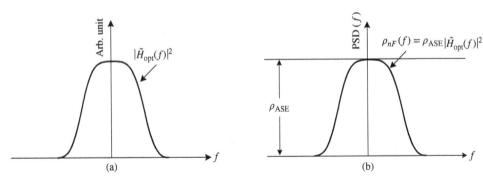

Figure 6.6 Impact of the optical filter on noise: (a) absolute square of the filter transfer function, and (b) PSD of the noise at the filter output.

Now Eq. (6.25) may be written as

$$I = I_0 + I_{s-sp} + I_{sp-sp}.$$ (6.29)

Here, I_0 is a constant photocurrent, $I_{s-sp}(t)$ is the noise current due to *signal–ASE beating*, and I_{sp-sp} is the noise current due to *ASE–ASE beating* $(n_F(t)n_F^*(t))$. Since I_0 is the deterministic current, its variance is zero and it only alters the mean of the photocurrent. Using Eq. (6.29), the mean and variance of I can be calculated as

$$< I >=< I_0 > + < I_{s-sp} > + < I_{sp-sp} >,$$ (6.30)

$$\sigma^2 = < I^2 > - < I >^2$$

$$= \sigma_{s-sp}^2 + \sigma_{sp-sp}^2 + 2R_{s-sp,sp-sp},$$ (6.31)

where

$$\sigma_{s-sp}^2 = < I_{s-sp}^2 > - < I_{s-sp} >^2,$$ (6.32)

$$\sigma_{sp-sp}^2 = < I_{sp-sp}^2 > - < I_{sp-sp} >^2,$$ (6.33)

$$R_{s-sp,sp-sp} = < I_{s-sp}I_{sp-sp} > - < I_{s-sp} >< I_{sp-sp} > .$$ (6.34)

In Eq. (6.31), the first term and second term on the right-hand side represent the variance of signal–ASE beat noise and ASE–ASE beat noise, respectively, and the last term represents the correlation between the two. It can be shown that these two noise processes are uncorrelated and, therefore, the last term of Eq. (6.31) is zero. Total variance may be written as

$$\sigma^2 = \sigma_{s-sp}^2 + \sigma_{sp-sp}^2.$$ (6.35)

In the following subsections, we develop expressions for each of the terms.

6.5.1 Signal–ASE Beat Noise

In this section, we develop an analytical expression for the variance of $I_{s-sp}(t)$. Let us first consider the case in which the electrical filter bandwidth is much larger than B_o and, hence, its impact can be ignored. Let

$$\psi_{out} = |\psi_{out}| \exp(i\theta_{out}),$$ (6.36)

$$n_F = |n_F| \exp(i\theta_F).$$ (6.37)

We obtain

$$\psi_{out}n_F^* + \psi_{out}^*n_F = |\psi_{out}||n_F|[\exp(i\Delta\theta) + \exp(-i\Delta\theta)]$$

$$= 2|\psi_{out}||n_F| \cos(\Delta\theta),$$ (6.38)

where $\Delta\theta = \theta_{out} - \theta_F$. $\Delta\theta$ is a random variable with uniform distribution in the interval $[0, 2\pi]$. Now, using Eq. (6.38) in Eq. (6.27), the mean of $I_{s-sp}(t)$ is given by

$$< I_{s-sp}(t) >= 2R|\psi_{out}| < |n_F| >< \cos(\Delta\theta) >= 0,$$ (6.39)

since cos $(\Delta\theta)$ takes positive values with the same probability as it takes negative values. The variance is

$$\sigma_{s-sp}^2 = <I_{s-sp}^2(t)> = 4R^2|\psi_{\text{out}}|^2 <|n_F|^2><\cos^2(\Delta\theta)> . \tag{6.40}$$

Using the following relation:

$$\cos^2(\Delta\theta) = \frac{1 + \cos(2\Delta\theta)}{2}, \tag{6.41}$$

and making use of Eq. (6.21), Eq. (6.40) becomes

$$\sigma_{s-sp}^2 = 4R^2 P_{\text{out}} P_{\text{ASE}} < \frac{1}{2} + \frac{1}{2}\cos(2\Delta\theta) >$$
$$= 2R^2 P_{\text{out}} P_{\text{ASE}}. \tag{6.42}$$

Eq. (6.42) is an important result. We will use this result later in Section 7.4 to evaluate the performance of a fiber-optic system consisting of a chain of amplifiers. Using Eq. (6.21) in Eq. (6.42), we find

$$\sigma_{s-sp}^2 = 2R^2 P_{\text{out}} \rho_{\text{ASE}} B_o. \tag{6.43}$$

Eq. (6.43) may be rewritten as

$$\sigma_{s-sp}^2 = 2I_{\text{out}} I_{\text{ASE}}, \tag{6.44}$$

where I_{ASE} and I_{out} are the noise current due to ASE and signal current, respectively, given by

$$I_{\text{ASE}} = RP_{\text{ASE}} = R\rho_{\text{ASE}} B_o, \tag{6.45}$$

$$I_{\text{out}} = RP_{\text{out}}. \tag{6.46}$$

Next, let us consider the case in which the optical filter is absent, but the electrical filter is present. Let the current $I(t)$ pass through an electrical filter with the transfer function $\tilde{H}_e(f)$. The signal–ASE noise current before the electrical filter is given by Eq. (6.27),

$$I_{s-sp} = R[\psi_{\text{out}} n^*(t) + \psi_{\text{out}}^* n(t)]. \tag{6.47}$$

Since the optical filter is absent, $n_F(t)$ of Eq. (6.27) is replaced by $n(t)$. Suppose $n(t)$ is the input of the electrical filter, its output would be

$$\tilde{n}_{EF}(f) = \tilde{n}(f)\tilde{H}_e(f). \tag{6.48}$$

Therefore, if $I_{s-sp}(t)$ is the input of the electrical filter, its output is

$$I_{s-sp,EF}(t) = R[\psi_{\text{out}} n_{EF}^*(t) + \psi_{\text{out}}^* n_{EF}(t)]. \tag{6.49}$$

Eq. (6.49) is the same as Eq. (6.38) except that $n_F(t)$ is replaced by $n_{EF}(t)$. So, proceeding as before, the variance may be written as

$$\sigma^2 = 2R^2 P_{\text{out}} <|n_{EF}|^2> . \tag{6.50}$$

The mean noise power after the electrical filter can be calculated as follows. The power spectral density after the electrical filter is

$$\rho_{n_{EF}}(f) = \rho_{\text{ASE}}|\tilde{H}_e(f)|^2. \tag{6.51}$$

The noise power is

$$< |n_{EF}|^2 > = \int_{-\infty}^{\infty} \rho_{n_{EF}}(f)df$$

$$= 2\rho_{ASE}B_e, \tag{6.52}$$

where

$$B_e = \frac{1}{2} \int_{-\infty}^{\infty} |\tilde{H}_e(f)|^2 df. \tag{6.53}$$

Substituting Eq. (6.52) in Eq. (6.50), we find

$$\sigma_{s-sp}^2 = 4R^2 P_{out}\rho_{ASE}B_e. \tag{6.54}$$

Eq. (6.54) may be rewritten as

$$\sigma_{s-sp}^2 = 2I_{out}I_{ASE}, \tag{6.55}$$

where I_{out} and I_{ASE} are the signal current and noise current in the electrical bandwidth of B_e, respectively, given by

$$I_{out} = RP_{out}, \tag{6.56}$$

$$I_{ASE} = 2R\rho_{ASE}B_e. \tag{6.57}$$

When the electrical filter is an ideal low-pass filter with cutoff frequency f_e,

$$\tilde{H}_e(f) = 1 \quad \text{for} \quad |f| < f_e$$

$$= 0 \quad \text{otherwise}, \tag{6.58}$$

Eq. (6.53) becomes

$$B_e = \frac{1}{2} \int_{-\infty}^{\infty} df = f_e. \tag{6.59}$$

Next, let us consider the case in which the bandwidths of optical and electrical filters are comparable. When the optical filter is an ideal band-pass filter with full bandwidth f_o and the electrical filter is an ideal low-pass filter with cutoff frequency f_e, the variance is (see Example 6.8)

$$\sigma_{s-sp}^2 = 4R^2 P_{out}\rho_{ASE}B_{eff}, \tag{6.60}$$

$$B_{eff} = \min\{f_o/2, f_e\}. \tag{6.61}$$

Example 6.2

The ASE PSD of an amplifier ρ_{ASE} is 1.3×10^{-16} W/Hz. The gain of the amplifier $G = 20$ dB and the input power of the amplifier is $10\,\mu$W. The output of the amplifier is incident on a photodetector with responsivity $R = 0.8$ A/W. Calculate the variance of the signal–ASE beat noise. Assume that the receiver can be modeled as an ideal low-pass filter with cutoff frequency $f_e = 7$ GHz. Ignore the optical filter.

Solution:

$$G(\text{dB}) = 20 \text{ dB} = 10 \log_{10}(G).$$

The gain in the linear unit is

$$G = 10^{G(\text{dB})/10} = 100.$$

The output power of the amplifier is

$$P_{\text{out}} = GP_{\text{in}} = 100 \times 10 \times 10^{-6} \text{ W} = 1 \text{ mW}.$$

The variance of the signal–ASE beat noise current is

$$\sigma_{s-sp}^2 = 4R^2 P_{\text{out}} \rho_{\text{ASE}} f_e$$
$$= 4 \times 0.8^2 \times 1 \times 10^{-3} \times 1.3 \times 10^{-16} \times 7 \times 10^9 \text{ A}^2$$
$$= 2.32 \times 10^{-9} \text{ A}^2.$$

6.5.2 ASE–ASE Beat Noise

The absolute square of the ASE noise leads to the noise current I_{sp-sp}, known as the ASE–ASE beat noise current given by Eq. (6.28). Analytical expressions for the variance of ASE–ASE beat noise for arbitrary transfer functions of optical and electrical filters is discussed in Example 6.11. Here we consider a few special cases. When the optical filter bandwidth B_o is much smaller than the electrical filter bandwidth B_e, the impact of the electrical filter can be ignored and, in this case, the mean and variance are calculated as (see Example 6.10)

$$< I_{sp-sp} >= R\rho_{\text{ASE}} B_o, \tag{6.62}$$

$$\sigma_{sp-sp}^2 = R^2 \rho_{\text{ASE}}^2 B_o^2. \tag{6.63}$$

When the optical filter is an ideal band-pass filter with full bandwidth f_o and the electrical filter is an ideal low-pass filter with cutoff frequency f_e, the mean and variance are given by [1] (see Example 6.11),

$$< I_{sp-sp} >= R\rho_{\text{ASE}} f_o, \tag{6.64}$$

$$\sigma_{sp-sp}^2 = R^2 \rho_{\text{ASE}}^2 (2f_o - f_e) f_e \quad \text{if} \quad f_e < f_o,$$
$$= R^2 \rho_{\text{ASE}}^2 f_o^2 \quad \text{otherwise.} \tag{6.65}$$

When $f_e \ll f_o$, Eq. (6.65) becomes

$$\sigma_{sp-sp}^2 = 2R^2 \rho_{\text{ASE}}^2 f_o f_e. \tag{6.66}$$

6.5.3 Total Mean and Variance

Let us assume that the optical and electrical filters are ideal filters with bandwidths f_o and f_e, respectively. Since the mean of the signal–ASE beat current is zero, the total mean current is the sum of the deterministic photocurrent given by Eq. (6.26) and the mean of the ASE–ASE beat current,

$$< I >= I_0 + R\rho_{\text{ASE}} f_o. \tag{6.67}$$

Without loss of generality, we can assume that the signal field at the receiver is polarized in the x-direction with a suitably chosen reference axis so that $\psi_{\text{out},y} = 0$. Now, Eq. (6.74) becomes

$$P = |\psi_{\text{out},x} + n_{F,x}|^2 + |n_{F,y}|^2. \tag{6.75}$$

The photocurrent is

$$I = RP \tag{6.76}$$

$$= R\{|\psi_{\text{out},x}|^2 + \psi_{\text{out},x} n_{F,x}^* + \psi_{\text{out},x}^* n_{F,x} + |n_{F,x}|^2 + |n_{F,y}|^2\} \tag{6.77}$$

$$= I_0 + I_{s-sp} + I_{sp-sp}, \tag{6.78}$$

where

$$I_0 = RP_{\text{out}} = R|\psi_{\text{out},x}|^2, \tag{6.79}$$

$$I_{s-sp} = R(\psi_{\text{out},x} n_{F,x}^* + \psi_{\text{out},x}^* n_{F,x}), \tag{6.80}$$

$$I_{sp-sp} = R(|n_{F,x}|^2 + |n_{F,y}|^2). \tag{6.81}$$

Since Eq. (6.80) is the same as Eq. (6.27) with n_F replaced by $n_{F,x}$, we have

$$< I_{s-sp} >= 0, \tag{6.82}$$

$$\sigma_{s-sp}^2 = 4R^2 \rho_{\text{ASE}} P_{\text{out}} B_{\text{eff}}. \tag{6.83}$$

Here, ρ_{ASE} is the PSD of $n_{F,x}$. $n_{F,x}$ and $n_{F,y}$ are two independent noise processes. So, the mean ASE–ASE beat noise current is doubled, i.e.,

$$< I_{sp-sp} >= 2 < |n_{F,x}|^2 >= 2R\rho_{\text{ASE}} f_o. \tag{6.84}$$

From Eq. (6.81), we find

$$< I_{sp-sp}^2 >= R^2(< |n_{F,x}|^4 > + < |n_{F,y}|^4 > +2 < |n_{F,x}|^2 >< |n_{F,y}|^2 >). \tag{6.85}$$

Since $< |n_{F,x}|^m >=< |n_{F,y}|^m >, m = 1, 2, ..., 4$, Eq. (6.85) reduces to

$$< I_{sp-sp}^2 >= 2R^2[< |n_{F,x}|^4 > + < |n_{F,x}|^2 >^2], \tag{6.86}$$

$$\sigma_{sp-sp}^2 = < I_{sp-sp}^2 > - < I_{sp-sp} >^2 = 2R^2[< |n_{F,x}|^4 > - < |n_{F,x}|^2 >^2] \tag{6.87}$$

$$= 2R^2 \rho_{\text{ASE}}^2 (2f_o - f_e)f_e \quad \text{if} \quad f_e < f_o \tag{6.88}$$

$$= 2R^2 \rho_{\text{ASE}}^2 f_o^2 \quad \text{otherwise.} \tag{6.89}$$

Note that the variance of the signal–noise beating with two polarizations given by Eq. (6.83) is the same as that obtained earlier for the case of a single polarization. This is because the noise field can be split into two polarization components: one aligned with the signal and the other orthogonal to the signal. The noise component orthogonal to the signal does not interfere with it and hence the variance of the signal–noise beating obtained in the two cases is identical. However, the mean and variance of the noise–noise beating current are doubled compared with the single-polarization case since the amplifier noise power in both polarizations leads to noise current. In Sections 6.5.1 to 6.5.3, we assumed that there is only one polarization. Such an assumption is valid if a polarizer aligned with the signal polarization is placed before the photodetector that removes the noise in the orthogonal polarization.

6.5.5 Amplifier Noise Figure

The noise figure is commonly used to characterize the noise added by an amplifier. It is defined as the ratio of the electrical SNR at the amplifier input to that at the amplifier output [3, 4],

$$F_n = \frac{(\text{SNR})_{\text{in}}}{(\text{SNR})_{\text{out}}}. \tag{6.90}$$

See Fig. 6.7. Since $(\text{SNR})_{\text{out}}$ can never exceed $(\text{SNR})_{\text{in}}$, the noise figure is greater than unity. Note that the SNRs appearing in the above equation are measured in the electrical domain using photodetectors at the input and output of the amplifiers, and measuring the electrical signal and noise powers. To minimize the parameters of the measurement unit entering into the definition of F_n, ideal photodetectors with 100% quantum efficiency are used, and thermal noise is ignored. Let us first consider $(\text{SNR})_{\text{in}}$. When the incident power is P_{in}, the photocurrent is

$$I_{\text{in}} = RP_{\text{in}}, \tag{6.91}$$

where responsivity is given by Eq. (5.17),

$$R = \frac{\eta}{hf_0} = \frac{q}{hf_0}, \tag{6.92}$$

assuming $\eta = 1$. Here, f_0 is the carrier frequency. The electrical signal power delivered to a load resistor R_L is

$$S_{\text{in}} = I_{\text{in}}^2 R_L. \tag{6.93}$$

We assume that there is no noise in the optical signal before the amplifier. The noise power at the output of PD_1 is due to the shot noise and is given by Eq. (5.79),

$$N_{\text{shot}} = 2qI_{\text{in}}R_L B_e. \tag{6.94}$$

Here, we have ignored the dark current. Therefore, we have

$$\text{SNR}_{\text{in}} = \frac{S_{\text{in}}}{N_{\text{shot}}} = \frac{RP_{\text{in}}}{2qB_e}. \tag{6.95}$$

Next, consider SNR_{out}. The output optical power of the amplifier is

$$P_{\text{out}} = GP_{\text{in}}, \tag{6.96}$$

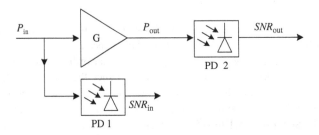

Figure 6.7 Measurement of the amplifier noise figure.

and the corresponding photocurrent and electrical signal power are

$$I_{out} = RGP_{in},$$ (6.97)

$$S_{out} = (RGP_{in})^2 R_L,$$ (6.98)

respectively. The noise power delivered to a resistor R_L consists mainly of two components. They are due to shot noise and signal−ASE beating noise. In this analysis, we ignore the ASE−ASE beating noise and assume that the optical filter is absent. The total noise power can be obtained by adding the shot noise given by Eq. (6.94) and the signal−ASE beating noise power given by Eq. (6.60),

$$N_{out} = N_{shot} + N_{s-sp}$$

$$= 2qRP_{out}R_L B_e + 4R^2 \rho_{ASE} P_{out} B_e R_L.$$ (6.99)

Note that the power spectral density of the ASE noise in Eq. (6.99) is in single polarization. Although the amplifier adds noise in both polarizations, the noise in the polarization orthogonal to the signal polarization does not interfere with the signal to generate signal−ASE beat noise. The SNR at the output of PD_2 can be written as

$$SNR_{out} = \frac{S_{out}}{N_{out}} = \frac{(RGP_{in})^2 R_L}{[q + 2R\rho_{ASE}]2RGP_{in} B_e R_L}$$

$$= \frac{RGP_{in}}{(q + 2R\rho_{ASE})2B_e}.$$ (6.100)

Substituting Eqs. (6.95) and (6.100) in Eq. (6.90), we find

$$F_n = \frac{RP_{in}}{2qB_e} \frac{(q + 2R\rho_{ASE})2B_e}{RGP_{in}}$$

$$= \frac{q + 2R\rho_{ASE}}{Gq}.$$ (6.101)

Using Eq. (6.92), Eq. (6.101) can be written as

$$\rho_{ASE} = (GF_n - 1)hf_0/2.$$ (6.102)

The power spectral density ρ_{ASE} can also be expressed in terms of n_{sp} (see Eq. (6.17)),

$$\rho_{ASE} = n_{sp}(G - 1)hf_0.$$ (6.103)

Equating the right-hand sides of Eqs. (6.102) and (6.103), we find an expression that relates the amplifier noise figure and spontaneous emission factor n_{sp},

$$F_n = \frac{2n_{sp}(G - 1)}{G} + \frac{1}{G}.$$ (6.104)

When $G \gg 1$,

$$F_n \cong 2n_{sp}.$$ (6.105)

Since the minimum value of n_{sp} is 1, the lowest achievable noise figure is 2. The noise figure is expressed in dB units as

$$F_n(dB) = 10 \log_{10} F_n.$$ (6.106)

When $n_{sp} = 1$, $F_n = 3$ dB, which corresponds to an ideal amplifier with the lowest ASE noise.

Example 6.4

An optical amplifier at 1550 nm has a noise figure of 4.5 dB. The signal output of the amplifier is 0 dBm, which is incident on a photodetector. Calculate the amplifier gain if the standard deviation of the signal–ASE beat noise current is 0.066 mA. Assume $R = 0.9$ A/W, $B_e = 7.5$ GHz, and the optical filter is absent.

Solution:

$$P_{out}(dBm) = 0 \text{ dBm},$$

$$P_{out} = 10^{P_{out}(dBm)/10} \text{ mW} = 1 \text{ mW}.$$

From Eq. (6.54), we have

$$\rho_{ASE} = \frac{\sigma_{s-sp}^2}{4R^2 B_e P_{out}} = \frac{(0.066 \times 10^{-3})^2}{4 \times 0.9^2 \times 7.5 \times 10^9 \times 1 \times 10^{-3}} \text{ W/Hz}$$

$$= 1.79 \times 10^{-16} \text{ W/Hz},$$

$$F_n = 10^{F_n(dB)/10} = 2.818,$$

$$f_0 = \frac{c}{\lambda} = 193.55 \text{ THz}.$$

Using Eq. (6.102), we find

$$G = \frac{1}{F_n}\left(\frac{2\rho_{ASE}}{h f_0} + 1\right) = \frac{1}{2.818}\left(\frac{2 \times 1.79 \times 10^{-16}}{6.626 \times 10^{-34} \times 193.55 \times 10^{12}} + 1\right) = 992.$$

6.5.6 Optical Signal-to Noise Ratio

The noise added by an amplifier is characterized by the noise figure, which is the ratio of electrical SNRs at the input and output of the amplifier. The noise added by the amplifier may also be characterized by the *optical signal-to-noise ratio* (OSNR), defined as

$$\text{OSNR} = \frac{\text{mean signal power}}{\text{mean noise power in a bandwidth of 0.1 nm}}. \tag{6.107}$$

At 1550 nm, 0.1 nm corresponds to $B_{opt} = 12.49$ GHz and the mean noise power in the bandwidth of B_{opt} is

$$P_{ASE} = 2\rho_{ASE}B_{opt}, \tag{6.108}$$

$$\text{OSNR} = \frac{P_{out}}{P_{ASE}}. \tag{6.109}$$

Or in decibels,

$$\text{OSNR (dB)} = 10 \log_{10}\text{OSNR}, \tag{6.110}$$

The factor 2 is introduced in Eq. (6.108) to account for two polarizations. Note that B_{opt} is not the same as the effective bandwidth of the optical filter B_o defined in Eq. (6.22). B_{opt} is a reference bandwidth used in the definition of OSNR, which may or may not be equal to B_o.

Example 6.5

An amplifier operating at 1545 nm has a gain $G = 25$ dB, $F_n = 6$ dB, and input power $P_{in} = -22$ dBm. Calculate the OSNR in a bandwidth of 12.49 GHz.

Solution:

$$P_{out} = GP_{in},$$

$$P_{out}(dBm) = G(dB) + P_{in}(dBm)$$

$$= 25 \text{ dB} - 22 \text{ dBm}$$

$$= 3 \text{ dBm},$$

$$P_{out} = 10^{P_{out}(dBm)/10} \text{ mW}$$

$$= 2 \text{ mW}.$$

From Eq. (6.102), we have

$$\rho_{ASE} = (GF_n - 1)\frac{hf_0}{2},$$

$$F_n = 10^{F_n(dB)/10} = 3.98,$$

$$G = 10^{G(dB)/10} = 316.22,$$

$$f_0 = \frac{c}{\lambda_0} = \frac{3 \times 10^8}{1545 \times 10^{-9}} = 194.17 \text{ THz},$$

$$\rho_{ASE} = (316.22 \times 3.98 - 1) \times 6.626 \times 10^{-34} \times 194.17 \times 10^{12}/2 \text{ W/Hz}$$

$$= 8.09 \times 10^{-17} \text{ W/Hz}$$

$$B_{opt} = 12.49 \text{ GHz},$$

$$\text{OSNR} = \frac{P_{out}}{2\rho_{ASE} B_{opt}} = \frac{2 \times 10^{-3}}{2 \times 8.09 \times 10^{-17} \times 12.49 \times 10^9} = 989,$$

$$\text{OSNR(dB)} = 10 \log_{10}(\text{OSNR}) = 29.95 \text{ dB}.$$

6.6 Semiconductor Optical Amplifiers

A semiconductor optical amplifier (SOA) or semiconductor laser amplifier (SLA) is nothing but a laser operating slightly below threshold. The optical field incident on one facet is amplified at the other accompanied by the ASE. The SOAs can be divided into two types: (i) cavity-type SOA or Fabry–Perot amplifier (FPA), (ii) traveling wave amplifier (TWA).

6.6.1 Cavity-Type Semiconductor Optical Amplifiers

Let the power reflectivities of mirrors M_1 and M_2, shown in Fig. 6.8, be R_1 and R_2, respectively. Assuming that the power is conserved at each mirror, the corresponding power transmittivities at mirror M_j are given by

$$T_j = 1 - R_j, \quad j = 1, 2. \tag{6.111}$$

The optical field transmitted at A is $t_1 \psi_{\text{in}}$, where $\psi_{\text{in}} = \sqrt{P_{\text{in}}}$ and $|t_j| = \sqrt{T_j}$, $j = 1, 2$. Let g be the gain coefficient and α_{int} be the cavity internal loss. The net gain coefficient is $g_s = \Gamma g - \alpha_{\text{int}}$, where Γ is the overlap factor introduced in Eq. (3.109). As shown in Fig. 6.9, the partial optical field ψ_0 at B after a single pass is

$$\psi_0 = \psi_{\text{in}} t_1 t_2 \sqrt{G_s} \exp(i\phi_0), \tag{6.112}$$

where $\phi_0 = 2\pi n L / \lambda$ is the phase-shift due to propagation, n is the refractive index of the gain medium, and λ is the free-space wavelength. In the small-signal limit, the single-pass gain is

$$G_s = \exp(g_s L). \tag{6.113}$$

A fraction of the optical field is reflected at the mirror M_2 and then at M_1. After one round trip, the partial field at B is (see Fig. 6.9)

$$\psi_1 = \psi_{\text{in}} t_1 r_2 r_1 t_2 [\sqrt{G_s} \exp(i\phi_0)]^3, \tag{6.114}$$

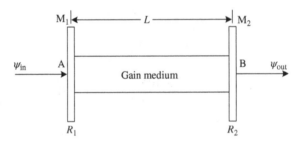

Figure 6.8 Cavity-type semiconductor optical amplifier.

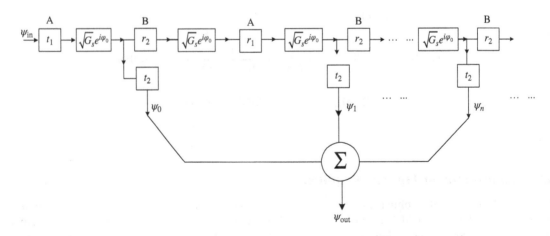

Figure 6.9 The optical signal output of the amplifier is the sum of the partial fields due to repeated reflections.

where $|r_j| = \sqrt{R_j}, j = 1, 2$. After j round trips, the partial field at B is

$$\psi_j = \psi_{in} t_1 t_2 (r_1 r_2)^j [\sqrt{G_s} \exp(i\phi_0)]^{2j+1}. \tag{6.115}$$

The total field output at B is the superposition of partial fields,

$$\psi_{out} = \sum_{j=0}^{\infty} \psi_j = t_1 t_2 \psi_{in} \sqrt{G_s} \exp(i\phi_0) \sum_{j=0}^{\infty} h^j, \tag{6.116}$$

where

$$h = r_1 r_2 G_s \exp(i2\phi_0). \tag{6.117}$$

The summation in Eq. (6.116) is a geometric series and if

$$|h| < 1, \tag{6.118}$$

we have

$$\sum_{j=0}^{\infty} h^j = \frac{1}{1-h}. \tag{6.119}$$

Therefore, Eq. (6.116) becomes

$$\psi_{out} = \frac{\psi_{in} t_1 t_2 \sqrt{G_s}}{1-h} \exp(i\phi_0). \tag{6.120}$$

The overall gain G is defined as

$$G(f) = \frac{|\psi_{out}|^2}{|\psi_{in}|^2} = \frac{|t_1|^2 |t_2|^2 G_s(f)}{[1-h(f)][1-h^*(f)]}. \tag{6.121}$$

Using Eq. (6.117), Eq. (6.121) can be rewritten as

$$G(f) = \frac{(1-R_1)(1-R_2)G_s(f)}{1 + R_1 R_2 G_s^2(f) - 2\sqrt{R_1 R_2} G_s(f) \cos(2\phi_0)}. \tag{6.122}$$

Using the relation

$$\cos(2\phi_0) = 1 - 2\sin^2 \phi_0, \tag{6.123}$$

Eq. (6.122) can be put in a different form:

$$G(f) = \frac{(1-R_1)(1-R_2)G_s(f)}{(1-RG_s)^2 + 4RG_s \sin^2(2\pi n f L/c)}, \tag{6.124}$$

where $R = \sqrt{R_1 R_2}$ is the geometric mean of facet reflectivities. From Eq. (6.124), we see that the peak gain occurs when

$$\frac{2\pi n f L}{c} = m\pi, \quad m = 0, \pm 1, \pm 2, \ldots \tag{6.125}$$

or

$$f_m = \frac{mc}{2nL}, \tag{6.126}$$

which is the same as the resonant frequency given by Eq. (3.44). Therefore, the cavity-type optical amplifier amplifies any input signal whose frequency is matched to the resonant frequency f_m of the cavity. The peak gain occurs when the signal frequency is equal to one of the resonant frequencies given by Eq. (6.126), and it is given by

$$G_{\text{peak}} = \frac{(1 - R_1)(1 - R_2)G_s(f)}{(1 - RG_s)^2}. \tag{6.127}$$

The separation between two peaks is known as the *free spectral range* (FSR):

$$\text{FSR} = f_{m+1} - f_m = \frac{c}{2nL}. \tag{6.128}$$

Fig. 6.10 shows the gain of the amplifier as a function of the frequency of the input signal. When the phase accumulated in a round trip is $2m\pi$, the partial fields add up coherently, leading to signal amplification. As the frequency of the input field deviates from the resonant frequency, $mc/2nL$, the gain decreases. Fig. 6.11 shows the gain within a free spectral range for different values of R. As can be seen, as R decreases, the bandwidth increases and the peak gain decreases. For example, on a hot day, if you leave your car outside with all the windows closed, it becomes too hot (large gain) because of the repeated reflections of radiation within the car. If you open one of the windows, the reflectivity is reduced, which lowers the gain. The common characteristic of cavity-type amplifiers is that there exists a trade-off between gain and bandwidth. To see that, let us first define the half-width at half-maximum (HWHM) as the frequency deviation Δf from f_m at which the $G(f)$ becomes half of the peak gain, $G(f_m)$, i.e.,

$$G(f_m + \Delta f) = 0.5G_{\text{peak}}. \tag{6.129}$$

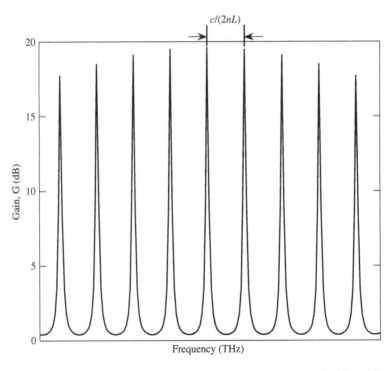

Figure 6.10 Gain of a cavity-type SOA vs. frequency. When the signal frequency coincides with one of the resonant frequencies of the cavity, the gain is maximum.

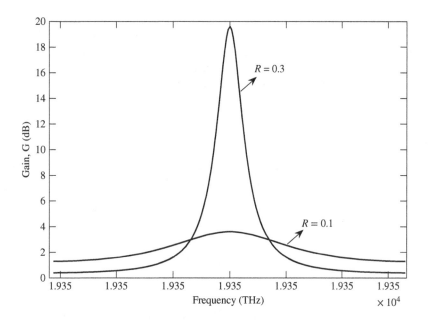

Figure 6.11   Gain–bandwidth trade-off in cavity-type SOA.

Using Eq. (6.127), Eq. (6.124) can be written as

$$G(f) = \frac{G_{\text{peak}}}{1 + 4RG_s \sin^2(2\pi nfL/c)/(1 - RG_s)^2}. \qquad (6.130)$$

Using Eqs. (6.129) and (6.130) and noting that $2\pi n f_{\text{m}} L/c = m\pi$, we obtain

$$\sin^2\left(\frac{2\pi \Delta f L n}{c}\right) = \frac{(1 - RG_s)^2}{4RG_s}. \qquad (6.131)$$

The FWHM is defined as $2\Delta f$. The FWHM is also known as the 3-dB bandwidth. From Eq. (6.131), we have

$$f_{3\text{ dB}} = 2\Delta f = \frac{c}{\pi Ln} \sin^{-1}\left\{\frac{1 - RG_s}{2\sqrt{RG_s}}\right\}. \qquad (6.132)$$

In obtaining Eq. (6.132), we have assumed that the single-pass gain G_s varies much more slowly with frequency compared with variations of G, and G_s can be treated as a constant. From Eq. (3.39), it follows that at the threshold the total cavity loss equals the net gain, i.e., $RG_s = 1$. Since $|h| = RG_s$, Eq. (6.118) corresponds to the situation where the amplifier is biased below threshold and the expression for the gain given by Eq. (6.124) is valid only if $RG_s < 1$. Typically, the amplifier is biased slightly below threshold and $1 - RG_s$ is much smaller than unity. Using $\sin x \cong x$, for $x \ll 1$, Eq. (6.132) can be approximated as

$$f_{3\text{ dB}} \approx \frac{c}{2\pi Ln} \frac{1 - RG_s}{\sqrt{RG_s}}. \qquad (6.133)$$

From Eq. (6.127), we find that the peak gain increases as RG_s approaches unity. However, from Eq. (6.133), it follows that the 3-dB bandwidth becomes quite small under this condition. Thus, there is a trade-off between gain and bandwidth. As an example, consider $R_1 = R_2 = 0.3$, $G_s = 2.5$, and $L = 300$ μm. Using Eqs. (6.127) and (6.132), we find $G_{peak} = 20$ and $f_{3\ dB} = 36$ GHz. For optical communication systems, the desired amplifier bandwidth is typically greater than 1 THz and, therefore, a cavity-type semiconductor amplifier is unsuitable as an in-line amplifier for high-bit-rate optical communication systems. In addition, they are very sensitive to fluctuations in bias current, temperature, and polarization of the incident field [5].

Example 6.6

In a cavity-type SOA, the cavity length is 500 μm, $R_1 = R_2 = 0.32$, and the peak gain is 15 dB. Find the single-pass gain and the 3-dB bandwidth. Assume $n = 3.2$.

Solution:

$$G_{peak} = 15 \text{ dB},$$

$$G_{peak} = 10^{G_{peak}(dB)/10} = 31.62.$$

From Eq. (6.127), we have

$$(1 - RG_s)^2 G_{peak} = (1 - R_1)(1 - R_2)G_s,$$

$$1 + 0.32^2 G_s^2 - 2 \times 0.32 G_s = 0.0146 G_s,$$

or

$$0.1024 G_s^2 - 0.6546 G_s + 1 = 0.$$

This quadratic equation has solutions

$$G_s = 2.52 \quad \text{or} \quad 3.86.$$

Eq. (6.127) is derived under the assumption that $|h| < 1$ or $RG_s < 1$. When $G_s = 3.86$, $RG_s > 1$, and therefore it is not consistent with Eq. (6.127) and this solution is rejected.
The 3-dB bandwidth is given by Eq. (6.132),

$$f_{3\ dB} = \frac{c}{\pi Ln} \sin^{-1}\left(\frac{1 - RG_s}{2\sqrt{RG_s}}\right)$$

$$= \frac{3 \times 10^8}{\pi \times 500 \times 10^{-6} \times 3.2} \sin^{-1}\left(\frac{1 - 0.32 \times 2.52}{2(0.32 \times 2.52)^{1/2}}\right)$$

$$= 6.44 \text{ GHz}.$$

6.6.2 Traveling-Wave Amplifiers

From Fig. 6.11, it can be seen that as R decreases, the bandwidth increases. In the limiting case of $R = 0$, Eq. (6.124) becomes $G = G_s$, i.e., the overall gain is equal to the single-pass gain G_s. This should be expected since there are no partial fields due to round trips when $R = 0$. Such an amplifier is known as a traveling-wave amplifier (TWA).

It is useful to calculate the ratio of the peak gain to the minimum gain. From Eq. (6.124), we find that the gain is minimum when $\sin^2\phi_0 = 1$, and it is given by

$$G_{\min} = \frac{(1 - R_1)(1 - R_2)G_s(f)}{(1 + RG_s)^2},$$

(6.134)

and $G_{\max} = G_{\text{peak}}$. Using Eqs. (6.127) and (6.134), we find the gain ripple as

$$\Delta G = \frac{G_{\max}}{G_{\min}} = \frac{(1 + RG_s)^2}{(1 - RG_s)^2}.$$

(6.135)

Or in decibels,

$$\Delta G(\text{dB}) = G_{\max}(\text{dB}) - G_{\min}(\text{dB}).$$

(6.136)

Fig. 6.12 shows the gain as a function of frequency of the input optical field and the gain ripple ΔG is the separation between the points corresponding to the maximum and minimum gains. For example, when $RG_s = 0.9$, ΔG is 25.5 dB. The fluctuations in gain as a function of frequency are undesirable for wide-band amplifiers. To keep the gain ripple quite small, $RG_s \ll 1$, which can be achieved by reducing the reflectivities of the end facets. To have $\Delta G < 3$ dB, $RG_s \leq 0.17$, which can be achieved by reducing the facet reflectivities.

From Eq. (6.135), we find that the gain ripple ΔG of an ideal TWA ($R = 0$) is 0 dB and it has a large bandwidth determined solely from the characteristics of the gain medium. However, in practice, even with the best antireflection (AR) coatings, there is some residual reflectivity. Therefore, some authors [5, 6] use the term nearly traveling-wave amplifier (NTWA) to denote an amplifier with $RG_s \leq 0.17$. For a NTWA, the gain ripple $\Delta G \leq 3$ dB. A NTWA has been fabricated with $R = 4 \times 10^{-4}$ [7] and it has a 3-dB bandwidth

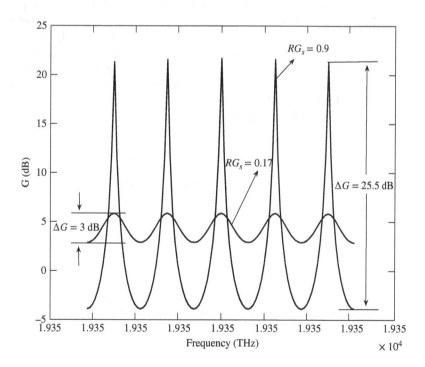

Figure 6.12   The gain ripple ΔG increases as RG_s approaches unity.

about 9 THz and a gain ripple ΔG of about 1.5 dB. Owing to the non-resonant character of the NTWA, it was found to be less sensitive to temperature, bias current, and input signal polarization fluctuations compared with the FPA.

In order to reduce the reflectivities of the end facets, AR coatings may be applied to the end facets. The optical field transmission at a dielectric interface is accompanied by reflection. However, if we deposit an intermediate layer (AR coating) between the two dielectric media, it is possible to avoid reflections. In the following section, the principle behind the AR coating is discussed.

6.6.3 AR Coating

The purpose of an AR coating is to avoid the reflection occurring at the interface between two dielectric media. This reflection can be avoided by introducing an AR coating between the dielectrics with different permittivities. Consider a three-layered dielectric medium as shown in Fig. 6.13. The optical field incident from the left gets partially reflected at the interface 1 and the rest is transmitted. Let the reflected field at the interface 1 be ψ_1. At the interface 2, the field gets partially reflected again. Let the reflected field at the interface 2 arriving at interface 1 be ψ_2. If the reflected fields add destructively, there will be no reflection to the left of the interface 1, i.e., $\psi_1 + \psi_2 = 0$. For destructive interference, the phase accumulated over a round trip in the middle layer should be $(2l + 1)\pi, l = 0, \pm1, \pm2, \ldots$ The acquired phase in a round trip is

$$\phi = 2k_2w = \frac{4\pi n_2 w}{\lambda_0} = (2l + 1)\pi \quad \text{or} \quad w = \frac{(2l + 1)\lambda_0}{4n_2}, \tag{6.137}$$

where k_2 is the propagation constant of the middle layer with refractive index n_2, λ_0 is the free-space wavelength, and w is the width of the middle layer. Thus, by introducing a coating of width $(2l + 1)\lambda_0/4n_2$ between two media of refractive indices n_1 and n_3, the backward reflection can be avoided and therefore the power of optical signal transmitted to layer 3 is the same as the incident power to layer 1. The middle layer is known as the AR coating. In the absence of the AR coating, the transmitted power is less than the incident power.

In some other application, it may be desired to have the power of the incident wave the same as the reflected wave at interface 1 and no optical field should be transmitted to the medium with refractive index n_3. To make this happen, the field reflected at interface 2 should add in-phase with the field reflected at interface 1. In this case, the phase accumulated over a round trip in the middle layer should be $2l\pi, l = 0, \pm1, \pm2, \ldots$, i.e.,

$$\phi = \frac{4\pi w n_2}{\lambda_0} = 2l\pi \quad \text{or} \quad w = \frac{l\lambda_0}{2n_2}. \tag{6.138}$$

In this case, the structure shown in Fig. 6.13 acts as a dielectric mirror.

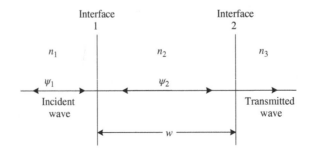

Figure 6.13 Three-layered dielectric structure.

The reflectivity of the AR coating is quite sensitive to the width w and refractive index n_2 of the AR coating. To increase the tolerance, a multilayer AR coating can be used. The experimental results of Ref. [8] show that the tolerances in w and n_2 of a single-layer AR coating for realizing the power reflectivity $R \leq 1 \times 10^{-3}$ are about $\pm 60^{\circ}$ A and ± 0.05, respectively. The double-layer AR coating has larger tolerances, $\pm 90^{\circ}$ A in widths and ± 0.3 in refractive index, for the same power reflectivity as in the single-layer AR coating. In this section, we have assumed that the optical field is a plane wave and as a result, we obtained a simple expression for the width of the AR coating. However, in a waveguide, plane waves should be replaced by the modes of the waveguide and the reflectivity should be calculated [8].

6.6.4   Gain Saturation

As the input signal power increases beyond a certain threshold, the gain G decreases for both cavity-type FPA and TWA. This is known as gain saturation. This phenomenon can be explained as follows. When the population inversion is achieved, the stimulated emission dominates the absorption. Since the stimulated emission rate is proportional to photon density, a larger input signal power enhances the stimulated emission and, therefore, the excited carriers are depleted and the gain decreases. Under steady-state conditions, we can set dN_e/dt to zero in Eq. (3.123) to obtain

$$G_0(N_e - N_{e,0})N_{ph} + \frac{N_e}{\tau_e} = \frac{I}{qV}, \tag{6.139}$$

where

$$G_0 = \Gamma \sigma_g v. \tag{6.140}$$

Simplifying Eq. (6.139), we find

$$N_e = \frac{I/qV + G_0 N_{e,0} N_{ph}}{G_0 N_{ph} + 1/\tau_e}, \tag{6.141}$$

$$g = \sigma_g(N_e - N_{e,0}) = \frac{(I/qV - N_{e,0}/\tau_e)\tau_e \sigma_g}{G_0 N_{ph}\tau_e + 1}. \tag{6.142}$$

Eq. (6.142) can be rewritten as

$$\Gamma g = \frac{\Gamma g_0}{1 + N_{ph}/N_{ph,\text{sat}}}, \tag{6.143}$$

where

$$g_0 = (I/qV - N_{e,0}/\tau_e)\sigma_g \tau_e, \tag{6.144}$$

$$N_{ph,\text{sat}} = \frac{1}{\Gamma \sigma_g v \tau_e}. \tag{6.145}$$

The optical power P and photon density N_{ph} are related by (Eq. (3.136))

$$P = N_{ph}\hbar\omega_0 vA. \tag{6.146}$$

So, Eq. (6.143) may be rewritten in terms of P as

$$\Gamma g = \frac{\Gamma g_0}{1 + P/P_{\text{sat}}}, \tag{6.147}$$

$$P_{\text{sat}} = \frac{\hbar\omega_0 A}{\Gamma \sigma_g \tau_e}. \tag{6.148}$$

P_{sat} is known as the saturation power. When $P \ll P_{sat}$, $g \approx g_0$. Therefore, g_0 is known as a small signal gain. When P is comparable with P_{sat}, g decreases. This is because, as the photon density increases, the stimulated emission rate increases, which depletes the carriers in the conduction band and from Eq. (6.142) it follows that g decreases as N_e decreases. The evolution of power in the amplifier is given by

$$\frac{dP}{dz} = \left(\frac{\Gamma g_0}{1 + P/P_{sat}} - \alpha \right) P. \tag{6.149}$$

The single-pass gain is

$$G_s = \frac{P(L)}{P(0)}. \tag{6.150}$$

When the internal loss ($\alpha \approx 0$) is small, Eq. (6.149) may be rewritten as

$$\left(\frac{1 + P/P_{sat}}{P} \right) dP = \Gamma g_0 dz. \tag{6.151}$$

Integrating Eq. (6.151) from 0 to amplifier length L, we obtain

$$\int_{P(0)}^{P(L)} \left[\frac{1}{P} + \frac{1}{P_{sat}} \right] dP = \int_0^L \Gamma g_0 dz,$$

$$\ln \frac{P(L)}{P(0)} + \frac{P(L) - P(0)}{P_{sat}} = \Gamma g_0 L. \tag{6.152}$$

Let

$$G_{s_0} = \exp(\Gamma g_0 L) \tag{6.153}$$

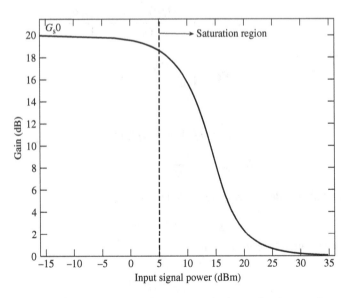

Figure 6.14 Dependence of gain on the input signal power. As the input signal power increases, the gain becomes smaller. $P_{sat} = 0$ dBm, $G_{s_0} = 20$ dB, and the internal loss is ignored.

be the small signal gain. Using Eqs. (6.150) and (6.153), Eq. (6.152) can be expressed as

$$G_s = G_{s_0} \exp\left[-\frac{(G_s - 1)P(L)}{G_s P_{sat}}\right]. \tag{6.154}$$

When $P(0)(= P(L)/G_s) \ll P_{sat}$, the exponent in Eq. (6.154) is close to zero and $G_s \approx G_{s_0}$. Fig. 6.14 shows the gain G_s as a function of the input signal power $P(0)$. When $P(0)$ exceeds P_{sat}, the gain G_s decreases from its unsaturated value G_{s0}.

For a TWA, the overall gain $G \approx G_s$. For a cavity-type SOA, the gain is given by Eq. (6.124). For both types of amplifier, G decreases as the input power increases due to gain saturation. The saturation power for a TWA is higher than that for a cavity-type SOA [7]. This is because the electron lifetime τ_e is lower at higher carrier density and from Eq. (6.147), we see that P_{sat} is inversely proportional to τ_e. The carrier density is higher for a TWA since G_s can be much larger and still $RG_S < 1$.

Example 6.7

A 1530-nm TWA has the following parameters:
Effective area of mode A = 5 μm^2
Active volume = 7.5×10^{-16} m^3
Carrier lifetime = 1 ns
Gain cross-section $\sigma_g = 7.3 \times 10^{-20}$ m^2
$N_{e,0} = 3.5 \times 10^{23}$ m^{-3}
Overlap factor $\Gamma = 0.3$

Calculate (a) the saturation power and (b) the bias current I to have the small signal gain coefficient $g_0 = 4.82 \times 10^3$ m^{-1}.

Solution:
The saturation power is given by Eq. (6.148),

$$P_{sat} = \frac{hf_o A}{\Gamma \sigma_g \tau_e}, \tag{6.155}$$

$$f_0 = \frac{c}{\lambda_0} = \frac{3 \times 10^8}{1530 \times 10^{-9}} = 196.08 \text{ THz}, \tag{6.156}$$

$$P_{sat} = \frac{6.626 \times 10^{-34} \times 196.08 \times 10^{12} \times 5 \times 10^{-12}}{0.3 \times 7.3 \times 10^{-20} \times 1 \times 10^{-9}} \text{ W}$$

$$= 29.7 \text{ mW}. \tag{6.157}$$

The relation between g_0 and I is given by Eq. (6.144),

$$g_0 = \left(\frac{I}{qV} - \frac{N_{e,0}}{\tau_e}\right)\sigma_g \tau_e, \tag{6.158}$$

$$I = \left[\frac{g_0}{\sigma_g \tau_e} + \frac{N_{e,o}}{\tau_e}\right] qV$$

$$= \left[\frac{4.82 \times 10^3}{7.3 \times 10^{-20} \times 1 \times 10^{-9}} + \frac{3.5 \times 10^{23}}{1 \times 10^{-9}}\right] \times 1.602 \times 10^{-19} \times 7.5 \times 10^{-16} \text{ A}$$

$$= 50 \text{ mA}.$$

6.7   Erbium-Doped Fiber Amplifier

Optical fibers have the lowest loss around 1550 nm, and therefore the operating wavelength of long-haul fiber-optic systems is around 1550 nm. It is desirable to have the signal amplification around 1550 nm. It is found that Er^{3+} ions have an excited state which is separated from the ground state by the energy difference corresponding to a wavelength of $\sim$ 1530 nm and, therefore, an information-bearing signal at the carrier wavelength around 1550 nm can be amplified by means of stimulated emission if the population inversion is achieved. Optical pumps at the wavelength of 980 nm or 1480 nm are typically used to achieve the population inversion. Erbium-doped fibers (EDFs) are the silica optical fibers doped with erbium. Sometimes they are co-doped with aluminum to increase the solubility of the Er^{3+}. Fig. 6.15 shows the structure of a typical EDFA. The signal to be amplified is combined with the pump beam at either 1480 nm or 980 nm using a wavelength-selective coupler (WSC). Semiconductor laser diodes are used as the optical pumps. For ruby lasers or Nd–Yag lasers, a flash light with a broad spectrum can be used as the optical pump whereas for EDFA, a flash light can not be used because the linewidth of the energy band that absorbs the pump is quite narrow and therefore a semiconductor laser with narrow linewidth is used as the pump. In the EDFA, the signal is amplified and the pump is attenuated. An optical isolator is used at the output end so that the reflections occurring at various points along the fiber-optic transmission lines after the EDFA should not interfere with the signal inside the amplifier. The configuration shown in Fig. 6.15 is known as a forward-pumping scheme. There are other configurations in which the pump propagates backward, or two pumps with one pump propagate forward and the other propagates backward [1, 9].

6.7.1   Gain Spectrum

An EDFA has a very broad spectrum ($\sim$ 30 nm). The broad bandwidth of EDFAs makes them useful in WDM systems (see Chapter 9). If the excited level 2 does not have any energy sublevels, the optical signal corresponding to the energy difference between the ground level and level 2 would be the only frequency

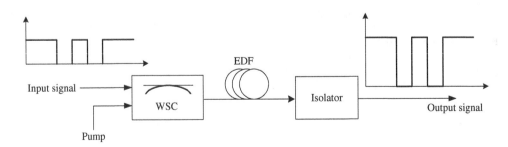

Figure 6.15   Erbium-doped fiber amplifier. WSC = wavelength selective coupler, EDF = erbium-doped fiber.

component amplified. This corresponds to the very narrow (impulse) gain spectrum. However, level 2 is a collection of sublevels. Broadening of the energy levels occurs when the erbium ions are incorporated into the glass of optical fibers and, thus, the gain spectrum is also broadened. This broadening is both *homogeneous* (all erbium ions exhibit the identical broadened spectrum) and *inhomogeneous* (different ions in different glass locations exhibit different spectra). Homogeneous broadening is due to the interactions with photons of the glass, whereas inhomogeneous broadening is caused by differences in the glass sites where different ions are hosted.

6.7.2 Rate Equations*

Consider a three-level system as shown in Fig. 6.16. An optical pump beam of frequency ω_p causes upward transitions from level 1 to level 3. Let the population densities of the level j be $N_j, j = 1, 2, 3$. The erbium ions excited to level 3 relax to level 2 by spontaneous emission and non-radiative processes. In practice, it is mostly non-radiative. The stimulated emission occurring between level 2 and level 1 is responsible for the signal amplification. Let the lifetime associated with spontaneous emission and non-radiative processes between any levels j and k be τ_{jk}. First consider the gain and loss rates for level 3. The population density of level 3 increases because of the net absorption of pump photons and it decreases because of the non-radiative emission,

$$\frac{dN_3}{dt} = R_{abs} + R_{stim} + R_{nr} + R_{sp}. \tag{6.159}$$

Consider the pump absorption. From the Einstein relation (see Eqs. (3.2) and (3.30)), we have

$$R_{abs} = B_{13} N_1 u_p, \tag{6.160}$$

where u_p is the energy density of the pump. We will write Eq. (6.160) in a slightly different form. The pump intensity $\mathcal{I}_p$ and energy density are related by (Eq. (3.50))

$$u_p = \frac{\mathcal{I}_p}{\upsilon}, \tag{6.161}$$

where υ is the speed of light in the medium. The photon flux density is defined as the mean number of photons per unit area per unit time. In other words, if n_p photons cross the area A over the time interval Δt, the photon flux density is

$$\phi_p = \frac{n_p}{A \Delta t}. \tag{6.162}$$

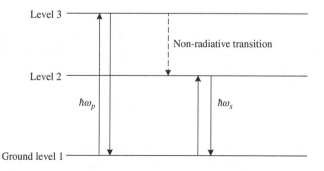

Figure 6.16   A three-level system. Signal amplification in erbium-doped fiber.

If a light wave of energy E_p crosses the area A over the time interval Δt, the optical intensity is

$$\mathcal{I}_p = \frac{E_p}{A\Delta t}. \tag{6.163}$$

Since the energy $E_p = n_p\hbar\omega_p$, where ω_p is the frequency of the pump wave, we find

$$\phi_p = \frac{\mathcal{I}_p}{\hbar\omega_p}. \tag{6.164}$$

Using Eqs. (6.161) and (6.164) in Eq. (6.160), we find

$$R_{abs} = \sigma_{13}N_1\phi_p, \tag{6.165}$$

where

$$\sigma_{13} = \frac{\hbar\omega_p B_{13}}{v} \tag{6.166}$$

is known as the *absorption cross-section* associated with the transition from level 1 to level 3. The physical meaning of σ_{13} is as follows. The optical power absorbed by an erbium ion is proportional to the optical intensity $\mathcal{I}_p$ of the light wave incident,

$$P_{abs} = k\mathcal{I}_p, \tag{6.167}$$

where k is a constant of proportionality that depends on the medium. Since $P_{abs}/\hbar\omega_p$ is the number of photons absorbed per unit time by an erbium ion (photon absorption rate) and $\mathcal{I}_p/\hbar\omega_p$ is the photon flux density (Eq. (6.164)), dividing Eq. (6.167) by $\hbar\omega_p$, we find

$$\frac{P_{abs}}{\hbar\omega_p} = k\phi_p. \tag{6.168}$$

If there are N_1 erbium ions per unit volume in the ground state, the total absorption rate is

$$R_{abs} = kN_1\phi_p, \tag{6.169}$$

which is the same as Eq. (6.165) if $k = \sigma_{13}$. Thus, the absorption cross-section can be imagined as an effective area that "captures" a fraction of the incident photons [1]. Similarly, the stimulated emission rate from level 3 to level 1 is given by

$$R_{stim} = -\sigma_{31}N_3\phi_p, \tag{6.170}$$

where $\sigma_{jk} = B_{jk}\hbar\omega_{jk}/v$ is the cross-section associated with the transition from j to k and ω_{jk} is the energy difference between the levels j and k. Since the transition from level 3 to level 2 is mostly non-radiative, absorption and stimulated emission between level 3 and level 2 can be ignored. Using Eqs. (6.165) and (6.170) in Eq. (6.159), we find

$$\frac{dN_3}{dt} = (\sigma_{13}N_1 - \sigma_{31}N_3)\phi_p - \frac{N_3}{\tau_{32}}. \tag{6.171}$$

In the case of the two-level atomic system discussed in Chapter 3, Section 3.2, we found that $B_{12} = B_{21}$, which implies that the emission and absorption cross-section are equal. However, in general, they could be different if the two levels are combinations of sublevels that are populated to various extents depending on the thermal distribution [1]. It is straightforward to write the rate equations for population densities N_1 and N_2 as

$$\frac{dN_2}{dt} = \frac{N_3}{\tau_{32}} - (N_2\sigma_{21} - N_1\sigma_{12})\phi_s - \frac{N_2}{\tau_{21}}, \tag{6.172}$$

$$\frac{dN_1}{dt} = (N_2\sigma_{21} - N_1\sigma_{12})\phi_s + (N_3\sigma_{31} - N_1\sigma_{13})\phi_p + \frac{N_2}{\tau_{21}}, \tag{6.173}$$

where ϕ_s is the photon flux density of the signal. Consider the rate equation (6.171). As soon as the pump photons cause transition from level 1 to level 3, erbium ions relax to level 2 by the non-radiative processes involving interaction with phonons of the glass matrix. Therefore, the second term $N_3\sigma_{31}\phi_p$ in Eq. (6.171) that accounts for the stimulated emission can be ignored. Under steady-state conditions, $dN_3/dt = 0$ and from Eq. (6.171), we obtain

$$\frac{N_3}{\tau_{32}} = (N_1\sigma_{13} - N_3\sigma_{31})\phi_p \approx N_1\sigma_{13}\phi_p. \tag{6.174}$$

Substituting Eq. (6.174) into Eqs. (6.172) and (6.173) and ignoring $N_3\sigma_{31}\phi_p$, we obtain [1]

$$\frac{dN_2}{dt} = N_1\sigma_{13}\phi_p - \frac{N_2}{\tau_{21}} - (N_2\sigma_{21} - N_1\sigma_{12})\phi_s, \tag{6.175}$$

$$\frac{dN_1}{dt} = (N_2\sigma_{21} - N_1\sigma_{12})\phi_s - N_1\sigma_{13}\phi_p + \frac{N_2}{\tau_{21}}. \tag{6.176}$$

Note that Eq. (6.175) is similar to Eq. (3.89), corresponding to the two-level system with $R_{\text{pump}} = N_1\sigma_{13}\phi_p$. The erbium ions are excited to level 2 from level 1 by an alternate route, i.e., first they make an upward transition to level 3 from level 1 by absorbing pump photons and they relax to level 2 by means of non-radiative processes. If the population inversion is achieved ($\sigma_{21}N_2 > \sigma_{12}N_1$), the energy of the pump is transferred to the signal.

Adding Eqs. (6.175) and (6.176), we find

$$\frac{d(N_1 + N_2)}{dt} = 0 \quad \text{or} \quad N_1 + N_2 = N_T \quad \text{(a constant)}. \tag{6.177}$$

Here, N_T denotes the erbium ion density. The steady-state solution of Eqs. (6.175) and (6.176) can be obtained by setting

$$\frac{dN_1}{dt} = \frac{dN_2}{dt} = 0. \tag{6.178}$$

Using Eq. (6.177) in Eq. (6.175), we obtain under the steady-state condition

$$N_2 = \frac{N_T[\sigma_{13}\phi_p + \sigma_{12}\phi_s]\tau_{21}}{1 + \sigma_{13}\tau_{21}\phi_p + (\sigma_{12} + \sigma_{21})\phi_s\tau_{21}}. \tag{6.179}$$

The photon flux density ϕ_p and optical power P_p are related by Eq. (6.164),

$$\phi_p = \frac{\mathcal{I}}{\hbar\omega_p} = \frac{P_p}{A_{\text{eff}}\hbar\omega_p}, \tag{6.180}$$

where A_{eff} is the cross-section of the erbium ion distribution. Similarly, we have

$$\phi_s = \frac{P_s}{A_{\text{eff}}\hbar\omega_s}. \tag{6.181}$$

Substituting Eqs. (6.180) and (6.181) in Eq. (6.179), we obtain

$$N_2 = \frac{N_T[P'_s + P'_p]}{1 + P'_p + P'_s(1 + \eta)}, \tag{6.182}$$

$$N_1 = \frac{N_T \eta P'_s}{1 + P'_p + P'_s(1 + \eta)}, \tag{6.183}$$

$$P'_s = \frac{P_s}{P_s^{\text{th}}}, \quad P'_p = \frac{P_p}{P_p^{\text{th}}}, \tag{6.184}$$

$$\eta = \frac{\sigma_{21}}{\sigma_{12}}, \tag{6.185}$$

where P_s^{th} and P_p^{th} are threshold powers given by

$$P_s^{\text{th}} = \frac{A_{\text{eff}} \hbar \omega_s}{\sigma_{12} \tau_{21}}, \quad P_p^{\text{th}} = \frac{A_{\text{eff}} \hbar \omega_p}{\sigma_{13} \tau_{21}}. \tag{6.186}$$

The evolution of the signal beam due to stimulated emission, absorption, and scattering is similar to that of a semiconductor laser (see Eq. (3.32)),

$$\frac{dP_s}{dz} = \Gamma_s g_s P_s - \alpha_s P_s, \tag{6.187}$$

where

$$g_s = N_2 \sigma_{21} - N_1 \sigma_{12}, \tag{6.188}$$

Γ_s is the overlap factor that accounts for the fraction of the optical mode cross-section of the signal that overlaps with the erbium ion transverse distribution profile, and α_s is the internal loss coefficient of the erbium-doped fiber at the signal wavelength. Similarly, we have

$$\frac{dP_p}{dz} = \Gamma_p g_p P_p - \alpha_p P_p, \tag{6.189}$$

where

$$g_p = -N_1 \sigma_{13}. \tag{6.190}$$

Note that g_p is negative and the pump power is attenuated, whereas g_s could be positive indicating signal amplification. Using Eqs. (6.180) and (6.183) in Eqs. (6.187) and (6.188), we obtain

$$g_p = \frac{-N_T \sigma_{13}[1 + \eta P'_s]}{1 + P'_P + P'_s(1 + \eta)}, \tag{6.191}$$

$$g_s = \frac{N_T \sigma_{12}[\eta P'_p - 1]}{1 + P'_P + P'_s(1 + \eta)}. \tag{6.192}$$

Eqs. (6.187) and (6.189) together with Eqs. (6.191) and (6.192) form a coupled nonlinear differential equation which governs the growth of signal and pump powers in the EDFA. Fig. 6.17 shows the numerical solution of Eqs. (6.187) and (6.189) assuming the typical EDFA parameters. As can be seen, the signal is amplified whereas the pump is attenuated. The small signal gain can be found analytically from Eqs. (6.187) and (6.189) when $P_s(z) \ll P_s^{\text{th}}$ and $P_p(z) \gg P_p^{\text{th}}$ at any z, i.e., the signal is weak and the pump is strong. Under these conditions, Eq. (6.192) reduces to

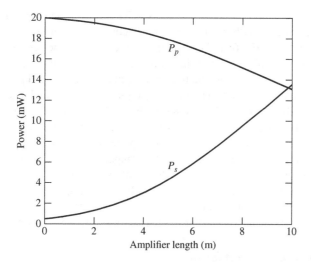

Figure 6.17   Evolution of signal and pump power in EDFA.

$$g_s = N_T \sigma_{12} \eta, \tag{6.193}$$

and from Eq. (6.187) we obtain

$$P_s(L) = P_s(0) \exp\left[(\Gamma_s N_T \sigma_{12} \eta - \alpha_s)L\right]. \tag{6.194}$$

Typically, $\alpha_s \ll \Gamma_s N_T \sigma_{12} \eta$ and therefore, signal power grows exponentially with distance. Fig. 6.18 shows the amplifier gain G as a function of the amplifier length for various pump powers. For the given pump power,

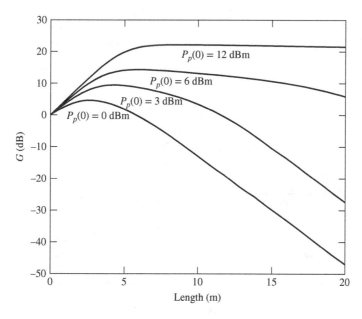

Figure 6.18   Amplifier gain vs. length in EDFA. The amplifier gain becomes maximum at a certain amplifier length which is a function of pump power.

gain increases with distance initially and then it decreases. This is because the pump decays as it propagates through the EDF and at a certain length, the pump power becomes less than or equal to P_p^{th} (Eq. (6.194) is no longer valid in this case) and from Eq. (6.192), we find that g_s could become zero or negative, which indicates that the signal is attenuated. Physically, when the pump power is less than a certain threshold, erbium ions pumped to level 2 (via level 3) are not adequate to cause population inversion. From a practical standpoint, it is desirable to have a low pump threshold so that the population inversion can be achieved at relatively lower pump powers. From Eq. (6.186), we see that the pump threshold is inversely proportional to the product of pump absorption cross-section σ_{13} and lifetime associated with the transition from level 2 to level 1, τ_{21}. The larger pump absorption cross-section enables higher pump absorption. As a result, more erbium ions make the transition to level 3 and consequently to level 2. The longer lifetime τ_{21} implies that erbium ions are in the excited level 2 for a longer time. For the erbium-doped silica fiber, the lifetime τ_{21} is very large (~ 10 ms) and as result, the population inversion can be achieved with a low pump power.

Fig. 6.19 shows the dependence of the gain on the input signal power for various pump powers. The gain saturates at large signal powers, which is similar to the case of semiconductor optical amplifiers.

6.7.3 Amplified Spontaneous Emission

So far we have ignored the impact of spontaneous emission. The Er^{3+} ions in the excited level spontaneously emit photons. These photons are amplified as they propagate down the fiber leading to ASE. The population density of the excited level is depleted because of the ASE and, therefore, the amplifier gain decreases. In the case of EDFA, the fiber-loss coefficient is much smaller than the gain coefficient g and Eq. (6.5) can be used to calculate the spontaneous noise factor n_{sp} with a slight modification. Eq. (6.5) is valid for a non-degenerate system. It is straightforward to modify Eq. (6.5) for the case of a degenerate system, as

$$n_{sp} = \frac{\sigma_{21}N_2}{\sigma_{21}N_2 - \sigma_{12}N_1}. \tag{6.195}$$

Since the population densities N_2 and N_1 vary along the fiber length, Eq. (6.187) has to be solved numerically to obtain the ASE power. Typically, the noise figure ($\approx 2n_{sp}$) of an EDFA is in the range of 4–8 dB. Spontaneous emission occurs at random and in all directions. The optical field due to spontaneous emission

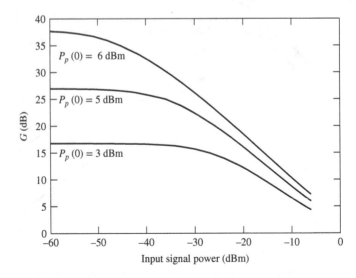

Figure 6.19 Gain saturation in EDFA.

can be expressed as the superposition of the guided mode and radiation modes of the erbium-doped fiber. The radiation modes escape to the cladding and do not degrade the system performance. Similarly, the ASE propagating as a guided mode in the backward direction does not a effect the system performance. However, backward-propagating ASE and radiation modes lead to degradation of the amplifier performance since the ASE reduces the gain of the amplifier.

6.7.4   Comparison of EDFA and SOA

One of the disadvantages of SOAs is their polarization sensitivity. The amplifier gain G depends on whether the mode in the waveguide is transverse electric (TE) or transverse magnetic (TM). Therefore, the incident optical field with arbitrary polarization experiences different amounts of gain for its x- and y- polarization components. This is known as *polarization-dependent gain* (PDG), which is undesirable for light-wave systems since it alters the polarization state of the light wave. Fiber amplifiers such as EDFA provide a uniform gain for x- and y- polarization components because of the circular symmetry of the fiber.

Another drawback of the SOA is the presence of *interchannel cross-talk* for WDM systems (see Chapter 9). Consider the signal corresponding to two channels of a WDM system,

$$q(t,z) = s_1(t,z)\exp(i\omega_1 t) + s_2(t,z)\exp(i\omega_2 t), \tag{6.196}$$

where ω_j and $s_j(t,z)$ are the optical carrier frequency and field envelope of the jth channel. The total signal power is

$$P = |q|^2 = |s_1|^2 + |s_2|^2 + 2|s_1||s_2|\cos(\Delta\omega t + \theta_1 - \theta_2), \tag{6.197}$$

where $\Delta\omega = |\omega_1 - \omega_2|$ is the channel separation and $\theta_j = \text{Arg}[s_j]$. Substituting Eqs. (6.197) in Eqs. (3.117) and (3.120), we obtain

$$\frac{dN_e}{dt} = \frac{I}{qV} - \frac{N_e}{\tau_e} - \frac{G_0(N_e - N_0)}{\hbar\omega}[|s_1|^2 + |s_2|^2 + 2|s_1||s_2|\cos(\Delta\omega t + \theta_1 - \theta_2)]. \tag{6.198}$$

On the right-hand side of Eq. (6.198), we have a term oscillating at the beat frequency $\Delta\omega$. This leads to the carrier population density N_e oscillating at the beat frequency $\Delta\omega$. Since the gain coefficient is related to N_e by Eq. (6.142), the gain is also modulated at frequency $\Delta\omega$ and from Eq. (3.32), we have

$$\frac{\partial P}{\partial z} = g(z,t)P. \tag{6.199}$$

Since the gain coefficient $g(z,t)$ depends on the instantaneous channel powers $|s_1(t,z)|^2$ and $|s_2(t,z)|^2$, the amplifier gain changes with time depending on the bit patterns in channels 1 and 2. Owing to the randomly changing bit patterns, the noise in the system is enhanced, leading to performance degradations. This is known as interchannel cross-talk. This cross-talk can be avoided if the SOA operates in the unsaturated regime. However, in the WDM system, saturation occurs quickly because there are many channels and the signal saturation power is the sum of the powers of each channel.

The carrier lifetime τ_e ($\tau_e \sim 0.5$ ns) of the SOA is much shorter than the lifetime τ_{21} ($\tau_{21} \sim 10$ ms) associated with the excited state of erbium ions. When the bit interval T_b is much shorter than the lifetime τ_{21}, erbium ions do not follow the fast variations of the signal, but they respond only to the average power of the signal. Therefore, in this case, all pulses experience the same gain in the case of an EDFA. For example, when the bit rate is 2.5 Gb/s, $T_b = 0.4$ ns $\ll \tau_{21}$ and the EDFA gain does not change from bit to bit. However, for SOA, T_b and τ_e are comparable and the gain experienced by the current bit depends on the signal power of the previous bits.

6.8   Raman Amplifiers

Distributed Raman amplifiers have become a viable alternative to EDFAs because of their relatively lower ASE [10, 13–15]. Raman amplifiers are based on stimulated Raman scattering, which occurs in fibers at high powers (see Chapter 10 for more details). As an intense pump beam of frequency ω_p propagates down the fiber, an optical wave of lower frequency ω_s is generated due to SRS. The frequency difference, $\omega_p - \omega_s = \Omega$, is known as *Stokes's shift*. If a signal field of frequency ω_s (Stokes wave) is incident at the input of the fiber along with the pump beam, the signal field gets amplified due to SRS. As shown in Fig. 6.20, the pump photons cause transitions to the excited level 3 from level 1, and silica molecules relax to one of the vibrational levels in band 2; the energy difference $\hbar(\omega_p - \omega_s)$ appears as molecular vibrations or optical phonons. If a signal photon corresponding to the energy difference between level 3 and one of the levels in band 2 is present, the molecules are stimulated to emit signal photons of the same kind, leading to the amplification of the signal photons, which is known as SRS. The silica molecule could also make a transition to band 2 from level 3 by spontaneous emission, whether or not the signal beam is present. This is known as spontaneous Raman scattering and is the source of noise in Raman amplifiers. Band 2 is a collection of vibrational states of silicon molecules. In other words, part of the pump energy is converted into signal energy and the rest is dissipated as molecular vibrations. Quantum mechanically, a pump photon of energy $\hbar\omega_p$ is annihilated to create a signal photon of lower energy $\hbar\omega_s$ and an optical phonon of energy $\hbar\Omega$. A semiclassical description of the Raman scattering is provided in Section 10.11. Fig. 6.21 shows the typical Raman gain spectrum as a function of the frequency shift for a silica-core single-mode fiber. The frequency shift shown in Fig. 6.21 refers to the frequency deviation of the Stokes wave from the pump. The Raman gain curve has a peak around a frequency shift, Ω of about 14 THz. In amorphous materials such as fused silica, molecular vibrational states form a continuum [14] shown by crossed lines in Fig. 6.20 and, therefore, the Raman gain occurs over a broad range of frequencies up to 40 THz. Fig. 6.22 shows a schematic of the Raman amplifier with co-propagating pump.

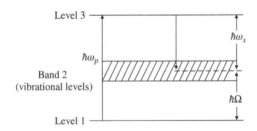

Figure 6.20   Energy levels of silica.

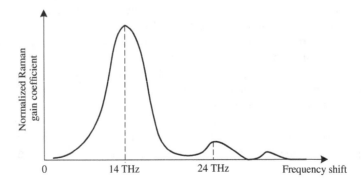

Figure 6.21   Typical Raman gain spectrum of silica fibers.

The signal and pump are combined using a fiber coupler and the combined signal is launched into the fiber. For a signal wavelength at 1550 nm, the pump wavelength should be about 1450 nm to ensure the highest gain (corresponding to a frequency difference of about 14 THz). To achieve a gain flatness over a broad range of signal frequencies, multiple pumps are usually used in practical systems [10].

Fig. 6.23 shows a schematic of a Raman amplifier with counter-propagating pump. The advantage of a counter-propagating pump scheme is that the transfer of power fluctuations from the pump to the signal can be reduced compared with the co-propagating scheme. The lifetime associated with the excited state of silica is in the range of 3 to 6 fs. Because of such a short lifetime, the transfer of power from the pump to the signal is almost instantaneous, leading to the transfer of pump fluctuations to the signal. However, if the pump is counter-propagating, the interaction time is equal to the transit time (= fiber length/speed of light) through the fiber, which would be the effective lifetime. For an 80-km fiber length, the transit time is about 0.4 ms, which is much larger than the actual lifetime (in the range of femtoseconds). In the scheme of co-propagating pumps, the pump lasers must be very quiet, i.e., they must have very low-intensity fluctuations [10]. Some light-wave systems use both co-propagating and counter-propagating pumps.

Raman amplifiers can be divided into two types: distributed and lumped. Distributed Raman amplifiers utilize the existing transmission fiber as a gain medium, whereas in lumped Raman amplifiers a dedicated short-span fiber is used to provide amplification. Typically, the length of the lumped amplifier is less than 15 km. In the case of lumped amplifiers, a highly nonlinear fiber with very small effective area can be used so that the pump intensity (= power/area) and gain can be maximized. In contrast, in the case of a distributed Raman amplifier, the fiber parameters can not be optimized to achieve the maximum gain since the nonlinear effects are enhanced in small-effective-area fibers which leads to performance degradation (see Chapter 10).

6.8.1 Governing Equations

Assuming the signal and pump beams are CW, the evolution of the signal and pump powers for the forward-pumping scheme is governed by (see Section 10.11)

$$\frac{dP_s}{dz} = \frac{g_R P_p P_s}{A_p} - \alpha_s P_s, \tag{6.200}$$

$$\frac{dP_p}{dz} = -\frac{\omega_p}{\omega_s} \frac{g_R P_p P_s}{A_s} - \alpha_p P_p, \tag{6.201}$$

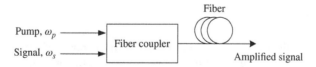

Figure 6.22 Schematic of the Raman amplifier. The pump co-propagates with the signal.

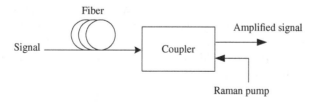

Figure 6.23 Schematic of the Raman amplifier in which the pump is counter-propagating.

where g_R is the Raman gain coefficient; P, α, and A denote the power, loss coefficient, and effective cross-section, respectively; and the subscripts p and s denote the pump and the signal, respectively. In general, these equations have to be solved numerically to calculate the amplifier gain. But more insight can be gained if we do the undepleted pump approximation. Under this approximation, the depletion of the pump due to the transfer of power to the signal (first term in Eq. (6.201)) is ignored. The solution of Eq. (6.201) under this approximation is

$$P_p(z) = P_p(0)\exp(-\alpha_p z). \tag{6.202}$$

Substituting Eq. (6.202) in Eq. (6.200) and rearranging, we obtain

$$\frac{dP_s}{P_s} = \frac{g_R P_p(0)\exp(-\alpha_p z)}{A_p} - \alpha_s. \tag{6.203}$$

Integrating Eq. (6.203) from 0 to L, we find

$$\ln\ P_s(L) - \ln\ P_s(0) = \frac{g_R P_p(0)L_{\text{eff}}}{A_p} - \alpha_s L \tag{6.204}$$

or

$$P_s(L) = P_s(0)\exp\left[\frac{g_R P_p(0)L_{\text{eff}}}{A_p} - \alpha_s L\right], \tag{6.205}$$

where

$$L_{\text{eff}} = \frac{1 - \exp(-\alpha_p L)}{\alpha_p} \tag{6.206}$$

is the effective fiber length over which the pump power is significant. The Raman amplifier gain may be defined as

$$G = \frac{P_s(L)}{P_s(0)\exp(-\alpha_s L)} = \exp\left[\frac{g_R P_p(0)L_{\text{eff}}}{A_p}\right]. \tag{6.207}$$

When $\alpha_p L \gg 1$, $L_{\text{eff}} \approx 1/\alpha_p$ and Eq. (6.207) can be approximated as

$$G \approx \exp\left[\frac{g_R P_p(0)}{A_p \alpha_p}\right]. \tag{6.208}$$

From Eq. (6.208), we see that the gain increases exponentially with the pump launch power and it is independent of the fiber length (when $\alpha_p L \gg 1$).

Fig. 6.24 shows the gain as a function of the fiber length obtained by solving Eqs. (6.200) and (6.201) numerically. As can be seen, the gain becomes approximately independent of length for $L > 40$ km. The solid and dotted lines in Fig. 6.25 show the evolution of signal power in the presence and absence of the Raman pump, respectively. From Fig. 6.26, we see that the pump is almost depleted at the end of the fiber due to SRS and fiber internal loss. Fig. 6.27 shows the gain as a function of the pump power for various signal powers. As can be seen, the gain reduces as the signal launch power increases. This saturation effect is similar to that in EDFA or SOA, and it can be understood from Eq. (6.201). When the signal launch power is large, the pump depletes rapidly since the first term on the right-hand side is proportional to the signal power. A reduction in pump power leads to a lower level of population inversion and the gain decreases. Nevertheless, the gain saturation of Raman amplifiers occurs at much higher signal powers than that of SOA and EDFA signal powers, which make them attractive for multi-channel amplification. Typically, the pump power required to achieve

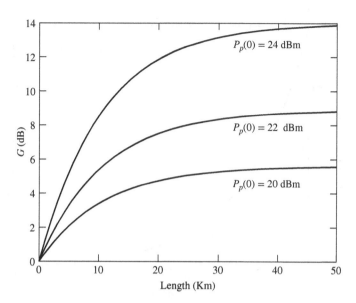

Figure 6.24   The gain of the Raman amplifier becomes roughly independent of length for large L. Signal launch power $=$ -6 dBm, $g_R = 6 \times 10^{-14}$ m/W, $A_p = 40$ μm², $A_s = 50$ μm², $\alpha_p = 9.21 \times 10^{-5}$ m^{-1}, $\alpha_s = 4.605 \times 10^{-5}$ m^{-1}.

a certain gain for Raman amplifiers is larger than that needed for EDFA in the unsaturated regime. However, as the signal power increases because of the increase in number of channels of a WDM system, the gain of a Raman amplifier is greater than that of an EDFA [10]. This is because of the higher saturation power of the Raman amplifier.

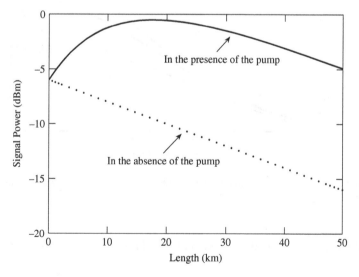

Figure 6.25   Evolution of signal power in the Raman amplifier with and without the pump. $P_p(0) = 23$ dBm for the solid line. The parameters are the same as those of Fig. 6.24.

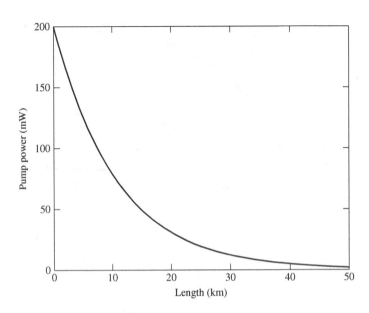

Figure 6.26   Decay of the Raman pump. The parameters are the same as those of Fig. 6.24.

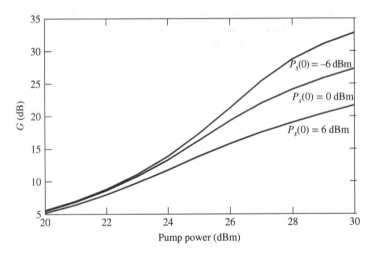

Figure 6.27   Dependence of gain on the pump power. Gain saturates as the signal power increases. The parameters are the same as those of Fig. 6.24.

For the backward-pumping scheme, the evolution of signal and pump powers is given by

$$\frac{dP_s}{dz} = \frac{g_r}{A_p} P_p P_s - \alpha_s P_s, \tag{6.209}$$

$$-\frac{dP_p}{dz} = -\frac{\omega_p}{\omega_s} \frac{g_R}{A_s} P_p P_s - \alpha_p P_p. \tag{6.210}$$

Under the undepleted pump approximations, Eqs. (6.209) and (6.210) can be solved as before since the pump is injected at $z = L$, the pump power at $z = L$, $P_p(L)$ is known. Ignoring the first term on the right-hand side of Eq. (6.210), the solution of Eq. (6.210) is

$$P_p(z) = P_P(L) \exp[-\alpha_p(L - z)]. \tag{6.211}$$

Substituting Eq. (6.211) in Eq. (6.209) and proceeding as before, we obtain the same expression for $P_s(L)$ as in Eq. (6.205). Thus, the gain for the forward- and backward-pumping scheme is the same under the undepleted pump approximations.

6.8.2 Noise Figure

Spontaneous Raman scattering occurs randomly over the entire bandwidth of the amplifier and spontaneous emission photons are amplified by SRS. The spontaneous emission factor, n_{sp}, is nearly unity since a Raman system acts as a fully inverted system with the ground-state population density $N_1 \approx 0$. Therefore, the noise figure of the Raman amplifier is close to 3 dB, whereas that of the EDFA is typically in the range of 4 to 8 dB. Distributed Raman amplifiers can be imagined as tiny amplifiers placed throughout the fiber transmission line with very small amplifier spacing. Because of the distributed nature of the amplification, the OSNR of the distributed Raman amplifiers is higher than that of the lumped amplifiers such as EDFA (see Section 7.4.2).

6.8.3 Rayleigh Back Scattering

One of the primary sources of noise in Raman amplifiers is double Rayleigh back scattering (DRBS). Consider a signal propagating in the forward direction and ASE propagating backward in a distributed Raman amplifier as shown in Fig. 6.28. Because of the microscopic non-uniformity of the silica composition, ASE gets reflected and, therefore, it interferes with the signal, leading to performance degradation. This is known as *single Rayleigh back scattering* (SRBS). Consider the signal and ASE both propagating in the forward direction, as shown in Fig. 6.29. The ASE is reflected backward by a scatterer and it is reflected again by another scatterer,

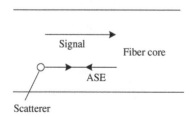

Figure 6.28   Single Rayleigh back scattering.

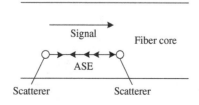

Figure 6.29   Double Rayleigh back scattering.

so that now the ASE and signal are both propagating in the forward direction. This is known as *double Rayleigh back scattering* (DRBS). The DRBS occurs not only for ASE, but also for the signal modulated by the data, i.e., a part of the modulated signal undergoes the DRBS process and interferes with the current bit. Since the Rayleigh back scattering can occur anywhere along the fiber, the part of the signal that undergoes DRBS has a random delay and it acts as noise on the current bit, leading to performance degradation. DRBS also occurs in fibers without distributed amplification. However, in the presence of Raman amplification, the back-scattered signal and ASE are amplified by SRS in both directions and, therefore, DRBS is one of the primary sources of noise in the distributed Raman amplifiers.

6.9 Additional Examples

Example 6.8

In Fig. 6.30, transfer functions of the optical and electrical filters are $\tilde{H}_{opt}(f)$ and $\tilde{H}_e(f)$, respectively. Show that the variance of the signal–ASE beat noise is

$$\sigma^2_{s-sp} = 4R^2 P_{out} \rho_{ASE} B_{eff},$$

where

$$B_{eff} = \frac{1}{2} \int_{-\infty}^{\infty} |\tilde{H}_{opt}(f)|^2 |\tilde{H}_e(f)|^2 df.$$

Solution:
From Eq. (6.27), we have

$$I_{s-sp}(t) = R[\psi_{out} n_F^*(t) + \psi_{out}^* n_F(t)].$$

The signal–ASE beat noise current passes through the electrical filter, as shown in Fig. 6.30. Let the corresponding output of the electrical filter be $I_F(t)$, i.e.,

$$\tilde{I}_F(f) = \tilde{I}_{s-sp}(f)\tilde{H}_e(f). \tag{6.212}$$

Let

$$\mathcal{F}[n_F(t)] = \tilde{n}_F(f), \tag{6.213}$$

$$\mathcal{F}[n_F^*(t)] = \tilde{n}_F^*(-f). \tag{6.214}$$

Taking the Fourier transform of Eq. (6.27) and using Eqs. (6.212), (6.213), and (6.214), we obtain

$$\tilde{I}_F(f) = R[\psi_{out}\tilde{n}_F^*(-f) + \psi_{out}^*\tilde{n}_F(f)]\tilde{H}_e(f). \tag{6.215}$$

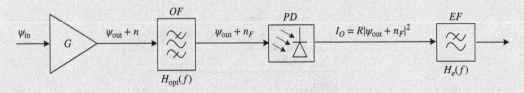

Figure 6.30 Impact of optical and electrical filter on ASE. OF = optical filter, PD = photodetector, EF = electrical filter.

Using Parseval's theorem, the mean noise power is

$$N_{s-sp} = \lim_{T \to \infty} \frac{1}{T} \int_{-T/2}^{T/2} <I_F^2(t)> dt$$

$$= \lim_{T \to \infty} \frac{1}{T} \int_{-\infty}^{\infty} <|\tilde{I}_F(f)|^2> df. \tag{6.216}$$

From Eq. (6.215), we have

$$<|\tilde{I}_F(f)|^2> = R^2|\tilde{H}_e(f)|^2[|\psi_{\text{out}}|^2 <|\tilde{n}_F(f)|^2> +|\psi_{\text{out}}|^2 <|\tilde{n}_F(-f)|^2>$$

$$+(\psi_{\text{out}}^*)^2 <\tilde{n}_F(f)\tilde{n}_F(-f)> +\psi_{\text{out}}^2 <\tilde{n}_F^*(-f)\tilde{n}_F^*(f)>]. \tag{6.217}$$

The contribution of the third and fourth terms on the right-hand side of Eq. (6.217) is zero. This can be shown by writing

$$n_F(f) = |n_F(f)| \exp[i\theta_F(f)], \tag{6.218}$$

$$\psi_{\text{out}} = |\psi_{\text{out}}| \exp[i\theta_{\text{out}}], \tag{6.219}$$

$$(\psi_{\text{out}}^*)^2 <\tilde{n}_F(f)\tilde{n}_F(-f)> = <|\tilde{n}_F(f)||\tilde{n}_F(-f)| \exp(i[\theta_F(f) + \theta_F(-f) - 2\theta_{\text{out}}]>)$$

$$\psi_{\text{out}}^2 <\tilde{n}_F^*(-f)\tilde{n}_F^*(f)> = <|\tilde{n}_F(f)||\tilde{n}_F(-f)| \exp(i[-\theta_F(f) - \theta_F(-f) + 2\theta_{\text{out}}]) >, \tag{6.220}$$

$$(\psi_{\text{out}}^*)^2 <\tilde{n}_F(f)\tilde{n}_F(-f)> +\psi_{\text{out}}^2 <\tilde{n}_F^*(-f)\tilde{n}_F^*(f)> = 2|\psi_{\text{out}}|^2|\tilde{n}_F(f)||\tilde{n}_F(-f)|$$

$$<\cos[\theta_F(f) + \theta_F(-f) - 2\theta_{\text{out}}]>$$

$$= 0, \tag{6.221}$$

since $\theta_F(f)$ is a random variable with uniform distribution in the interval $[0, 2\pi]$. By definition, the power spectral densities are

$$\rho_{\text{ASE}} = \lim_{T \to \infty} \frac{<|\tilde{n}(f)|^2>}{T}, \tag{6.222}$$

$$\rho_{n_F} = \lim_{T \to \infty} \frac{<|\tilde{n}_F(f)|^2>}{T} = \rho_{\text{ASE}}|\tilde{H}_{\text{opt}}(f)|^2, \tag{6.223}$$

$$\rho_{I_F} = \lim_{T \to \infty} \frac{<|\tilde{I}_F(f)|^2>}{T}. \tag{6.224}$$

Using Eqs. (6.217), (6.223), and (6.224), we find

$$\rho_{I_F} = 2R^2|\tilde{H}_e(f)|^2|\psi_{\text{out}}|^2\rho_{n_F}$$

$$= 2R^2P_{\text{out}}\rho_{\text{ASE}}|\tilde{H}_{\text{eff}}(f)|^2, \tag{6.225}$$

where

$$\tilde{H}_{\text{eff}}(f) = \tilde{H}_{\text{opt}}(f)\tilde{H}_e(f) \tag{6.226}$$

is the transfer function of an effective filter, which is the cascade of optical and electrical filters. From Eqs. (6.216), (6.224), and (5.79), we have

$$N_{s-sp} = \sigma_{s-sp}^2 R_L = R_L \int_{-\infty}^{\infty} \rho_{I_F}(f)df. \qquad (6.227)$$

Therefore,

$$\sigma_{s-sp}^2 = 2R^2 P_{\text{out}} \rho_{\text{ASE}} \int_{-\infty}^{\infty} |\tilde{H}_{\text{eff}}(f)|^2 df$$

$$= 4R^2 P_{\text{out}} \rho_{\text{ASE}} B_{\text{eff}}, \qquad (6.228)$$

where B_{eff} is the effective bandwidth of the filter obtained by cascading the optical and electrical filters. Let us consider two limiting cases of Eq. (6.228). When the electrical filter bandwidth is much larger than the optical filter bandwidth, i.e., $\tilde{H}_e(f) \cong 1$, from Eq. (6.226), we have $\tilde{H}_{\text{eff}}(f) = \tilde{H}_{\text{opt}}(f)$ and Eq. (6.228) reduces to Eq. (6.42). When the optical filter bandwidth is much larger than the electrical filter bandwidth, i.e., $\tilde{H}_{\text{opt}}(f) \cong 1$, $\tilde{H}_{\text{eff}}(f) = \tilde{H}_e(f)$, and Eq. (6.228) becomes

$$\sigma_{s-sp}^2 = 4R^2 P_{\text{out}} \rho_{\text{ASE}} B_e, \qquad (6.229)$$

where

$$B_e = \frac{1}{2} \int_{-\infty}^{\infty} |\tilde{H}_e(f)|^2 df. \qquad (6.230)$$

When the optical filter is an ideal band-pass filter with full bandwidth f_o and the electrical filter is an ideal low-pass filter with bandwidth f_e, it is easy to see that

$$B_{\text{eff}} = \min(f_e, f_o/2). \qquad (6.231)$$

Example 6.9

An amplifier has an output signal power of 0 dBm and a noise power per polarization per unit frequency interval (single-sided) of -126 dBm/Hz. Calculate the variance of the signal–ASE beat noise current. Assume $B_o = 20$ GHz, $B_e \gg B_o$, and $R = 0.9$ A/W.

Solution:

$$P_{\text{out}} = 10^{P_{\text{out}}(\text{dBm})/10} = 1 \text{ mW}.$$

The noise power per polarization per unit frequency interval is the single-sided power spectral density,

$$\rho_{\text{ASE},sp} = 10^{\rho_{\text{ASE}}(\text{dBm/Hz})/10} \text{ mW/Hz}$$

$$= 1.58 \times 10^{-13} \text{ mW/Hz}.$$

From Eq. (6.17), we have $\rho_{\text{ASE}} = 1.58 \times 10^{-13}$ mW/Hz. Noise power $P_{\text{ASE}} = \rho_{\text{ASE}} B_o = 3.17$ mW. The variance of the signal–ASE beat noise current is

$$\sigma_{s-sp}^2 = 2R^2 P_{\text{out}} P_{\text{ASE}} = 5.13 \times 10^{-9} \text{ A}^2.$$

Example 6.10

Find an analytical expression for the mean and variance of ASE–ASE beat noise. Assume that the optical filter is an ideal band-pass filter with bandwidth B_o and the samples of ASE noise $n(t)$ are identically distributed complex Gaussian random variables. Ignore the electrical filter.

Solution:

From Eqs. (6.21) and (6.28), it follows that

$$< I_{sp-sp} > = R < |n_F(t)|^2 >= R\rho_{\text{ASE}}B_o, \tag{6.232}$$

$$< I_{sp-sp}^2 > = R^2 < |n_F(t)|^4 > . \tag{6.233}$$

Let

$$n_F = n_{Fr} + in_{Fi}. \tag{6.234}$$

With the assumption that the samples of $n(t)$ are identically distributed complex random variables, it follows that

$$< n_{Fr}^2 > = < n_{Fi}^2 >= \frac{< |n_F|^2 >}{2} = \frac{\rho_{\text{ASE}}B_o}{2}, \tag{6.235}$$

$$< n_{Fr}n_{Fi} > = 0, \tag{6.236}$$

$$< |n_F|^4 > = < (n_{Fr}^2 + n_{Fi}^2)^2 >=< n_{Fr}^4 > + < n_{Fi}^4 > +2 < n_{Fr}^2 n_{Fi}^2 > . \tag{6.237}$$

For Gaussian random variables $N_1, N_2, N_3,$ and N_4, from the moment theorem, we have

$$< N_1 N_2 N_3 N_4 >=< N_1 N_2 >< N_3 N_4 > + < N_1 N_3 >< N_2 N_4 > + < N_1 N_4 >< N_2 N_3 > . \tag{6.238}$$

If we choose $N_i = n_{Fr}, i = 1, 2, 3, 4$, using Eqs. (6.235) and (6.236), we find

$$< n_{Fr}^4 >= 3(< n_{Fr}^2 >)^2 = 3\rho_{\text{ASE}}^2 B_o^2/4. \tag{6.239}$$

Similarly,

$$< n_{Fi}^4 >= 3(< n_{Fi}^2 >)^2 = 3\rho_{\text{ASE}}^2 B_o^2/4. \tag{6.240}$$

If we choose $N_1 = N_2 = n_{Fr}$ and $N_3 = N_4 = n_{Fi}$, we find

$$< n_{Fr}^2 n_{Fi}^2 >= \rho_{\text{ASE}}^2 B_o^2/4. \tag{6.241}$$

Substituting Eqs. (6.239)–(6.241) in Eq. (6.237), and using Eq. (6.233), we obtain

$$< I_{sp-sp}^2 > = 2R^2 \rho_{\text{ASE}}^2 B_o^2, \tag{6.242}$$

$$\sigma_{sp-sp}^2 = < I_{sp-sp}^2 > - < I_{sp-sp}>^2, \tag{6.243}$$

$$\sigma_{sp-sp}^2 = R^2 \rho_{\text{ASE}}^2 B_o^2. \tag{6.244}$$

Example 6.11

Show that the variance of ASE–ASE beat noise current for the case of arbitrary optical and electrical filters (see Fig. 6.31) and for the case of a single polarization is given by

$$\sigma_{sp-sp}^2 = R^2 \rho_{ASE}^2 B_{oe}^2,$$ (6.245)

where

$$B_{oe}^2 = \int \int H_{opt}^2(t''-t')H_e(t')H_e(t'')dt'dt''.$$ (6.246)

Further, when the optical filter is an ideal band-pass filter with full bandwidth f_o and the electrical filter is an ideal low-pass filter with cutoff frequency f_e, show that

$$B_{oe}^2 = (2f_o - f_e)f_e \quad \text{if} \quad f_e < f_o,$$ (6.247)

$$= f_o^2 \quad \text{otherwise}.$$ (6.248)

Solution: From Eq. (6.25), the current before the electrical filter is

$$I_{in}(t) = R\{|\psi_{out}|^2 + |n_F(t)|^2 + 2\text{Re}[\psi_{out}n_F^*(t)]\}.$$ (6.249)

Let

$$I_{in,sp-sp}(t) = R|n_F(t)|^2.$$ (6.250)

$$I_{out,sp-sp}(t) = \mathcal{F}^{-1}[\tilde{I}_{in,sp-sp}(f)\tilde{H}_e(f)]$$

$$= \int_{-\infty}^{\infty} I_{in,sp-sp}(t')H_e(t-t')dt'.$$ (6.251)

Without loss of generality, let us assume that the decision is based on the sample at $t = 0$,

$$I_{out,sp-sp}(0) = \int_{-\infty}^{\infty} I_{in,sp-sp}(t')H_e(-t')dt'.$$ (6.252)

$$< I_{out,sp-sp}(0) > = R \int_{-\infty}^{\infty} < |n_F(t')|^2 > H_e(-t')dt'$$

$$= R\rho_{ASE}B_o \int_{-\infty}^{\infty} H_e(t)dt,$$ (6.253)

where we have used Eq. (6.21). Since

$$\tilde{H}_e(f=0) = \int_{-\infty}^{\infty} H_e(t)dt,$$ (6.254)

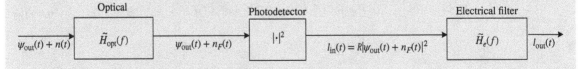

Figure 6.31 Impact of optical and electrical filters on ASE.

the mean noise current due to ASE–ASE noise beating is

$$< I_{sp-sp} > = < I_{\text{out},sp-sp}(0) > = R\rho_{\text{ASE}}B_o\tilde{H}_e(f = 0). \tag{6.255}$$

If the electrical filter is an ideal low-pass filter given by

$$H_e(f) = \text{rect}\left(\frac{f}{2f_e}\right), \tag{6.256}$$

Eq. (6.255) becomes

$$< I_{sp-sp} > = R\rho_{\text{ASE}}B_o. \tag{6.257}$$

Squaring Eq. (6.252) and then averaging, we find

$$< I^2_{\text{out},sp-sp}(0) > = \int_{-\infty}^{\infty}\int_{-\infty}^{\infty} < I_{\text{in},sp-sp}(t')I_{\text{in},sp-sp}(t'')H_e(-t')H_e(-t'') > dt'\,dt''$$

$$= R^2 \int_{-\infty}^{\infty}\int_{-\infty}^{\infty} < [n^2_{Fr}(t') + n^2_{Fi}(t')][n^2_{Fr}(t'') + n^2_{Fi}(t'')] >$$

$$\times H_e(-t')H_e(-t'')dt'\,dt''. \tag{6.258}$$

Using the moment theorem (see Eq. (6.238)), we obtain

$$< n^2_{Fr}(t')n^2_{Fr}(t'') > = < n^2_{Fr}(t') >< n^2_{Fr}(t'') > +2 < n_{Fr}(t')n_{Fr}(t'')>^2, \tag{6.259}$$

$$< n^2_{Fi}(t')n^2_{Fi}(t'') > = < n^2_{Fi}(t') >< n^2_{Fi}(t'') > +2 < n_{Fi}(t')n_{Fi}(t'')>^2, \tag{6.260}$$

$$< n^2_{Fr}(t')n^2_{Fi}(t'') > = < n^2_{Fr}(t') >< n^2_{Fi}(t'') > +2 < n_{Fr}(t')n_{Fi}(t'')>^2. \tag{6.261}$$

$$< n^2_{Fr}(t') > = < n^2_{Fi}(t') > = \rho_{\text{ASE}}B_o/2, \tag{6.262}$$

$$< n_{Fr}(t')n_{Fr}(t'') > = < n_{Fi}(t')n_{Fi}(t'') > = \frac{\rho_{\text{ASE}}}{2}\mathcal{H}_{\text{opt}}(t' - t''), \tag{6.263}$$

$$< n_{Fr}(t')n_{Fi}(t'') > = 0, \tag{6.264}$$

where

$$\mathcal{F}[\mathcal{H}_{\text{opt}}(t)] = |\tilde{H}_{\text{opt}}(f)|^2. \tag{6.265}$$

Using Eqs. (6.259)–(6.264) in Eq. (6.258), we find

$$< [n^2_{Fr}(t') + n^2_{Fi}(t')][n^2_{Fr}(t'') + n^2_{Fi}(t'')] > = \rho^2_{\text{ASE}}B^2_o + \rho^2_{\text{ASE}}\mathcal{H}^2_{\text{opt}}(t' - t''), \tag{6.266}$$

$$< I^2_{\text{out},sp-sp}(0) > = R^2\rho^2_{\text{ASE}}[B^2_o\mathcal{H}_e + B^2_{oe}], \tag{6.267}$$

$$B^2_{oe} = \int\int \mathcal{H}^2_{\text{opt}}(t'' - t')H_e(t')H_e(t'')dt'\,dt'', \tag{6.268}$$

$$\mathcal{H}_e = \left[\int H_e(t')dt'\right]^2, \tag{6.269}$$

$$\sigma^2_{sp-sp} = < I^2_{\text{out},sp-sp} > - < I_{\text{out},sp-sp}>^2. \tag{6.270}$$

Using Eqs. (6.253) and (6.267), we obtain

$$\sigma^2_{sp-sp} = R^2 \rho^2_{\text{ASE}} B^2_{oe}. \qquad (6.271)$$

Next, let us consider the case in which the optical filter is an ideal band-pass filter with full bandwidth f_o and the electrical filter is an ideal low-pass filter with cutoff frequency f_e.

$$\tilde{H}_{\text{opt}}(f) = \text{rect}\left(\frac{f}{f_o}\right), \qquad (6.272)$$

$$\tilde{H}_e(f) = \text{rect}\left(\frac{f}{2f_e}\right), \qquad (6.273)$$

$$\tilde{H}_{\text{opt}}(f) = |H_{\text{opt}}(f)|^2 = [\text{rect}(f/f_o)]^2 \qquad (6.274)$$

$$= \text{rect}(f/f_o), \qquad (6.275)$$

$$\mathcal{H}_{\text{opt}}(t) = \mathcal{F}^{-1}[\text{rect}(f/f_o)] \qquad (6.276)$$

$$= f_o \text{sinc}(f_o t), \qquad (6.277)$$

where

$$\text{sinc}(x) = \frac{\sin(\pi x)}{\pi x}, \qquad (6.278)$$

$$\mathcal{H}^2_{\text{opt}}(t) = f_o^2 \text{sinc}^2(f_o t). \qquad (6.279)$$

Eq. (6.268) may be rewritten as

$$B^2_{oe} = \int u(t'') H_e(t'') dt'', \qquad (6.280)$$

$$u(t'') = \mathcal{H}^2_{\text{opt}}(t'') * H_e(t''). \qquad (6.281)$$

Here $*$ denotes the convolution. Since the convolution in the time domain becomes a multiplication in the frequency domain, the Fourier transform of Eq. (6.281) is

$$\tilde{u}(f) = \tilde{H}_{\text{opt2}}(f) H_e(f). \qquad (6.282)$$

$$\tilde{H}_{\text{opt2}}(f) = \mathcal{F}[\mathcal{H}^2_{\text{opt}}(t'')] \qquad (6.283)$$

$$= \mathcal{F}[f_o^2 \text{sinc}^2(f_o t)] \qquad (6.284)$$

$$= \text{triang}(f/f_o) f_o, \qquad (6.285)$$

where

$$\text{triang}(f/f_o) = 1 - |f|/f_o \quad \text{if} \quad |f| < f_o, \qquad (6.286)$$

$$= 0 \quad \text{otherwise}. \qquad (6.287)$$

Substituting Eq. (6.283) in Eq. (6.282), we find

$$\tilde{u}(f) = \text{triang}\,(f/f_o)\text{rect}(f/(2f_e))f_o.$$

(6.288)

The product in Eq. (6.280) may be rewritten as a convolution in the frequency domain at $f = 0$ as

$$B_{oe}^2 = \int \tilde{u}(f)\tilde{H}_e(-f)df,$$

(6.289)

$$\tilde{u}(f)\tilde{H}_e(-f) = \text{triang}\,(f/f_o)\text{rect}\left(\frac{f}{2f_e}\right)\text{rect}\left(\frac{f}{2f_e}\right)f_o,$$

(6.290)

$$= \text{triang}\,(f/f_o)\text{rect}\left(\frac{f}{2f_e}\right)f_o.$$

(6.291)

First let us consider the case $f_e < f_o$. The integral in Eq. (6.289) is equal to the area of the shaded region shown in Fig. 6.32.

$$B_{oe}^2 = (2f_o - f_e)f_e.$$

(6.292)

If $f_e \geq f_o$, B_{oe}^2 equals the area of the triangle,

$$B_{oe}^2 = f_o^2.$$

(6.293)

Substituting Eqs. (6.292) and (6.293) in Eq. (6.271), we find

$$\sigma_{sp-sp}^2 = R^2\rho_{\text{ASE}}^2(2f_o - f_e)f_e \quad \text{if} \quad f_e \leq f_o,$$

(6.294)

$$= R^2\rho_{\text{ASE}}^2 f_o^2 \quad \text{otherwise.}$$

(6.295)

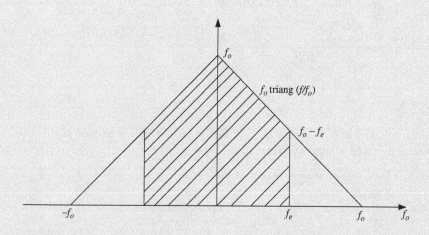

Figure 6.32 Calculation of the variance of ASE–ASE beat noise for the case of ideal optical and electrical filters.

Example 6.12

The electrical SNRs at the amplifier input and output are 30 dB and 25 dB, respectively. The signal power at the input and output of the amplifier are -13 dBm and 2 dBm, respectively. Find the ASE power spectral density per polarization, ρ_{ASE}. Assume that the carrier frequency $f_0 = 195$ THz.

Solution:

$$F_n = \frac{\text{SNR}_{\text{in}}}{\text{SNR}_{\text{out}}}$$

$$F_n(\text{dB}) = (\text{SNR})_{\text{in}}(\text{dB}) - (\text{SNR})_{\text{out}}(\text{dB})$$

$$= (30 - 25)\ \text{dB}$$

$$= 5\ \text{dB},$$

$$F_n = 10^{F_n(\text{dB})/10} = 3.1623.$$

The amplifier gain is

$$G = \frac{P_{\text{out}}}{P_{\text{in}}},$$

$$G(\text{dB}) = P_{\text{out}}(\text{dBm}) - P_{\text{in}}(\text{dBm}) = 2\ \text{dBm} - (-13)\ \text{dBm} = 15\ \text{dBm},$$

$$G = 10^{G(\text{dB})/10} = 31.62.$$

The ASE PSD is given by Eq. (6.102),

$$\rho_{\text{ASE}} = \frac{h f_0 (G F_n - 1)}{2}$$

$$= 6.626 \times 10^{-34} \times 195 \times 10^{12} \times (31.62 \times 3.162 - 1)/2\ \text{W/Hz}$$

$$= 6.39 \times 10^{-18}\ \text{W/Hz}.$$

Example 6.13

A Fabry–Perot amplifier has a peak gain G_{max} of 20 dB and a single-pass gain G_s of 5 dB. Calculate the geometric mean of the facet reflectivity R. Assume $R_1 = R_2$.

Solution:
From Eq. (6.127), we have

$$\sqrt{G_{\text{max}}} = \frac{(1 - R)\sqrt{G_s}}{1 - R G_s}.$$

$$10\log_{10} G_{max} = 20,$$

$$G_{max} = 10^2 = 100,$$

$$G_s = 10^{5/10} = 3.16,$$

$$(1 - 3.16R) \times 10 = (1 - R)\sqrt{3.16},$$

$$R = 0.276.$$

Example 6.14

A Fabry–Perot amplifier has to be designed such that its full 3-dB bandwidth is greater than 25 GHz. Calculate the upper bound on the single-pass gain G_s. Assume $R_1 = R_2 = 0.3$, refractive index $n = 3.5$, amplifier length $L = 200$ μm.

Solution:
From Eq. (6.128), we have

$$\text{FSR} = \frac{c}{2nL} = \frac{3 \times 10^8}{2 \times 3.5 \times 200 \times 10^{-6}} = 214 \text{ GHz}.$$

From Eq. (6.132), the 3-dB bandwidth is

$$f_{3\,\text{dB}} = \frac{2\,\text{FSR}}{\pi} \sin^{-1}\left\{ \frac{1 - RG_s}{(4RG_s)^{1/2}} \right\},$$

or

$$\sin\left(\frac{\pi f_{3\,\text{dB}}}{2\text{FSR}} \right) = \frac{1 - RG_s}{(4RG_s)^{1/2}}. \tag{6.296}$$

If $f_{3\,\text{dB}} = 25$ GHz, the left-hand side of Eq. (6.296) becomes 0.1805:

$$(1 - 0.3G_s)^2 = 1.2G_s \times (0.1805)^2.$$

Solving the above quadratic equation, we find

$$G_s = 4.79 \quad \text{or} \quad 2.31.$$

If $G_s = 4.79$, $(1 - RG_s) < 0$, which corresponds to biasing above the threshold and violates the condition given by Eq. (6.118) and, therefore, it has to be rejected. To have $f_{3\,\text{dB}} > 25$ GHz, G_s has to be less than 2.31.

Example 6.15

For a Fabry–Perot amplifier, show that the round-trip amplitude gain is

$$RG_s = \frac{\sqrt{G_{max}} - \sqrt{G_{min}}}{\sqrt{G_{max}} + \sqrt{G_{min}}},$$

where G_{max} and G_{min} are the maximum and minimum values of G.

Solution:
From Eq. (6.135), we have

$$\frac{\sqrt{G_{max}}}{\sqrt{G_{min}}} = \frac{1 + RG_s}{1 - RG_s} = x,$$

or

$$(1 - RG_s)x = (1 + RG_s),$$

$$RG_s = \frac{x - 1}{x + 1}$$

$$= \frac{\sqrt{G_{max}} - \sqrt{G_{min}}}{\sqrt{G_{max}} + \sqrt{G_{min}}}.$$

Exercises

6.1 Explain the meaning of (a) signal–ASE beating noise and (b) ASE–ASE beating noise.

6.2 An optical amplifier operating at 1300 nm has a mean noise power per unit frequency interval per polarization (single-sided) of -125 dBm/Hz. Calculate the noise figure. Assume $G = 30$ dB.

(Ans: 6 dB.)

6.3 The output of an amplifier passes through an ideal optical band-pass filter of bandwidth (f_o) 20 GHz, a photodetector of responsivity 0.9 A/W, and an ideal electrical low-pass filter of bandwidth (f_e) 8.5 GHz. The amplifier input power is -23 dBm and its gain is 20 dB. The OSNR at the amplifier output (in a bandwidth of 12.49 GHz) is 15 dB. A polarizer is placed just before the photodetector, which blocks one of the polarization modes. Find (a) the variance of the signal–ASE beat noise current and (b) the variance of the ASE–ASE beat noise current. Repeat (a) and (b) if the polarizer is absent.

(Ans: (a) 8.75×10^{-9} A^2; (b) 8.73×10^{-11} A^2. When the polarizer is absent, (a) 8.75×10^{-9} A^2; (b) 1.746×10^{-10} A^2.)

6.4 State the difference between OSNR and electrical SNR.

6.5 The OSNR (within a bandwidth of 0.1 nm) at the output of an amplifier operating at 1550 nm is 22 dB. The output of the amplifier passes through an ideal photo-detector ($\eta = 1$). Calculate the electrical SNR at the photodetector. Ignore the thermal noise and ASE–ASE beat noise. Assume that the input power of the amplifier is -20 dBm, gain $G = 23$ dB, $B_e \ll B_o$, $R = 0.8$ A/W, and $B_e = 8$ GHz.

(Ans: 20.92 dB.)

6.6 In a cavity-type SOA, $R_1 = R_2 = 0.3$, FSR $= 30$ GHz, and the single-pass gain $G_s = 4.75$ dB. Find (a) the peak gain and (b) the gain G at frequency $\Delta f = 5$ GHz, where Δf is the frequency shift from the resonant frequency. Assume $n = 3.3$.

(Ans: (a) 21.28 dB; (b) 2.07 dB.)

6.7 Explain the gain–bandwidth trade-off in semiconductor amplifiers.

6.8 In a cavity-type SOA, the maximum and minimum gains are 20.78 dB and 4.43 dB, respectively. The geometric mean of the reflectivities, R, is 0.32. Calculate the single-pass gain G_s.

(Ans: 4.47 dB.)

6.9 In a cavity-type SOA, FSR $= 300$ GHz, refractive index $n = 3.3$, $G_s = 4.3$ dB. (a) Calculate the peak gain G_{peak} and the 3-dB bandwidth if $R_1 = R_2 = 0.3$, (b) repeat if $R_1 = R_2 = 0.1$.

(Ans: (a) $G_{\text{peak}}(\text{dB}) = 17.62$ dB and $f_{3\,\text{dB}} = 16.06$ GHz; (b) $G_{\text{peak}}(\text{dB}) = 6.4$ dB and $f_{3\,\text{dB}} = 141.8$ GHz.)

6.10 Explain how population inversion is achieved in an EDFA.

6.11 Explain the meaning of absorption cross-section.

6.12 Solve Eqs. (6.187) and (6.189) numerically and plot the signal power as a function of amplifier length for various pump powers, $P_p(0) = 10$ mW and $P_s(0) = 10$ µW. Assume $N_T = 1.1 \times 10^{25}$ m^{-3}, $\Gamma_s = 0.4$, $\Gamma_p = 0.64$, $\sigma_{13} = 2.7 \times 10^{-25}$ m^{-2}, $\sigma_{12} = 1.8 \times 10^{-25}$ m^{-2}, $\tau_{21} = 12$ ms, A$_{\text{eff}} = 3.4 \times 10^{-12}$ m^{-2}.

6.13 Explain the difference between spontaneous Raman scattering and stimulated Raman scattering.

6.14 Solve Eqs. (6.200) and (6.201) numerically. Plot the gain as a function of the length for pump powers $P_p(0) = 200$ mW and $P_s(0) = 1$ mW. Plot the gain obtained by the undepleted pump approximation given by Eq. (6.207) and compare the analytical result (Eq. (6.207)) and that obtained by the numerical solution of Eqs. (6.200) and (6.201). Assume $\alpha_s = 0.2$ dB/km and $\alpha_p = 0.5$ dB/km.

6.15 Provide an explanation as to why gain saturates for large signal powers in any type of amplifier.

6.16 In a distributed Raman amplifier system, the pump power of the input $= 250$ mW, effective area of the pump mode $= 30$ µm^2, loss coefficient at the pump wavelength $= 9.5 \times 10^{-5}$ m^{-1}, Raman gain coefficient $g_R = 6 \times 10^{-14}$ m/W, and length $= 50$ km. Calculate the gain of the amplifier under the undepleted pump approximation.

(Ans: 7.17 dB.)

6.17 In a hybrid Raman/EDFA amplified system, the fiber loss at the signal wavelength = 0.18 dB/km, EDFA gain = 14 dB. The hybrid Raman/EDFA compensates for the fiber loss exactly. Calculate the Raman pump power. Assume the following parameters: $A_p = 25$ μm^2, $\alpha_p = 9 \times 10^{-5}$ m^{-1}, $g_R = 5.8 \times 10^{-14}$ m/W, fiber length = 100 km.

(Ans: 96.99 mW.)

6.18 Explain the difference between single Rayleigh scattering and double Rayleigh scattering.

Further Reading

A.E. Siegman, *Lasers*. University Science Books, Mill Valley, CA, 1986.

E. Desurvire, *Erbium-doped Fiber Amplifiers* John Wiley & Sons, Hoboken, NJ, 2002.

P.C. Becker, N.A. Olsson, and J.R. Simpson, *Erbium-doped Fiber Amplifiers, Fundamentals and Technology*. Academic Press, SanDiego, 1999.

P.W. Milonni and J.H. Eberly, *Laser Physics*. John Wiley & Sons, Hoboken, NJ, 2010.

G.P. Agrawal and N.K. Datta, *Semiconductor Lasers*, 2nd edn., Van Nostrand Reinhold, New York, 1993.

B.E.A. Saleh and M.C. Teich, *Fundamentals of Photonics*, 2nd edn., John Wiley & Sons, Hoboken, NJ, 2007.

J.M. Senior, *Optical Fiber Communications*, 2nd edn., Prentice-Hall, London, 1992.

G. Keiser, *Optical Fiber Communications*, 4th edn., McGraw-Hill, New York, 2011.

G. Lachs, *Fiber Optic Communications*. McGraw-Hill, New York, 1998.

References

[1] G. Keiser, *Optical Fiber Communications*, 4th edn. McGraw-Hill, New York, 2011, chapter 11.

[2] P.C. Becker, N.A. Olsson, and J.R. Simpson, *Erbium-doped Fiber Amplifiers, Fundamentals and Technology*. Academic Press, San Diego, 1999.

[3] H. Kogelnik and A. Yariv, *Proc. IEEE*, vol. **52**, p. 165, 1964.

[4] A. Yariv, *Opt. Lett.*, vol. **15**, p. 1064, 1990.

[5] G. Eisenstein and R.M. Jopson, *Int. J. Electron.*, vol. **60**, p. 113, 1986.

[6] M.J. Mahonny, *J. Lightwave Technol.*, vol. **6**, p. 531, 1988.

[7] T. Saitoh and T. Mukai, *J. Lightwave Technol.*, vol. **6**, p. 1656, 1988.

[8] T. Saitoh, T. Mukai, and O. Mukami, *J. Lightwave Technol.*, vol. **LT3**, p. 288, 1985.

[9] E. Desurvire, *Erbium-doped Fiber Amplifiers*. John Wiley & Sons, Hoboken, NJ, 2002.

[10] M.N. Islam, *IEEE J. Select. Top. Quant. Electron.*, vol. **8**, p. 548, 2002.

[11] K. Rottwitt and A.J. Stentz, Raman amplification in lightwave communication systems. In I.P. Kaminow and L. Tingye (eds), *Optical Fiber Telecommunications IV A*. Academic Press, San Diego, 2002, p. 217.

[12] A.F. Evans, A. Kobyakov, and M. Vasilyev, Distributed Raman transmission: applications and fiber issues. In M.N. Islam (ed.), *Raman Amplifiers in Telecommunications 2: Sub-Systems and Systems*. Springer-Verlag, New York, 2004, p. 383.

[13] Y. Emori, S Kado, and S. Namiki, *Opt. Fiber Technol.*, vol. **8**, p. 383, 2002.

[14] R. Shuker and R.W. Gammon, *Phys. Rev. Lett.*, vol. **25**, p. 222, 1970.

[15] G.P. Agrawal, *Nonlinear Fiber Optics*, 4th edn. Academic Press, New York, 2007, chapter 8.

7

Transmission System Design

7.1 Introduction

So far, we have discussed photonic/optoelectronic components such as lasers, modulators, optical fibers, optical amplifiers, and receivers. In this chapter, we put together these components to form a fiber-optic transmission system. Critical system/signal parameters that affect the performance are identified and design guidelines are provided. In Section 7.2, the performance of a simple fiber-optic system consisting of a transmitter, a fiber, a preamplifier, and a receiver is analyzed. The transmission performance advantage of a coherent receiver over the direct detection receiver for this unrepeated system is discussed. Section 7.3 covers the dispersion-induced limitations and provides a simple design rule relating the bit rate, dispersion coefficient, and reach. In Section 7.4, optical amplifier noise-induced limitations are discussed. For a long-haul fiber-optic system, optical amplifier noise is one of the dominant impairments. A design rule pertaining to amplifier spacing, number of amplifiers, and total reach is also discussed in Section 7.4.

7.2 Fiber Loss-Induced Limitations

Let us consider an unrepeated direct detection system based on OOK, as shown in Fig. 7.1. Let P_{in} be the transmitted power when '1' is sent. The received power is

$$P_r = \begin{cases} P_{1r} = P_{in} \exp\left(-\alpha L\right) & \text{when '1' is sent,} \\ P_{0r} = 0 & \text{when '0' is sent.} \end{cases} \tag{7.1}$$

The variances of shot noise and thermal noise are given by Eqs. (5.72) and (5.76). Ignoring the dark current, for bit '1', we have

$$\sigma^2_{1,\text{shot}} = 2qI_1B_e, \tag{7.2}$$

$$\sigma^2_{1,\text{thermal}} = 4K_BTB_e/R_L, \tag{7.3}$$

where

$$I_1 = RP_{in} \exp\left(-\alpha L\right) \tag{7.4}$$

is the mean photocurrent of bit '1'. The mean photocurrent of bit '0' is

$$I_0 = RP_{0r} = 0. \tag{7.5}$$

Fiber Optic Communications: Fundamentals and Applications, First Edition. Shiva Kumar and M. Jamal Deen.
© 2014 John Wiley & Sons, Ltd. Published 2014 by John Wiley & Sons, Ltd.

Figure 7.1 A simple fiber-optic system consisting of a transmitter, a receiver, and an optical fiber.

The total variance is

$$\sigma_1^2 = \sigma_{1,\text{shot}}^2 + \sigma_{1,\text{thermal}}^2 = 2qI_1B_e + 4K_BTB_e/R_L. \tag{7.6}$$

For bit '0', the mean photocurrent I_0 is zero and, therefore, the shot noise variance is negligible. The total noise variance is

$$\sigma_0^2 = 4K_BTB_e/R_L. \tag{7.7}$$

Fig. 7.2 shows a plot of current vs. time when the bit pattern is 1011. When the bit pattern is long, it is more convenient to superpose the signals in two bit slots and obtain the eye diagram as shown in Fig. 7.3. If there is no noise in the system, the lines overlap and the eye diagram would have four lines; the eye is then said to be wide open (see Fig. 7.3(a)). In the presence of noise, the current in each bit slot fluctuates and the eye would be partially closed (see Fig. 7.3(b)). If the difference between I_1 and I_0 is small, the eye opening is small and if there is noise, this would lead to poor system performance. Therefore, to assess the quality of a signal at the receiver, the Q-factor is defined as

$$Q = \frac{I_1 - I_0}{\sigma_1 + \sigma_0}. \tag{7.8}$$

Here, I_1 and I_0 are the mean currents at the upper level (bit '1') and lower level (bit '0') of the eye diagram, respectively, and σ_1 and σ_0 are the standard deviations of bit '1' and bit '0', respectively. The analytical expressions for these quantities are given by Eqs. (7.4)–(7.7). Physically, σ_j is a measure of the spread of levels of bit 'j', $j = 0, 1$, and I_j is the mean of the levels of bit 'j' in the eye diagram. When the difference

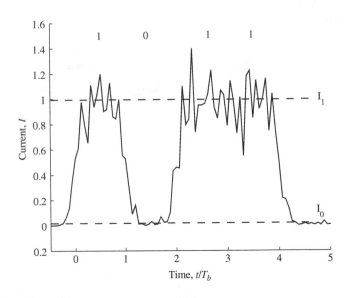

Figure 7.2 Time diagram of the current at the receiver.

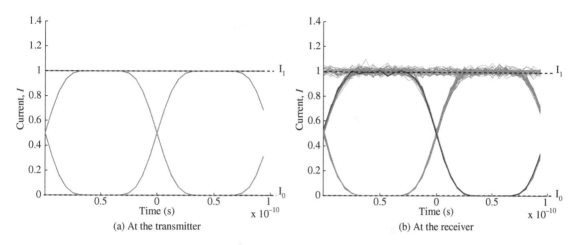

Figure 7.3   Eye diagrams. (a) At the transmitter and (b) at the receiver.

between the mean levels $I_1 - I_0$ is large and/or the spreads of the levels are small, the eye is wide open and the Q-factor is large. Using Eqs. (7.4)–(7.8), the Q-factor may be written as

$$Q = \frac{RP_{1r}}{\sqrt{aP_{1r} + b} + \sqrt{b}},\tag{7.9}$$

where

$$P_{1r} = P_{in} \exp{(-\alpha L)},\tag{7.10}$$

$$a = 2qRB_e,\tag{7.11}$$

$$b = 4k_B T B_e / R_L.\tag{7.12}$$

From Eqs. (7.9)–(7.10), we see that as the fiber loss increases, Q decreases. At the receiver, the samples of current are taken at $t = nT_b$ and if the current sample is higher than the threshold current, I_T, the decision circuit decides that a bit '1' is sent. Otherwise, a bit '0' is sent. In the presence of noise and distortion, when a bit '1' is sent, the received current sample could be lower than I_T, causing a bit error. Suppose there are N_e bit errors in a long bit sequence consisting of N bits; the bit error rate is defined as

$$BER = \lim_{N \to \infty} \frac{N_e}{N}.\tag{7.13}$$

If we assume that the noise is Gaussian distributed, BER can be related to the Q-factor by (see Chapter 8)

$$BER = \frac{1}{2}\text{erfc}\left(\frac{Q}{\sqrt{2}}\right) \approx \frac{\exp(-Q^2/2)}{\sqrt{2\pi}Q}.\tag{7.14}$$

When the variances of bit '1' and '0' are large or the difference between the means of '1' and '0' is small, Q is small and hence the BER becomes large. To achieve a BER of 10^{-9}, the required Q is 6. If $Q < 6$, BER $> 10^{-9}$. Therefore, the maximum transmission distance to achieve the fixed BER is determined by the total loss in the system. Fig. 7.4 shows the BER as a function of transmission distance L for a 10-Gb/s system. For fixed fiber length, the BER decreases as the received power (or the fiber launch power) increases. Suppose

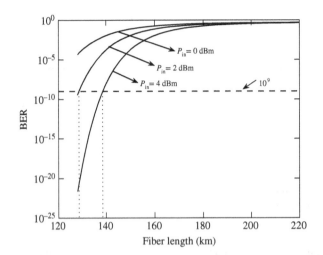

Figure 7.4 BER as a function of fiber length. $B = 10$ Gb/s, $B_e = 7.5$ GHz, $R_L = 1$ kΩ, $R = 1$ A/W, $T = 290$ K, fiber loss coefficient $\alpha = 0.2$ dB/km.

Figure 7.5 A fiber-optic system consisting of a transmitter, a fiber, an amplifier, and a receiver.

we introduce a preamplifier of gain G, as shown in Fig. 7.5. Now, the received power is $GP_{\text{in}} \exp(-\alpha L)$ when '1' is sent. The preamplifier adds ASE noise with the PSD per polarization given by Eq. (6.17). The mean current for bit '0' is given by Eq. (6.84),

$$I_0 = 2R\rho_{\text{ASE}}B_o. \tag{7.15}$$

We assume that the optical filter is an ideal band-pass filter with bandwidth f_o, the electrical filter is an ideal low-pass filter with bandwidth f_e, and $f_e < f_o$. In this case $B_o = f_o$. The variance of bit '0' is

$$\sigma_0^2 = \sigma_{\text{shot},0}^2 + \sigma_{\text{thermal},0}^2 + \sigma_{sp-sp}^2. \tag{7.16}$$

Using Eqs. (5.72), (5.76), and (6.87), we find

$$\sigma_0^2 = 2qI_0f_e + \frac{4k_BTf_e}{R_L} + 2R^2\rho_{\text{ASE}}^2(2f_o - f_e)f_e. \tag{7.17}$$

Similarly, the mean and variance for bit '1' are

$$I_1 = RGP_{\text{in}} + 2R\rho_{\text{ASE}}f_o, \tag{7.18}$$

$$\sigma_1^2 = \sigma_{\text{shot},1}^2 + \sigma_{\text{thermal},1}^2 + \sigma_{s-sp}^2 + \sigma_{sp-sp}^2. \tag{7.19}$$

Using Eqs. (5.72), (5.76), (6.83), and (6.87), we find

$$\sigma_1^2 = 2qI_1f_e + \frac{4k_BTf_e}{R_L} + 2R^2\rho_{\text{ASE}}[2P_{\text{out}}f_e + \rho_{\text{ASE}}(2f_o - f_e)f_e]. \tag{7.20}$$

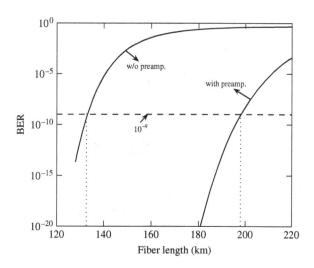

Figure 7.6 BER as a function of fiber length with and without preamplifier. $P_{in} = 2\,dBm$, $n_{sp} = 1.5$, and $G = 20\,dB$. Other parameters are the same as in Fig. 7.4.

Using Eqs. (7.15), (7.17), (7.18), and (7.20) in Eqs. (7.8) and (7.14), the BER is calculated and shown in Fig. 7.6. Using a preamplifier of 20-dB gain, the maximum transmission distance at a BER of 10^{-9} is about 200 km when the fiber launch power $P_{in} = 2\,dBm$. Note that for the same launch power, the maximum transmission distance is limited to about 130 km when no preamplifier is used. It can be increased to about 140 km if the launch power is 4 dBm (see Fig. 7.4). Here, we have ignored the fiber nonlinear effects. In the presence of fiber nonlinearity, as the launch power increases, nonlinear distortions limit the maximum achievable transmission distance (see Chapter 10).

Example 7.1

In the fiber-optic system of Fig. 7.1, it is desired that $Q \geq 6$ at the receiver. Fiber loss coefficient $\alpha = 0.046\,km^{-1}$ and length $= 130\,km$. Find the lower limit on the transmitter power. Assume $T = 25°C$, $R_L = 50\,\Omega$, $R = 1\,A/W$, and $B_e = 7\,GHz$.

Solution:
From Eq. (7.9), we have

$$Q = \frac{RP_{1r}}{\sqrt{aP_{1r} + b} + \sqrt{b}}, \tag{7.21}$$

$$P_{1r} = P_{in} \exp(-\alpha L), \tag{7.22}$$

$$a = 2qRB_e, \tag{7.23}$$
$$= 2 \times 1.602 \times 10^{-19} \times 1 \times 7 \times 10^9\,A^2/W$$
$$= 2.24 \times 10^{-9}\,A^2/W,$$

$$b = 4k_B T B_e / R_L$$

$$= \frac{4 \times 1.38 \times 10^{-23} \times 298 \times 7 \times 10^9}{50} \, \text{A}^2$$

$$= 2.3 \times 10^{-12} \text{A}^2. \tag{7.24}$$

Rearranging Eq. (7.21), we have

$$\sqrt{aP_{1r} + b} = \frac{RP_{1r}}{Q} - \sqrt{b}. \tag{7.25}$$

Squaring Eq. (7.25) and simplifying, we obtain

$$aP_{1r} = \left(\frac{RP_{1r}}{Q}\right)^2 - \frac{2RP_{1r}\sqrt{b}}{Q}. \tag{7.26}$$

or

$$P_{1r} = \frac{2\sqrt{b}}{RQ} + \frac{aQ^2}{R^2}. \tag{7.27}$$

When $Q = 6$,

$$P_{1r} = \frac{2 \times \sqrt{2.3 \times 10^{-12}}}{6} + 2.24 \times 10^{-9} \times 36$$

$$= 1.829 \, \text{mW}. \tag{7.28}$$

From Eq. (7.22),

$$P_{in} = P_{1r} \exp(\alpha L)$$

$$= 1.829 \times 10^{-2} \times \exp(0.046 \times 130)$$

$$= 7.23 \, \text{mW}. \tag{7.29}$$

The lower limit on the transmitter peak power is 7.23 mW. If $P_{in} < 7.23$ mW, $Q < 6$.

7.2.1 Balanced Coherent Receiver

Consider a fiber-optic system based on OOK. Let the output of the fiber-optic link be connected to a balanced homodyne coherent receiver as shown in Fig. 7.7. In this analysis, we ignore the LO phase noise, relative

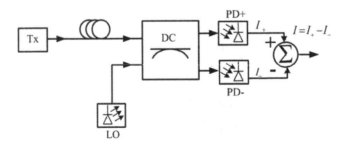

Figure 7.7 A fiber-optic system with a balanced coherent receiver.

intensity noise (RIN), and imperfections in 90° hybrid. In the presence of shot noise, Eqs. (5.109) and (5.110) are modified as (with $\omega_{\mathrm{IF}} = 0$, $\Delta\phi = \pi/2$)

$$I_+ = \frac{R}{2}\{A_r|s(t)|^2 + |A_{\mathrm{LO}}|^2 + 2A_rA_{\mathrm{LO}}\mathrm{Re}\{s(t)\}\} + n_{\mathrm{shot}+}, \tag{7.30}$$

$$I_- = \frac{R}{2}\{A_r|s(t)|^2 + |A_{\mathrm{LO}}|^2 - 2A_rA_{\mathrm{LO}}\mathrm{Re}\{s(t)\}\} + n_{\mathrm{shot}-}, \tag{7.31}$$

where $n_{\mathrm{shot}+}$ and $n_{\mathrm{shot}-}$ are the shot noise introduced by PD^+ and PD^-, respectively. Subtracting Eq. (7.30) from Eq. (7.31), we have

$$I = I_+ - I_- = 2RA_rA_{\mathrm{LO}}\mathrm{Re}\{s(t)\} + n_{\mathrm{shot}+} - n_{\mathrm{shot}-}. \tag{7.32}$$

Let

$$n_{\mathrm{shot}} = n_{\mathrm{shot}+} - n_{\mathrm{shot}-}. \tag{7.33}$$

Since $n_{\mathrm{shot}+}$ and $n_{\mathrm{shot}-}$ are statistically independent, the variance of n_{shot} is the sum of the variances of $n_{\mathrm{shot}+}$ and $n_{\mathrm{shot}-}$:

$$\begin{aligned} \sigma_{\mathrm{shot}}^2 &= \sigma_{\mathrm{shot}+}^2 + \sigma_{\mathrm{shot}-}^2 \\ &= 2qI_+B_e + 2qI_-B_e. \end{aligned} \tag{7.34}$$

Let

$$P_{\mathrm{LO}} = A_{\mathrm{LO}}^2, \tag{7.35}$$

$$P_r(t) = A_r^2 s^2(t). \tag{7.36}$$

Here, P_{LO} and P_r are the LO and receiver power, respectively. First consider the OOK. For bit '1' ($s(t) = 1$ within the bit slot), the mean and variances of the current are

$$I_1 = 2R\sqrt{P_{1r}P_{\mathrm{LO}}}, \tag{7.37}$$

$$P_{1r} = A_r^2, \tag{7.38}$$

$$\sigma_{1,\mathrm{shot}}^2 = 2qI_{1+}B_e + 2qI_{1-}B_e, \tag{7.39}$$

$$I_{1+} = \frac{R}{2}\left\{P_{1r} + P_{\mathrm{LO}} + 2\sqrt{P_{1r}P_{\mathrm{LO}}}\right\}, \tag{7.40}$$

$$I_{1-} = \frac{R}{2}\left\{P_{1r} + P_{\mathrm{LO}} - 2\sqrt{P_{1r}P_{\mathrm{LO}}}\right\}. \tag{7.41}$$

Using Eqs. (7.40) and (7.41), Eq. (7.39) becomes

$$\sigma_{1,\mathrm{shot}}^2 = 2qB_eR(P_{1r} + P_{\mathrm{LO}}). \tag{7.42}$$

The total noise variance of '1' is

$$\begin{aligned} \sigma_1^2 &= \sigma_{1,\mathrm{shot}}^2 + \sigma_{1,\mathrm{thermal}}^2 \\ &= 2qB_eR(P_{1r} + P_{\mathrm{LO}}) + 4k_BTB_e/R_L. \end{aligned} \tag{7.43}$$

Similarly, for bit '0', we have

$$I_0 = 0, \tag{7.44}$$

$$\sigma_{0,shot}^2 = 2qB_e RP_{LO}, \tag{7.45}$$

$$\sigma_0^2 = 2qB_e RP_{LO} + 4k_B TB_e/R_L. \tag{7.46}$$

The Q-factor is calculated as

$$Q_{OOK} = \frac{I_1 - I_0}{\sigma_1 + \sigma_0}. \tag{7.47}$$

Some approximations can be made to Eq. (7.47) to gain some insight. When $P_{LO} \gg P_{1r}$, from Eqs. (7.42) and (7.45) we have

$$\sigma_{1,shot}^2 = \sigma_{0,shot}^2 = 2qB_e RP_{LO}. \tag{7.48}$$

Let the photocurrent due to LO be I_{LO}. Eq. (7.48) may be rewritten as

$$\sigma_{shot,\ eff}^2 \equiv \sigma_{1,shot}^2 = 2qI_{LO}B_e. \tag{7.49}$$

Comparing Eq. (7.49) with Eq. (5.72), the effective PSD of shot noise in balanced detection is

$$\rho_{shot,\ eff} = qI_{LO} \tag{7.50}$$

If P_{LO} is sufficiently large, the shot noise will dominate the thermal noise and it may be ignored in Eqs. (7.43) and (7.46),

$$\sigma_1^2 = \sigma_0^2 = 2qB_e RP_{LO}. \tag{7.51}$$

Now Eq. (7.47) reduces to

$$Q_{OOK} = \frac{2R\sqrt{P_{1r}P_{LO}}}{2\sqrt{2qB_e RP_{LO}}},$$

$$= \sqrt{\frac{RP_{1r}}{2qB_e}}. \tag{7.52}$$

Note that the Q-factor is independent of P_{LO} under these conditions. Using Eq. (5.17), the Q-factor may be rewritten as

$$Q_{OOK} = \sqrt{\frac{\eta P_{1r}}{2h\bar{f}B_e}}. \tag{7.53}$$

where $\bar{f}$ is the mean frequency. The energy of a bit '1' at the receiver is $E_{1r} = P_{1r}T_b$. If the receiver filter is an ideal Nyquist filter ($B_e = 1/(2T_b)$), Eq. (7.53) becomes

$$Q_{OOK} = \sqrt{\eta N_{1r}}, \tag{7.54}$$

where $N_{1r} = E_{1r}/h\bar{f}$ is the number of signal photons of bit '1'. For an OOK signal, the mean number of received photons per bit, $\bar{N}_{rec} = N_{1r}/2$. So, Eq. (7.54) becomes [1]

$$Q_{OOK} = \sqrt{2\eta \bar{N}_{rec}}. \tag{7.55}$$

For an ideal photondetector, $\eta = 1$. To have a BER of 10^{-9}, $Q = 6$ and from Eq. (7.55), we see that the average number of signal photons per bit, $\bar{N}_{rec}$, should be 18. In other words, if the mean number of signal photons is less than 18 for an OOK system, the BER exceeds 10^{-9}. Since $P_r = P_{in} \exp(-\alpha L)$, the maximum transmission distance at which the mean number of photons becomes equal to 18 can easily be calculated. For example, when the peak fiber launch power $P_{in} = 2\,\text{dBm}$ and bit rate $= 10\,\text{Gb/s}$, the peak received power is

$$P_r(\text{dBm}) = P_{in}(\text{dBm}) - \text{loss(dB)}, \tag{7.56}$$

$$P_r = \frac{N_{1r}h\overline{f}}{T_b}, \tag{7.57}$$

$$P_r(\text{dBm}) = 10\log_{10}(2\overline{N}_{rec}h\overline{f}B) \tag{7.58}$$

$$= -43.3\,\text{dBm},$$

where $\overline{N}_{rec} = 18$ and $\overline{f} = 193.54\,\text{THz}$ is used. From Eq. (7.56), we find

$$\text{loss(dB)} = P_{in}(\text{dBm}) - P_r(\text{dBm}) \tag{7.59}$$

$$= 2\,\text{dBm} - (-43.3)\,\text{dBm}$$

$$= 45.3\,\text{dB}.$$

With fiber loss coefficient $\alpha = 0.2\,\text{dB/km}$, the maximum transmission distance to reach the BER of 10^{-9} is about 225 km. This should be compared with the results shown in Fig. 7.4 for the case of a direct detection system, in which the transmission distance is limited to about 130 km for the same launch power. Fig. 7.8 shows the BER as a function of the length for various launch powers. Solid line and × marks show the BER using the exact Q (Eq. (7.47)) and the approximate Q (Eq. (7.52)), respectively. As can be seen, there is a good agreement between the two. However. if the LO power is not sufficiently large, the thermal noise could dominate the shot noise and Eq. (7.52) would not be accurate. Fig. 7.9 shows the dependence of BER on the LO power for different load resistances, calculated using Eq. (7.47). As can be seen, when the LO power is small, the BER is large due to thermal noise. When the LO power is large, the shot noise dominates over the thermal noise and the BER becomes roughly equal for a range of load resistances.

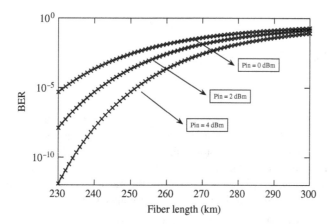

Figure 7.8 BER as a function of the fiber length for a fiber-optic system based on OOK with balanced coherent detection. Parameters: $P_{LO} = 100\,\text{mW}$, $T = 290\,\text{K}$, $R_L = 100\,\Omega$, and $\eta = 1$. Laser phase noise is ignored.

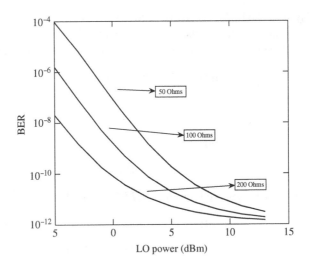

Figure 7.9 BER as a function of LO power. $L = 230$ km, other parameters are the same as those of Fig. 7.8 except for LO power and R_L.

Next, let us consider PSK. For bit '1', the mean and variances are the same as those of OOK given by Eqs. (7.37)–(7.46). For bit '0',

$$I_0 = -2R\sqrt{P_{1r}P_{LO}}, \tag{7.60}$$

$$\sigma_0^2 = \sigma_1^2. \tag{7.61}$$

When the P_{LO} is sufficiently large, the Q-factor can be calculated as before:

$$Q_{PSK} = \sqrt{\frac{2\eta P_{1r}}{hf B_e}}$$

$$= 2\sqrt{\eta N_{1r}}. \tag{7.62}$$

For PSK, $\overline{N}_{rec} = N_{1r}$. So, Eq. (7.62) becomes

$$Q_{PSK} = 2\sqrt{\eta \overline{N}_{rec}}. \tag{7.63}$$

To have a BER of 10^{-9}, the average number of signal photons per bit should be 9 assuming $\eta = 1$ [1, 2]. Comparing Eqs. (7.55) and (7.63), we see that the receiver sensitivity can be improved by 3 dB using PSK for the fixed number of mean received photons. Fig. 7.10 shows the theoretical limit on the achievable BER for a shot noise-limited system. As can be seen, for the given mean received power, the PSK outperforms the OOK. In other words, to achieve a given BER, the mean received power for OOK should be 3 dB higher than that for PSK. The reason for the superior performance of the PSK is that constellation points are separated by $2\sqrt{\overline{P}_{rec}}$ for PSK, whereas the corresponding separation for OOK is $\sqrt{2\overline{P}_{rec}}$ ($\overline{P}_{rec} = P_{1,rec}/2$). In [3], the receiver sensitivity close to the shot noise limit is experimentally demonstrated in a 10-Gb/s PSK system.

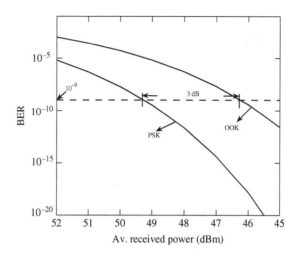

Figure 7.10   BER as a function of the mean received power. Thermal noise is ignored.

Example 7.2

In a 1.55-μm coherent fiber-optic system as shown in Fig. 7.7, the parameters are as follows. Mean fiber launch power = 1 dBm, fiber loss = 0.2 dB/km, fiber length = 240 km, quantum efficiency $\eta = 0.7$, $T = 290\,\mathrm{K}$, $R_L = 100\,\Omega$, $P_{LO} = 10\,\mathrm{dBm}$, and $B_e = 7.5\,\mathrm{GHz}$. Calculate exact and approximate Q-factor if the signal is (a) OOK, (b) PSK. Assume that the NRZ format with rectangular pulses is used.

Solution:
The mean frequency

$$\bar{f} = \frac{c}{\lambda} = \frac{3 \times 10^8}{1.55 \times 10^{-6}} = 193.54\,\mathrm{THz}. \tag{7.64}$$

The responsivity is given by Eq. (5.17),

$$R = \frac{\eta q}{h\bar{f}} = \frac{0.7 \times 1.602 \times 10^{-9}}{6.626 \times 10^{-34} \times 193.54 \times 10^{12}}\,\mathrm{A/W}.$$

$$= 0.874\,\mathrm{A/W}. \tag{7.65}$$

(a) For OOK, the peak power is twice the average power $\bar{P}_{in}$ when the NRZ with rectangular pulses is used:

$$P_{in} = 2\bar{P}_{in}. \tag{7.66}$$

Eq. (7.66) may be rewritten in dB units as

$$P_{in}(\mathrm{dBm}) = 10\log_{10}2 + \bar{P}_{in}(\mathrm{dBm}) \tag{7.67}$$

$$= 4\,\mathrm{dBm}.$$

The received peak power is

$$P_{1r} = P_{in}\exp(-\alpha L), \tag{7.68}$$

$$P_{1r}(\text{dBm}) = P_{\text{in}}(\text{dBm}) - \text{loss}(\text{dB}) \qquad (7.69)$$

$$= 4 - 0.2 \times 240 = -44\,\text{dBm}$$

$$= 10^{P_{1r}(\text{dBm})/10}\,\text{mW} = 3.98 \times 10^{-5}\,\text{mW},$$

$$P_{\text{LO}} = 10^{P_{\text{LO}}(\text{dBm})/10}\,\text{mW} = 10\,\text{mW}. \qquad (7.70)$$

The mean of bit '1' is

$$I_1 = 2R\sqrt{P_{1r}P_{\text{LO}}} \qquad (7.71)$$

$$= 2 \times 0.874\sqrt{3.98 \times 10^{-5} \times 10}\,\text{mA}$$

$$= 3.48 \times 10^{-2}\,\text{mA}.$$

The total noise variance of '1' is given by Eq. (7.43),

$$\sigma_1^2 = 2qB_e R(P_{1r} + P_{\text{LO}}) + 4k_B T B_e/R_L \qquad (7.72)$$

$$= 2 \times 1.602 \times 10^{-19} \times 7.5 \times 10^9 \times 0.874 \times (3.98 \times 10^{-8} + 10^{-2})$$

$$+ 4 \times 1.38 \times 10^{-23} \times 290 \times 7.5 \times 10^9 / 100\,\text{A}^2$$

$$= 2.22 \times 10^{-11}\,\text{A}^2.$$

The mean and variance of '0' are calculated as follows:

$$I_0 = 0. \qquad (7.73)$$

Since $P_{1r} \ll P_{\text{LO}}$, comparing Eqs. (7.43) and (7.46), we find

$$\sigma_1^2 = \sigma_0^2 = 2.22 \times 10^{-11}\,\text{A}^2, \qquad (7.74)$$

$$Q = \frac{I_1 - I_0}{\sigma_1 + \sigma_0} \qquad (7.75)$$

$$= \frac{3.48 \times 10^{-5}}{2 \times \sqrt{2.22 \times 10^{-11}}} = 3.7.$$

The approximate Q-factor is given by Eq. (7.53),

$$Q = \sqrt{\frac{\eta P_{1r}}{2h\bar{f}B_e}} = 3.8. \qquad (7.76)$$

(b) For PSK, the peak power is the same as the average power,

$$P_{\text{in}} = \overline{P}_{\text{in}}. \qquad (7.77)$$

$$P_{\text{in}}(\text{dBm}) = 1\,\text{dBm}. \qquad (7.78)$$

The peak power at the receiver is

$$P_{1r}(\text{dBm}) = P_{\text{in}}(\text{dBm}) - \text{loss}(\text{dB}) \qquad (7.79)$$

$$= 1 - 0.2 \times 240$$

$$= -47\,\text{dBm}, \qquad (7.80)$$

$$P_{1r} = 1.99 \times 10^{-5}\,\text{mW}. \tag{7.81}$$

The mean of bit '1' is

$$I_1 = 2R\sqrt{P_{1r}P_{\text{LO}}}, \tag{7.82}$$

$$= 2.46 \times 10^{-2}\,\text{mA}. \tag{7.83}$$

The noise variance of '1' may be calculated as before,

$$\sigma_1^2 = 2qB_e R(P_{1r} + P_{\text{LO}}) + 4k_B T B_e / R_L \tag{7.84}$$

$$= 2.22 \times 10^{-11}\,\text{A}^2. \tag{7.85}$$

For bit '0',

$$I_0 = -I_1 \tag{7.86}$$

$$= -2.46 \times 10^{-2}\,\text{mA},$$

$$\sigma_0 = \sigma_1. \tag{7.87}$$

Therefore, the Q-factor is

$$Q = \frac{I_1}{\sigma_1} = 5.23. \tag{7.88}$$

The approximate Q-factor for PSK is given by Eq. (7.62),

$$Q_{\text{PSK}} = \sqrt{\frac{2\eta P_{1r}}{hf B_e}} \tag{7.89}$$

$$= 5.38. \tag{7.90}$$

7.3 Dispersion-Induced Limitations

Consider a simple fiber-optic system consisting of an OOK transmitter, a receiver, and a fiber, as shown in Fig. 7.11. Fig. 7.12 shows the input and output bit patterns. As can be seen, a pulse corresponding to bit '1' broadens and occupies the adjacent bit slot corresponding to bit '0'. This is known as inter-symbol interference (ISI), and it leads to performance degradation. To estimate the maximum achievable transmission distance with negligible ISI for the given bit rate, let us consider a single Gaussian pulse launched to the fiber,

$$u(t,0) = A_{\text{in}} \exp\left(-\frac{t^2}{2T_0^2}\right). \tag{7.91}$$

Figure 7.11 A simple fiber-optic system.

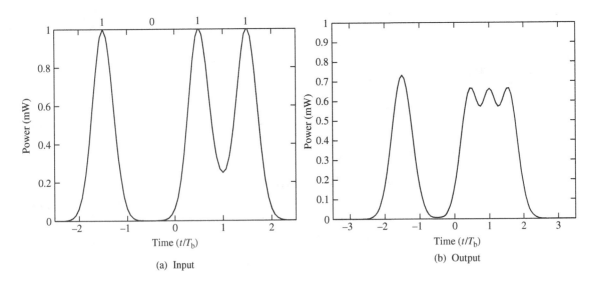

(a) Input   (b) Output

Figure 7.12   Input and output of the fiber. $\beta_2 = -21 \text{ ps}^2/\text{km}$, $L = 40 \text{ km}$, FWHM $= 50 \text{ ps}$, bit rate $= 10 \text{ Gb/s}$. Fiber loss is ignored. (a) Input and (b) output.

After propagating a distance L, the power distribution is given by Eq. (2.161),

$$P(t, L) = P_{\text{in}} \exp\left(-\frac{t^2}{T_L^2}\right), \tag{7.92}$$

where

$$T_L^2 = \frac{T_0^4 + \beta_2^2 L^2}{T_0^2}. \tag{7.93}$$

The FWHM at the transmitter and the receiver is $1.665T_0$ and $1.665T_L$, respectively. From Eq. (7.93), we see that if we choose very small T_0, T_L becomes very large since T_0^2 appears in the denominator. If we choose very large T_0, T_L could become large when $T_0^4 \gg \beta_2^2 L^2$. Therefore, for the given $|\beta_2| L$, T_0 has to be optimized. The optimum T_0 can be found by setting

$$\frac{dT_L}{dT_0} = 0. \tag{7.94}$$

Using Eq. (7.93) in Eq. (7.94), we find the optimum T_0 as

$$T_0^{\text{opt}} = \sqrt{\beta_2 L}. \tag{7.95}$$

The r.m.s. width of a Gaussian pulse is related to T_0 by [4]

$$\sigma(z = 0) \equiv \sigma_0 = T_0/\sqrt{2}, \tag{7.96}$$

$$\sigma(z = L) \equiv \sigma_L = T_L/\sqrt{2}. \tag{7.97}$$

Now, Eq. (7.93) may be rewritten as

$$\sigma_L^2 = \frac{4\sigma_0^4 + \beta_2^2 L^2}{4\sigma_0^2}. \tag{7.98}$$

To have negligible ISI, the pulse at the receiver should remain within its bit slot. The commonly used criterion is [4]

$$\sigma_L \le T_B/4, \tag{7.99}$$

where T_B is the bit interval. Using Eqs. (7.98) and (7.99), we obtain

$$\left[\frac{4\sigma_0^4 + \beta_2^2 L^2}{4\sigma_0^2} \right]^{1/2} B \le \frac{1}{4}. \tag{7.100}$$

If we choose the optimum pulse width T_0^{opt}, the corresponding σ_0^{opt} is $\sqrt{\beta_2 L/2}$. Using this value of σ_0 in Eq. (7.100), we find [4]

$$B(|\beta_2|L)^{1/2} \le \frac{1}{4}. \tag{7.101}$$

For fixed $|\beta_2|$, as the bit rate increases linearly, the maximum transmission distance decreases as $L^{-1/2}$. To undo the pulse broadening due to fiber dispersion, a DCF may be used (see Chapter 2) or it may be compensated in the electrical domain using a DSP (see Chapter 11). In some applications such as metro/access networks, it would be expensive to use DCF or coherent receivers. For such applications, Eq. (7.101) provides a simple design rule relating the reach, bit rate, and dispersion.

Example 7.3

A fiber-optic system is upgraded to operate at 10 Gb/s from 2.5 Gb/s. The maximum transmission distance at 2.5 Gb/s at a BER of 10^{-9} was 100 km. Find the corresponding distance at 10 Gb/s. Assume that the transmission fiber is unchanged and the penalty due to fiber dispersion is the same in both systems.

Solution:
Let

$$B_1 = 2.5 \, \text{Gb/s}, \ B_2 = 10 \, \text{Gb/s},$$

$$L_1 = 160 \, \text{km}.$$

From Eq. (7.101), we have

$$B_1 L_1^{1/2} = B_2 L_2^{1/2},$$

$$L_2 = 10 \, \text{km}.$$

7.4 ASE-Induced Limitations

To transmit an optical signal over a long distance, amplifiers have to be introduced along the transmission line. Otherwise, the received power could be too low to detect. In Section 7.2.1, we found that 9 photons/bit is required at the receiver for a PSK signal to achieve a BER of 10^{-9}. The number of photons/bit at the receiver can be increased by introducing amplifiers. However, amplification by stimulated emission is always accompanied by ASE, which enhances the noise in the system. In a long-haul fiber-optic system consisting of a chain of amplifiers, ASE builds up over many amplifiers, which degrades the transmission performance.

Figure 7.13   A long-haul fiber-optic system consisting of a transmitter, a receiver, N fibers, and N amplifiers.

In this section, we ignore the fiber dispersion and consider only the fiber loss, amplifier gain, and ASE. Fig. 7.13 shows a fiber-optic system consisting of transmission fibers and in-line amplifiers. Let H_j and G_j, $j = 1, \dots, N$ be the fiber loss and amplifier gain of the jth stage, respectively. The power spectral density of ASE introduced by the jth amplifier per polarization is

$$\rho_{\text{ASE},j} = n_{sp}h\overline{f}(G_j - 1). \tag{7.102}$$

Let us first consider the signal propagation in the absence of noise. Let $\overline{P}_{\text{in}}$ be the mean transmitter output power. In this section, we assume that the transmitter output is CW. Later, in Sections 7.4.3 and 7.4.4, we consider the OOK/PSK modulation formats. The received power is

$$P_r = \prod_{j=1}^{N} H_j G_j \overline{P}_{\text{in}}. \tag{7.103}$$

Next, let us consider the propagation of ASE due to the first amplifier. Let the full bandwidth of the optical filter be Δf. The noise power per polarization within the filter bandwidth immediately after the first amplifier is $n_{sp}h\overline{f}(G - 1)\Delta f$. Therefore, the mean noise power at the receiver due to the first amplifier is

$$P_{1,\text{ASE}} = n_{sp}h\overline{f}(G_1 - 1)\Delta f \prod_{j=2}^{N} H_j G_j. \tag{7.104}$$

The mean noise power at the receiver due to the nth amplifier is

$$P_{n,\text{ASE}} = n_{sp}h\overline{f}(G_n - 1)\Delta f \prod_{j=n+1}^{N} H_j G_j. \tag{7.105}$$

The total mean noise power at the receiver due to all the amplifiers is

$$P_{\text{ASE}} = \sum_{n=1}^{N} P_{n,\text{ASE}} = n_{sp}h\overline{f}\Delta f \sum_{n=1}^{N}(G_n - 1) \prod_{j=n+1}^{N} H_j G_j. \tag{7.106}$$

When an in-line amplifier fully compensates for the fiber loss, we have $G_j = 1/H_j$. From now on, we assume that amplifiers (and fibers) are identical and $G_j = 1/H_j$. Now, Eq. (7.106) can be simplified as

$$P_{\text{ASE}} = N n_{sp}h\overline{f}\Delta f(G - 1), \tag{7.107}$$

where $G = G_j, j = 1, 2, \dots, N$. The cascade of in-line amplifiers and fibers is equivalent to a single amplifier with unity gain and power spectral density of ASE,

$$\rho_{\text{ASE}}^{\text{eq}} = N n_{sp}h\overline{f}(G - 1). \tag{7.108}$$

In this section, we assume ideal in-line amplifiers. If EDFAs are used as in-line amplifiers, the saturation power of each successive amplifier has to be increased slightly to compensate for the gain saturation caused by the build-up of the ASE [5]. The optical signal-to-noise ratio is defined as (Eq. (6.107))

$$OSNR = \frac{\text{mean signal power}}{\text{mean noise power in a bandwidth of } B_{opt}}, \tag{7.109}$$

where B_{opt} is the reference bandwidth, typically chosen to be 12.5 GHz. The noise power in both polarizations is twice that given by Eq. (7.107). Using Eqs. (7.103) and (7.107), we find the OSNR at the receiver to be

$$OSNR = \frac{\overline{P}_{in}}{2Nn_{sp}h\overline{f}(G-1)B_{opt}}. \tag{7.110}$$

Sometimes, it is convenient to express OSNR in dB units. Assuming $G \gg 1$ and noise figure $F_n \cong 2n_{sp}$, and dividing the numerator and denominator of Eq. (7.110) by 1 mW, it can be written in dB units as

$$OSNR(dB) = \overline{P}_{in}[dBm] - N[dB] - G[dB] - F_n[dB] - 10\log_{10}\left(\frac{h\overline{f}B_{opt}}{1\,mW}\right)$$

$$= \overline{P}_{in}[dBm] - N[dB] - G[dB] - F_n[dB] + 58, \tag{7.111}$$

where we have used $\overline{f} = 194$ THz.

7.4.1 Equivalent Noise Figure

Fig. 7.14(a) shows a two-stage amplifier with loss element, such as a dispersion compensation module between the two stages. Fig. 7.14(b) shows the equivalent amplifier with gain G_{eq} and noise figure $F_{n,eq}$. From Fig. 7.14(a), it is easy to see that

$$P_{out} = G_1 H G_2 \overline{P}_{in}. \tag{7.112}$$

Therefore,

$$G_{eq} = G_1 H G_2. \tag{7.113}$$

The ASE PSD of the amplifier j is given by Eq. (6.102),

$$\rho_{ASE,j} = (G_j F_{n,j} - 1)h\overline{f}/2, \quad j = 1, 2. \tag{7.114}$$

The noise power per polarization due to the amplifier 1 in a bandwidth of Δf is

$$N_1 = \rho_{ASE,1}\Delta f. \tag{7.115}$$

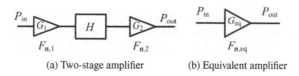

(a) Two-stage amplifier   (b) Equivalent amplifier

Figure 7.14 Two-stage amplifier with a loss element in between. (a) Two-stage amplifier and (b) Equivalent amplifier.

The ASE due to amplifier 1 is attenuated by the loss element and, then, it is amplified by the second amplifier. Therefore, the noise power due to amplifier 1 at the output end is

$$N_{out,1} = \rho_{ASE,1}\Delta f H G_2. \tag{7.116}$$

Similarly, the noise power due to amplifier 2 at the output end is $\rho_{ASE,2}\Delta f$, and the total noise power at the output is

$$N_{out} = (\rho_{ASE,1}HG_2 + \rho_{ASE,2})\Delta f. \tag{7.117}$$

Therefore, the PSD at the output is

$$\rho_{ASE,eq} = \frac{N_{out}}{\Delta f} = \rho_{ASE,1}HG_2 + \rho_{ASE,2}. \tag{7.118}$$

Using Eq. (7.114), we obtain

$$\rho_{ASE,eq} = [G_1 F_{n,1} G_2 H + G_2 F_{n,2} - HG_2 - 1]\,\overline{hf}/2. \tag{7.119}$$

For an equivalent amplifier shown in Fig. 7.14(b), we have

$$\rho_{ASE,eq} = (G_{eq}F_{n,eq} - 1)\overline{hf}/2. \tag{7.120}$$

Comparing Eqs. (7.119) and (7.120), we obtain

$$F_{n,eq} = F_{n,1} + \frac{F_{n,2}}{G_1 H} - \frac{1}{G_1}. \tag{7.121}$$

Note that in Eq. (7.121), the noise figure of the second amplifier is divided by the gain of the first amplifier and the loss, H. Therefore, in practice, an amplifier with higher noise figure is used as the second stage and/or an amplifier with higher gain is used as the first stage. Typically, $G_1 \gg 1$ and Eq. (7.121) reduces to

$$F_{n,eq} \cong F_{n,1} + \frac{F_{n,2}}{G_1 H}. \tag{7.122}$$

If two amplifiers are cascaded without a loss element in between, $H = 1$ and Eq. (7.122) becomes [4, 6]

$$F_{n,eq} \cong F_{n,1} + \frac{F_{n,2}}{G_1}. \tag{7.123}$$

The effective noise figure of a cascaded chain of k amplifiers is given by (Exercise 7.9)

$$F_{n,eq} = F_{n,1} + \frac{F_{n,2} - 1}{G_1} + \frac{F_{n,3} - 1}{G_1 G_2} + \cdots + \frac{F_{n,k} - 1}{G_1 G_2 \cdots G_{k-1}}. \tag{7.124}$$

7.4.2 Impact of Amplifier Spacing

Consider the long-haul fiber-optic system with identical amplifier gains and noise figures. The power spectral density of ASE at the output is given by (Eq. (7.108))

$$\rho_{ASE,1}^{eq} = n_{sp}\overline{hf}[\exp{(\alpha L_a)} - 1]N, \tag{7.125}$$

where L is the amplifier spacing and $G = \exp{(\alpha L_a)}$. Let the total transmission distance, L_{tot}, be NL. Eq. (7.125) can be rewritten as

$$\rho_{ASE,1}^{eq} = n_{sp}\overline{hf}[\exp{(\alpha L_a)} - 1]\frac{L_{tot}}{L_a}. \tag{7.126}$$

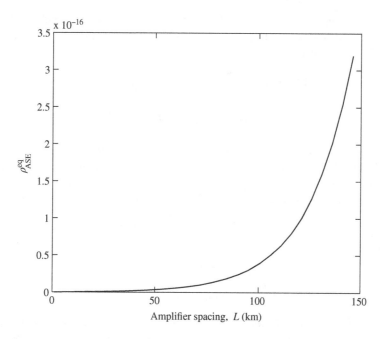

Figure 7.15 Dependence of the effective PSD on the amplifier spacing in a long-haul fiber-optic system.

Suppose we reduce the amplifier spacing by a factor of 2, but keep L_{tot} fixed. Now, the PSD is

$$\rho_{ASE,2}^{eq} = n_{sp} h\bar{f} [\exp(\alpha L_a/2) - 1]\frac{L_{tot}}{L_a/2}. \qquad (7.127)$$

By expanding the exponential function by a Taylor series, it is easy to show that $\rho_{ASE,2} < \rho_{ASE,1}$. In general, as the amplifier spacing is increased, the PSD of ASE at the output becomes large. This is because the PSD increases exponentially with L due to the first term in the square bracket of Eq. (7.126) when $\exp(\alpha L_a) \gg 1$, although it scales inversely with L_a due to the term in the denominator. In practice, $\exp(\alpha L_a) \gg 1$ and the exponential growth dominates the linear increment. Fig. 7.15 shows the PSD of ASE at the receiver as a function of amplifier spacing. As can be seen, the PSD of ASE increases almost exponentially with the amplifier spacing. From the theoretical standpoint, it is desirable to make the amplifier spacing as small as possible. However, due to practical limitations, the amplifier spacings are in the range of 60–125 km for long-haul terrestrial communication systems. If distributed amplification, such as Raman amplifiers, is used, the growth of ASE power can be substantially reduced.

7.4.3   Direct Detection Receiver

Let us consider the impact of ASE in a long-haul direct detection system consisting of a chain of amplifiers as shown in Fig. 7.13 based on OOK. When a bit '1' is sent, the optical power at the receiver is

$$P_{1r} = P_{in}, \qquad (7.128)$$

where P_{in} is the peak power. The mean currents are

$$I_1 = RP_{in} + 2R\rho_{ASE}^{eq}B_o, \qquad (7.129)$$

$$I_0 = 2R\rho_{ASE}^{eq}B_o. \qquad (7.130)$$

As in Section 7.2, we assume that the optical filter is an ideal band-pass filter with bandwidth $B_0 = f_0$, and the electrical filter is an ideal low-pass filter with bandwidth f_e. The variances of bit '0' and '1' are given by Eqs. (7.17) and (7.20), respectively as

$$\sigma_0^2 = 2qI_0 f_e + \frac{4k_B T f_e}{R_L} + 2R^2 \left(\rho_{\text{ASE}}^{\text{eq}} \right)^2 (2f_o - f_e) f_e, \tag{7.131}$$

$$\sigma_1^2 = 2qI_1 f_e + \frac{4k_B T f_e}{R_L} + 2R^2 \rho_{\text{ASE}}^{\text{eq}} [2P_{\text{in}} f_e + \rho_{\text{ASE}}^{\text{eq}} (2f_o - f_e) f_e]. \tag{7.132}$$

The long-haul fiber-optic systems are typically amplifier noise-limited and, hence, some approximations can be made while calculating the Q-factor. The variance of shot noise and thermal noise can be ignored compared with the variance of signal–spontaneous beat noise. Also, when the signal power is large, spontaneous–spontaneous beat noise can be ignored as well. Under these conditions,

$$\sigma_1^2 \cong 4R^2 P_{\text{in}} [Nn_{sp} h\bar{f}(G-1)] f_e, \tag{7.133}$$

$$\sigma_0^2 \cong 0, \tag{7.134}$$

$$Q \cong \sqrt{\frac{P_{\text{in}}}{4Nn_{sp} h\bar{f}(G-1)f_e}}. \tag{7.135}$$

From Eq. (7.135), we see that the Q-factor is independent of the responsivity R, under these approximations. Q can be increased by increasing P_{in} or decreasing the gain G. Since $G = 1/H$, Q can be increased by using low-loss fibers. From Eq. (7.135), we see that as the number of amplifiers (or n_{sp} or G) increases, P_{in} has to be increased to keep the Q-factor at a fixed value. The enhancement of the signal power to counter the noise increase is known as the *power penalty*. Suppose the number of amplifiers increases from N to $2N$, then the launched power should be doubled to keep the Q-factor fixed (or equivalently, BER fixed). In this case the power penalty is 3 dB. Using Eqs. (7.110) and (7.135), we find

$$\text{OSNR} = \frac{Q^2 f_e}{B_{\text{opt}}}, \tag{7.136}$$

where we have used $\bar{P}_{\text{in}} = P_{\text{in}}/2$ for OOK. Solid and dotted lines in Fig. 7.16 show the BER obtained using the exact Q-factor (Eqs. (7.131) and (7.132)) and approximate Q-factor (Eq. (7.135)), respectively. Since

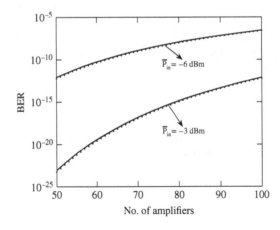

Figure 7.16 BER vs. number of amplifiers for direct detection system. Parameters: $n_{sp} = 2$, $\alpha = 0.2$ dB/km, amp. spacing $= 80$ km, gain $G = 16$ dB, $R_L = 1000\,\Omega$, $T = 200$ K, $R = 1$ A/W.

the difference between these curves is negligible, the approximation that the variance of receiver noise is much smaller than that of ASE is good. When spontaneous–spontaneous beat noise is comparable with signal–spontaneous beat noise, Eq. (7.136) needs to be modified [7–9]. Note that the Gaussian distribution is an approximation and the amplifier noise after the photodetector is actually chi-square distributed [7] (see Chapter 8).

Example 7.4

In a 1.55-μm long-haul fiber-optic system based on NRZ-OOK as shown in Fig. 7.13, 80 identical amplifiers are placed periodically with a spacing of 80 km. The mean fiber launch power $= -3$ dBm, fiber loss coefficient $\alpha = 0.0461$ km^{-1}, amplifier loss is fully compensated by the amplifiers, and $n_{sp} = 1.5$. Electrical filter bandwidth, $f_e = 7$ GHz and $f_e < f_0$. Calculate (a) OSNR in a reference bandwidth of 0.1 nm, (b) Q-factor. Ignore shot noise, thermal noise, and spontaneous–spontaneous beat noise.

Solution:
(a) Since

$$\bar{f} = \frac{c}{\lambda},$$

$$d\bar{f} = -\frac{c}{\lambda^2}d\lambda.$$

With $d\lambda = 0.1$ nm,

$$d\bar{f} = B_{\text{opt}} = \frac{-3 \times 10^8 \times (0.1 \times 10^{-9})}{(1.55 \times 10^{-6})^2} \, \text{Hz}$$

$$= 12.48 \, \text{GHz},$$

$$G = \exp(\alpha L_a)$$

$$= \exp(0.0461 \times 80)$$

$$= 39.96,$$

$$G(\text{dB}) = 10 \log_{10} G$$

$$= 16.01 \, \text{dB},$$

$$N(\text{dB}) = 10 \log_{10} 80$$

$$= 19.03 \, \text{dB},$$

$$F_n \cong 2n_{sp},$$

$$F_n(\text{dB}) = 4.77 \, \text{dB},$$

$$P_{\text{in}}(\text{dBm}) = -3 \, \text{dBm}.$$

Using Eq. (7.111), we find

$$\text{OSNR}(\text{dB}) = \bar{P}_{\text{in}}(\text{dBm}) - N(\text{dB}) - G(\text{dB}) - F_n(\text{dB}) + 58$$

$$= -3 - 19.03 - 16.01 - 4.77 + 58$$

$$= 15.19 \, \text{dB}.$$

The mean launch power

$$\overline{P}_{\text{in}} = -3\,\text{dBm}.$$

So the peak power is (assuming NRZ rectangular pulses)

$$P_{\text{in}} = 2\overline{P}_{\text{in}}$$
$$= 2 \times 10^{-(3/10)}\,\text{mW}$$
$$= 1\,\text{mW}.$$

Using Eq. (7.135), we find

$$Q = \sqrt{\frac{P_{\text{in}}}{4Nn_{sp}h\overline{f}(G-1)f_e}}$$
$$= \sqrt{\frac{1 \times 10^{-3}}{4 \times 80 \times 1.5 \times 6.626 \times 10^{-34} \times 193.5 \times 10^{12} \times (39.96 - 1) \times 7 \times 10^9}}$$
$$= 7.71.$$

7.4.4 Coherent Receiver

Consider the fiber-optic system shown in Fig. 7.13 with balanced coherent detection. Let the received field envelope be

$$q(t) = A_r s(t) + n_{\text{ASE}}(t), \tag{7.137}$$

where $A_r s(t)$ and $n_{\text{ASE}}(t)$ are the signal and ASE noise field envelopes, respectively. In this case, Eq. (7.32) is modified as

$$I = 2RA_{\text{LO}}\text{Re}\left\{A_r s(t) + n_{\text{ASE}}(t)\right\} + n_{\text{shot}+} - n_{\text{shot}-}. \tag{7.138}$$

Consider a bit '1' of the OOK system. Let us first ignore the shot noise and write Eq. (7.138) as

$$I = \overline{I} + \delta I, \tag{7.139}$$
$$\overline{I} = 2R\sqrt{P_{\text{LO}}P_{1r}}, \tag{7.140}$$
$$\delta I = 2R\sqrt{P_{\text{LO}}}\text{Re}\left\{n_{\text{ASE}}\right\}. \tag{7.141}$$

Let

$$n_{\text{ASE}} = |n_{\text{ASE}}|\exp(i\theta), \tag{7.142}$$
$$\text{Re}\left\{n_{\text{ASE}}\right\} = |n_{\text{ASE}}|\cos(\theta). \tag{7.143}$$

Since θ is a random variable with uniform distribution, it follows that

$$<\delta I> = 0, \tag{7.144}$$
$$<\delta I^2> = 4R^2 P_{\text{LO}} < |n_{\text{ASE}}|^2\cos^2\theta>. \tag{7.145}$$

Proceeding as in Section 6.5.1, Eq. (7.145) is simplified as

$$< \delta I^2 >= 2R^2 P_{LO} P_{ASE}, \tag{7.146}$$

where P_{ASE} is the mean noise power within the receiver bandwidth. The variance given by Eq. (7.146) represents the signal–spontaneous beat noise due to the interaction of the LO signal and ASE. Note that spontaneous–spontaneous beat noise is absent when the balanced coherent receiver is used. We assume that $P_{LO} \gg P_{1r}$. So the variances of '1' due to ASE, shot noise, and thermal noise are

$$\sigma_{1,ASE}^2 = 2R^2 P_{LO} P_{ASE}, \tag{7.147}$$

$$\sigma_{1,shot}^2 = 2qB_e R P_{LO}, \tag{7.148}$$

$$\sigma_{1,thermal}^2 = 4k_B T B_e / R_L, \tag{7.149}$$

$$\sigma_1^2 = \sigma_{1,ASE}^2 + \sigma_{1,shot}^2 + \sigma_{1,thermal}^2. \tag{7.150}$$

Similarly, for bit '0', we have

$$\sigma_{0,ASE}^2 = \sigma_{1,ASE}^2, \tag{7.151}$$

$$\sigma_{0,shot}^2 = \sigma_{1,shot}^2, \tag{7.152}$$

$$\sigma_{0,thermal}^2 = \sigma_{1,thermal}^2, \tag{7.153}$$

$$\sigma_0^2 = \sigma_1^2. \tag{7.154}$$

The mean currents of bit '1' and '0' are

$$I_1 = 2R\sqrt{P_{LO} P_{1r}}, \tag{7.155}$$

$$I_0 = 0. \tag{7.156}$$

The Q-factor is

$$Q_{OOK} = \frac{I_1}{2\sigma_1} \tag{7.157}$$

$$= \frac{R\sqrt{P_{LO} P_{1r}}}{\sqrt{2R^2 P_{LO} P_{ASE} + 2qB_e R P_{LO} + 4k_B T B_e / R_L}}. \tag{7.158}$$

For a long-haul fiber-optic system, the ASE noise due to amplifier chains is dominant. Hence, the shot noise and thermal noise can be ignored in Eq. (7.157) to obtain

$$Q_{OOK} = \sqrt{\frac{P_{1r}}{2P_{ASE}}}. \tag{7.159}$$

For a fiber-optic system with the loss fully compensated by amplifier gain, we have

$$P_{1r} = P_{in} = 2\overline{P}_{in}. \tag{7.160}$$

We assume that the electrical filter bandwidth B_e is smaller than the optical filter bandwidth, so that (see Eq. (6.52))

$$P_{ASE} = 2\rho_{ASE}^{eq} B_e. \tag{7.161}$$

Eq. (7.159) becomes

$$Q_{OOK} = \sqrt{\frac{\overline{P}_{in}}{2\rho_{ASE}^{eq} B_e}}. \tag{7.162}$$

Note that the Q-factor is independent of the responsivity R and LO power P_{LO} when the shot noise and thermal noise are ignored. For PSK, $I_0 = -I_1$ and Eq. (7.159) is modified as

$$Q_{PSK} = \sqrt{\frac{\overline{P}_{in}}{\rho_{ASE}^{eq} B_e}}. \tag{7.163}$$

Fig. 7.17 shows the BER as a function of the number of amplifiers. Solid lines show the exact Q-factors obtained by including shot noise and thermal noise and the × marks show the approximate Q-factors obtained using Eqs. (7.159) and (7.163). As can be seen, PSK outperforms OOK. For a fixed BER, the transmission reach can be doubled by using PSK compared with OOK. Eq. (7.163) can be cast into another form by setting

$$N_s = \frac{P_{in} T_b}{h\overline{f}}, \tag{7.164}$$

$$.N_n = \frac{P_{ASE} T_b}{h\overline{f}}. \tag{7.165}$$

Here, N_s and N_n denote the mean number of signal photons and noise photons, respectively. Using Eqs. (7.164) and (7.165) in Eq. (7.163) and with $B_e = 1/(2T_b)$, we find

$$Q_{PSK}^2 = \frac{2N_s}{N_n}. \tag{7.166}$$

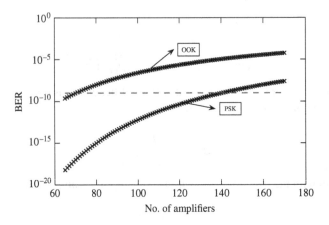

Figure 7.17 BER as a function of the number of amplifiers for a coherent fiber-optic system. Parameters: $n_{sp} = 2$, α = 0.2 dB/km, amp. spacing = 80 km, gain $G = 16$ dB, $R_L = 1000\ \Omega$, $T = 200$ K, $R = 1$ A/W, $\overline{P}_{in} = -6$ dBm, and $P_{LO} = 10$ mW.

To have $Q = 6$, the signal photon-to-noise photon ratio N_s/N_n should be 18. When N_n is very small, the amplifier noise variance becomes comparable with the shot noise and Eq. (7.166) becomes less accurate. When the shot noise is included, Eq. (7.166) is modified as (see Example 7.8)

$$Q_{PSK}^2 = \frac{2N_s}{N_n + 1/2}. \tag{7.167}$$

Example 7.5

In a 1.55-μm coherent long-haul fiber-optic system based on PSK as shown in Fig. 7.13, fiber loss = 0.2 dB/km, amplifier spacing $L_a = 100$ km. Fiber loss is fully compensated by the amplifier placed periodically along the transmission line. The mean fiber launch power = -2 dBm, $n_{sp} = 1.4$, and $B_e = 5$ GHz. Find the transmission distance at which the BER becomes equal to 10^{-9}. Ignore shot noise, thermal noise, and spontaneous–spontaneous beat noise.

Solution:

$$\text{loss(dB)} = 0.2 \times 100 = 20\,\text{dB}.$$

When the loss is fully compensated by the gain, we have

$$G(\text{dB}) = \text{loss(dB)}$$
$$= 20\,\text{dB},$$
$$G = 10^{20/10} = 100.$$

The ASE power spectral density

$$\rho_{ASE}^{eq} = N n_{sp} h\bar{f}(G - 1)$$
$$= N \times 1.4 \times 6.626 \times 10^{-34} \times 193.54 \times 10^{12} \times (100 - 1)$$
$$= N \times 1.77 \times 10^{-17}\,\text{W/Hz}.$$

From Eq. (7.163), we have

$$Q_{PSK} = \sqrt{\frac{\bar{P}_{in}}{\rho_{ASE}^{eq} B_e}} = \sqrt{\frac{\bar{P}_{in}}{N n_{sp} h\bar{f}(G - 1)B_e}},$$

$$\bar{P}_{in} = 10^{-2/10}\,\text{mW} = 6.309 \times 10^{-4}\,\text{mW}.$$

To have a BER of 10^{-9}, Q should be 6,

$$6 = \sqrt{\frac{6.309 \times 10^{-4}}{N \times 1.77 \times 10^{-17} \times 5 \times 10^9}},$$

or

$$N = \text{floor}(197.24) = 197.$$

$$\text{Total transmission distance} = 197 \times 100\,\text{km}$$
$$= 19,700\,\text{km}.$$

7.4.5   Numerical Experiments

Consider a 10-Gb/s fiber-optic direct detection system based on OOK. The schematic of the system is shown in Fig. 7.13. We assume that the gains of the amplifiers are identical and fiber lengths are the same. Let us first consider the impact of ASE only by ignoring the fiber dispersion. Fig. 7.18(a) shows the eye diagram at the receiver for system #1 when the amplifier spacing = 100 km and number of amplifiers = 30, so that the total transmission distance = 3000 km. Fig. 7.18(b) shows the eye diagram for system #2 when the amplifier spacing = 25 km and number of amplifiers = 120, i.e., we reduce the amplifier spacing by a factor of 4 and increase the number of amplifiers by the same factor. Comparing Fig. 7.18(a) and 7.18(b), we find that the impact of ASE is much less in system #2 than in system #1. This is because the variance of signal–ASE beating, σ^2_{s-sp}, increases, linearly with the number of amplifiers (Eq. (7.133)). But, as the number of amplifiers is increased, the amplifier spacing is reduced, which reduces the fiber loss and also the gain G. The gain G is

$$G = \exp(\alpha L_a), \tag{7.168}$$

where α is the fiber loss coefficient and L_a is the amplifier spacing. When L_a is reduced by a factor of 4, but the total transmission distance is fixed, G reduces exponentially whereas the number of amplifiers increases linearly. Since the exponential decrease dominates the linear increase, the net effect is to lower σ^2_{s-sp} and the performance increases if we choose a smaller amplifier spacing.

Next, we consider the impact of fiber dispersion and ignore ASE. Fig. 7.19(a) and (b) shows the eye diagrams when $\beta_2 = -1 \text{ ps}^2/\text{km}$ and $\beta_2 = -5 \text{ ps}^2/\text{km}$, respectively. As can be seen, as the dispersion coefficient increases the pulses broaden, leading to performance degradation. Fig. 7.20(a) and (b) shows the eye diagrams when ASE is turned on and $\beta_2 = -1 \text{ ps}^2/\text{km}$ and $\beta_2 = -5 \text{ ps}^2/\text{km}$, respectively.

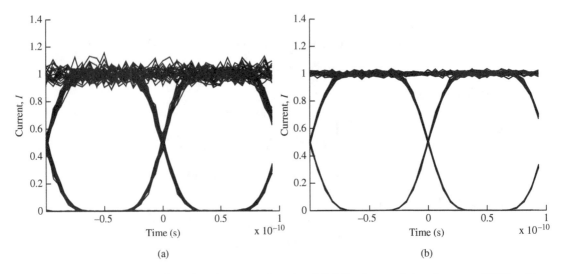

Figure 7.18   Eye diagrams for a direct detection system based on OOK. Total transmission distance = 3000 km, bit rate = 10 Gb/s, and $n_{sp} = 1.5$. (a) System #1. Amplifier spacing = 100 km, number of amplifiers = 30 and (b) System #2. Amplifier spacing = 25 km, number of amplifiers = 120.

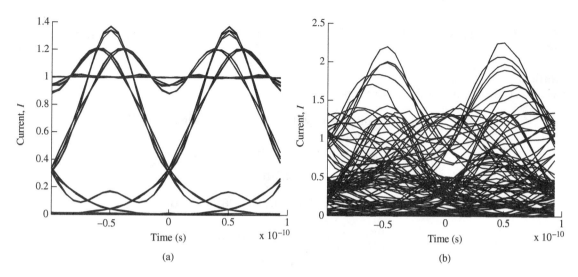

Figure 7.19 Eye diagrams for a direct detection system. Amplifier spacing $= 100$ km, number of amplifiers $= 10$. ASE is turned off. (a) $\beta_2 = -1\,\text{ps}^2/\text{km}$ and (b) $\beta_2 = -5\,\text{ps}^2/\text{km}$.

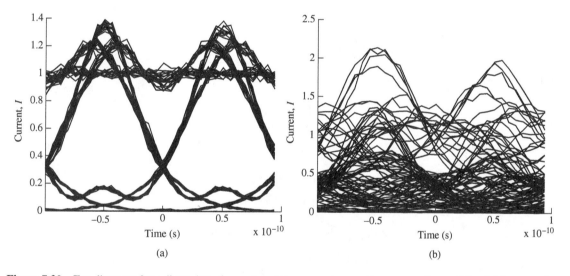

Figure 7.20 Eye diagrams for a direct detection system. The parameters are the same as those of Fig. 7.19, except that ASE is turned on. $n_{sp} = 1.5$, number of amplifiers $= 10$. (a) $\beta_2 = -1\,\text{ps}^2/\text{km}$ and (b) $\beta_2 = -5\,\text{ps}^2/\text{km}$.

7.5   Additional Examples

Example 7.6

In the fiber-optic system of Fig. 7.5, $\alpha = 0.18$ dB/km and fiber length $L = 190$ km, $T = 298$ K, $P_{in} = 1$ dBm, $R = 1.1$ A/W, $R_L = 200\,\Omega$, $f_e = 7.5$ GHz, and $f_o = 20$ GHz. The noise figure and gain of the preamplifier are

4.5 dB and 20 dB, respectively. Find the Q-factor at the receiver, assuming that the operating wavelength is 1.55 μm.

Solution:
The optical power at the receiver after the preamplifier is

$$P_{out}(\text{dBm}) = P_{in}(\text{dBm}) - \text{loss(dB)} + \text{gain of the preamplifier(dB)},$$

$$\text{loss(dB)} = 0.18 \times 190\,\text{dB}$$

$$= 34.2\,\text{dB}.$$

$$\text{Gain of the preamplifier(dB)} = G(\text{dB}) = 20\,\text{dB},$$

$$P_{out}(\text{dBm}) = 1 - 34.2 + 20\,\text{dBm}$$

$$= -13.2\,\text{dBm}.$$

$$P_{out} = 10^{-13.2/10}\,\text{mW} = 4.78 \times 10^{-2}\,\text{mW},$$

$$\rho_{\text{ASE}} = h\overline{f}(G - 1)n_{sp},$$

$$\overline{f} = \frac{c}{\lambda} = \frac{3 \times 10^8}{1.55 \times 10^{-6}} = 193.54\,\text{THz}.$$

From Eq. (6.105), we have

$$F_n \cong 2n_{sp},$$

$$F_n(\text{dB}) = 4.5\,\text{dB},$$

$$F_n = 10^{4.5/10} = 2.818,$$

$$n_{sp} = 1.409,$$

$$G = 10^{G(\text{dB})/10} = 100,$$

$$\rho_{\text{ASE}} = 6.626 \times 10^{-34} \times 193.54 \times 10^{12} \times (100 - 1) \times 1.409$$

$$= 1.78 \times 10^{-17}\,\text{W/Hz},$$

$$I_1 = RP_{out} + 2\rho_{\text{ASE}}f_o$$

$$= 1.1 \times 4.7 \times 10^{-5} + 2 \times 1.78 \times 10^{-17} \times 20 \times 10^9\,\text{A}$$

$$= 5.32 \times 10^{-2}\,\text{mA},$$

$$I_0 = 2R\rho_{\text{ASE}}f_o$$

$$= 2 \times 1.1 \times 1.78 \times 10^{-17} \times 20 \times 10^9\,\text{A}$$

$$= 7.83 \times 10^{-4}\,\text{mA},$$

$$\sigma_1^2 = 2qI_1f_e + \frac{4k_BTf_e}{R_L} + 2R^2\rho_{\mathrm{ASE}}[2P_{\mathrm{out}}f_e + \rho_{\mathrm{ASE}}(2f_o - f_e)f_e]$$

$$= 2 \times 1.602 \times 10^{-19} \times 5.32 \times 10^{-5} \times 7.5 \times 10^9 + \frac{4 \times 1.38 \times 10^{-23} \times 298 \times 7.5 \times 10^9}{200}$$

$$+ 2 \times 1.1^2 \times 1.78 \times 10^{-17} \times [2 \times 4.78 \times 10^{-5} \times 7.5 \times 10^9$$

$$+ 1.78 \times 10^{-17} \times (2 \times 20 \times 10^9 - 7.5 \times 10^9) \times 7.5 \times 10^9]$$

$$= 3.18 \times 10^{-11}\,\mathrm{A}^2,$$

$$\sigma_0^2 = 2qI_0f_e + \frac{4k_BTf_e}{R_L} + 2R^2\rho_{\mathrm{ASE}}^2 B(2f_o - f_e)f_e$$

$$= 2 \times 1.602 \times 10^{-19} \times 7.83 \times 10^{-7} \times 7.5 \times 10^9 + \frac{4 \times 1.38 \times 10^{-23} \times 298 \times 7.5 \times 10^9}{200}$$

$$+ 2 \times 1.1^2 \times (1.78 \times 10^{-17})^2 \times (2 \times 20 \times 10^9 - 7.5 \times 10^9) \times 7.5 \times 10^9$$

$$= 8.05 \times 10^{-13}\,\mathrm{A}^2,$$

$$Q = \frac{I_1 - I_0}{\sigma_1 + \sigma_0}$$

$$= \frac{5.32 \times 10^{-5} - 7.83 \times 10^{-7}}{\sqrt{3.18 \times 10^{-11}} + \sqrt{8.05 \times 10^{-13}}}$$

$$= 8.031.$$

Example 7.7

A two-stage amplifier with a DCF between the stages needs to be designed. The insertion loss of the DCF is 7 dB. There are two amplifiers Amp1 and Amp2 with gains $G_1 = 8\,\mathrm{dB}$ and $G_2 = 16\,\mathrm{dB}$, respectively. The noise figures of the amplifiers are $F_{n,1} = 7\,\mathrm{dB}$ and $F_{n,2} = 5.5\,\mathrm{dB}$. Find the optimum amplifier configuration.

Solution:
Since Amp2 has the lower noise figure, let us first choose Amp2 as the first amplifier. From Eq. (7.122) with the indices reversed, we have

$$F_{n,\mathrm{eq}} = F_{n,2} + \frac{F_{n,1}}{G_2H},$$

$$F_{n,1} = 10^{F_{n,1}(\mathrm{dB})/10} = 5.01,$$

$$F_{n,2} = 10^{F_{n,2}(\mathrm{dB})/10} = 3.54,$$

$$G_2 = 10^{G_2(\mathrm{dB})/10} = 39.81,$$

$$H = 10^{H(\mathrm{dB})/10} = 5.01,$$

$$F_{n,eq} = 3.54 + \frac{5.01}{39.81 \times 5.01} = 3.56,$$

$$F_{n,eq}(dB) = 5.52\,dB.$$

Thus, the F_n of the equivalent amplifier is roughly the same as that of Amp2. The net gain is

$$G_{eq} = G_2 H G_1,$$

$$G_{eq}(dB) = G_2(dB) + H(dB) + G_1(dB)$$

$$= 16 - 7 + 8\,dB$$

$$= 17\,dB.$$

If we choose Amp1 as the first amplifier, from Eq. (7.122) we find

$$F_{n,eq} = F_{n,1} + \frac{F_{n,2}}{G_1 H}$$

$$= 5.01 + \frac{3.54}{6.309 \times 5.52}$$

$$= 5.11,$$

$$F_{n,eq}(dB) = 7.08\,dB.$$

In this case, the equivalent noise figure is roughly the same as that of Amp1. Therefore, the optimum configuration is the one in which the first amplifier is Amp2.

Example 7.8

In the presence of ASE noise and shot noise, show that the Q-factor of a fiber-optic system based on PSK with balanced detection is given by

$$Q^2 = \frac{2N_s}{N_n + 1/2}, \tag{7.169}$$

where N_s and N_n are the mean number of signal photons and noise photons, respectively. Assume $\eta = 1$.

Solution:
For PSK, $I_1 = -I_0$. Ignoring the thermal noise, Eq. (7.158) is modified as

$$Q_{PSK} = \frac{2R\sqrt{P_{LO}P_{1r}}}{\sqrt{2R^2 P_{LO}P_{ASE} + 2qB_e R P_{LO}}}. \tag{7.170}$$

Since $R = \eta q/h\bar{f}$ and $\eta = 1$, Eq. (7.170) becomes

$$Q_{PSK} = \sqrt{\frac{2P_{1r}}{P_{ASE} + h\bar{f}B_e}}. \tag{7.171}$$

Using a Nyquist filter,

$$B_e = \frac{1}{2T_b}. \tag{7.172}$$

Eq. (7.171) becomes

$$Q_{PSK} = \sqrt{\frac{2P_{1r}T_b}{P_{ASE}T_b + h\bar{f}/2}}. \tag{7.173}$$

Using Eqs. (7.164) and (7.165), Eq. (7.173) becomes

$$Q_{PSK}^2 = \frac{2N_s}{N_n + 1/2}. \tag{7.174}$$

Example 7.9

A 1.55-μm long-haul fiber-optic system based on OOK that uses a direct detection receiver is shown in Fig. 7.21. The output of the TF passes through a two-stage amplifier with gain $G_1 = 16\,dB$ and G_2 determined by the condition that the output power of Amp2 is the same as that at the transmitter output. The noise figures of Amp1 and Amp2 are 5.5 dB and 7.5 dB, respectively. The loss and dispersion coefficients of the TF are 0.18 dB/km and $-21\,ps^2/km$, respectively, and the corresponding coefficients of the DCF are 0.5 dB/km and 145 ps^2/km, respectively. Other parameters: mean transmitter output power = $-2\,dBm$, length of the TF = 100 km, number of spans = 70, $f_e = 7\,GHz$. Calculate (a) the length of the DCF so that the DCF compensates for 90% of the accumulated dispersion of the TF, (b) the gain G_2, and (c) the Q-factor. Ignore the shot noise, thermal noise, and spontaneous–spontaneous beat noise.

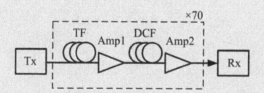

Figure 7.21 A long-haul fiber-optic system based on OOK.

Solution:
(a) Total accumulated dispersion of the single-span TF:

$$s_{TF} = \beta_{2,TF}L_{TF}$$
$$= -21 \times 10^{-27} \times 100 \times 10^3 \, s^2$$
$$= -2.1 \times 10^{-21} \, s^2.$$

Accumulated dispersion of the DCF:

$$s_{DCF} = -0.9 s_{TF} = 1.89 \times 10^{-21} \, s^2,$$

$$L_{\text{DCF}} = \frac{s_{\text{DCF}}}{\beta_{2,\text{DCF}}}$$

$$= \frac{1.89 \times 10^{-21}}{145 \times 10^{-27}} = 13.03 \text{ km}.$$

(b) Since the amplifiers compensate for the loss due to TF and DCF exactly, we have

$$G_1(\text{dB}) + G_2(\text{dB}) = H_{\text{TF}}(\text{dB}) + H_{\text{DCF}}(\text{dB}),$$

$$H_{\text{TF}}(\text{dB}) = 0.18 \times 100 = 18 \text{ dB},$$

$$H_{\text{DCF}}(\text{dB}) = 0.5 \times 13.03 = 6.517 \text{ dB},$$

$$G_1(\text{dB}) = 16,$$

$$G_2(\text{dB}) = 18 + 6.517 - 16 = 8.517 \text{ dB}.$$

(c) A two-stage amplifier with a DCF in between can be replaced by an equivalent amplifier with gain

$$G_{\text{eq}}(\text{dB}) = G_1(\text{dB}) + G_2(\text{dB}) - H_{\text{DCF}}(\text{dB}) = 18 \text{ dB},$$

$$G_{\text{eq}} = 10^{18/10} = 63.09,$$

$$F_{n,\text{eq}} = F_{n,1} + \frac{F_{n,2}}{G_1 H_{\text{DCF}}} - \frac{1}{G_1},$$

$$F_{n,1}(\text{dB}) = 5.5 \text{ dB},$$

$$F_{n,1} = 10^{5.5/10} = 3.548,$$

$$F_{n,2}(\text{dB}) = 7.5 \text{ dB},$$

$$F_{n,2} = 10^{7.5/10} = 5.62,$$

$$H_{\text{DCF}} = 10^{-H_{\text{DCF}}(\text{dB})/10} = 0.223,$$

$$G_1 = 10^{16/10} = 39.81,$$

$$F_{n,\text{eq}} = 3.548 + \frac{5.62}{39.81 \times 0.223} - \frac{1}{39.81} = 4.156.$$

Since we have 70 identical spans, the PSD of the ASE at the receiver is

$$\rho_{\text{ASE,eq}} = 70 \times h\bar{f}(G_{\text{eq}}F_{n,\text{eq}} - 1)/2$$

$$= 70 \times 6.626 \times 10^{-34} \times 193.54 \times 10^{12} \times (63.09 \times 4.156 - 1)/2 \text{ W/Hz}$$

$$= 1.17 \times 10^{-15} \text{ W/Hz}.$$

For OOK, the peak power is twice the average power:

$$P_{\text{in}} = 2 \times 10^{\bar{P}_{\text{in}}(\text{dBm})/10} \text{ mW}$$

$$= 1.2 \times 10^{-3} \text{ W}.$$

Using Eq. (7.135), we find

$$Q = \sqrt{\frac{P_{in}}{4\rho_{ASE,eq}f_e}}$$

$$= \sqrt{\frac{1.26 \times 10^{-3}}{4 \times 1.17 \times 10^{-15} \times 7 \times 10^{9}}}$$

$$= 6.2$$

Exercises

7.1 In a fiber-optic system based on OOK as shown in Fig. 7.1, fiber loss = 0.21 dB/km, length $L = 120$ km, peak power at the transmitter = 2 dBm, $T = 23°C$, $R_L = 100\,\Omega$, $B_e = 7$ GHz, and $R = 1.1$ A/W. Find (a) the peak power at the receiver, (b) the Q-factor, and (c) the BER.

(Ans: (a) −23.3 dBm, (b) 4.89, (c) 4.84×10^{-7}.)

7.2 In a 1.55-μm fiber-optic system based on OOK as shown in Fig. 7.5, the peak transmitter power $P_{in} = 2$ dBm, fiber loss coefficient $\alpha = 0.2$ dB/km, $T = 290$ K, $R_L = 1000\,\Omega$, $R = 1$ A/W, $B_o = 20$ GHz, $B_e = 7.5$ GHz, gain and n_{sp} of the preamplifier are 25 dB and 1.5, respectively. Find the maximum achievable transmission distance to have a BER of 10^{-9}.

(Ans: 190.5 km.)

7.3 In a 1.55-μm coherent fiber-optic system based on PSK, as shown in Fig. 7.7, find the lower limit on the LO power such that the shot noise dominates the thermal noise and the difference between the Q-factor given by Eq. (7.62) and the exact Q-factor is $\leq 2.5\%$. The mean received power = −45 dBm, $T = 293$ K, $\eta = 0.8$, $R_L = 200\,\Omega$, and $B_e = 7$ GHz.

(Ans: 5 mW.)

7.4 The received signal of an unrepeatered coherent fiber-optic system passes through a preamplifier of gain G and spontaneous noise factor n_{sp}, as shown in Fig. 7.22. Develop a mathematical expression for the Q-factor including the LO–spontaneous beat noise and shot noise.

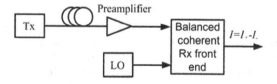

Figure 7.22 A balanced coherent receiver with a preamplifier.

7.5 Find the maximum transmission distance of a metro network operating at 10 Gb/s in which the fiber dispersion is not compensated in the optical or electrical domain if the transmission fiber dispersion β_2 is (a) 5 ps²/km, (b) −21 ps²/km. Use the criteria given by Eq. (7.101).

(Ans: (a) 125 km, (b) 29.76 km.)

7.6 For a long-haul fiber-optic system consisting of 10 identical amplifiers with gain $G = 30\,\text{dB}$ and noise figure $F_n = 4.5\,\text{dB}$, it is desirable to have a OSNR of 15 dB (in 0.1 nm bandwidth) at the receiver. Find the transmitter power launched to the fiber.

(Ans: $-1.5\,\text{dBm}$.)

7.7 A 20,000-km transmission system at 10 Gb/s based on OOK with direct detection needs to be designed. The required Q-factor at the receiver should be ≥ 5. Fiber loss = 0.18 dB/km and the loss is exactly compensated by periodically placed amplifiers. Write a program to find the maximum amplifier spacing allowed. Assume $B_e = 7.5\,\text{GHz}$, $R = 1\,\text{A/W}$. Ignore shot noise and thermal noise.

7.8 Explain how a larger amplifier spacing deteriorates the system performance in a long-haul fiber-optic system.

7.9 For a cascaded chain of k amplifiers, show that

$$F_{n,\text{eq}} = F_{n,1} + \frac{F_{n,2} - 1}{G_1} + \frac{F_{n,3} - 1}{G_1 G_2} + \cdots + \frac{F_{n,k} - 1}{G_1 G_2 \cdots G_{k-1}}, \tag{7.175}$$

where $F_{n,j}$ and G_j are the noise figure and gain of the jth amplifier, respectively.

7.10 Write a program to calculate the OSNR and BER of a long-haul fiber-optic system with direct/coherent detection. Include shot noise, thermal noise, signal–spontaneous beat noise and spontaneous–spontaneous beat noise (if applicable). Compare the BERs obtained using the exact Q-factor and approximate Q-factor obtained by ignoring shot noise, thermal noise, and spontaneous–spontaneous beat noise.

Further Reading

G. Keiser, *Optical Fiber Communications*, 4th edn. McGraw-Hill, New York, 2011.
G.P. Agrawal, *Fiber-optic Communication Systems*. John Wiley & Sons, Hoboken, NJ 2010.
K.P. Ho, *Phase Modulated Optical Communication Systems*. Springer-Verlag, Berlin, 2005.
J. M. Senior, *Optical Fiber Communications*, 2nd edn. Prentice-Hall, London, 1992.
S. Betti, G. DeMarchis, and E. Iannone, *Coherent Optical Communication Systems*. John Wiley & Sons, New York, 1995.
T. Okoshi and K. Kikuchi, *Coherent Optical Fiber Communications*. Kluwer Academic, Dordrecht, 1998.

References

[1] K.P. Ho, *Phase Modulated Optical Communication Systems*. Springer-Verlag, Berlin, 2005.
[2] S. Betti, G. De Marchis, and E. Iannone, *Coherent Optical Communication Systems*. John Wiley, Hoboken, NJ, 1995.
[3] K. Kikuchi and S. Tsukamoto, *J. Lightwave Technol.*, vol. 26, p. 1817, 2008.
[4] G.P. Agrawal, *Fiber-optic Communication Systems*. John Wiley & Sons, Hoboken, NJ, 2010.
[5] C.R. Giles and E. Desurvire, *J. Lightwave Technol.*, vol. 9, p. 147, 1991.
[6] Y. Yamamoto and T. Mukai, *Opt. Quant. Electron.*, vol. 21, p. S1, 1989.
[7] P.A. Humblet and M. Azizoglu, *J. Lightwave Technol.*, vol. 9, p. 1576, 1991.
[8] D. Marcuse, *J. Lightwave Technol.*, vol. 8, p. 1816, 1990.
[9] D. Marcuse, *J. Lightwave Technol.*, vol. 9, p. 505, 1991.

8

Performance Analysis

8.1 Introduction

In Chapter 4, various types of digital modulation schemes such as PSK, OOK, and FSK were introduced and in Chapter 5, different receiver architectures such as direct detection, homodyne, and heterodyne detections were discussed. In this chapter, the performances of these modulation schemes with the different receiver architectures are investigated. Firstly, the concept of a matched filter is introduced in Section 8.2. In practice, optical matched filters are rarely used in optical communication systems due to the difficulties involved in fabricating such a matched filter. Nevertheless, the performance is optimum when the matched filters are used, and the expressions for the error probability developed in this Chapter using the matched filters provide a lower bound on the achievable BER.

8.2 Optimum Binary Receiver for Coherent Systems

In this section, we consider the generalized model for the optimum binary receivers (See Fig. 8.1) and later, we apply this model to various detection schemes. Let $x_1(t)$ and $x_2(t)$ be the real optical signals used to transmit bits '1' and '0', respectively:

$$x(t) = \begin{cases} x_1(t) & \text{when the message = '1'} \\ x_0(t) & \text{when the message = '0'}. \end{cases} \tag{8.1}$$

Here, $x_j(t), j = 0, 1$ are arbitrary pulses of duration $\leq T_b$, where T_b is the bit interval. We assume that the channel can be modeled as an *additive white Gaussian noise* (AWGN) channel, which means that the power spectral density of the noise is constant and the probability distribution of the noise process is Gaussian. The output of the channel may be written as

$$y(t) = x(t) + n(t), \tag{8.2}$$

where $n(t)$ is the noise added by the channel and

$$\rho_n = N_0/2 \tag{8.3}$$

is the power spectral density of $n(t)$. Let the Fourier transform of $x_j(t)$ be

$$\tilde{x}_j(\omega) = \mathcal{F}[x_j(t)], j = 0, 1. \tag{8.4}$$

The channel output is passed through a filter. The purpose of this filter is to alter the ratio of the signal power and noise power so that the best performance can be attained. The filter multiplies the signal spectrum by

Fiber Optic Communications: Fundamentals and Applications, First Edition. Shiva Kumar and M. Jamal Deen.
© 2014 John Wiley & Sons, Ltd. Published 2014 by John Wiley & Sons, Ltd.

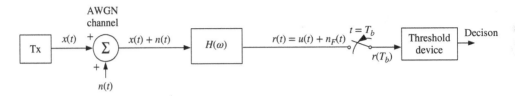

Figure 8.1   A generalized model for the optimum binary receivers.

$H(\omega)$ and, therefore, after passing through the filter, the signal component is

$$
u(t) = \begin{cases} u_1(t) & \text{when bit '1' is transmitted} \\ u_0(t) & \text{when bit '0' is transmitted} \end{cases} \tag{8.5}
$$

with

$$
u_j(t) = \frac{1}{2\pi} \int_{-\infty}^{\infty} \tilde{x}_j(\omega) H(\omega) \exp\left(-i\omega t\right) d\omega,
$$
$$
j = 0, 1. \tag{8.6}
$$

After passing through the filter, the noise variance is given by

$$
\sigma^2 = \frac{1}{2\pi} \int_{-\infty}^{\infty} \frac{N_0}{2} |H(\omega)|^2 \, d\omega. \tag{8.7}
$$

The received signal $r(t)$ can be written as the superposition of the signal and noise at the filter output. The decision is based on samples of $r(t)$:

$$
r(t) = u(t) + n_F(t), \tag{8.8}
$$

where $n_F(t)$ is the noise at the filter output. To determine if the message is bit '0' or bit '1', the received signal $r(t)$ is sampled at intervals of T_b. Since the noise sample $n_F(T_b)$ is a Gaussian random variable with zero mean and variance σ^2, the received signal sample $r(T_b)$ is a Gaussian random variable with mean $u(T_b)$ and variance σ^2. Its pdf is given by

$$
p(r) = \frac{1}{\sqrt{2\pi}\sigma} \exp\left\{ -\frac{[r - u(T_b)]^2}{2\sigma^2} \right\}. \tag{8.9}
$$

Let r_T be the threshold. If $r(T_b) > r_T$, the threshold device decides that the bit '1' is transmitted. Otherwise, the bit '0' is transmitted. When a bit '1' is transmitted, $u(T_b) = u_1(T_b)$. In this case, the conditional pdf is

$$
p(r|\text{'1' sent}) \equiv p_1(r) = \frac{1}{\sqrt{2\pi}\sigma} \exp\left\{ -\frac{[r - u_1(T_b)]^2}{2\sigma^2} \right\}. \tag{8.10}
$$

Fig. 8.2 shows the conditional pdf $p_1(r)$. The area of the shaded region in Fig. 8.2 is the chance that the received signal $r(T_b) < r_T$ when bit '1' is transmitted. A bit error is made if the decision device chooses a bit '0' when a bit '1' is transmitted. This happens if $r(T_b) < r_T$. Therefore, the probability of mistaking a bit '1' as a bit '0' is the area under the curve $p_1(r)$ from $-\infty$ to r_T and is given by

$$
P(0|1) = \frac{1}{\sqrt{2\pi}\sigma} \int_{-\infty}^{r_T} \exp\left\{ -\frac{[r - u_1(T_b)]^2}{2\sigma^2} \right\} dr. \tag{8.11}
$$

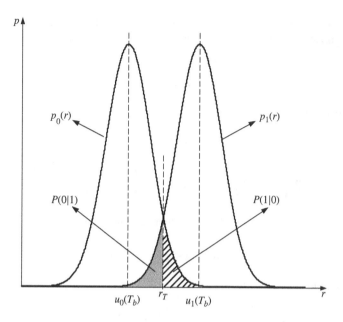

Figure 8.2 Conditional pdfs. $P(0|1)$ is the chance of mistaking bit '1' as bit '0'. $P(1|0)$ is the chance of mistaking bit '0' as bit '1'.

Similarly, when a bit '1' is sent, the conditional pdf is

$$p(r|\text{'0' sent}) \equiv p_0(r) = \frac{1}{\sqrt{2\pi}\sigma} \exp\left\{-\frac{[r - u_0(T_b)]^2}{2\sigma^2}\right\}. \tag{8.12}$$

The probability of mistaking a bit '0' as a bit '1' is the area under the curve $p_0(r)$ from r_T to ∞ (shown as slanted lines in Fig. 8.2):

$$P(1|0) = \frac{1}{\sqrt{2\pi}\sigma} \int_{r_T}^{\infty} \exp\left\{-\frac{[r - u_0(T_b)]^2}{2\sigma^2}\right\} dr. \tag{8.13}$$

The total BER is given by

$$P_b = P(0|1)P(1) + P(1|0)P(0), \tag{8.14}$$

where $P(j)$ is the probability of sending bit 'j', $j = 0, 1$. Assuming that the bits '1' and '0' are equally probable, we obtain

$$P_b = \frac{1}{2}[P(0|1) + P(1|0)]. \tag{8.15}$$

Substituting Eqs. (8.13) and (8.11) in Eq. (8.15), we find

$$P_b = \frac{1}{2}\left[\int_{-\infty}^{r_T} p_1(r)dr + \int_{r_T}^{\infty} p_0(r)dr\right] \tag{8.16}$$

$$= \frac{1}{2\sqrt{2\pi}\sigma}\left\{\int_{-\infty}^{r_T} \exp\left\{-\frac{[r - u_1(T_b)]^2}{2\sigma^2}\right\} dr + \int_{r_T}^{\infty} \exp\left\{-\frac{[r - u_0(T_b)]^2}{2\sigma^2}\right\} dr. \tag{8.17}$$

To find the minimum BER, the threshold r_T and the filter transfer function $H(\omega)$ should be optimized. Let us first consider the optimization of threshold r_T. P_b is minimum or maximum when

$$\frac{\partial P_b}{\partial r_T} = 0. \tag{8.18}$$

Differentiating Eq. (8.16) with respect to r_T and setting it to zero, we find

$$p_1(r_T) = p_0(r_T), \tag{8.19}$$

$$\exp\left\{-\frac{[r_T - u_1(T_b)]^2}{2\sigma^2}\right\} = \exp\left\{-\frac{[r_T - u_0(T_b)]^2}{2\sigma^2}\right\}. \tag{8.20}$$

Thus, the optimum threshold r_T corresponds to the inter section of curves $p_0(r)$ and $p_1(r)$ in Fig. 8.2. From Eq. (8.20), we see that

$$r_T - u_1(T_b) = \pm[r_T - u_0(T_b)]. \tag{8.21}$$

Taking the negative sign in Eq. (8.21), we find

$$r_T = [u_0(T_b) + u_1(T_b)]/2. \tag{8.22}$$

If we choose the positive sign, it would lead to $u_0(T_b) = u_1(T_b)$, which is not true in our case. The optimum threshold condition given by Eq. (8.19) is valid for arbitrary pdfs, while that given by Eq. (8.22) holds true for Gaussian distributions. From Eq. (8.22), we see that the optimum threshold r_T is at the middle of $u_0(T_b)$ and $u_1(T_b)$. Since the conditional pdfs $p_1(r)$ and $p_0(r)$ are symmetrically located with respect to the optimum threshold r_T (Fig. 8.2), $P(0|1)$ and $P(1|0)$ should be equal. Therefore, Eq. (8.17) can be rewritten as

$$P_b = \frac{1}{\sqrt{2\pi}\sigma} \int_{r_T}^{\infty} \exp\left\{-\frac{[r - u_0(T_b)]^2}{2\sigma^2}\right\} dr. \tag{8.23}$$

Let

$$z = \frac{r - u_0(T_b)}{\sqrt{2}\sigma}, \tag{8.24}$$

$$P_b = \frac{1}{\sqrt{\pi}} \int_{[r_T - u_0(T_b)]/\sqrt{2}\sigma}^{\infty} \exp\left(-z^2\right) dz$$

$$= \frac{1}{2}\text{erfc}\left(\frac{r_T - u_0(T_b)}{\sqrt{2}\sigma}\right), \tag{8.25}$$

where erfc($\cdot$) is the complementary error function defined as

$$\text{erfc}(z) = \frac{2}{\sqrt{\pi}} \int_z^{\infty} \exp\left(-y^2\right) dy. \tag{8.26}$$

Using Eq. (8.22), Eq. (8.25) becomes

$$P_b = \frac{1}{2}\text{erfc}\left(\sqrt{\frac{v}{8}}\right), \tag{8.27}$$

where

$$v = \frac{[u_1(T_b) - u_0(T_b)]^2}{\sigma^2} \tag{8.28}$$

$$= \frac{\left[\int_{-\infty}^{\infty} [\tilde{x}_1(\omega) - \tilde{x}_0(\omega)] \, H(\omega) \exp\left(-i\omega T_b\right) d\omega\right]^2}{\pi N_0 \int_{-\infty}^{\infty} |H(\omega)|^2 \, d\omega}. \tag{8.29}$$

From Fig. 8.3, we see that as v increases, P_b decreases and therefore, to minimize P_b, v should be maximized. v can be maximized by the proper choice of filter transfer function $H(\omega)$. If the filter is too wide (Fig. 8.4(a)), the variance of noise given by Eq. (8.7) increases since the variance is proportional to the area under the

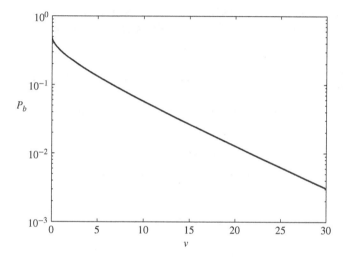

Figure 8.3   Dependence of the BER on v.

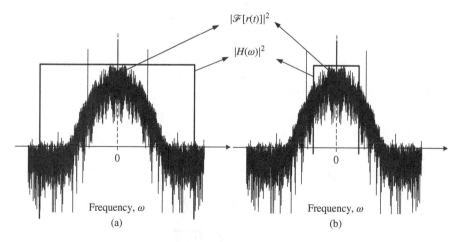

Figure 8.4   Received signal spectrum and the receiver filter transfer function: (a) wide-bandwidth filter; (b) narrow-bandwidth filter.

curve $|H(\omega)|^2$ and, therefore, v decreases. If the filter is too narrow (Fig. 8.4(b)), a significant fraction of the signal component is truncated by the filter and, therefore, the numerator of Eq. (8.29) becomes too small. The optimum filter transfer function can be obtained by setting the variation of v with respect to $H(\omega)$ and $H^*(\omega)$ to zero:

$$\frac{\delta v}{\delta H} = 0 \quad \text{and} \quad \frac{\delta v}{\delta H^*} = 0. \tag{8.30}$$

To find the variations given by Eq. (8.30), let us replace the integrals of Eq. (8.29) by summations:

$$v = \lim_{\Delta\omega_n \to 0} \frac{\left[\sum_n [\tilde{x}_1(\omega_n) - \tilde{x}_0(\omega_n)] H(\omega_n) \exp(-i\omega_n T_b) \Delta\omega_n\right]^2}{\pi N_0 \sum_n |H(\omega_n)|^2 \Delta\omega_n}. \tag{8.31}$$

v can be optimized by setting its partial derivatives with respect to $H(\omega_n)$ and $H^*(\omega_n)$ to zero. Note that $H(\omega_n)$ and $H^*(\omega_n)$ are independent variables. Alternatively, $\text{Re}[H(\omega_n)]$ and $\text{Im}[H(\omega_n)]$ can also be chosen as independent variables:

$$\frac{\partial v}{\partial H(\omega_n)} = \frac{2\sqrt{N}[\tilde{x}_1(\omega_n) - \tilde{x}_0(\omega_n)] \exp(-i\omega_n T_b)}{D} - \pi N_0 H^*(\omega_n) \frac{N}{D^2} = 0, \tag{8.32}$$

where N and D denote the numerator and denominator of Eq. (8.29), respectively. Simplifying Eq. (8.32), we obtain

$$H(\omega_n) = k[\tilde{x}_1^*(\omega_n) - \tilde{x}_0^*(\omega_n)] \exp(i\omega_n T_b),$$

$$H(\omega) = k[\tilde{x}_1^*(\omega) - \tilde{x}_0^*(\omega)] \exp(i\omega T_b), \tag{8.33}$$

where

$$k = \left(\frac{2D}{\sqrt{N}}\right) \frac{1}{\pi N_0} \tag{8.34}$$

is an arbitrary constant, which we set to unity from now on. The same result can be obtained by setting the variation of v with respect to $H^*(\omega)$ to zero. The filter with the transfer function given by Eq. (8.33) is called a *matched filter*. Using Eq. (8.33) in Eq. (8.29), we obtain

$$v_{\max} = \frac{1}{\pi N_0} \int_{-\infty}^{\infty} |\tilde{x}_1(\omega) - \tilde{x}_0(\omega)|^2 \, d\omega$$

$$= \frac{2}{N_0} \int_0^{T_b} [x_1(t) - x_0(t)]^2 \, dt, \tag{8.35}$$

where we have made use of Parseval's relations. Let

$$E_{jk} = \int_0^{T_b} x_j(t) x_k(t) \, dt, \quad j = 1, 0, \tag{8.36}$$

$$E_{jj} \equiv E_j. \tag{8.37}$$

Using Eqs. (8.36) and (8.37) in Eqs. (8.35) and (8.27), we obtain

$$v_{\max} = \frac{2}{N_0}[E_1 + E_0 - 2E_{10}], \tag{8.38}$$

$$P_{b,\min} = \frac{1}{2}\text{erfc}\left(\sqrt{\frac{v_{\max}}{8}}\right). \tag{8.39}$$

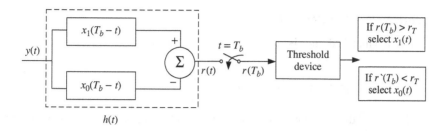

Figure 8.5   The matched filter as a parallel combination of two filters.

The transfer function of the matched filter may be rewritten as

$$H(\omega) = \tilde{x}_1^*(\omega) \exp(i\omega T_b) - \tilde{x}_0^*(\omega) \exp(i\omega T_b). \tag{8.40}$$

Substituting Eq. (8.40) in Eq. (8.6), the signal sample at $t = T_b$ is

$$u(T_b) = \frac{1}{2\pi} \int_{-\infty}^{\infty} \tilde{x}(\omega)[\tilde{x}_1^*(\omega) - \tilde{x}_0^*(\omega)]\, d\omega. \tag{8.41}$$

Taking the inverse Fourier transform of Eq. (8.40), we obtain the impulse response of the matched filter as

$$h(t) = x_1(T_b - t) - x_0(T_b - t). \tag{8.42}$$

This filter can be implemented as a parallel combination of two filters, as shown in Fig. 8.5. The optimum threshold is given by Eq. (8.22),

$$r_T = \left[\frac{u_1(T_b) + u_0(T_b)}{2} \right]$$

$$= \frac{1}{4\pi} \int_{-\infty}^{\infty} [\tilde{x}_1(\omega) + \tilde{x}_0(\omega)] H(\omega) \exp(-i\omega T_b)\, d\omega. \tag{8.43}$$

Using Eq. (8.40), Eq. (8.43) becomes

$$r_T = \frac{1}{4\pi} \int_{-\infty}^{\infty} [|\tilde{x}_1(\omega)|^2 - |\tilde{x}_0(\omega)|^2 + \tilde{x}_0(\omega)\tilde{x}_1^*(\omega) - \tilde{x}_0^*(\omega)\tilde{x}_1(\omega)]\, d\omega. \tag{8.44}$$

Since $x_j(t)$ is real, we have

$$\tilde{x}_j^*(\omega) = \tilde{x}_j(-\omega), \quad j = 1, 2. \tag{8.45}$$

So, the last two terms of Eq. (8.44) become

$$I = \int_{-\infty}^{\infty} \tilde{x}_0(\omega)\tilde{x}_1(-\omega)\, d\omega - \int_{-\infty}^{\infty} \tilde{x}_0(-\omega)\tilde{x}_1(\omega)\, d\omega. \tag{8.46}$$

After substituting $\omega = -\omega'$ in the first integral, we find $I = 0$. So, using Eq. (8.36) and Parseval's relations, Eq. (8.44) becomes

$$r_T = \frac{1}{2}[E_1 - E_0]. \tag{8.47}$$

Thus, the optimum threshold is half of the energy difference.

8.2.1   Realization of the Matched Filter

Suppose the input to the matched filter is $y(t) (= x(t) + n(t))$. The decision is based on the signal $r(t)$, which is given by

$$r(t) = [\mathcal{F}^{-1}\{\tilde{y}(\omega)H(\omega)\}] = \int_{-\infty}^{\infty} y(\tau)h(t - \tau)\, d\tau. \tag{8.48}$$

In Eq. (8.48), we have used the fact that the product in the spectral domain becomes convolution in the time domain. From Eq. (8.42), we have

$$h(t) = x_1(T_b - t) - x_0(T_b - t), \tag{8.49}$$

$$h(t - \tau) = x_1[T_b - (t - \tau)] - x_0[T_b - (t - \tau)]. \tag{8.50}$$

Hence,

$$r(t) = \int_{-\infty}^{\infty} y(\tau)[x_1(T_b + \tau - t) - x_0(T_b + \tau - t)]\, d\tau. \tag{8.51}$$

The decision is made based on the sample of $r(t)$ at $t = T_b$:

$$r(T_b) = \int_{-\infty}^{\infty} y(\tau)[x_1(\tau) - x_0(\tau)]d\tau = r_1(T_b) - r_0(T_b), \tag{8.52}$$

where

$$r_j(T_b) = \int_0^{T_b} y(\tau)x_j(\tau)d\tau, \quad j = 0, 1. \tag{8.53}$$

In Eq. (8.53), we have made use of the fact that $x_j(t)$ is zero when $t < 0$ and $t > T_b$. Thus, the matched filter can be realized by the correlation receiver shown in Fig. 8.6. If the energies of the signal $u_1(t)$ and $u_0(t)$ are equal, i.e., $E_1 = E_0$, from Eq. (8.47) we have $r_T = 0$. In this case, the equivalent form of correlation receiver is shown in Fig. 8.7. If $x_0(t) = -x_1(t)$, a simplified form of correlation receiver as shown in Fig. 8.8 may be used.

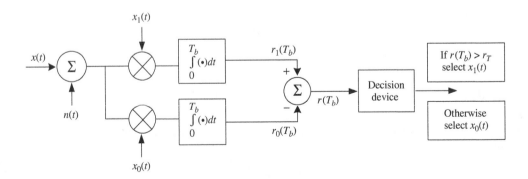

Figure 8.6   Realization of the matched filter using correlators.

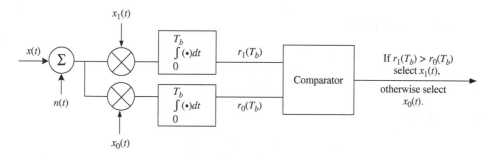

Figure 8.7   Realization of the matched filter using a correlator when $E_1 = E_0$.

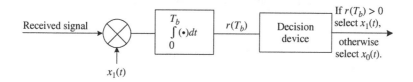

Figure 8.8   A realization of the matched filter when $x_1(t) = -x_0(t)$.

Example 8.1

Find the matched filter and its output at $t = T_b$ for the following signals.

(a)

$$x(t) = \begin{cases} \pm A & 0 < t < T_B \\ 0 & \text{otherwise.} \end{cases} \qquad (8.54)$$

(b)

$$x(t) = \begin{cases} \pm A \exp\left[-\frac{(t-T_B/2)^2}{2T_0^2}\right] & 0 < t < T_B \\ 0 & \text{otherwise.} \end{cases} \qquad (8.55)$$

Assume $T_0 \ll T_B$.

Solution:

(a) To transmit '1' ('0'), A ($-A$) is sent over the interval $0 < t < T_B$. Suppose '1' is transmitted. The Fourier transform of $x(t)$ is

$$\tilde{x}_1(\omega) = A \int_0^{T_B} \exp(-i\omega t)dt = \frac{2A \sin(\omega T_b/2)}{\omega} e^{-i\omega T_b/2} \qquad (8.56)$$

$$= -\tilde{x}_0(\omega). \qquad (8.57)$$

From Eq. (8.33), the matched filter is

$$H(\omega) = [\tilde{x}_1^*(\omega) - \tilde{x}_0^*(\omega)]\exp(-i\omega T_B)$$

$$= \frac{4A\sin(\omega T_b/2)}{\omega}\exp(-i\omega T_b/2). \tag{8.58}$$

Using Eq. (8.6), the signal sample at $t = T_B$ is

$$u(T_B) = \pm\frac{8A^2}{2\pi}\int_{-\infty}^{\infty}\frac{\sin^2(\omega T_B/2)}{\omega^2}\,d\omega$$

$$= \pm 2A^2 T_B. \tag{8.59}$$

Note that $A^2 T_b$ is the pulse energy, E_1. Since $x_0(t) = -x_1(t)$, the matched filter can also be realized using a correlation receiver as shown in Fig. 8.8:

$$r(T_B) = \int_0^{T_B} y(t)x_1(t)\,dt, \tag{8.60}$$

$$u(T_B) = \int_0^{T_B} x(t)x_1(t)\,dt = \pm A^2 T_b, \tag{8.61}$$

$$r(T_b) = u(T_b) + n_F(T_b). \tag{8.62}$$

If $r(T_B)$ is positive (negative), the threshold device selects bit '1' ('0'). Note that the signal value given by Eq. (8.61) is half of that given by Eq. (8.59). As far as the BER is concerned, this makes no difference since the noise sample corresponding to Fig. 8.8 is half of that corresponding to Fig. 8.5.

(b) Since $T_0 \ll T_B$, the signal power outside the bit interval is negligible. So, we approximate Eq. (8.55) as

$$x(t) = s(t - T_B/2), \tag{8.63}$$

$$s(t) = \pm A\exp\left(-\frac{t^2}{2T_0^2}\right). \tag{8.64}$$

From Eq. (2.152), we have

$$\tilde{s}(\omega) = \pm\frac{A}{a}\exp\left[-\frac{\omega^2}{4\pi a^2}\right], \tag{8.65}$$

$$a = \frac{1}{\sqrt{2\pi}T_0}. \tag{8.66}$$

Using the time-shifting property, we find

$$\tilde{x}(\omega) = \tilde{s}(\omega)e^{i\omega T_b/2}, \tag{8.67}$$

$$\tilde{x}_1(\omega) = -\tilde{x}_0(\omega) = \frac{A}{a}\exp\left[-\frac{\omega^2}{4\pi a^2} + i\omega T_b/2\right], \tag{8.68}$$

$$H(\omega) = \frac{2A}{a}\exp\left[-\frac{\omega^2}{4\pi a^2} + \frac{i\omega T_b}{2}\right]. \tag{8.69}$$

Substituting Eqs. (8.67) and (8.69) in Eq. (8.6), we find

$$u(T_b) = \pm \frac{A^2}{\pi a^2} \int_{-\infty}^{\infty} e^{-\frac{\omega^2}{2\pi a^2}} \, d\omega$$

$$= \pm 2A^2 \sqrt{\pi} T_0, \tag{8.70}$$

where $A^2 \sqrt{\pi} T_0$ is the pulse energy E_1.

8.2.2 *Error Probability with an Arbitrary Receiver Filter*

From Eqs. (8.27) and (8.28), we have

$$P_b = \frac{1}{2} \text{erfc} \left(\sqrt{\frac{v}{8}} \right), \tag{8.71}$$

$$v = \frac{[u_1(T_b) - u_0(T_b)]^2}{\sigma^2}, \tag{8.72}$$

where $u_j(T_b)$ and σ^2 are given by Eqs. (8.6) and (8.7), respectively. Eqs. (8.71) and (8.72) are valid for arbitrary filter shapes. From Eq. (7.8), we have

$$Q = \frac{I_1 - I_0}{\sigma_1 + \sigma_0}. \tag{8.73}$$

Since the mean of bit '1' ('0') is $u_1(T_b)(u_0(T_b))$, Eq. (8.72) becomes

$$v = \left(\frac{I_1 - I_0}{\sigma} \right)^2. \tag{8.74}$$

When $\sigma_1 = \sigma_0 \equiv \sigma$, from Eqs. (8.74) and (8.73), we find

$$v = 4Q^2, \tag{8.75}$$

$$P_b = \frac{1}{2} \text{erfc} \left(\frac{Q}{\sqrt{2}} \right). \tag{8.76}$$

8.3 Homodyne Receivers

Consider a fiber-optic transmission system with a homodyne balanced receiver, as shown in Fig. 8.9. Its mathematical representation is shown in Fig. 8.10. Let the transmitted optical field distribution be

$$q_s = s(t) \exp(-i\omega_c t), \tag{8.77}$$

where $s(t)$ is the complex field envelope, ω_c is the angular frequency of the optical carrier. We make the following assumptions to find the best achievable performance.

(1) Fiber dispersion, PMD, and nonlinearity are absent so that the fiber channel can be modeled as an AWGN channel. In fact, fiber dispersion and PMD cause distortion, and nonlinearity causes both distortion and noise enhancement, which will be discussed in Chapters 10 and 11.

(2) The frequency, phase, and polarization of the local oscillator are exactly aligned with the received signal.

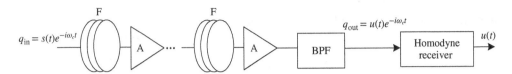

Figure 8.9 A fiber-optic transmission system with a homodyne receiver. F = fiber, A = amplifier, BPF = band-pass filter.

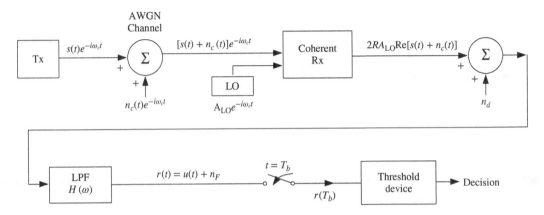

Figure 8.10 Mathematical representation of fiber-optic link with homodyne coherent receiver. LO = local oscillator, LPF = low-pass filter.

The fiber-optic channel noise can be written as

$$q_n = n_c(t) \exp(-i\omega_c t), \tag{8.78}$$

where $n_c(t)$ is the complex field envelope of the noise field with its PSD given by Eq. (6.17),

$$\rho_{\text{ASE}} = n_{sp} h\bar{f}(G - 1). \tag{8.79}$$

where $\bar{f}$ is the mean frequency. The output of the fiber-optic link is the sum of the signal field and the noise field,

$$q_{\text{out}} = [s(t) + n_c(t)] \exp(-i\omega_c t). \tag{8.80}$$

When the phase and frequency of the LO are aligned with those of the received signal, the current output of the balanced homodyne receiver may be written as (see Eq. (5.112))

$$I_s(t) = 2RA_{\text{LO}}\text{Re}[s(t) + n_c(t)] = 2RA_{\text{LO}}[s(t) + n_{cr}(t)]. \tag{8.81}$$

Here, the subscript r denotes the real part and $s(t)$ is assumed to be real. Let $n_d(t)$ be the noise introduced at the detection stage. We assume that the LO power is sufficiently large so that the shot noise dominates the thermal noise. Ignoring the thermal noise, the PSD of $n_d(t)$ is given by Eq. (7.50),

$$\rho_{n_d} = \rho_{\text{shot,eff}} = qRA_{\text{LO}}^2. \tag{8.82}$$

The total current is

$$I_{\text{tot}} = I_s(t) + n_d(t) = 2RA_{\text{LO}}[s(t) + n(t)], \tag{8.83}$$

where

$$n(t) = n_{cr}(t) + \frac{n_d(t)}{2RA_{LO}} \tag{8.84}$$

is a white noise process with Gaussian distribution. Its PSD is

$$\rho_n = \frac{N_0^{\text{homo}}}{2} = \frac{\rho_{\text{ASE}}}{2} + \frac{\rho_{\text{shot,eff}}}{4A_{LO}^2 R^2} \tag{8.85}$$

$$= \frac{\rho_{\text{ASE}}}{2} + \frac{q}{4R}. \tag{8.86}$$

The factor $1/2$ is introduced in the first term of Eq. (8.85) since the PSD of the real part of $n_c(t)$ is half of that of $n_c(t)$. The scaling factor $2RA_{LO}$ appearing in Eq. (8.83) multiplies both signal and noise and, hence, it is of no consequence in evaluating the performance. Dropping this term, we write the normalized signal used for decision as

$$I_d = s(t) + n(t). \tag{8.87}$$

8.3.1 PSK: Homodyne Detection

The optical field envelope may be written as

$$s(t) = \begin{cases} s_1(t) & \text{for bit '1'} \\ s_0(t) = -s_1(t) & \text{for bit '0'}. \end{cases} \tag{8.88}$$

We assume that $s(t)$ is real and the filter shown in Fig. 8.6 is matched to $s(t)$. Replacing $x(t)$ by $s(t)$ in Eq. (8.36), we obtain

$$E_1 = \int_0^{T_b} s_1^2(t)dt = E_0, \tag{8.89}$$

$$E_{10} = -\int_0^{T_b} s_1^2(t)dt = -E_1, \tag{8.90}$$

$$v_{\max} = \frac{2}{N_0^{\text{homo}}}(E_1 + E_0 - 2E_{10}) = \frac{8E_1}{N_0^{\text{homo}}}. \tag{8.91}$$

Since the bits '1' and '0' are equally probable, the average energy transmitted is

$$E_{av} = \frac{E_1 + E_0}{2} = E_1. \tag{8.92}$$

The average energy forms a basis for comparison of various modulation formats and Eq. (8.91) can be written as

$$v_{\max} = \frac{8E_{av}}{N_0^{\text{homo}}}. \tag{8.93}$$

The matched filter is given by (Eq. (8.40))

$$H(\omega) = [\tilde{s}_1^*(\omega) - \tilde{s}_0^*(\omega)] \exp(-i\omega T_b) = 2\tilde{s}_1^*(\omega) \exp(-i\omega T_b), \tag{8.94}$$

and the threshold r_T is (Eq. (8.47))

$$r_T = \frac{E_1 - E_0}{2} = 0. \tag{8.95}$$

Using Eq. (8.93) in Eq. (8.39), we obtain

$$P_b = \frac{1}{2}\text{erfc}(\sqrt{\gamma^{\text{homo}}}),$$

(8.96)

$$\gamma^{\text{homo}} = \frac{E_{\text{av}}}{N_0^{\text{homo}}}.$$

(8.97)

The parameter γ^{homo} represents the normalized energy per bit, which serves as a figure of merit in digital communication.

When γ^{homo} is much larger than unity, Eq. (8.96) can be approximated as

$$P_b \cong \frac{\exp(-\gamma^{\text{homo}})}{2\sqrt{\pi\gamma^{\text{homo}}}}.$$

(8.98)

8.3.1.1   Relation between Q-factor and BER

The BER and Q-factor are related as follows. From Eq. (7.8), we have

$$Q = \frac{I_1 - I_0}{\sigma_1 + \sigma_0}.$$

(8.99)

For a PSK signal, $I_1 = -I_0$ and $\sigma_1 = \sigma_0$. So,

$$Q = \frac{I_1}{\sigma_1}.$$

(8.100)

Suppose that the correlator shown in Fig. 8.8 is used as the matched filter. The mean of bit '1' after the correlator is (after setting the scaling factor $2RA_{\text{LO}}$ to unity in Eq. (8.83))

$$I_1 = \int_0^{T_b} s_1^2(t)\,dt$$

$$= E_{\text{av}}.$$

(8.101)

In this case, we have

$$h(t) = s_1(T_b - t),$$

(8.102)

$$H(\omega) = s_1^*(\omega)\exp(-i\omega T_b).$$

(8.103)

The variance of bit '1' (or bit '0') after the correlator is (see Eq. (8.7))

$$\sigma_1^2 = \frac{N_0^{\text{homo}}}{2}\frac{1}{2\pi}\int_{-\infty}^{\infty}|H(\omega)|^2 d\omega = \frac{N_0^{\text{homo}}}{2}\int_0^{T_b} s_1^2(t)dt = \frac{N_0^{\text{homo}}}{2}E_{\text{av}}.$$

(8.104)

Here, we have used Parseval's relations. Substituting Eqs. (8.101) and (8.104) in Eq. (8.100), we find

$$Q = \sqrt{\frac{2E_{\text{av}}}{N_0^{\text{homo}}}}.$$

(8.105)

From Eqs. (8.97) and (8.96), we have

$$\gamma^{\text{homo}} = \frac{Q^2}{2},$$

(8.106)

$$P_b = \frac{1}{2}\text{erfc}\left(\frac{Q}{\sqrt{2}}\right).$$

(8.107)

Eq. (8.107) holds true even when the matched filter is not used (see Section 8.2.2).

8.3.2   On–Off Keying

In this case, the optical field envelope may be written as

$$s(t) = s_1(t) \qquad \text{for bit '1'} \tag{8.108}$$

$$= 0 \qquad \text{for bit '0'} \tag{8.109}$$

Therefore,

$$E_0 = 0 \quad \text{and} \quad E_{10} = 0. \tag{8.110}$$

The average energy is

$$E_{av} = E_1/2. \tag{8.111}$$

Using Eqs. (8.111) and (8.110), Eqs. (8.38) and (8.39) can be written as

$$\upsilon_{max} = 4E_{av}/N_0^{homo} = 4\gamma^{homo} \tag{8.112}$$

and

$$P_b = \frac{1}{2}\text{erfc}\left(\sqrt{\frac{\gamma^{homo}}{2}}\right) \tag{8.113}$$

$$\cong \frac{\exp(-\gamma^{homo}/2)}{\sqrt{2\pi\gamma^{homo}}} \quad \text{when} \quad \gamma^{homo} \gg 1. \tag{8.114}$$

Fig. 8.11 shows the error probability as a function of the parameter γ^{homo}. Comparing Eqs. (8.113) and (8.96), we see that to achieve a fixed BER, the average energy should be doubled for the systems based on OOK compared with the systems based on PSK when the noise power of the channel is fixed. Alternatively, when the average energy of the transmitted signal is fixed, the system based on PSK can tolerate twice the noise power compared with the systems based on OOK to achieve the same BER.

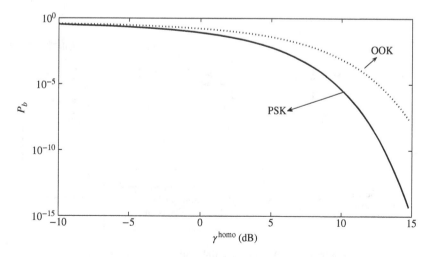

Figure 8.11   Plot of error probability vs. $10\log_{10}\gamma^{homo}$.

8.4   Heterodyne Receivers

The photocurrent in a heterodyne receiver is given by Eq. (5.94),

$$I(t) = 2RA_{\text{LO}}\text{Re}\{[s(t) + n_c(t)]\exp[-i(\omega_{\text{IF}}t + \Delta\phi)]\}. \tag{8.115}$$

Ignoring the thermal noise, the PSD of the noise current $n_d(t)$ introduced at the detection stage is

$$\rho_{n_d} = \rho_{\text{shot,eff}} = qRA_{\text{LO}}^2. \tag{8.116}$$

The total current is

$$I_{\text{tot}}(t) = 2RA_{\text{LO}}[s(t) + n_{cI}(t)]\cos(\omega_{\text{IF}}t + \Delta\phi)$$

$$+ 2RA_{\text{LO}}n_{cQ}(t)\sin(\omega_{\text{IF}}t + \Delta\phi) + n_d(t), \tag{8.117}$$

where

$$n_{cI}(t) = \text{Re}[n_c(t)], \quad n_{cQ}(t) = \text{Im}[n_c(t)], \tag{8.118}$$

and $s(t)$ is real. See Fig. 8.12. The total current may be rewritten as a summation of signal current and noise current:

$$I_{\text{tot}}(t) = I_s(t) + n_{\text{het}}(t), \tag{8.119}$$

where

$$I_s(t) = 2RA_{\text{LO}}s(t)\cos(\omega_{\text{IF}}t + \Delta\phi), \tag{8.120}$$

$$n_{\text{het}}(t) = 2RA_{\text{LO}}\left\{n_{cI}\cos\left(\omega_{\text{IF}}t + \Delta\phi\right) + n_{cQ}\sin\left(\omega_{\text{IF}}t + \Delta\phi\right) + \frac{n_d(t)}{2RA_{\text{LO}}}\right\}. \tag{8.121}$$

Note that $2RA_{\text{LO}}$ appears as a scaling factor both in the signal and noise components. As before, we drop this scaling factor from now on. The PSD of $n_{\text{het}}(t)$ is (see Example 8.5)

$$\frac{N_0^{\text{het}}}{2} = \frac{\rho_{\text{ASE}}}{4} + \frac{\rho_{\text{shot,eff}}}{4R^2A_{\text{LO}}^2} \tag{8.122}$$

$$= \frac{\rho_{\text{ASE}}}{4} + \frac{q}{4R}. \tag{8.123}$$

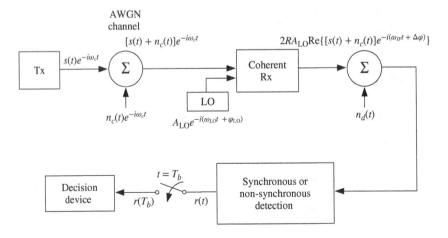

Figure 8.12   A fiber-optic link with heterodyne receiver.

For signals such as OOK, the information is contained only in the amplitude and if a detection scheme ignores the phase of the received signal, it does not lead to the loss of information. Such a scheme is known as *asynchronous detection*. For example, if the received signal passes through an envelope detector, the phase information is lost and the transmitted information is retrieved asynchronously without having to track the phase of the received signal. In non-optical communication, asynchronous receivers are known as non-coherent receivers [1, 2]. In contrast, a detector can detect the phase of the transmitted signal by carefully aligning the phase of the microwave oscillator (or equivalently synchronizing the timings of the oscillator output) with the received signal and such a scheme is known as *synchronous detection*. For PSK signals, a synchronous detector has to be used. For homodyne receivers, asynchronous detection schemes can be realized by introducing an envelope detector in the DSP unit. The performance of modulation schemes with homodyne asynchronous detection is similar to the corresponding heterodyne receivers.

8.4.1 PSK: Synchronous Detection

The received signal in the absence of noise can be written as

$$I(t) = \begin{cases} I_1(t) & \text{for '1'} \\ I_0(t) & \text{for '0'} \end{cases} \tag{8.124}$$

where

$$I_j(t) = s_j(t)\cos(\omega_{\mathrm{IF}}t + \Delta\phi), \quad j = 0, 1, \tag{8.125}$$

with $s_0(t) = -s_1(t)$. We assume that $s(t)$ is real. The filter matched to $I(t)$ is (see Eq. (8.40))

$$H_I(\omega) = [\tilde{I}_1^*(\omega) - \tilde{I}_0^*(\omega)]\exp(i\omega T_b)$$

$$= [\tilde{s}_1^*(\omega - \omega_{\mathrm{IF}})e^{i\Delta\phi} + \tilde{s}_1^*(\omega + \omega_{\mathrm{IF}})e^{-i\Delta\phi}]\exp(i\omega T_b). \tag{8.126}$$

This matched filter can be realized as a correlator, as shown in Fig. 8.13. In this case, Eq. (8.52) is modified as

$$r(T_b) = \int_0^{Tb} y(\tau)[I_1(\tau) - I_0(\tau)]\,d\tau$$

$$= 2\int_0^{Tb} y(\tau)s_1(\tau)\cos(\omega_{\mathrm{IF}}\tau + \Delta\phi)\,d\tau. \tag{8.127}$$

Using Eq. (8.35), we have

$$v_{\max} = \frac{2}{N_0^{\mathrm{het}}}\int_0^{T_b}[I_1(t) - I_0(t)]^2\,dt = \frac{8}{N_0^{\mathrm{het}}}\int_0^{Tb}s_1^2(t)\cos^2(\omega_{\mathrm{IF}}t + \Delta\phi)\,dt = \frac{4E_1}{N_0^{\mathrm{het}}}. \tag{8.128}$$

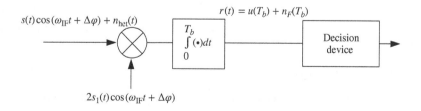

Figure 8.13   Matched filter for a PSK signal in a fiber-optic system with heterodyne receiver.

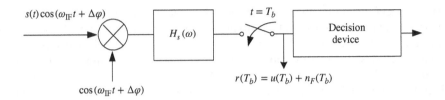

Figure 8.14   Matched filter for the baseband signal after down-conversion.

Alternatively, the photocurrent can be down-converted to the baseband by demodulating synchronously and the result is applied to a filter $H_s(\omega)$, matched to the baseband signal $s(t)$, as shown in Fig. 8.14. In this case, we find that (see Example 8.7)

$$v_{\max} = \frac{4E_1}{N_0^{\text{het}}}. \tag{8.129}$$

In both cases, we obtain the same $v_{\max}$ and, using Eq. (8.129) in Eq. (8.39), we obtain

$$P_b = \frac{1}{2} \operatorname{erfc}\left(\sqrt{\frac{\gamma^{\text{het}}}{2}}\right) \tag{8.130}$$

where

$$\gamma^{\text{het}} = \frac{E_{\text{av}}}{N_0^{\text{het}}}. \tag{8.131}$$

8.4.1.1   ASE Limited Systems

When $\rho_{\text{ASE}} \gg \rho_{\text{shot}}/(4A_{\text{LO}}^2 R^2)$, the second term of Eq. (8.122) can be ignored. In this case,

$$N_0^{\text{het}} = \frac{\rho_{\text{ASE}}}{2}. \tag{8.132}$$

Now, Eq. (8.130) reduces to

$$P_b^{\text{het}} = \frac{1}{2} \operatorname{erfc}\left(\sqrt{\frac{E_{\text{av}}}{\rho_{\text{ASE}}}}\right). \tag{8.133}$$

Under the same conditions, from Eq. (8.85), we have

$$N_0^{\text{homo}} = \rho_{\text{ASE}} \tag{8.134}$$

and

$$P_b^{\text{homo}} = \frac{1}{2} \operatorname{erfc}\left(\sqrt{\frac{E_{\text{av}}}{\rho_{\text{ASE}}}}\right). \tag{8.135}$$

Thus, the performances of the homodyne and heterodyne receivers are the same for a PSK signal when the ASE is dominant.

8.4.1.2   Shot Noise-Limited Systems

In this case, $\rho_{\text{shot}} \gg \rho_{\text{ASE}} A_{\text{LO}}^2 R^2$ and we ignore the first term of Eq. (8.122) and use Eq. (8.82) to obtain

$$N_0^{\text{het}} = \frac{\rho_{\text{shot,eff}}}{2R^2 A_{\text{LO}}^2} = \frac{h\bar{f}}{2\eta} \tag{8.136}$$

and

$$P_b^{\text{het}} = \frac{1}{2} \text{erfc}\left(\sqrt{\frac{\eta E_{\text{av}}}{h\bar{f}}}\right).$$
(8.137)

The corresponding expressions for homodyne detection are

$$N_0^{\text{homo}} = \frac{h\bar{f}}{2\eta},$$
(8.138)

$$P_b^{\text{homo}} = \frac{1}{2} \text{erfc}\left(\sqrt{\frac{2\eta E_{\text{av}}}{h\bar{f}}}\right).$$
(8.139)

Comparing Eqs. (8.137) and (8.139), we find that the homodyne receiver has a 3-dB advantage over the heterodyne receiver when the shot noise is dominant [3]–[5]. In other words, to reach a fixed error probability, the received signal power in the case of a heterodyne receiver should be twice that of a homodyne receiver. This can be explained as follows. In the case of the heterodyne receiver, the signal is modulated by a carrier at a frequency ω_{IF}. Since the average energy of the signal $s(t)\cos(\omega_{\text{IF}}t + \Delta\phi)$ is half of that of the transmitted signal $s(t)$, the heterodyne receivers have a 3-dB disadvantage over the homodyne receivers. However, for long-haul systems, the noise due to in-line amplifiers is dominant and, hence, the performances of homodyne and heterodyne receivers are roughly the same.

Sine $E_{\text{av}}/h\bar{f}$ is the mean number of received signal photons N_s, Eqs. (8.137) and (8.139) may be rewritten as

$$P_b^{\text{het}} = \frac{1}{2} \text{erfc}(\sqrt{\eta N_s}),$$
(8.140)

$$P_b^{\text{homo}} = \frac{1}{2} \text{erfc}(\sqrt{2\eta N_s}).$$
(8.141)

For example, when $\eta = 1$ and $N_s = 9$, $P_b^{\text{homo}} = \frac{1}{2}\text{erfc}(\sqrt{18}) = 10^{-9}$, in agreement with Eq. (7.63). When $N_s = 18$, $P_b^{\text{het}} = 10^{-9}$ [4].

8.4.2 OOK: Synchronous Detection

The received signal in the absence of noise can be written as

$$I(t) \begin{cases} = I_1(t) & \text{for '1'} \\ = I_0(t) & \text{for '0'} \end{cases}$$
(8.142)

$$I_j(t) = s_j(t)\cos(\omega_{\text{IF}}t), \quad j = 0, 1,$$
(8.143)

with $s_0(t) = 0$. Here $\Delta\phi$ is set to zero for simplicity. Proceeding as in Section 8.4.1, we find

$$P_b = \frac{1}{2} \text{erfc}\left(\sqrt{\frac{\gamma^{\text{het}}}{4}}\right).$$
(8.144)

Example 8.2

In a 10-Gb/s unrepeatered fiber-optic system based on PSK, rectangular NRZ pulses are transmitted with a peak power of 5 dBm. The fiber loss is 50 dB. The gain and n_{sp} of the pre-amplifier used at the receiver are 30 dB and 1.5, respectively. $R = 0.9$ A/W. Find the error probability if the receiver is (a) a balanced homodyne or (b) a balanced heterodyne. Ignore thermal noise. Repeat this example if the signal is OOK with the same peak power.

Solution:
Signal calculation

Launch peak power = P_0(dBm) = 5 dBm
Fiber loss(dB) = 50 dB
Preamplifier gain, G(dB) = 30 dB
The peak power of the received signal is

$$P_r(\text{dBm}) = P_0(\text{dBm}) - \text{fiber loss(dB)} + G(\text{dB})$$

$$= 5 \text{ dBm} - 50 \text{ dB} + 30 \text{ dB}$$

$$= -15 \text{ dBm}, \tag{8.145}$$

$$P_r = 10^{P_r(\text{dBm})/10} \text{ mW} = 3.16 \times 10^{-5} \text{ W}. \tag{8.146}$$

The bit interval is

$$T_b = \frac{1}{10 \times 10^9} = 10^{-10} \text{ s}. \tag{8.147}$$

The energy of bit '1' is

$$E_1 = P_r T_b$$

$$= 3.16 \times 10^{-5} \times 10^{-10} \text{ J}$$

$$= 3.16 \times 10^{-15} \text{ J}. \tag{8.148}$$

For PSK, the average energy $E_{av} = E_1$.

Noise calculation
The PSD of ASE is given by Eq. (6.17),

$$\rho_{\text{ASE}} = n_{sp} h \bar{f}(G - 1), \tag{8.149}$$

$$n_{sp} = 1.5, \tag{8.150}$$

$$h = 6.626 \times 10^{-34} \text{ Js}, \tag{8.151}$$

$$\bar{f} = \frac{c}{\lambda} = \frac{3 \times 10^8}{1550 \times 10^{-9}} = 193.54 \text{ THz}, \tag{8.152}$$

$$G = 10^{G(\text{dB})/10} = 1000, \tag{8.153}$$

$$\rho_{\text{ASE}} = 1.5 \times 6.626 \times 10^{-34} \times 193.54 \times (1000 - 1)$$

$$= 1.921 \times 10^{-16} \text{ W/Hz.} \tag{8.154}$$

The PSD of effective shot noise is given by Eq. (7.50),

$$\rho_{\text{shot,eff}} = qI_{\text{LO}}$$

$$= qRA_{\text{LO}}^2. \tag{8.155}$$

From Eq. (8.85), we have

$$N_0^{\text{homo}} = \rho_{\text{ASE}} + \frac{\rho_{\text{shot,eff}}}{2R^2 A_{\text{LO}}^2}$$

$$= \rho_{\text{ASE}} + \frac{q}{2R}. \tag{8.156}$$

Electron charge $q = 1.602 \times 10^{-19}$ C,

$$R = 0.9 \text{ A/W,} \tag{8.157}$$

$$N_0^{\text{homo}} = 1.921 \times 10^{-16} + \frac{1.602 \times 10^{-19}}{2 \times 0.9} \text{ W/Hz}$$

$$= 1.922 \times 10^{-16} \text{ W/Hz,} \tag{8.158}$$

From Eq. (8.122), we have

$$N_0^{\text{het}} = \frac{\rho_{\text{ASE}}}{2} + \frac{\rho_{\text{shot,eff}}}{2R^2 A_{\text{LO}}^2}$$

$$= 9.617 \times 10^{-17} \text{ W/Hz,} \tag{8.159}$$

Error Probability

(a) For a balanced homodyne receiver with PSK signal, the error probability is given by Eq. (8.96),

$$P_b^{\text{PSK}} = \frac{1}{2}\text{erfc}\left(\sqrt{\frac{E_{\text{av}}}{N_0^{\text{homo}}}}\right) = \frac{1}{2}\text{erfc}\left(\sqrt{\frac{3.16 \times 10^{-15}}{1.922 \times 10^{-16}}}\right)$$

$$= 4.86 \times 10^{-9}. \tag{8.160}$$

If the signal is OOK, from Eq. (8.113) we have

$$P_b^{\text{OOK}} = \frac{1}{2}\text{erfc}\left(\sqrt{\frac{E_{\text{av}}}{2N_0^{\text{homo}}}}\right), \tag{8.161}$$

$$E_{\text{av}} = \frac{E_1}{2} = 1.58 \times 10^{-15} \text{ J,} \tag{8.162}$$

$$P_b^{\text{OOK}} = \frac{1}{2}\text{erfc}\left(\sqrt{\frac{1.58 \times 10^{-15}}{2 \times 1.922 \times 10^{-16}}}\right)$$

$$= 2.06 \times 10^{-3}. \tag{8.163}$$

(b) For a balanced heterodyne receiver with PSK signal, the error probability is given by Eq. (8.130),

$$P_b^{\mathrm{PSK}} = \frac{1}{2}\mathrm{erfc}\left(\sqrt{\frac{E_{\mathrm{av}}}{2N_0^{\mathrm{het}}}}\right)$$

$$= \frac{1}{2}\mathrm{erfc}\left(\sqrt{\frac{3.16\times10^{-15}}{2\times9.617\times10^{-17}}}\right)$$

$$= 4.901\times10^{-9}. \tag{8.164}$$

If the signal is OOK, from Eq. (8.144) we have

$$P_b^{\mathrm{OOK}} = \frac{1}{2}\mathrm{erfc}\left(\sqrt{\frac{E_{\mathrm{av}}}{4N_0^{\mathrm{het}}}}\right)$$

$$= 2.07\times10^{-3}. \tag{8.165}$$

8.4.3   FSK: Synchronous Detection

To transmit bit '1' ('0'), the frequency of the optical carrier is shifted by $\Delta\omega/2$ $(-\Delta\omega 2)$. The complex field envelopes corresponding to bits '1' and '0' are

$$s_1(t) = A\exp\left(\frac{-i\Delta\omega t}{2}\right),$$

$$s_0(t) = A\exp\left(\frac{i\Delta\omega t}{2}\right), \tag{8.166}$$

for a duration of T_b. The photocurrents are

$$I = I_1(t) \quad \text{for bit '1'} \tag{8.167}$$

$$= I_0(t) \quad \text{for bit '0'} \tag{8.168}$$

where

$$I_1(t) = \mathrm{Re}[s_1(t)\exp(-i\omega_{\mathrm{IF}}t)] = A\cos\left[\left(\omega_{\mathrm{IF}} + \frac{\Delta\omega}{2}\right)t\right],$$

$$I_0(t) = A\cos\left[\left(\omega_{\mathrm{IF}} - \frac{\Delta\omega}{2}\right)t\right]. \tag{8.169}$$

Here, we have ignored the constant factor $2RA_{\mathrm{LO}}$ and $\Delta\phi$ is set to zero. The filter matched to $I(t)$ can be realized as a correlator, as shown in Fig. 8.15. Replacing $x_1(t)$ and $x_0(t)$ of Eq. (8.36) by $I_1(t)$ and $I_0(t)$, respectively, we obtain

$$E_{10}^e = \int_0^{T_b} I_1(t)I_0(t)\,dt$$

$$= \frac{A^2}{2}\int_0^{T_b}[\cos(\Delta\omega t) + \cos(\omega_{\mathrm{IF}}t)]\,dt$$

$$= \frac{A^2 T_b}{2}[\mathrm{sinc}(\Delta\omega T_b/\pi) + \mathrm{sinc}(\omega_{\mathrm{IF}}T_b/\pi)], \tag{8.170}$$

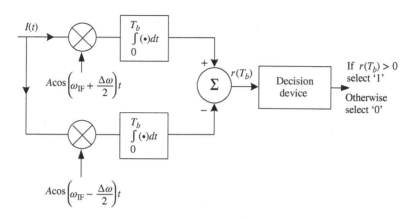

Figure 8.15   Matched filter for FSK signal in a fiber-optic system with heterodyne receiver.

where

$$\text{sinc}(x) = \frac{\sin(\pi x)}{\pi x}. \tag{8.171}$$

The superscript e is introduced to indicate that E represents the energy of a bit in the electrical domain. When $\omega_{IF} \gg 1/T_b$, the second term on the right-hand side of Eq. (8.170) can be ignored. Similarly,

$$E_0^e = E_1^e = A^2 \int_0^{T_b} \cos^2\left[\left(\omega_{IF} + \frac{\Delta\omega}{2}\right)t\right] dt$$

$$= \frac{A^2}{2} \int_0^{T_b} \left\{1 + \cos\left[2\left(\omega_{IF} + \frac{\Delta\omega}{2}\right)t\right]\right\} dt$$

$$= \frac{A^2 T_b}{2}. \tag{8.172}$$

The contribution from the second term on the right-hand side of Eq. (8.172) is negligible since $\omega_{IF} \gg 1/T_b$. Here, E_j^e denotes the normalized energy of the bit 'j', $j = 0, 1$ in the electrical domain. The corresponding energy in the optical domain is

$$E_1^{opt} = \int_0^{T_b} |s_1(t)|^2 dt = A^2 T_b = E_{av}. \tag{8.173}$$

Substituting Eqs. (8.172) and (8.170) in Eq. (8.38), we obtain

$$v_{max} = \frac{2E_{av}}{N_0^{het}}[1 - \text{sinc}(\Delta\omega T_b/\pi)], \tag{8.174}$$

$$P_b = \frac{1}{2}\text{erfc}\left(\sqrt{\frac{v_{max}}{8}}\right). \tag{8.175}$$

The above equation is valid for arbitrary frequency difference $\Delta\omega$. To find the minimum achievable BER, we need to minimize P_b with respect to $\Delta\omega$ or equivalently maximize v_{max}. In other words, the sinc function should be minimum. From Fig. 8.16, we see that the minimum value is -0.217 at $\Delta\omega T_b = 1.43\pi$. Choosing

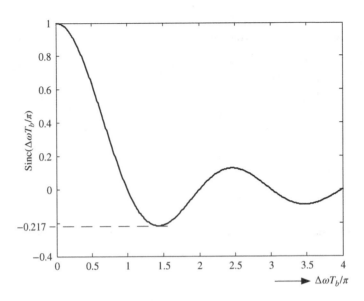

Figure 8.16 Sinc function.

this value of $\Delta\omega$ in Eq. (8.174), we obtain

$$P_b = \frac{1}{2}\text{erfc}\left(\sqrt{\frac{1.217\gamma^{\text{het}}}{4}}\right). \tag{8.176}$$

Comparing Eqs. (8.144) and (8.176), we find that a system based on OOK requires an average energy 1.217 times that of a system based on FSK to achieve the given BER.

8.4.3.1 Orthogonal FSK

The signals $I_1(t)$ and $I_0(t)$ are said to be *orthogonal* if

$$E_{10}^e = \int_0^{T_b} I_1(t)I_0(t)\,dt = 0. \tag{8.177}$$

From Eq. (8.170), we find that E_{10}^e is zero if

$$\Delta\omega T_b = n\pi, \quad n = 1, 2, 3, \ldots \tag{8.178}$$

or

$$\Delta f = \frac{n}{2T_b}. \tag{8.179}$$

As Δf increases, the signal bandwidth increases too. The minimum frequency separation occurs when $n = 1$:

$$\Delta f_{\min} = \frac{1}{2T_b}. \tag{8.180}$$

The FSK scheme using the frequency separation given by Eq. (8.180) is called *minimum shift keying* (MSK). Since E_{10}^e is zero for orthogonal FSK,

$$v_{\max} = \frac{2}{N_0^{\text{het}}}(E_1^e + E_0^e) = \frac{2E_{\text{av}}}{N_0^{\text{het}}} \tag{8.181}$$

and

$$P_b = \frac{1}{2}\text{erfc}\left(\sqrt{\frac{\gamma^{\text{het}}}{4}}\right). \tag{8.182}$$

Comparing Eqs. (8.182) and (8.144), we see that the expression for P_b for systems based on OOK is the same as that for systems based on orthogonal FSK when synchronous detection is used in both systems.

8.4.4   OOK: Asynchronous Receiver

In Section 8.4.2, we discussed the error probability for the OOK signal with the receiver filter matched to $I(t)$ exactly. Here, we consider the case in which the receiver filter is matched to $I(t)$ except for the phase. The transmitted optical field envelope can be written as

$$s(t) = \begin{cases} s_1(t) & \text{for bit '1'} \\ 0 & \text{for bit '0'} \end{cases} \tag{8.183}$$

Without loss of generality, we assume that $s(t)$ is real. The received signal passes through a filter matched to $s_1(t)\cos(\omega_{\text{IF}}t)$. From Eq. (8.120), the photocurrent is

$$I_1(t) = s_1(t)\cos(\omega_{\text{IF}}t + \Delta\phi),$$

$$\tilde{I}_1(\omega) = \frac{[\tilde{s}_1(\omega - \omega_{\text{IF}})\exp(-i\Delta\phi) + \tilde{s}_1(\omega + \omega_{\text{IF}})\exp(i\Delta\phi)]}{2}. \tag{8.184}$$

As before, we have ignored the scaling factor $2RA_{\text{LO}}$. The matched filter need not match the phase $\Delta\phi$ of $I_1(t)$. One possible way of realizing this type of matched filter is to use the envelope detector which detects only the envelope and ignores the phase, i.e., it is matched to $s_1(t)\cos(\omega_{\text{IF}}t)$. The transfer function of the matched filter is

$$H_I(\omega) = \{\mathcal{F}[s_1(t)\cos(\omega_{\text{IF}}t)]\}^* \exp(i\omega T_b)$$

$$= \frac{[\tilde{s}_1^*(\omega - \omega_{\text{IF}}) + \tilde{s}_1^*(\omega + \omega_{\text{IF}})]\exp(i\omega T_b)}{2}. \tag{8.185}$$

The impulse response of the matched filter can be found from Eq. (8.42) by replacing $x_1(t)$ with $s_1(t)\cos(\omega_{\text{IF}}t)$ and $x_0(t)$ with zero:

$$h_I(t) = s_1(T_b - t)\cos[\omega_{\text{IF}}(T_b - t)]. \tag{8.186}$$

The matched filter can be realized as a correlator (see Fig. 8.6), and the signal component of its output is (Eq. (8.48))

$$I_F(t) = \int_0^{T_b} s(\tau)\cos(\omega_{\text{IF}}\tau + \Delta\phi)s_1(T_b + \tau - t)\cos[\omega_{\text{IF}}(T_b + \tau - t)]d\tau \tag{8.187}$$

$$= \frac{1}{2}\int_0^{T_b} s(\tau)s_1(T_b + \tau - t)\{\cos[\omega_{\text{IF}}(t - T_b) + \Delta\phi]$$

$$+ \cos(\omega_{\text{IF}}(2\tau + T_b - t) + \Delta\phi)\}d\tau. \tag{8.188}$$

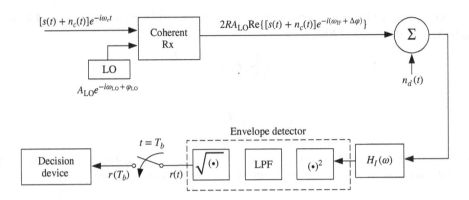

Figure 8.17 A heterodyne receiver with an envelope detector and a matched filter $H_I(\omega)$ for OOK.

In Eq. (8.188), the second term on the right-hand side corresponds to the Fourier transform of $s(\tau)s_1(\tau + T_b - t)$ at $\omega = \pm 2\omega_{IF}$. Since the spectral width of $s(\tau)$ is much smaller than ω_{IF}, the second term can be ignored. Therefore,

$$I_F(t) = \underbrace{\frac{s_F(t)}{2}}_{\text{envelope}} \cos[\omega_{IF}(t - T_b) + \Delta\phi], \tag{8.189}$$

where

$$s_F(t) = \int_0^{T_b} s(\tau)s_1(T_b + \tau - t)\,d\tau. \tag{8.190}$$

The output of the matched filter passes through an envelope detector which can be imagined as a cascade of squarer, low-pass filter, and square-rootor, as shown in Fig. 8.17. When we square $I_F(t)$, we obtain a term proportional to $\cos[2\omega_{IF}(t - T_b) + \Delta\phi]$ which is rejected by the low-pass filter. The signal output of the envelope detector is the envelope of $I_F(t)$ (shown in Eq. (8.189)), which is given by

$$u(t) = s_F(t)/2, \tag{8.191}$$

$$u(T_b) = \frac{s_F(T_b)}{2} = \frac{1}{2}\int_0^{T_b} s(\tau)s_1(\tau)\,d\tau, \tag{8.192}$$

$$u(T_b) = E_1/2 \quad \text{when '1' is sent}$$
$$= 0 \qquad \text{when '0' is sent.} \tag{8.193}$$

Next, consider the noise component before the matched filter given by Eq. (8.121),

$$n(t) = n_{cI}\cos(\omega_{IF}t + \Delta\phi) + n_{cQ}\sin(\omega_{IF}t + \Delta\phi) + \frac{n_d(t)}{2RA_{LO}}. \tag{8.194}$$

Here, we have dropped the scaling factor $2RA_{LO}$. Since the ASE is expressed as the modulated noise process, it is convenient to express the detector noise as the modulated noise process as well, i.e.,

$$n_d(t) = n_{dI}\cos(\omega_{IF}t + \Delta\phi) + n_{dQ}\sin(\omega_{IF}t + \Delta\phi), \tag{8.195}$$

where n_{dI} and n_{dQ} are the in-phase and quadrature components of the detector noise. Substituting Eq. (8.195) in Eq. (8.194), we find

$$n(t) = n_I \cos(\omega_{IF} t + \Delta\phi) + n_Q \sin(\omega_{IF} t + \Delta\phi), \tag{8.196}$$

where

$$n_I = n_{cI} + n_{dI}/(2RA_{LO}), \tag{8.197}$$

$$n_Q = n_{cQ} + n_{dQ}/(2RA_{LO}). \tag{8.198}$$

We assume that $n(t)$ is a narrow-band Gaussian noise process with zero mean and it is band-limited to the frequency interval $f_{IF} - B \leq |f| \leq f_{IF} + B$. First consider $n_I \cos(\omega_{IF} t + \Delta\phi)$. After passing through the matched filter, it becomes (see Eq. (8.188))

$$\frac{1}{2} \int_0^{T_b} n_I(\tau) s_1(T_b + \tau - t)\{\cos[\omega_{IF}(t - T_b) + \Delta\phi] + \cos(\omega_{IF}(2\tau + T_b - t) + \Delta\phi)\}\, d\tau. \tag{8.199}$$

As before, the second term on the right-hand side can be ignored. So, it becomes

$$n_{FI} \cos[\omega_{IF}(t - T_b) + \Delta\phi], \tag{8.200}$$

where

$$n_{FI} = \frac{1}{2} \int_0^{T_b} n_I(\tau) s_1(T_b + \tau - t)\, d\tau. \tag{8.201}$$

Similarly, the second term of Eq. (8.196) becomes

$$n_{FQ} \sin[\omega_{IF}(t - T_b) + \Delta\phi], \tag{8.202}$$

where

$$n_{FQ} = \frac{1}{2} \int_0^{T_b} n_Q(\tau) s_1(T_b + \tau - t)\, d\tau. \tag{8.203}$$

Combining Eqs. (8.200) and (8.202), the noise output of the matched filter is

$$n_F(t) = [n_{FI} \cos(\omega_{IF}(t - T_b) + \Delta\phi) + n_{FQ} \sin(\omega_{IF}(t - T_b) + \Delta\phi)], \tag{8.204}$$

where $n_{FI}(t)$ and $n_{FQ}(t)$ are the in-phase and quadrature components of $n_F(t)$. The PSD of $n(t)$ is $N_0^{het}/2$. From Eq. (8.7), we have

$$\sigma_F^2 = \langle n_F^2 \rangle$$

$$= \frac{N_0^{het}}{2} \frac{1}{2\pi} \int_{-\infty}^{\infty} |H_I(\omega)|^2\, d\omega \tag{8.205}$$

$$= \frac{N_0^{het}}{16\pi} \int_{-\infty}^{\infty} [|\tilde{s}_1(\omega - \omega_{IF})|^2 + |\tilde{s}_1(\omega + \omega_{IF})|^2\, d\omega]. \tag{8.206}$$

In Eq. (8.205) we ignore cross-products such as $\tilde{s}_1^*(\omega - \omega_{IF})s_1(\omega + \omega_{IF})$. This is because $\tilde{s}_1(\omega - \omega_{IF})$ and $\tilde{s}_1^*(\omega + \omega_{IF})$ represent frequency components centered around ω_{IF} and $-\omega_{IF}$, respectively. If the spectral width of $s_1(t)$ is smaller than ω_{IF}, these frequency components do not overlap. Noting that the contributions from the first and second terms on the right-hand side of Eq. (8.206) are the same, we find

$$\sigma_F^2 = \frac{N_0^{het}}{8\pi} \int_{-\infty}^{\infty} |\tilde{s}_1(\omega)|^2 d\omega = \frac{N_0^{het} E_1}{4}. \tag{8.207}$$

The total output of the matched filter is

$$
I_F(t) + n_F(t) = \left(\frac{s_F(t)}{2} + n_{FI}(t) \right) \cos \theta + n_{FQ}(t) \sin \theta
$$

$$
= \underbrace{\sqrt{\left[\frac{s_F(t)}{2} + n_{FI}(t) \right]^2 + n_{FQ}^2(t)}} \cdot \cos(\theta - \phi), \tag{8.208}
$$

$$
\text{envelope}
$$

where

$$
\theta = \omega_{\mathrm{IF}}(t - T_b) + \Delta\phi, \tag{8.209}
$$

$$
\phi = \tan^{-1} \left\{ \frac{n_{FQ}(t)}{s_F(t)/2 + n_{FI}(t)} \right\}. \tag{8.210}
$$

After passing through the envelope detector, the output sample at $t = T_b$ is proportional to the envelope:

$$
r(T_b) = \sqrt{[s_F(T_b)/2 + n_{FI}(T_b)]^2 + n_{FQ}^2(T_b)}. \tag{8.211}
$$

When a bit '0' is transmitted, $s_F(T_b) = 0$. Therefore,

$$
r(T_b) = \sqrt{n_{FI}^2(T_b) + n_{FQ}^2(T_b)}. \tag{8.212}
$$

For a narrow-band noise process, it can be shown that the variances of the in-phase component $n_{FI}(t)$ and the quadrature component $n_{FQ}(t)$ are the same as for the narrow-band noise $n_F(t)$ [2]. Therefore, $n_{FI}(T_b)$ and $n_{FQ}(T_b)$ are Gaussian random variables with variance σ_F^2 given by Eq. (8.207). The pdf of the envelope when '0' is transmitted is given by the Rayleigh distribution [6],

$$
p(r|\text{'0' sent}) = \frac{r}{\sigma_F^2} \exp\left(-\frac{r^2}{2\sigma_F^2} \right). \tag{8.213}
$$

When a bit '1' is transmitted, $r(t)$ is an envelope of a cosine wave in the presence of Gaussian noise (Eq. (8.208)), its amplitude $s_F(T_b)/2 = E_1/2$ (see Eq. (8.193)) and, therefore, the pdf of $r(t)$ is given by the Rician distribution [6]

$$
p(r|\text{'1' sent}) = \frac{r}{\sigma_F^2} \exp\left(-\frac{r^2 + E_1^2/4}{2\sigma_F^2} \right) I_0\left(\frac{rE_1}{2\sigma_F^2} \right), \tag{8.214}
$$

where $I_0(x)$ is the modified zero-order Bessel function of the first kind. The threshold is determined by the intersection of two curves $p(r|\text{'1' sent})$ and $p(r|\text{'0' sent})$ (see Eq. (8.19)):

$$
p(r|\text{'1' sent}) = p(r|\text{'0' sent}), \tag{8.215}
$$

$$
\exp\left(-\frac{E_1^2/4}{2\sigma_F^2} \right) I_0\left(\frac{r_T E_1}{2\sigma_F^2} \right) = 1. \tag{8.216}
$$

This equation is satisfied to a close approximation [1]

$$r_T = \frac{E_1}{4} \sqrt{1 + \frac{8\sigma_F^2}{E_1^2/4}} \tag{8.217}$$

$$= \frac{E_1}{4} \sqrt{1 + \frac{8N_0^{\text{het}}}{E_1}}, \tag{8.218}$$

where we have used Eq. (8.207) for σ_F^2. When a bit '0' is transmitted, if $r > r_T$, '0' is mistaken as '1' and this probability is

$$P(1|0) = \int_{r_T}^{\infty} p(r|\text{'0' sent})dr = \frac{1}{\sigma_F^2} \int_{r_T}^{\infty} r \exp\left(-\frac{r^2}{2\sigma_F^2}\right) dr \tag{8.219}$$

$$= \exp\left(-\frac{r_T^2}{2\sigma_F^2}\right) \cong \exp\left[-\frac{\gamma^{\text{het}}}{4}\left(1 + \frac{4}{\gamma^{\text{het}}}\right)\right], \tag{8.220}$$

where we have used Eq. (8.218) for r_T and

$$\gamma^{\text{het}} = \frac{E_{\text{av}}}{N_0^{\text{het}}}. \tag{8.221}$$

For OOK, $E_{\text{av}} = E_1/2$, so

$$\gamma^{\text{het}} = \frac{E_1}{2N_0^{\text{het}}}. \tag{8.222}$$

When $\gamma^{\text{het}} \gg 1$,

$$P(1|0) \cong \exp\left(-\frac{\gamma^{\text{het}}}{4}\right). \tag{8.223}$$

Similarly, the probability of mistaking '1' as '0' is

$$P(0|1) = \int_0^{r_T} p(r|\text{'1' sent})dr = \frac{1}{\sigma_F^2} \int_0^{r_T} r \exp\left(-\frac{r^2 + E_1^2/4}{2\sigma_F^2}\right) I_0\left(\frac{rE_1}{2\sigma_F^2}\right) dr. \tag{8.224}$$

Let

$$x = \frac{r}{\sigma_F}, \tag{8.225}$$

$$a = \frac{E_1}{2\sigma_F} = \sqrt{2\gamma^{\text{het}}}. \tag{8.226}$$

Now, Eq. (8.224) becomes

$$P(0|1) = \int_0^{r_T/\sigma_F} x \exp\left(-\frac{a^2 + x^2}{2}\right) I_0(ax) \, dx$$

$$= \left[1 - Q_1\left(a, \frac{r_T}{\sigma_F}\right)\right], \tag{8.227}$$

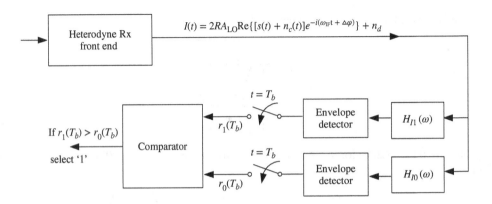

Figure 8.18 A heterodyne receiver with envelope detectors and matched filters for FSK.

where $Q_1(a, r_T/\sigma_F)$ is the generalized Marcum Q-function defined as [6]

$$Q_1(a, b) = \int_b^\infty x \exp\left(-\frac{a^2 + x^2}{2}\right) I_0(ax)\, dx$$

$$= \exp\left(-\frac{a^2 + b^2}{2}\right) \sum_{k=0}^\infty \left(\frac{a}{b}\right)^k I_k(ab), \quad b > a > 0. \tag{8.228}$$

Substituting Eqs. (8.226) and (8.217) in Eq (8.227), we obtain

$$P(0|1) = \left[1 - Q_1\left(\sqrt{2\gamma^{\text{het}}}, \sqrt{\frac{\gamma^{\text{het}}}{2}\left(1 + \frac{4}{\gamma^{\text{het}}}\right)}\right)\right]. \tag{8.229}$$

Combining Eqs. (8.224) and (8.229), we find

$$P_b = \frac{1}{2}[P(1|0) + P(0|1)]$$

$$= \frac{1}{2}\left\{\exp\left[-\frac{\gamma^{\text{het}}}{4}\left(1 + \frac{4}{\gamma^{\text{het}}}\right)\right] + 1 - Q_1\left(\sqrt{2\gamma^{\text{het}}}, \sqrt{\frac{\gamma^{\text{het}}}{2}\left(1 + 4/\gamma^{\text{het}}\right)}\right)\right\}. \tag{8.230}$$

8.4.5 FSK: Asynchronous Detection

The transmitted signals $s_1(t)$ and $s_0(t)$ are the same as those in Section 8.4.3 (Eq. (8.166)). The output of the heterodyne receiver front end in the absence of noise may be written as

$$I(t) = \begin{cases} I_1(t) & \text{when } s_1(t) \text{ is transmitted} \\ I_0(t) & \text{when } s_0(t) \text{ is transmitted}, \end{cases} \tag{8.231}$$

where

$$I_1(t) = \begin{cases} 2RA_{\text{LO}}A \cos\left[\left(\omega_{\text{IF}} + \frac{\Delta\omega}{2}\right)t + \Delta\phi\right] & \text{for } 0 < t \leq T_b \\ 0 & \text{otherwise}, \end{cases} \tag{8.232}$$

$$I_0(t) = \begin{cases} 2RA_{LO}A\cos\left[\left(\omega_{IF} - \frac{\Delta\omega}{2}\right)t + \Delta\phi\right] & \text{for } 0 < t \leq T_b \\ 0 & \text{otherwise.} \end{cases} \tag{8.233}$$

As before, we ignore the scaling factor $2RA_{LO}$. The matched filter $H_{Ij}(\omega)$ is matched to $I_j(t), j = 0, 1$ except for the phase factor as in Section 8.4.4 (See Fig. 8.18). Suppose $s_1(t)$ is transmitted so that $I(t) = I_1(t)$. $H_{Ij}(\omega)$ can be realized as correlators and their outputs in the absence of noise can be written as

$$I_{1F}(t) = A^2 \int_0^{T_b} I_1(\tau)\cos\left[\left(\omega_{IF} + \frac{\Delta\omega}{2}\right)(\tau + T_b - t)\right] d\tau, \tag{8.234}$$

$$I_{0F}(t) = A^2 \int_0^{T_b} I_1(\tau)\cos\left[\left(\omega_{IF} - \frac{\Delta\omega}{2}\right)(\tau + T_b - t)\right] d\tau. \tag{8.235}$$

Eq. (8.234) is similar to Eq. (8.188). Ignoring the frequency component centered around $2\omega_{IF}$ and simplifying Eq. (8.234), we obtain

$$I_{1F}(t) = \frac{E_1}{2}, \tag{8.236}$$

where $E_1 = A^2 T_b$. Similarly, from Eq. (8.235), we obtain

$$\begin{aligned} I_{0F}(t) &= \frac{A^2}{2} \int_0^{T_b} \cos\left[\Delta\omega\tau + \left(\omega_{IF} - \frac{\Delta\omega}{2}\right)(t - T_b) + \Delta\phi\right] d\tau \\ &= \frac{A^2}{2} \int_0^{T_b} [\cos(\Delta\omega\tau)\cos\theta(t) - \sin(\Delta\omega\tau)\sin\theta(t)] d\tau \\ &= \frac{A^2}{2} \left[\frac{\sin(\Delta\omega T_b)}{\Delta\omega}\right]\cos\theta(t) - \left[\frac{\cos(\Delta\omega T_b) - 1}{\Delta\omega}\right]\sin\theta(t), \end{aligned} \tag{8.237}$$

where

$$\theta(t) = \left(\omega_{IF} - \frac{\Delta\omega}{2}\right)(t - T_b) + \Delta\phi. \tag{8.238}$$

If

$$2\pi\Delta f T_b = 2n\pi, \quad n = 1, 2, \ldots,$$
$$\Delta f = \frac{n}{T_b}, \tag{8.239}$$

from Eq. (8.237) we find that $I_{0F}(t) = 0$. The signals are *orthogonal* for asynchronous detection if the output of the filter $H_{Ij}(\omega)$ is zero when $s_k, k \neq j$ is transmitted. Comparing Eqs. (8.178) and (8.239), we find that the minimum frequency difference to achieve orthogonality for asynchronous detection is twice that for synchronous detection. In this section, we assume that $\Delta f = 1/T_b$ so that the output of the filter $H_{I0}(\omega)$ is zero (ignoring noise) when $s_1(t)$ is transmitted. In this case, the outputs of the envelope detectors can be written as

$$r_1(T_b) = \sqrt{\left[\frac{E_1}{2} + n_{1FI}(T_b)\right]^2 + n_{1FQ}^2(T_b)}, \tag{8.240}$$

$$r_0(T_b) = \sqrt{n_{0FI}^2(T_b) + n_{0FQ}^2(T_b)}, \tag{8.241}$$

where n_{jFI} and n_{jFQ} are the in-phase and quadrature components of the noise output of the matched filter, $H_{Ij}(\omega)$, respectively. The variance of n_{jFI} and n_{jFQ}, $j = 1, 0$ is given by Eq. (8.207)

$$\sigma_F^2 \equiv \sigma_{jFI}^2 = \sigma_{jFQ}^2 = \frac{N_0^{het} E_1}{4}, \quad j = 0, 1. \tag{8.242}$$

The pdf of the envelope $r_0(T_b)$ when $s_1(t)$ is transmitted is given by the Rayleigh distribution

$$p_{r0}(r_0|\text{'1' sent}) = \frac{r_0}{\sigma_F^2} \exp\left(-\frac{r_0^2}{2\sigma_F^2}\right). \tag{8.243}$$

The pdf of the envelope $r_1(T_b)$ is given by the Rician distribution

$$p_{r1}(r_1|\text{'1' sent}) = \frac{r_1}{\sigma_F^2} \exp\left(-\frac{r^2 + E_1^2/4}{2\sigma_F^2}\right) I_0\left(\frac{rE_1}{2\sigma_F^2}\right). \tag{8.244}$$

If $r_1(T_b) > r_0(T_b)$, it will be decided that '1' is transmitted. Therefore, an error is made if $r_1(T_b) < r_0(T_b)$ when $s_1(t)$ is transmitted. So, the probability of mistaking '1' as '0' is

$$P(0|\text{'1' sent}) = P(r_1(T_b) < r_0(T_b)|\text{'1' sent}). \tag{8.245}$$

The probability that $r_1(T_b) < r_0(T_b)$ can be found as follows. Since $r_1(T_b)$ and $r_0(T_b)$ are independent random variables, the joint pdf of $r_1(T_b)$ and $r_0(T_b)$ can be written as

$$p_{r_1 r_0}(r_1, r_0|\text{'1' sent}) = p_{r_0}(r_0|\text{'1' sent})p_{r_1}(r_1|\text{'1' sent}). \tag{8.246}$$

The chance that $r_1(T_b) < r_0(T_b)$ is the same as that $r_1(T_b)$ has a value r_1 in the range $0 < r_1 < \infty$ and $r_0(T_b)$ has a value greater than r_1,

$$P(r_1(T_b) < r_0(T_b)|\text{'1' sent}) = \int_{r_1}^{\infty} \left\{ \int_0^{\infty} p_{r_1 r_0}(r_1, r_0|\text{'1' sent}) dr_1 \right\} dr_0. \tag{8.247}$$

Using Eqs. (8.243), (8.244), and (8.246), Eq. (8.247) can be simplified as

$$
\begin{aligned}
P(0|\text{'1' sent}) &= \int_0^{\infty} p_{r_1}(r_1) \left\{ \int_{r_1}^{\infty} p_{r_0}(r_0) dr_0 \right\} dr_1 \\
&= \int_0^{\infty} p_{r_1}(r_1) \exp\left(-\frac{r_1^2}{2\sigma_F^2}\right) dr_1 \\
&= \frac{1}{\sigma_F^2} \int_0^{\infty} r_1 \exp\left(-\frac{r_1^2 + E_1^2/8}{\sigma_F^2}\right) I_0\left(\frac{r_1 E_1}{2\sigma_F^2}\right) dr_1.
\end{aligned} \tag{8.248}
$$

Let

$$r_1' = r_1\sqrt{2}, \tag{8.249}$$

$$E_1' = E_1/\sqrt{2}. \tag{8.250}$$

Eq. (8.248) becomes

$$P(0|\text{`1' sent}) = \frac{1}{2\sigma^2} \exp\left(-\frac{E_1^2}{16\sigma_F^2}\right) \int_0^\infty r_1' \exp\left(-\frac{r_1'^2 + E_1'^2/4}{2\sigma_F^2}\right) I_0\left(\frac{r_1' E_1'}{2\sigma_F^2}\right) dr_1'$$

$$= \frac{1}{2} \exp\left(-\frac{E_1^2}{16\sigma_F^2}\right) \underbrace{\int_0^\infty p_{r_1}(r_1|\text{`1' sent})dr_1}_{=1}$$

$$= \frac{1}{2} \exp\left(-\frac{\gamma^{\text{het}}}{4}\right). \tag{8.251}$$

Owing to the symmetry of the problem, $P(1|\text{`0' sent})$ is the same as $P(0|\text{`1' sent})$. Therefore,

$$P_b = P(0|\text{`1' sent}) = \frac{1}{2} \exp\left(-\frac{\gamma^{\text{het}}}{4}\right). \tag{8.252}$$

8.4.6 Comparison of Modulation Schemes with Heterodyne Receiver

Fig. 8.19 shows the error probability as a function of γ^{het} for various modulation schemes with the heterodyne receiver. First consider the synchronous detection. OOK requires 3 dB more γ^{het} (equivalently 3 dB more power or 3 dB less noise) than PSK to reach the same BER. FSK outperforms OOK by roughly 0.85 dB. Next consider the asynchronous detection. OOK performs slightly better than FSK. However, from the practical

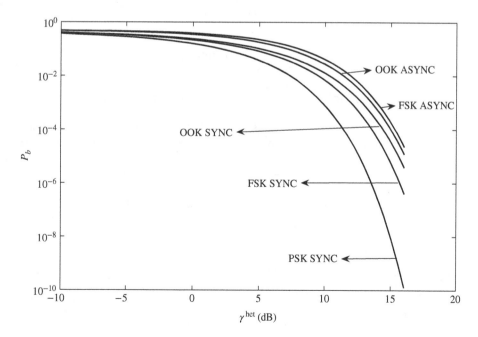

Figure 8.19 Error probability in heterodyne receiver as a function of γ^{het}. SYNC = synchronous detection. ASYNC = asynchronous detection.

stand point, FSK is preferred over OOK since the optimum threshold of FSK is fixed, whereas the optimum threshold of OOK depends on γ^{het} (see Eq. (8.217)). The performance with synchronous detection is always better than that with asynchronous detection, but synchronous detection is more sensitive to laser phase noise.

8.5   Direct Detection

In a direct detection system, the photocurrent is directly proportional to the optical signal power. Hence, the phase information of the optical field is lost. In other words, it is not possible to encode information on the phase of the optical carrier (PSK) although it is possible to encode information as the phase change of a current bit relative to the previous bit (DPSK). In this section, we analyze the performances of OOK, FSK, and DPSK.

8.5.1   OOK

Let $s(t)$ and $n_c(t)$ be the complex field envelope of the transmitted signal and channel noise, respectively (See Fig. 8.20). At the receiver, the signal passes through an optical filter with transfer function $H(\omega)$ that matches the transmitted signal. As in the case of asynchronous detection for OOK, the matched filter need not be phase synchronized with the received optical signal, but it can differ by an arbitrary phase factor ϕ. Using Eq. (8.40) and taking $x_0(t) = 0$, we have

$$H(\omega) = \tilde{x}_1^*(\omega) \exp(i\omega T_b + i\phi)$$
$$= \tilde{s}_1^*(\omega - \omega_c) \exp(i\omega T_b + i\phi). \tag{8.253}$$

Here, we have replaced $x(t)$ of Section 8.2 by $s(t)e^{-i\omega_c t}$. Let the optical filter output be

$$r(t) = [s_F(t) + n_F(t)]e^{-i\omega_c t}, \tag{8.254}$$

with

$$s_F(t)e^{-i\omega_c t} = \frac{1}{2\pi} \int_{-\infty}^{\infty} \tilde{x}(\omega)H(\omega) \exp(-i\omega t)\, d\omega, \tag{8.255}$$

$$n_F(t)e^{-i\omega_c t} = \frac{1}{2\pi} \int_{-\infty}^{\infty} \tilde{n}_c(\omega)H(\omega) \exp(-i\omega t)\, d\omega, \tag{8.256}$$

where $s_F(T_b)$ and $n_F(T_b)$ are the base-band signal and noise field envelopes, respectively. When a bit '1' is sent, $\tilde{x}(\omega) = \tilde{s}_1(\omega - \omega_c)$ and when a bit '0' is sent, $\tilde{x}(\omega) = 0$. Using Eq. (8.253) in Eq. (8.255), we obtain

$$s_F(T_b) = \frac{\exp(i\omega_c T_b + i\phi)}{2\pi} \int_{-\infty}^{\infty} |s_1(\omega - \omega_c)|^2\, d\omega$$
$$= \exp[i(\omega_c T_b + \phi)]E_1 \quad \text{when '1' is sent}$$
$$= 0 \quad \text{when '0' is sent}. \tag{8.257}$$

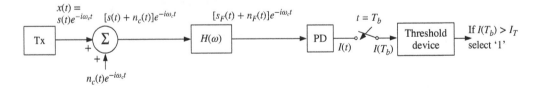

Figure 8.20   Direct detection receiver for OOK.

The output of the matched filter passes through the photodetector and the signal component of the photocurrent is

$$I_s(t) = R|s_F(T_b)|^2$$
$$= RE_1^2 \quad \text{when '1' is sent}$$
$$= 0 \quad \text{otherwise.} \tag{8.258}$$

The total current at the decision instant T_b is

$$I(T_b) = R|s_F(T_b) + n_F(T_b)|^2$$
$$= R[(s_{FI}(T_b) + n_{FI}(T_b))^2 + (s_{FQ}(T_b) + n_{FQ}(T_b))^2], \tag{8.259}$$

where $y_I = \text{Re}[y]$ and $y_Q = \text{Im}[y]$, $y = s_F, n_F$. Without loss of generality, we can assume $s_F(T_b)$ to be real so that $s_{FQ}(T_b) = 0$ and $s_{FI}(T_b) = E_1$. $\sqrt{R}n_{FI}(T_b)$ and $\sqrt{R}n_{FQ}(T_b)$ are independent Gaussian random variables with zero mean and variance,

$$\sigma^2 = R < n_{FI}^2 >= R < n_{FQ}^2 >= \frac{R}{2} < |n_F|^2 >$$

$$= \frac{R}{4\pi} \int_{-\infty}^{\infty} \rho_{\text{ASE}} |H(\omega)|^2 d\omega = \frac{\rho_{\text{ASE}}}{2} E_1 R. \tag{8.260}$$

When a bit '0' is transmitted, $s_{FI}(T_b) = s_{FQ}(T_b) = 0$ and in this case the pdf of I is given by the central chi-square distribution

$$p(I|\text{'0' sent}) \equiv p_0(I) = \frac{1}{2\sigma^2} \exp\left(-\frac{I}{2\sigma^2}\right). \tag{8.261}$$

When a bit '1' is transmitted, the current in the absence of noise is RE_1^2 and in this case the pdf is given by the non-central chi-square distribution

$$p(I|\text{'1' sent}) \equiv p_1(I) = \frac{1}{2\sigma^2} \exp\left(-\frac{RE_1^2 + I}{2\sigma^2}\right) I_0\left(\frac{\sqrt{IRE_1^2}}{\sigma^2}\right). \tag{8.262}$$

The threshold current is determined by the interSection of two curves $p_0(I)$ and $p_1(I)$:

$$p_1(I_T) = p_0(I_T) \tag{8.263}$$

or

$$\exp\left(-\frac{RE_1^2}{2\sigma^2}\right) I_0\left(\frac{\sqrt{I_T RE_1^2}}{\sigma^2}\right) = 1. \tag{8.264}$$

This equation is satisfied to a close approximation [1]

$$I_T = \frac{RE_1^2}{4}\left(1 + \frac{8\sigma^2}{RE_1^2}\right)$$

$$= \frac{RE_1^2}{4}\left(1 + \frac{4\rho_{\text{ASE}}}{E_1}\right). \tag{8.265}$$

When a bit '0' is transmitted, if $I > I_T$, the bit '0' is mistaken as the bit '1' and this probability is

$$P(1|0) = \int_{I_T}^{\infty} p_0(I)dI = \frac{1}{2\sigma^2} \int_{I_T}^{\infty} \exp\left(-\frac{I}{2\sigma^2}\right) dI$$

$$= \exp\left(-\frac{I_T}{2\sigma^2}\right). \tag{8.266}$$

Using Eqs. (8.260) and (8.265), we obtain

$$P(1|0) = \exp\left[-\frac{\gamma^{DD}}{2}\left(1 + \frac{2}{\gamma^{DD}},\right)\right] \tag{8.267}$$

where γ^{DD} is given by

$$\gamma^{DD} = \frac{E_{av}}{\rho_{ASE}}. \tag{8.268}$$

For OOK, $E_{av} = E_1/2$. So

$$\gamma^{DD} = \frac{E_1}{2\rho_{ASE}}, \tag{8.269}$$

When $\gamma^{DD} \gg 1$, $P(1|0) \cong \exp(-\gamma^{DD}/2)$.
    Similarly, the probability of mistaking bit '1' as bit '0' is

$$P(0|1) = \int_0^{I_T} p_1(I)dI$$

$$= \frac{1}{2\sigma^2} \int_0^{I_T} \exp\left(-\frac{RE_1^2 + I}{2\sigma^2}\right) I_0\left(\frac{\sqrt{IRE_1^2}}{\sigma^2}\right) dI. \tag{8.270}$$

Changing the variable of integration from I to x, where

$$x^2 = \frac{I}{\sigma^2}, \tag{8.271}$$

and letting $a^2 = RE_1^2/\sigma^2$, Eq. (8.270) becomes

$$P(0|1) = \int_0^{\sqrt{I_T}/\sigma} x \exp\left(-\frac{a^2 + x^2}{2}\right) I_0(ax) dx$$

$$= 1 - Q_1\left(a, \frac{\sqrt{I_T}}{\sigma}\right) \tag{8.272}$$

where $Q_1(a, \sqrt{I_T}/\sigma)$ is the generalized Marcum's Q-function given by Eq. (8.228). Using Eqs. (8.260), (8.265), and (8.268), Eq. (8.272) can be rewritten as

$$P(0|1) = 1 - Q_1\left(2\sqrt{\gamma^{DD}}, \sqrt{\gamma^{DD}\left(1 + \frac{2}{\gamma^{DD}}\right)}\right) \tag{8.273}$$

Combining Eqs. (8.267) and (8.273), we obtain

$$P_b = \frac{1}{2}[P(1|0) + P(0|1)]$$

$$= \frac{1}{2}\left\{\exp\left[-\frac{\gamma^{DD}}{2}\left(1+\frac{2}{\gamma^{DD}}\right)\right] + 1 - Q_1\left(2\sqrt{\gamma^{DD}}, \sqrt{\gamma^{DD}\left(1+\frac{2}{\gamma^{DD}}\right)}\right)\right\}. \qquad (8.274)$$

Note that Eq. (8.274) is the same as Eq. (8.230) obtained for the case of a heterodyne receiver if we replace γ^{het} by $2\gamma^{DD}$. In this analysis, we have ignored the receiver noise mechanisms such as shot noise and thermal noise and assumed that the optical filter is a matched filter. Without these approximations and assumptions, the analysis is quite cumbersome. When the optical filter is not matched to the transmitted signal, analytical expressions can be obtained using the approaches in Refs. [7]–[9]. In a simplified approach, chi-square distributions are approximated by Gaussian distributions and the BER can be estimated by calculating the Q-factor as in Chapter 7. This Gaussian approximation gives reasonably accurate results for OOK, although it is found to be inaccurate for DPSK signals with direct detection [7].

8.5.2 FSK

For FSK with direct detection, the transmitted signals $s_1(t)$ and $s_0(t)$ are the same as those in Section 8.4.3. (Eq. (8.166)). Since the energies of signals $s_1(t)$ and $s_0(t)$ are equal, we use the matched filters shown in Fig. 8.21 (similar to Fig. 8.7). The matched filters can be realized as a bank of band-pass filters. As before, the matched filters need not be synchronized with the received signal, but can differ by a phase factor ϕ. The signal field $u_j(t)$ and noise field $n_{F_j}(t)$ at the output of the matched filters are given by

$$u_j(t) = \frac{1}{2\pi}\int_{-\infty}^{\infty}\tilde{x}(\omega)H_j(\omega)e^{-i\omega t}d\omega, \quad j = 0, 1, \qquad (8.275)$$

$$n_{Fj}(t) = \frac{1}{2\pi}\int_{-\infty}^{\infty}\tilde{n}_c(\omega)H_j(\omega)e^{-i\omega t}d\omega, \quad j = 0, 1. \qquad (8.276)$$

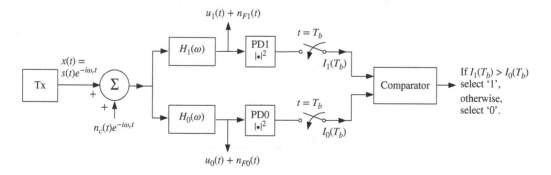

Figure 8.21   Direct detection receiver for FSK.

The transfer function of the matched filter is

$$H_j(\omega) = x_j^*(\omega)e^{i\phi+i\omega T_b}, \quad j = 0, 1$$

$$= \tilde{s}_j^*(\omega - \omega_c)e^{i\phi+i\omega T_b}, \tag{8.277}$$

where ϕ is an arbitrary phase factor. The output of the photodetector j at $t = T_b$ is

$$I_j(T_b) = R[|u_j(T_b) + n_{Fj}(T_b)|^2], \quad j = 0, 1. \tag{8.278}$$

First let us consider the photodetector outputs in the absence of noise. Suppose a bit '1' is transmitted so that $x(t) = x_1(t)$. The output of PD1 is

$$I_1(T_b) = R[|u_1(T_b)|^2]. \tag{8.279}$$

From Eqs. (8.275) and (8.277), we have

$$u_1(T_b) = \frac{e^{i\phi}}{2\pi} \int_{-\infty}^{\infty} |x_1(\omega)|^2 d\omega = E_1 e^{i\phi}, \tag{8.280}$$

$$I_1(T_b) = RE_1^2. \tag{8.281}$$

For this case, the output of PD0 is

$$I_0(T_b) = R|u_0(T_b)|^2$$

$$= R\left|\frac{1}{2\pi} \int_{-\infty}^{\infty} \tilde{x}_1(\omega)\tilde{x}_0^*(\omega) \, d\omega\right|^2. \tag{8.282}$$

Using Parseval's relations,

$$\frac{1}{2\pi} \int_{-\infty}^{\infty} \tilde{x}_1(\omega)\tilde{x}_0^*(\omega)d\omega = \int_{-\infty}^{\infty} x_1(t)x_0^*(t) \, dt, \tag{8.283}$$

Eq. (8.282) may be rewritten as

$$I_0(T_b) = R\left|\int_0^{T_b} x_1(t)x_0^*(t) \, dt\right|^2$$

$$= R\left|\int_0^{T_b} s_1(t)s_0^*(t) \, dt\right|^2$$

$$= R\left|\int_0^{T_b} \exp(-i2\pi\Delta ft) \, dt\right|^2$$

$$= R\left|\frac{\exp(-i\pi\Delta fT_b)\sin(\pi\Delta fT_b)}{\pi\Delta f}\right|^2$$

$$= \frac{R\sin^2(\pi\Delta fT_b)}{\pi^2\Delta f^2}. \tag{8.284}$$

$I_0(T_b)$ is zero if

$$\pi \Delta f T_b = n\pi, \quad n = 1, 2, \ldots \tag{8.285}$$

or

$$\Delta f = \frac{n}{T_b}. \tag{8.286}$$

Comparing Eqs. (8.286) and (8.239), we find that the orthogonality conditions for the asynchronous receiver and direct detection receiver are the same. In this section, we assume that the orthogonality condition is satisfied so that the output of PD0 (PD1) is zero when bit '1' (bit '0') is transmitted in the absence of noise. Since the photo-detector output is not sensitive to the phase factor ϕ, we ignore it from now on.

Case (i): bit '1' transmitted. Expanding Eq. (8.278), we obtain

$$I_1(T_b) = R[u_1^2(T_b) + 2u_1(T_b)n_{F_{1r}}(T_b) + n_{F_{1r}}^2(T_b) + n_{F_{1i}}^2(T_b)]$$

$$= R\{[E_1 + n_{F_{1r}}(T_b)]^2 + n_{F_{1i}}^2(T_b)\}, \tag{8.287}$$

$$I_0(T_b) = R[n_{F_{0r}}^2(T_b) + n_{F_{0i}}^2(T_b)], \tag{8.288}$$

where we have used $u_1(T_b) = E_1$ and subscripts r and i denote the real and imaginary parts, respectively. Let us first consider $I_1(T_b)$. $\sqrt{R}[E_1 + n_{F_{1r}}(T_b)]$ and $\sqrt{R}n_{F_{1i}}$ are Gaussian random variables with means $\sqrt{R}E_1$ and zero, respectively. The variances of these two random variables are equal and given by

$$\sigma^2 = \frac{R}{2\pi} \int_{-\infty}^{\infty} \frac{\rho_{ASE}}{2} |H_1(\omega)|^2 d\omega = \frac{R\rho_{ASE}E_1}{2}. \tag{8.289}$$

The pdf of $I_1(T_b)$ is given by the non-central chi-square distribution

$$p_1(I_1) = \frac{1}{2\sigma^2} \exp\left[-\frac{(RE_1^2 + I_1)}{2\sigma^2}\right] I_0\left(\frac{\sqrt{I_1 RE_1^2}}{\sigma^2}\right). \tag{8.290}$$

The pdf of $I_0(T_b)$ is given by the central chi-square distribution

$$p_1(I_0) = \frac{1}{2\sigma^2} \exp\left(-\frac{I_0}{2\sigma^2}\right). \tag{8.291}$$

If $I_1(T_b) > I_0(T_b)$, it will be decided that the bit '1' is transmitted. Therefore, an error is made if $I_1(T_b) < I_0(T_b)$ when $s_1(t)$ is transmitted:

$$P(0|\text{'1' sent}) = P(I_0(T_b) > I_1(T_b)|\text{'1' sent}). \tag{8.292}$$

The chance that $I_1(T_b) < I_0(T_b)$ can be found as follows. Since $I_1(T_b)$ and $I_0(T_b)$ are independent random variables, the joint pdf of I_1 and I_0 can be written as

$$p_1(I_1, I_0) = p_1(I_1)p_1(I_0). \tag{8.293}$$

The chance that $I_1(T_b) < I_0(T_b)$ is the same as that $I_1(T_b)$ has a value i_1 and $I_0(T_b)$ has a value i_0 greater than i_1. Since $I_1(T_b)$ can take any value in the range $(0, \infty)$, we have

$$P(I_0(T_b) > I_1(T_b)|\text{'1' sent}) = \int_{i_1}^{\infty} \left\{ \int_0^{\infty} p_1(i_1, i_0) \, di_1 \right\} di_0. \tag{8.294}$$

Using Eqs. (8.290), (8.291), and (8.293), we obtain

$$P(0|1) = \int_0^\infty p_1(i_1)di_1 \int_{i_1}^\infty p_1(i_0)di_0$$

$$= \frac{1}{(2\sigma^2)^2} \int_0^\infty \exp\left[-\frac{RE_1^2 + i_1}{2\sigma^2}\right] I_0\left(\frac{\sqrt{i_1 RE_1^2}}{\sigma^2}\right)di_1 \int_{i_1}^\infty \exp\left(-\frac{i_0}{2\sigma^2}\right)di_0$$

$$= \frac{1}{2\sigma^2} \int_0^\infty \exp\left[-\frac{RE_1^2 + 2i_1}{2\sigma^2}\right] I_0\left(\frac{\sqrt{i_1 RE_1^2}}{\sigma^2}\right)di_1. \tag{8.295}$$

Let $i_1' = 2i_1$ and $x = RE_1^2/2$. Now, Eq. (8.295) becomes

$$P(0|1) = \frac{1}{4\sigma^2} \int_0^\infty \exp\left[-\frac{2x + i_1'}{2\sigma^2}\right] I_0\left(\frac{\sqrt{i_1' x}}{\sigma^2}\right)di_1'$$

$$= \frac{1}{4\sigma^2} e^{-\frac{x}{2\sigma^2}} \int_0^\infty \exp\left[-\frac{x + i_1'}{2\sigma^2}\right] I_0\left(\frac{\sqrt{i_1' x}}{\sigma^2}\right)di_1'$$

$$= \frac{1}{2} e^{-\frac{x}{2\sigma^2}} \int_0^\infty p_1(i_1)di_1$$

$$= \frac{1}{2} \exp\left(-\frac{E_1}{2\rho_{\text{ASE}}}\right). \tag{8.296}$$

For FSK, $E_1 = E_0 = E_{\text{av}}$. Using Eq. (8.268), Eq. (8.296) may be rewritten as

$$P(0|1) = \frac{1}{2} \exp\left(-\frac{\gamma^{DD}}{2}\right). \tag{8.297}$$

Case (ii): bit '0' transmitted. Owing to the symmetry of the problem, $P(0|1)$ is same as $P(1|0)$. The error probability is

$$P_b = \frac{1}{2}[P(0|1) + P(1|0)] = \frac{1}{2} \exp\left(-\frac{\gamma^{DD}}{2}\right) \tag{8.298}$$

Note that this error probability is the same as that given by Eq. (8.251) for asynchronous detection if we replace γ^{het} by $2\gamma^{DD}$. If we ignore shot noise, we see that two expressions are identical.

8.5.3 DPSK

In the case of PSK, the information is transmitted as the *absolute* phase of the complex field envelope $s(t)$. But in the case of DPSK, the information is transmitted as the phase of the field envelope *relative* to the previous bit. To estimate the absolute phase of the transmitted PSK signal, a reference is required at the receiver. This reference is provided by the local oscillator whose phase should be synchronized with that of the optical carrier (or it should be post-corrected using the DSP). In contrast, for systems based on DPSK, the transmitted signal of the previous bit interval acts as a reference and, therefore, there is no need for the local oscillator and the phase synchronization. However, one of the drawbacks is that the phase of the previous bit is noisy and, therefore, this leads to performance degradation for DPSK compared with PSK.

Let $s_1(t)$ and $s_0(t)$ be the optical field envelopes of duration $\leq T_b$ with $s_0(t) = -s_1(t)$. We assume that $s(t)$ is real. To send a bit '1', the pulse in the current bit slot is the same as that in the previous bit slot and to send bit '0', the pulse in the current bit slot has a phase of $\pm\pi$ radians *relative* to the previous bit slot. Using $s_1(t)$ and $s_0(t)$, let us construct orthogonal signals over the period $2T_b$:

$$
\left.
\begin{aligned}
s_1'(t) &= s_1(t) + s_1(t - T_b) \quad \text{set I} \\
\text{or} & \\
s_1'(t) &= s_0(t) + s_0(t - T_b) \quad \text{set II}
\end{aligned}
\right\} \text{to send bit '1';}
\tag{8.299}
$$

$$
\left.
\begin{aligned}
s_0'(t) &= s_1(t) + s_0(t - T_b) \quad \text{set I} \\
\text{or} & \\
s_0'(t) &= s_0(t) + s_1(t - T_b) \quad \text{set II}
\end{aligned}
\right\} \text{to send bit '0'.}
\tag{8.300}
$$

Since $s_0(t) = -s_1(t)$, the signals corresponding to set I are negative of the signals corresponding to set II and we could use either of these sets. Figs. 8.22 and 8.23 show the signals $s_j(t)$ and $s_j'(t), j = 0, 1$, using Gaussian pulses. From Fig. 8.23, we see that $s_0'(t)$ is antisymmetric with respect to the point T_b while $s_1'(t)$ is symmetric; therefore, they are orthogonal over a period of $2T_b$:

$$
\left| \int_0^{2T_b} s_1'(t) s_0'(t) dt \right|^2 = 0.
\tag{8.301}
$$

In the case of direct detection of orthogonal FSK, we have seen that the optical receiver consists of matched filters, square-law detectors (i.e., photodetectors), and a comparator. The results of direct detection orthogonal FSK are applicable for DPSK signals as well, since the signals $s_1'(t)$ and $s_0'(t)$ are orthogonal. Therefore, the schematic of the optical receiver is the same as that of the direct detection FSK if we use the signals $s_1'(t)$ and

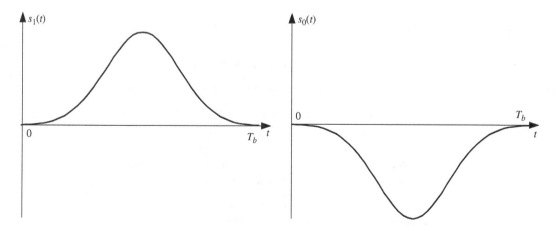

Figure 8.22   Optical complex field envelopes $s_1(t)$ and $s_0(t)$.

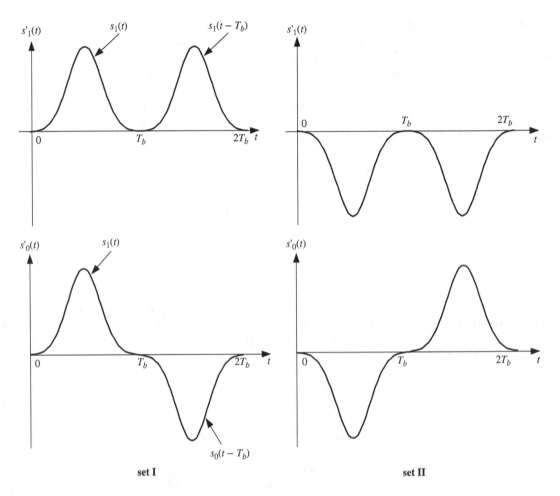

Figure 8.23 Orthogonal signals $s'_1(t)$ and $s'_0(t)$ constructed using $s_1(t)$ and $s_0(t)$.

$s'_0(t)$, and this is shown in Fig. 8.24. $H'_1(\omega)$ and $H'_0(\omega)$ are the filters matched to $s'_1(t)$ and $s'_0(t)$, except that they can have an arbitrary constant phase shift:

$$H'_j(\omega) = s'^*_j(\omega - \omega_c) \exp(i\omega T_b + i\phi), \quad j = 0, 1. \tag{8.302}$$

Taking the Fourier transform of Eq. (8.299) (set I) and using the shifting property of the Fourier transform,

$$H'_1(\omega) = [s^*_1(\omega - \omega_c) + s^*_1(\omega - \omega_c)\exp(i\omega T_b)]\exp(i\omega T_b + i\phi)$$

$$= H_1(\omega)[1 + \exp(i\omega T_b)], \tag{8.303}$$

where $H_1(\omega)$ is the filter matched to $s_1(t)e^{-i\omega_c t}$ except for the phase factor ϕ (see Eq. (8.277)). Fig. 8.25 shows the realization of $H'_1(\omega)$ using a delay-and-add filter. The second term of Eq. (8.303) corresponds to the delay by T_b. Similarly,

$$H'_0(\omega) = [s^*_1(\omega - \omega_c) + s^*_0(\omega - \omega_c)\exp(i\omega T_b)]\exp(i\omega T_b + i\phi)$$

$$= H_1(\omega)[1 - \exp(i\omega T_b)]. \tag{8.304}$$

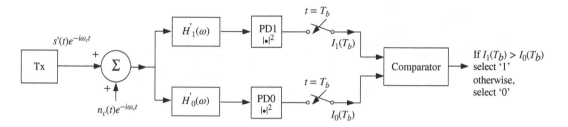

Figure 8.24 Direct detection receiver for DPSK.

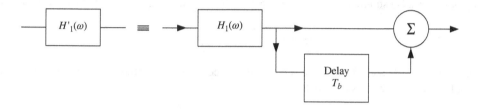

Figure 8.25 The filter matched to $s_1'(t)$.

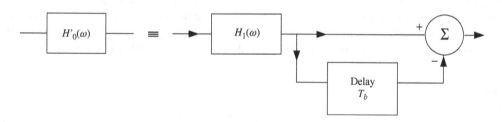

Figure 8.26 The filter matched to $s_0'(t)$.

Fig. 8.26 shows the realization of $H_0'(\omega)$ using a delay-and-subtract filter. Therefore, the schematic shown in Fig. 8.24 can be redrawn with signal $s(t)$ instead of $s'(t)$, as shown in Fig. 8.27.

The energy of the signal $s_1'(t)$ is

$$E_1' = \int_0^{2T_b} |s_1(t) + s_1(t - T_b)|^2 dt$$

$$= \int_0^{T_b} |s_1(t)|^2 dt + \int_{T_b}^{2T_b} |s_1(t - T_b)|^2 dt + 2\int_0^{2T_b} s_1(t) s_1(t - T_b)\, dt. \qquad (8.305)$$

Since $s_1(t)$ is a pulse that is zero outside the interval $[0, T_b]$, the last term in Eq. (8.305) vanishes and, therefore, we obtain

$$E_1' = 2E_1, \qquad (8.306)$$

where E_1 is the energy of the signal $s_1(t)$. A similar calculation shows that $E_0' = 2E_1$. Since the energies of the signals $s_1'(t)$ and $s_0'(t)$ are equal and they are orthogonal, the analytical expression derived for the case of

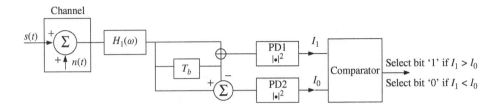

Figure 8.27 DPSK receiver.

the FSK with direct detection or asynchronous detection is applicable in this case except that the energy E_1 appearing in Eq. (8.297) should be replaced by $2E_1 (= 2E_{av})$, where E_{av} is the average energy of $s(t)$:

$$P_b = \frac{1}{2} \exp \left(-\frac{E_{av}}{\rho_{ASE}} \right) = \frac{1}{2} \exp \left(-\gamma^{DD} \right). \tag{8.307}$$

Comparing Eqs. (8.307) and (8.298), we see that the direct detection orthogonal FSK requires 3 dB more power than DPSK to reach the same BER.

Example 8.3

50% duty cycle rectangular RZ pulses are used in a direct detection long-haul 40-Gb/s DPSK system operating at 1550 nm. The peak transmitter power is 3 dBm. The fiber-optic link consists of N spans of 80-km standard SMF with a loss of 0.2 dB/km followed by an optical amplifier with $n_{sp} = 1$ and gain G being equal to the fiber loss. Find the maximum transmission distance so that the error probability is $<= 10^{-5}$. Ignore receiver noise.

Solution:
For 50% duty RZ pulses, we have

$$\text{average power} = \frac{\text{peak power}}{2}, \tag{8.308}$$

$$\text{average power (dBm)} = \text{peak power (dBm)} + 10 \log_{10}(1/2)$$

$$= 3 \, \text{dBm} - 3 \, \text{dB}$$

$$= 0 \, \text{dBm}, \tag{8.309}$$

$$\text{average power} = 10^{0/10} \, \text{mW} = 1 \, \text{mW}, \tag{8.310}$$

$$\text{average energy of a pulse} = E_{av} = \text{average power} \times \text{bit interval}. \tag{8.311}$$

Bit interval,

$$T_b = \frac{1}{40 \times 10^9} \, \text{s} = 25 \quad \text{ps}, \tag{8.312}$$

$$E_{av} = 1 \, \text{mW} \times 25 \, \text{ps} = 2.5 \times 10^{-14} \, \text{J}. \tag{8.313}$$

Operating frequency,

$$\bar{f} = \frac{3 \times 10^8}{1550 \times 10^{-9}} = 193.54 \, \text{THz}, \tag{8.314}$$

$$\text{fiber loss per span} = 0.2 \, \text{dB/km} \times 80 \, \text{km}$$

$$= 16 \, \text{dB}. \tag{8.315}$$

Amplifier gain,

$$G(\text{dB}) = 16 \, \text{dB}, \tag{8.316}$$

$$G = 10^{16/10} = 39.81. \tag{8.317}$$

PSD per amplifier,

$$\rho_{\text{ASE},1} = n_{sp} h\bar{f}(G - 1)$$

$$= 6.626 \times 10^{-34} \times 193.54 \times 10^{12} \times (39.81 - 1)$$

$$= 7.465 \times 10^{-18} \, \text{J}. \tag{8.318}$$

Total PSD due to all amplifiers,

$$\rho_{\text{ASE}}^{\text{tot}} = N\rho_{\text{ASE},1}. \tag{8.319}$$

From Eq. (8.307), we have

$$P_b = \frac{1}{2} \exp\left(-\frac{E_{\text{av}}}{\rho_{\text{ASE}}^{\text{tot}}}\right), \tag{8.320}$$

$$E_{\text{av}} = -\ln(2P_b)\rho_{\text{ASE}}^{\text{tot}}, \tag{8.321}$$

$$N = -\frac{E_{\text{av}}}{\ln(2P_b)\rho_{\text{ASE},1}}$$

$$= -\frac{2.5 \times 10^{-14}}{\ln(2 \times 10^{-5}) \times 7.465 \times 10^{-18}}$$

$$\cong 309. \tag{8.322}$$

Maximum achievable transmission distance so that $Pb <= 10^{-5}$,

$$L_{\text{max}} = 309 \times 80 \, \text{km}$$

$$= 24,720 \, \text{km}. \tag{8.323}$$

8.5.4   Comparison of Modulation Schemes with Direct Detection

Fig. 8.28 shows the performance of OOK, FSK, and DPSK in a direct detection system. The performance of OOK and FSK is roughly similar, and DPSK has a 3-dB advantage over FSK. To compare the performance of DPSK and PSK, let us ignore the shot noise. In this case

$$N_0^{\text{homo}} = \rho_{\text{ASE}} \tag{8.324}$$

and

$$\gamma^{\text{homo}} = \gamma^{DD} \equiv \gamma. \tag{8.325}$$

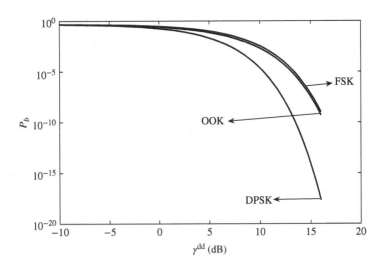

Figure 8.28 Error probability of various modulation formats in a direct detection system.

From Eq. (8.96), we have

$$P_b^{\text{PSK}} = \frac{1}{2}\text{erfc}(\sqrt{\gamma}). \tag{8.326}$$

When $\gamma \gg 1$,

$$P_b^{\text{PSK}} \cong \frac{\exp(-\gamma)}{2\sqrt{\pi\gamma}}. \tag{8.327}$$

From Eq. (8.307), we have

$$P_b^{\text{DPSK}} = \frac{1}{2}\exp(-\gamma). \tag{8.328}$$

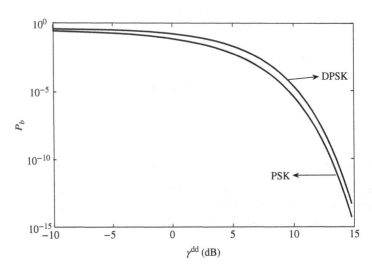

Figure 8.29 Comparison of the performances of homodyne PSK and direct detection DPSK.

Comparing Eqs. (8.327) and (8.328), we find that PSK and DPSK have a similar exponential dependence on γ when γ is large. Fig. 8.29 shows the performance of PSK with homodyne receiver and DPSK with direct detection receiver. To reach a BER of 10^{-3}, the DPSK requires roughly 1.2 dB more power than the PSK when the noise power is kept constant. However, to reach a BER of 10^{-9}, the DPSK requires only 0.5 dB more power than the PSK.

8.6   Additional Examples

Example 8.4

An optical signal passes through the single-mode fiber with dispersion coefficient β_2 and length L. The transmitted field envelope is

$$x_{tr}(t) = \begin{cases} \pm A \exp\left[-\frac{(t-T_B/2)^2}{2T_0^2}\right] & \text{for} \quad 0 < t < T_B \\ 0 & \text{otherwise.} \end{cases} \quad (8.329)$$

Find the filter matched to the received signal. Ignore β_1 and β_3. Assume $T_0 << T_B$.

Solution:
Since $T_0 \ll T_b$, we approximate that the signal is Gaussian in the range $[-\infty, \infty]$ since $T_0 << T_B$:

$$x_{tr}(t) = s_i(t - T_B/2)., \quad (8.330)$$

$$s_i(t) = \pm A \exp\left(-\frac{t^2}{2T_0^2}\right) \quad (8.331)$$

From Eq. (2.153), the Fourier transform of the signal $s_0(t)$ after the fiber transmission is

$$\tilde{s}_0(\omega) = \tilde{s}_i(\omega) H_f(\omega, L)$$

$$= \pm \frac{A}{a} \exp\left(-\frac{\omega^2}{4\pi b^2}\right), \quad (8.332)$$

$$\frac{1}{b^2} = \frac{1}{a^2} - i 2\pi \beta_2 L \quad (8.333)$$

$$a = \frac{1}{\sqrt{2\pi}} T_0. \quad (8.334)$$

The received signal is

$$x_o(t) = s_o(t - T_B/2), \quad (8.335)$$

$$x_o(\omega) = \tilde{s}_o(\omega) e^{i\omega T_b/2}. \quad (8.336)$$

Let the received signal corresponding to bit '1' and bit '0' be $x_{o,1}(t)$ and $x_{o,0}(t)$, respectively:

$$\tilde{x}_{o,1}(\omega) = -\tilde{x}_{o,0}(\omega) = \frac{A}{a} \exp\left[-\frac{\omega^2}{4\pi b^2} + \frac{i\omega T_b}{2}\right]. \quad (8.337)$$

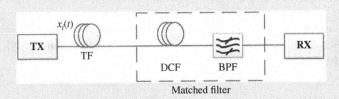

Matched filter

Figure 8.30 The filter matched to the received signal after the dispersive fiber channel. Tx = transmitter, TF = transmission fiber, DCF = dispersion compensation fiber.

The filter matched to the received signal is

$$H(\omega) = [\tilde{x}^*_{o,1}(\omega) - \tilde{x}^*_{o,0}(\omega)] \exp(i\omega T_b)$$

$$= \frac{2A}{a} \exp\left[-\frac{\omega^2}{4\pi(b^2)^*} + \frac{i\omega T_b}{2}\right]$$

$$= \frac{2A}{a} \underbrace{\exp\left[-\frac{\omega^2}{4\pi a^2}\right]}_{\text{band-pass filter}} \underbrace{\exp\left[-\frac{i\omega^2 \beta_2 L}{2}\right]}_{\substack{\text{dispersion} \\ \text{compensation} \\ \text{fiber}}} \exp\left[\frac{i\omega T_b}{2}\right]. \qquad (8.338)$$

As shown in Fig. 8.30, the matched filter can be realized by cascading a first-order Gaussian filter with a dispersion compensating fiber or fiber Bragg grating whose accumulated dispersion is equal in magnitude to that of the transmission fiber, but of opposite sign. The last term in Eq. (8.338) corresponds to a delay of $T_b/2$, which implies that the decision should be made at $t = T_b/2$ (instead of at $t = T_b$).

Example 8.5

Show that the PSD of noise in a balanced heterodyne receiver is

$$\rho_{n_{\text{het}}} = \frac{\rho_{\text{ASE}}}{4} + \frac{\rho_{\text{shot,eff}}}{4R^2 A^2_{\text{LO}}}. \qquad (8.339)$$

Solution:
Let

$$n_{\text{het}} = n_{\text{ASE}}(t) + n_d(t)/2RA_{\text{LO}}, \qquad (8.340)$$

$$n_{\text{ASE}} = n_{cI} \cos(\omega_{\text{IF}} t) + n_{cQ} \sin(\omega_{\text{IF}} t). \qquad (8.341)$$

The PSD of n_{cI} (or n_{cQ}) is $\rho_{\text{ASE}}/2$. Let us consider impact of multiplication by $\cos(\omega_{\text{IF}} t)$ (or $\sin(\omega_{\text{IF}} t)$). Let

$$n_1(t) = n_{cI}(t) \cos(\omega_{\text{IF}} t), \qquad (8.342)$$

$$\tilde{n}_1(\omega) = \frac{\tilde{n}_{cI}(\omega + \omega_{\text{IF}}) + \tilde{n}_{cI}(\omega - \omega_{\text{IF}})}{2}, \qquad (8.343)$$

$$\langle|\tilde{n}_1(\omega)|^2\rangle = \frac{1}{4}\{\langle|\tilde{n}_{cI}(\omega+\omega_{\mathrm{IF}})|^2\rangle + \langle|\tilde{n}_{cI}(\omega-\omega_{\mathrm{IF}})|^2\rangle$$

$$+ \langle\tilde{n}_{cI}(\omega+\omega_{\mathrm{IF}})\tilde{n}_{cI}^*(\omega-\omega_{\mathrm{IF}}) + \mathrm{c.c.}\rangle\}. \tag{8.344}$$

Typically, an optical filter of bandwidth B_o is introduced before the photodetectors. B_o is greater than the signal bandwidth, but much smaller than $\omega_{\mathrm{IF}}/2\pi$. In this case, $n_{cI}(t)$ and $n_{cQ}(t)$ are band-limited to $B_o/2$ and the spectra due to $\tilde{n}_{cI}(\omega+\omega_{\mathrm{IF}})$ and $\tilde{n}_{cI}(\omega-\omega_{\mathrm{IF}})$ do not overlap. Hence, the last two terms on the right-hand side of Eq. (8.344) are zero. Since the PSD of $n_1(t)$ is proportional to $\langle|\tilde{n}_1(\omega)|^2\rangle$, from Eq. (8.344) and Fig. 8.31(b), we find

$$\rho_{n_1}(f) = \frac{\rho_{\mathrm{ASE}}}{8} \quad \text{for} \quad |f-f_{\mathrm{IF}}| < B_o/2$$

$$= \frac{\rho_{\mathrm{ASE}}}{8} \quad \text{for} \quad |f+f_{\mathrm{IF}}| < B_o/2$$

$$= 0 \quad \text{elsewhere.} \tag{8.345}$$

Let

$$n_2(t) = n_{cQ}\sin(\omega_{\mathrm{IF}}t + \Delta\phi). \tag{8.346}$$

Proceeding as before, we find

$$\rho_{n_2}(f) = \frac{\rho_{\mathrm{ASE}}}{8} \quad \text{for} \quad |f-f_{\mathrm{IF}}| < B_o/2$$

$$= \frac{\rho_{\mathrm{ASE}}}{8} \quad \text{for} \quad |f+f_{\mathrm{IF}}| < B_o/2$$

$$= 0 \quad \text{elsewhere.} \tag{8.347}$$

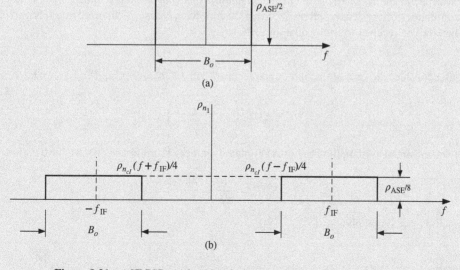

Figure 8.31 ASE PSD: (a) base band; (b) after multiplication by $\cos(\omega_{\mathrm{IF}}t)$.

From Eq. (8.341),

$$n_{\text{ASE}} = n_1 + n_2,$$

(8.348)

$$
\begin{aligned}
\rho_{n_{\text{ASE}}} &= \frac{\rho_{\text{ASE}}}{4} \quad \text{for} \quad |f - f_{\text{IF}}| < B_o/2 \\
&= \frac{\rho_{\text{ASE}}}{4} \quad \text{for} \quad |f + f_{\text{IF}}| < B_o/2 \\
&= 0 \quad \text{otherwise},
\end{aligned}
$$

(8.349)

where we have used the fact that n_1 and n_2 are statistically independent random processes. Since the signal spectrum is centered around f_{IF} extending from $f_{\text{IF}} - B_o/2$ to $f_{\text{IF}} + B_o/2$, n_{ASE} may be approximated as a white noise process over the band of interest,

$$\rho_{n_{\text{ASE}}} = \frac{\rho_{\text{ASE}}}{4}.$$

(8.350)

The PSD of $n_d(t)/2RA_{\text{LO}}$ is

$$\frac{\rho_{n_d}}{4R^2 A_{\text{LO}}} = \frac{\rho_{\text{shot,eff}}}{4R^2 A_{\text{LO}}^2}.$$

(8.351)

Combining Eqs. (8.350) and (8.351), we find the PSD of $n_{\text{het}}(t)$ is

$$\rho_{n_{\text{het}}} = \frac{N_0^{\text{het}}}{2} = \frac{\rho_{\text{ASE}}}{4} + \frac{\rho_{\text{shot,eff}}}{4R^2 A_{\text{LO}}^2}.$$

(8.352)

Example 8.6

To reach an error probability of 10^{-9}, find the mean number of signal photons required in a shot noise-limited coherent communication system based on OOK for the following cases: (i) balanced homodyne receiver; (ii) balanced heterodyne receiver. Assume quantum efficiency, $\eta = 1$.

Solution:
(i) Let us first consider the case of the homodyne receiver with OOK. From Eq. (8.113), we have

$$P_b = \frac{1}{2}\text{erfc}\left(\sqrt{\frac{E_{\text{av}}}{2N_0^{\text{homo}}}}\right).$$

(8.353)

For a shot noise-limited system, the PSD of ASE can be ignored. From Eq. (8.86), we have

$$N_0^{\text{homo}} = \frac{q}{2R} = \frac{h\bar{f}}{2\eta}.$$

(8.354)

The mean number of signal photons is

$$N_s = \frac{E_{\text{av}}}{h\bar{f}}.$$

(8.355)

Substituting Eqs. (8.354) and (8.355) in Eq. (8.353), we find

$$P_b = \frac{1}{2}\text{erfc}(\sqrt{\eta N_s})$$
$$= 10^{-9}. \tag{8.356}$$

When $\eta = 1$,

$$\sqrt{N_s} = \text{erfc}^{-1}(2 \times 10^{-9})$$
$$= 4.2411 \tag{8.357}$$

$$N_s \cong 18. \tag{8.358}$$

(ii) From Eq. (8.144), we have

$$P_b = \frac{1}{2}\text{erfc}\left(\sqrt{\frac{E_{av}}{4N_0^{het}}}\right). \tag{8.359}$$

From Eq. (8.123), we find

$$N_0^{het} = \frac{h\bar{f}}{2\eta}. \tag{8.360}$$

Substituting Eqs. (8.360) and (8.355) in Eq. (8.359), we obtain

$$P_b = \frac{1}{2}\text{erfc}(\sqrt{\eta N_s/2}) = 10^{-9} \tag{8.361}$$

$$\sqrt{N_s} = \text{erfc}^{-1}(2 \times 10^{-9})\sqrt{2} \tag{8.362}$$

$$N_s \cong 36. \tag{8.363}$$

Example 8.7

Show that the error probability in a fiber-optic system based on PSK that uses a heterodyne receiver with synchronous demodulator and a filter matched to the transmitted signal $s(t)$ (see Fig. 8.14) is given by

$$P_b = \frac{1}{2}\text{erfc}\left(\sqrt{\frac{E_{av}}{2N_0^{het}}}\right). \tag{8.364}$$

Solution:

Let the signal output of the synchronous demodulator be

$$x(t) = s(t)\cos^2(\omega_{IF}t + \Delta\phi)$$
$$= \frac{s(t)}{2}[1 + \cos(2\omega_{IF}t + 2\Delta\phi)]. \tag{8.365}$$

Since the filter is not matched to $x(t)$, Eq. (8.35) can not be used to find v_{max}. Instead, we use Eqs. (8.27) and (8.28) to calculate P_b. The Fourier transform of Eq. (8.365) is

$$\tilde{x}(\omega) = \frac{\tilde{s}(\omega)}{2} + \frac{\tilde{s}(\omega - 2\omega_{IF})e^{-i2\Delta\phi}}{4} + \frac{\tilde{s}(\omega + 2\omega_{IF})e^{i2\Delta\phi}}{4}. \tag{8.366}$$

The transfer function of the filter matched to $s(t)$ is

$$H_s(\omega) = [\tilde{s}_1^*(\omega) - \tilde{s}_0^*(\omega)] \exp(i\omega T_b)$$
$$= 2\tilde{s}_1^*(\omega) \exp(i\omega T_b). \tag{8.367}$$

Substituting Eqs. (8.366) and (8.367) into Eq. (8.6), we find

$$u_1(T_b) = \frac{1}{2\pi} \int_{-\infty}^{\infty} \frac{\tilde{s}_1(\omega)}{2} \cdot 2\tilde{s}_1^*(\omega)d\omega, \tag{8.368}$$

where we have ignored the overlap between the frequency components at $\omega \pm 2\omega_{IF}$ and ω since the bandwidth of $s(t)$ is assumed to be much smaller than ω_{IF}. From Parseval's relations, it follows that

$$u_1(T_b) = E_1 = -u_0(T_b). \tag{8.369}$$

Next, let us consider the noise propagation. Let the noise before the demodulator be

$$n_{het}(t) = n_{ASE}(t) + n_d(t), \tag{8.370}$$

$$n_{ASE} = n_{cI} \cos\theta + n_{cQ} \sin\theta, \tag{8.371}$$

$$n_d = n_{dI} \cos\theta + n_{dQ} \sin\theta, \tag{8.372}$$

$$\theta = \omega_{IF}t + \Delta\phi, \tag{8.373}$$

$$n_{het}(t) = n_{het,I} \cos\theta + n_{het,Q} \sin\theta, \tag{8.374}$$

$$n_{het,I}(t) = n_{cI} + n_{dI}, \tag{8.375}$$

$$n_{het,Q}(t) = n_{cQ} + n_{dQ}. \tag{8.376}$$

After the synchronous demodulator, the noise is

$$n(t) = n_{het}(t) \cos\theta$$
$$= n_{het,I}(t)\cos^2\theta + n_{het,Q}(t) \sin\theta \cos\theta$$
$$= \frac{n_{het,I}(t)}{2} + \frac{n_{het,I}(t) \cos 2\theta + n_{het,Q} \sin 2\theta}{2}, \tag{8.377}$$

$$\tilde{n}(\omega) = \frac{\tilde{n}_{het,I}(\omega)}{2} + \text{terms at } 2\omega_{IF}. \tag{8.378}$$

The components of $\tilde{n}(\omega)$ around $2\omega_{IF}$ are removed by the low-pass filter, $H_s(\omega)$ placed just before the decision. Hence, we ignore these terms and obtain

$$\langle|\tilde{n}(\omega)|^2\rangle = \frac{\langle|\tilde{n}_{het,I}(\omega)|^2\rangle}{4} \tag{8.379}$$

or

$$\rho_n(\omega) = \frac{\rho_{n_{\text{het},I}}(\omega)}{4}. \tag{8.380}$$

Since

$$\rho_{n_{\text{het},I}}(\omega) = 2\rho_{n_{\text{het}}} = N_0^{\text{het}} \tag{8.381}$$

then

$$\rho_n(\omega) = \frac{N_0^{\text{het}}}{4}. \tag{8.382}$$

Using Eq. (8.7), the variance of noise after the filter is

$$\sigma^2 = \frac{N_0^{\text{het}}}{4} \frac{1}{2\pi} \int_{-\infty}^{\infty} |H_s(\omega)|^2 \, d\omega. \tag{8.383}$$

Using Eq. (8.367) in Eq. (8.383), we obtain

$$\sigma^2 = N_0^{\text{het}} E_1. \tag{8.384}$$

Using Eqs. (8.369) and (8.384) in Eqs. (8.28) and (8.27), we find

$$v = \frac{4E_1^2}{N_0^{\text{het}} E_1} = \frac{4E_1}{N_0^{\text{het}}}, \tag{8.385}$$

$$P_b = \frac{1}{2}\text{erfc}\left(\sqrt{\frac{E_{\text{av}}}{2N_0^{\text{het}}}}\right). \tag{8.386}$$

Exercises

8.1 The transmitted signal is

$$x(t) = A \, \text{rect}\left(\frac{t - T_B/2}{T_B}\right), \tag{8.387}$$

where

$$\text{rect}(x) = 1 \quad \text{if } |x| < 1/2$$

$$= 0 \quad \text{Otherwise} \tag{8.388}$$

Show that the filter matched to the transmitted signal is an integrator with the limits of integration from 0 to T_B (integrate-and-dump filter).

8.2 Explain the meaning of a matched filter.

8.3 In a 25-Gb/s homodyne fiber-optic system operating at 1530 nm, the PSD of the ASE at the receiver ρ_{ASE} is 7.78×10^{-16} W/Hz. Find the average signal power required at the receiver to reach the BER of 10^{-9} if the signal is (a) PSK, (b) OOK. Assume that quantum efficiency η is 0.9 and rectangular NRZ pulses are used to transmit the data. Ignore thermal noise.

(Ans: (a) $-4.55\,$dBm; (b) $-1.55\,$dBm.)

8.4 Explain the difference between synchronous and asynchronous detection.

8.5 Show that the BER of DPSK with a heterodyne receiver and asynchronous detection is

$$P_b = \frac{1}{2}\exp\left(\frac{-\gamma^{\text{het}}}{2}\right). \tag{8.389}$$

8.6 Write a computer program to estimate the BER of the following modulation schemes with a heterodyne receiver and asynchronous detection: (a) OOK, (b) FSK, (c) DPSK.

8.7 N_s and N_n are the mean number of signal photons and noise photons at the receiver of the unrepeatered fiber-optic system with a preamplifier. The spontaneous noise factor and gain of the preamplifier are n_{sp} and G, respectively and $N_n = n_{sp}(G-1)$. Show that the error probabilities of FSK and DPSK are given by

$$P_b^{\text{FSK}} = \exp\left(\frac{-N_s}{2N_n}\right), \tag{8.390}$$

$$P_b^{\text{DPSK}} = \exp\left(\frac{-N_s}{N_n}\right), \tag{8.391}$$

respectively. Ignore shot noise and thermal noise.

8.8 Rectangular NRZ pulses are used in a direct detection 25-Gb/s DPSK system operating at 1540 nm. The average optical power at the receiver is 0 dBm. The fiber-optic link consists of 20 spans of identical fibers followed by amplifiers which exactly compensate the loss of fibers preceding. Each amplifier introduces ASE whose PSD is $\rho_{\text{ASE}}^{(1)}$. It is desirable that the BER $<= 2.1 \times 10^{-3}$. Find the upper limit on $\rho_{\text{ASE}}^{(1)}$.

(Ans: 3.654×10^{-16} W/Hz.)

References

[1] B.P. Lathi, *Modern Digital and Analog Communication Systems*, 3rd edn. Oxford University Press, New York, 1998.
[2] S. Haykin, *Communication Systems*, 4th edn. John Wiley & Sons, New York, 2001.
[3] S. Betti, G. De Marchis, and E. Iannone, *Coherent Optical Communication Systems*. John Wiley & Sons, New York, 1995.
[4] K.P. Ho, *Phase-modulated Optical Communication Systems*. Springer-Verlag, Berlin, 2005.
[5] G.P. Agrawal, *Fiber-optic Communication Systems*, 4th edn. John Wiley, & Sons, Hoboken, NJ, 2010.
[6] J.G. Proakis, *Digital Communications*, 4th edn. McGraw-Hill, New York, 2001, chapter 2.
[7] P.A. Humblet and M. Azizoglu, *J. Lightwave Technol.*, vol. **9**, p. 1576, 1991.
[8] D. Marcuse, *J. Lightwave Technol.*, vol. **8**, p. 1816 1990.
[9] D. Marcuse, *J. Lightwave Technol.*, vol. **9**, p. 505 1991.

9

Channel Multiplexing Techniques

9.1 Introduction

Typically, the single-channel symbol rates range from 10 Gsym/s to 40 Gsym/s. A symbol rate beyond 40 Gsym/s is hard to achieve in practice because of the speed of electronic components in transmitter and receiver circuits. In the low-loss region of the fiber (1530–1620 nm), it has a bandwidth greater than 10 THz. To utilize the full bandwidth of the fiber, several channels can be multiplexed and they can share the same fiber channel. An EDFA operating in C-band (1530–1565 nm) has a bandwidth of about 4.3 THz and, therefore, several channels can be amplified simultaneously by a single amplifier. The multiplexing techniques can be divided into three types: (i) polarization division multiplexing (PDM) or polarization multiplexing (PM), (ii) frequency or wavelength-division multiplexing (WDM), (iii) time-division multiplexing (TDM).

9.2 Polarization-Division Multiplexing

PDM is an effective technique to double the capacity. Commercial coherent systems make use of PDM and WDM to enhance the capacity. A single-mode fiber supports two polarization modes–one with the electric field aligned with the x-axis and the other aligned with the y-axis (see Section 2.7.5). Therefore, it is possible to transmit information using each of these polarization modes. A schematic of the PDM or PM is shown in Fig. 9.1. At the transmitter, a polarization beam splitter is used to split the x- and y-polarization components of the laser source. The x- (y-)polarization component of the laser is modulated by the electrical data m_x (m_y) using an optical modulator Mod x (Mod y). If the modulators are operating in the linear region, their outputs are (see Section 4.6.2.2)

$$\Psi_x = \mathbf{x}A_c m_x(t)e^{-i2\pi f_c t},$$

(9.1)

$$\Psi_y = \mathbf{y}A_c m_y(t)e^{-i2\pi f_c t},$$

(9.2)

where f_c is the laser frequency. The polarization beam combiner (PBC) combines these polarization components. The output of the PBC is

$$\Psi = \Psi_x + \Psi_y = A_c e^{-i2\pi f_c t}[m_x(t)\mathbf{x} + m_y(t)\mathbf{y}].$$

(9.3)

Fiber Optic Communications: Fundamentals and Applications, First Edition. Shiva Kumar and M. Jamal Deen.
© 2014 John Wiley & Sons, Ltd. Published 2014 by John Wiley & Sons, Ltd.

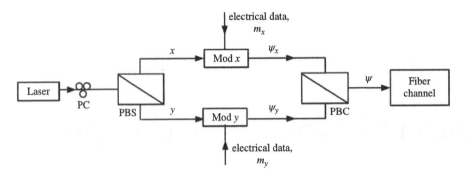

Figure 9.1   Polarization-division multiplexing. PC = polarization controller, PBS = polariaztion beam splitter, PBC = polarization beam combiner, Mod = optical modulator.

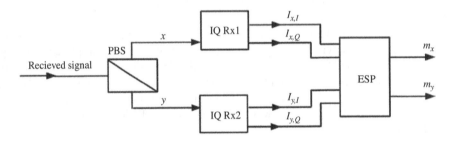

Figure 9.2   Block diagram of the polarization demultiplexing receiver. ESP = electrical signal processing.

These polarization components propagate as two polarization modes in a single-mode fiber. At the receiver, two IQ receivers are used to detect the x- and y-polarization components as shown in Fig. 9.2 (see Section 5.6.5). Let the complex currents corresponding to x- and y-polarization components be

$$I_x(t) = I_{x,I}(t) + iI_{x,Q}(t), \tag{9.4}$$

$$I_y(t) = I_{y,I}(t) + iI_{y,Q}(t). \tag{9.5}$$

Note that the x- and y-axes at the receiver may not be the same as those at the transmitter and, in addition, there is a coupling between the polarization modes during the fiber propagation due to random fluctuations in the refractive index. So, in the absence of noise, the complex currents in the frequency domain may be written as

$$\tilde{I}_x(\omega) = \tilde{m}_x(\omega)\tilde{M}_{xx}(\omega) + \tilde{M}_{xy}(\omega)\tilde{m}_y(\omega), \tag{9.6}$$

$$\tilde{I}_y(\omega) = \tilde{m}_x(\omega)\tilde{M}_{yx}(\omega) + \tilde{M}_{yy}(\omega)\tilde{m}_y(\omega) \tag{9.7}$$

or

$$\tilde{\mathbf{I}} = \tilde{\mathbf{M}}\tilde{\mathbf{m}}, \tag{9.8}$$

$$\tilde{\mathbf{I}} = \begin{bmatrix} \tilde{I}_x \\ \tilde{I}_y \end{bmatrix}, \tag{9.9}$$

$$\tilde{\mathbf{M}} = \begin{bmatrix} \tilde{M}_{xx}(\omega) & \tilde{M}_{xy}(\omega) \\ \tilde{M}_{yx}(\omega) & \tilde{M}_{yy}(\omega) \end{bmatrix}, \tag{9.10}$$

$$\tilde{\mathbf{m}} = \begin{bmatrix} \tilde{m}_x(\omega) \\ \tilde{m}_y(\omega) \end{bmatrix}. \tag{9.11}$$

Multiplying Eq. (9.8) by $\tilde{\mathbf{M}}^{-1}$ on both sides, we find

$$\tilde{\mathbf{m}} = \tilde{\mathbf{M}}^{-1}\tilde{\mathbf{I}}. \tag{9.12}$$

The digital signal processing of the coherent receiver can be used to compute $\tilde{\mathbf{M}}^{-1}$ (see Chapter 11) and, thus, the message signal vector $\tilde{\mathbf{m}}$ can be retrieved.

9.3 Wavelength-Division Multiplexing

In a WDM system, multiple optical carriers of different wavelengths are modulated by independent electrical data. Since wavelength λ and frequency f are related by $\lambda = c/f$, WDM may also be considered as frequency-division multiplexing (FDM). Fig. 9.3 shows the schematic of a WDM system. A CW laser operating at $\lambda_j, j = 1, 2, \ldots, N$ is modulated by electrical data j. The modulated signals are combined using a multiplexer and then launched to a fiber-optic link. At the end of the fiber-optic link, the channels are demultiplexed using a demultiplexer. If the data rate of a data stream modulating an optical carrier of wavelength λ_j is B, the total data rate is NB. Fig. 9.4 shows the WDM spectrum. Suppose that each channel is band-limited to f_s Hz. The spectrum of the channel j extends from $f_j - f_s/2$ to $f_j + f_s/2$, where $f_j = c/\lambda_j$. Typically, the optical carrier frequencies are equally spaced and the frequency difference between adjacent carriers is known as the *channel spacing* Δf. If the channel spacing Δf is smaller than the signal bandwidth f_s, the spectra of the neighboring channels overlap, leading to cross-talk and performance degradation. If the channel spacing Δf is much larger than f_s, it is a waste of fiber bandwidth. It is useful to define the spectral efficiency of a WDM system as

$$\eta = \frac{B}{\Delta f}, \tag{9.13}$$

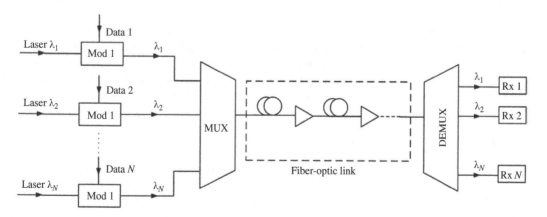

Figure 9.3 Schematic of a WDM system: Mod = modulator, MUX = multiplexer, DEMUX = demultiplexer.

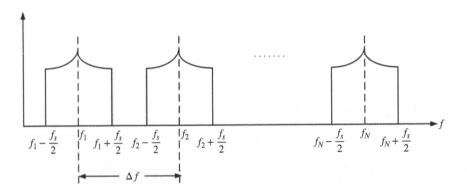

Figure 9.4   WDM spectrum.

where B is the data rate of a channel. If there are N channels, the total data rate is NB and the total bandwidth is about $N\Delta f$. Therefore, the spectral efficiency is also the ratio of the total data rate to the total bandwidth. For example, for a direct detection system based on NRZ-OOK, let the bit rate B be 10 Gb/s. If rectangular pulses are used, the first null of the NRZ spectrum occurs at 10 GHz (see Fig. 4.4) and the signal bandwidth f_s in a channel $\cong$ 20 GHz. If the channel spacing Δf is 20 GHz, the spectral efficiency η is 0.5 b/s/Hz. In this example, the overlap between the channel spectra is small and, hence, the cross-talk between the channels is negligible. The channel spacing is determined by the signal bandwidth in a channel. In this example, if the channel spacing is less than 20 GHz, there would be a significant overlap of spectra of adjacent channels, leading to cross-talk. However, if a Nyquist pulse is used instead, the signal bandwidth in a channel is 10 GHz (see Section 4.8) and in this case, the channel spacing can be reduced by a factor of two compared with the case of NRZ, which leads to an improvement in spectral efficiency by a factor of two. The spectral efficiency can also be considerably enhanced using coherent detection. For example, for a system based on QAM-16, let the symbol rate B_s be 25 Gsym/s. For QAM-16, the data rate B is $B_s \log_2 16 = 100$ Gb/s (see Section 4.9). If the channel spacing $\Delta f = 50$ GHz, the spectral efficiency = 2 b/s/Hz. The spectral efficiency can be increased using higher-order modulation formats such as QAM-64, but these signals suffer from distortions due to fiber nonlinear effects (see Chapter 10) limiting the maximum achievable transmission reach. Therefore, there is a trade-off between spectral efficiency and reach. When polarization multiplexing is used, the data rate is doubled for the given bandwidth and, therefore, the spectral efficiency is doubled compared with the case of single polarization.

In 2002, the International Telecommunication Union (ITU) standardized the channel wavelengths (or frequencies) of WDM systems on a 100-GHz (≈ 0.8 nm) grid in a wavelength range of 1528.77 nm to 1563.86 nm as defined by ITU-T G.694.1 [1]. However, for coherent communication systems with a symbol rate of 28 Gsym/s, such a large channel spacing leads to poor spectral efficiency. Recently, ITU standardized the WDM channels with a frequency spacing ranging from 12.5 GHz to 100 GHz and wider [1].

Example 9.1

Nyquist pulses are used in a single-polarization WDM system based on QAM-64. The symbol rate is 10 Gsym/s and the number of channels is 12. Calculate (a) the channel spacing to have a spectral efficiency of 6 b/s/Hz, (b) the signal bandwidth in a channel and the total bandwidth of the WDM signal, and (c) the total data rate.

Solution:
From Eq. (9.13), we have

$$\eta = \frac{B}{\Delta f}.$$

(9.14)

For QAM-64, from Section 4.9, we find

$$B = B_s \log_2 64 = 6 \times 10 \, \text{Gb/s}.$$

(9.15)

(a) The channel spacing is given by

$$\Delta f = \frac{B}{\eta} = \frac{60}{6} \, \text{GHz}$$

$$= 10 \, \text{GHz}.$$

(9.16)

(b) For a Nyquist pulse, the one-sided bandwidth $f_s/2$ is $B_s/2$. Therefore, the signal bandwidth in a channel $f_s = B_s = 10 \, \text{GHz}$.

$$\text{Total bandwidth of the WDM system} = (N-1)\Delta f + 2f_s/2$$

$$= (11 \times 10 + 10) \, \text{GHz}$$

$$= 120 \, \text{GHz}.$$

(9.17)

(c)
$$\text{Total data rate} = NB = 12 \times 60 \, \text{Gb/s}$$

$$= 720 \, \text{Gb/s}.$$

(9.18)

Example 9.2

A WDM system consists of 11 channels with a channel spacing of 100 GHz. The signal in each channel is band-limited to 50 GHz. The average power per channel is 0 dBm. The WDM signal is transmitted over a fiber of length 50 km. Fiber loss = 0.2 dB/km. Find the total power at the fiber output.

Solution:
Let the signal in channel k at the fiber input be

$$q_k(t) = \sum_n a_{n,k} f(t - nT_s) e^{i2\pi k \Delta f t}, \quad k = -5, -4, \dots, 5$$

$$= g_k(t) e^{i2\pi k \Delta f t},$$

(9.19)

where $\Delta f = 100 \, \text{GHz}$. The total signal field at the fiber input is

$$q_{\text{in}}(t) = \sum_{k=-5}^{5} q_k(t).$$

(9.20)

Taking the Fourier transform of Eq. (9.20), we find

$$\tilde{q}_{\text{in}}(f) = \sum_{k=-5}^{5} \tilde{q}_k(f) = \sum_{k=-5}^{5} \tilde{g}_k(f - k\Delta f),$$

(9.21)

where we have used the frequency-shifting property. Using Parseval's relation, the total energy is

$$
E_{\text{tot}} = \int_{-\infty}^{\infty} |q_{\text{in}}(t)|^2 \, dt = \int_{-\infty}^{\infty} |\tilde{q}_{\text{in}}(f)|^2 \, df
$$

$$
= \int_{-\infty}^{\infty} \tilde{q}_{\text{in}}(f)\tilde{q}_{\text{in}}^*(f)df
$$

$$
= \int_{-\infty}^{\infty} \sum_{k=-5}^{5} \tilde{g}_k(f - k\Delta f) \sum_{l=-5}^{5} \tilde{g}_l^*(f - l\Delta f)df
$$

$$
= \int_{-\infty}^{\infty} \left[\sum_{k=-5}^{5} |\tilde{g}_k(f - k\Delta f)|^2 + \sum_{\substack{k \\ k \neq l}} \sum_{l} \tilde{g}_k(f - k\Delta f)\tilde{g}_l^*(f - l\Delta f) \right] df. \tag{9.22}
$$

Consider the second term on the right-hand side of Eq. (9.22). $\tilde{g}_k(f - k\Delta f)$ corresponds to the channel k centered at $k\Delta f$ with a bandwidth of 50 GHz. Since $\Delta f > 50$ GHz, the overlap term $\tilde{g}_k(f - k\Delta f)\tilde{g}_l^*(f - l\Delta f)$ is zero when $k \neq l$. Therefore, the total energy is

$$
E_{\text{tot}} = \sum_{k=-5}^{5} \int_{-\infty}^{\infty} |\tilde{g}_k(f - k\Delta f)|^2 df
$$

$$
= \sum_{k=-5}^{5} \int_{-\infty}^{\infty} |q_k(t)|^2 dt. \tag{9.23}
$$

Thus, we see that the total energy is the sum of the energy of each channel. So, it follows that the total power is 11 times the power per channel:

$$
\text{power per channel} = 0\,\text{dBm}
$$

$$
= 10^{0.1 \times 0}\,\text{mW} = 1\,\text{mW}; \tag{9.24}
$$

$$
\text{total power} = 11\,\text{mW}
$$

$$
= 10\log_{10} 11\,\text{dBm}
$$

$$
= 10.413\,\text{dBm}; \tag{9.25}
$$

$$
\text{total fiber loss} = 0.2 \times 50 = 10\,\text{dB}; \tag{9.26}
$$

$$
\text{total power at the fiber output} = 10.413\,\text{dBm} - 10\,\text{dBm}
$$

$$
= 0.413\,\text{dBm}. \tag{9.27}
$$

9.3.1 WDM Components

Multiple wavelengths are combined using a multiplexer. The inverse operation of separating the wavelengths of a combined signal is achieved using a demultiplexer. The photonic device used as a multiplexer can also be used as a demultiplexer if the direction of propagation is reversed, because of the reciprocity property of optical field propagation. The simplest example for a multiplexer/demultiplexer is a prism which separates

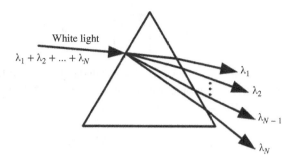

Figure 9.5 Wavelength separation using a prism.

(or combines) the different colors of white light, as shown in Fig. 9.5. But the angular separations provided by the prism are not large enough to separate the wavelengths of a WDM signal. The multiplexers can be divided into two categories: (i) interference-based multiplexers use Mach–Zehnder or other types of interferometer; (ii) diffraction-based multiplexers make use of diffraction to spatially separate the wavelengths. Examples include prisms and gratings.

9.3.1.1 Mach–Zehnder Interferometer-Based Demultiplexer

Mach–Zehnder interferometers can be cascaded to form a $1 \times N$ demultiplexer [2–5]. Let us first consider the theory of the 1×2 demultiplexer that separates two wavelengths. Fig. 9.6 shows a schematic of the demultiplexer. The 3-dB coupler is described by a matrix,

$$\mathbf{M}_{\text{coupler}} = \frac{1}{\sqrt{2}} \begin{bmatrix} 1 & i \\ i & 1 \end{bmatrix}. \tag{9.28}$$

The outputs of the 3-dB coupler 1 are

$$\mathbf{A}_{\text{out}}^{co1} = \mathbf{M}_{\text{coupler}} \mathbf{A}_{\text{in}}, \tag{9.29}$$

where

$$A_{\text{in}} = \begin{bmatrix} A_0 \\ 0 \end{bmatrix}, \tag{9.30}$$

A_0 is the input field envelope. Substituting Eqs. (9.28) and (9.30) in Eq. (9.29), we find

$$A_{\text{out},1}^{co1} = A_0/\sqrt{2}, \tag{9.31}$$

$$A_{\text{out},2}^{co1} = iA_0/\sqrt{2} \tag{9.32}$$

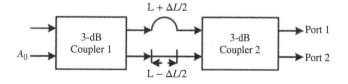

Figure 9.6 1×2 wavelength demultiplexer.

$$\mathbf{A}_{\text{out}}^{co1} = \begin{bmatrix} A_{\text{out,1}}^{co1} \\ A_{\text{out,2}}^{co1} \end{bmatrix}. \tag{9.33}$$

Optical fields in the upper and lower arms of the interferometer undergo phase shifts $k(L + \Delta L/2)$ and $k(L - \Delta L/2)$, respectively, where k is the propagation constant and ΔL is the path-length difference between two arms. Therefore, the inputs of the 3-dB coupler 2 can be written as

$$A_{\text{in,1}}^{co2} = \frac{A_0}{\sqrt{2}} \exp\left[ik(L + \Delta L/2)\right], \tag{9.34}$$

$$A_{\text{in,2}}^{co2} = \frac{iA_0}{\sqrt{2}} \exp\left[ik(L - \Delta L/2)\right]. \tag{9.35}$$

The outputs of the 3-dB coupler are

$$\mathbf{A}_{\text{out}}^{co2} = \mathbf{M}_{\text{coupler}} \mathbf{A}_{\text{in}}^{co2}, \tag{9.36}$$

$$A_{\text{out,1}}^{co2} = A_0 \exp(ikL) i \sin(k\Delta L/2), \tag{9.37}$$

$$A_{\text{out,2}}^{co2} = A_0 \exp(ikL) i \cos(k\Delta L/2). \tag{9.38}$$

The corresponding output powers are

$$P_{\text{out,1}} = |A_{\text{out,1}}^{co2}|^2 = A_0^2 \sin^2(k\Delta L/2), \tag{9.39}$$

$$P_{\text{out,2}} = |A_{\text{out,2}}^{co2}|^2 = A_0^2 \cos^2(k\Delta L/2). \tag{9.40}$$

Fig. 9.7 shows the power transmittances of ports 1 and 2. At a specific frequency, the power transmittance of port 1 is maximum and at the same frequency, the power transmittance of port 2 is zero. This implies that if the channel frequencies of a two-channel WDM system coincide with the frequencies corresponding to the peak power transmittances of ports 1 and 2, they can be separated.

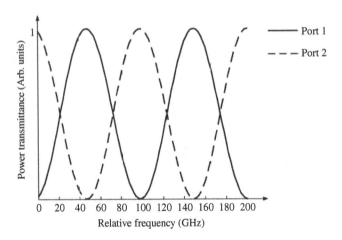

Figure 9.7   Power transmittance as a function of frequency deviation from the reference frequency of 194.8 THz.

Consider an optical wave with propagation constant $k_1 = 2\pi n/\lambda_1$, where n is the refractive index of the MZ interferometer. If

$$k_1 \Delta L = (2m+1)\pi, \quad m = 0, \pm 1, \pm 2, \tag{9.41}$$

we have

$$P_{out,1} = A_0^2,$$
$$P_{out,2} = 0. \tag{9.42}$$

So, all the input power appears in port 1. Given another optical wave with $k_2 = 2\pi n/\lambda_2$ and if

$$k_2 \Delta L = 2l\pi, \quad l = 0, \pm 1, \pm 2, \tag{9.43}$$

we find that all the input power appears in port 2. Therefore, if we choose the wavelengths λ_1 and λ_2 such that

$$\frac{n\Delta L}{\lambda_1} = \frac{(2m+1)}{2}, \tag{9.44}$$

$$\frac{n\Delta L}{\lambda_2} = l, \tag{9.45}$$

the optical fields with wavelengths λ_1 and λ_2 appear in ports 1 and 2, respectively. From Eqs. (9.44) and (9.45), we obtain

$$\lambda_2 - \lambda_1 = \frac{\lambda_1 \lambda_2 (2m'+1)}{2n\Delta L}, \quad m' = m - l = 0, \pm 1, \pm 2. \tag{9.46}$$

Since $f_j = c/\lambda_j, j = 1, 2$, from Eqs. (9.41) and (9.43), we obtain

$$\Delta f = \frac{(2m'+1)c}{2n\Delta L}, \tag{9.47}$$

where $\Delta f = f_1 - f_2$ is the channel spacing. For example, wavelengths $\lambda_1 = 1540\,\text{nm}$ and $\lambda_2 = 1540.4\,\text{nm}$ are multiplexed in a WDM system. At the receiver, these wavelengths can be separated if

$$\Delta L = \frac{\lambda_1 \lambda_2 (2m+1)}{(\lambda_2 - \lambda_1)2n}, \quad m = 0, \pm 1, \pm 2, \dots . \tag{9.48}$$

If we choose $m = 0$, we find $\Delta L = 2\,\text{mm}$.

A $1 \times N$ demultiplexer can be constructed by cascading the 1×2 demultiplexer of Fig. 9.6. Fig. 9.8 shows a schematic of a 1×4 demultiplexer. Suppose the input consists of four channels with wavelengths $\lambda_1, \lambda_2,$

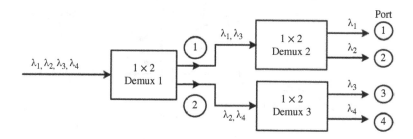

Figure 9.8 1×4 wavelength demultiplexer.

λ_3, and λ_4, equally spaced in frequency. Let the frequency spacing be Δf. The path-length difference ΔL of Demux 1 is chosen so that odd and even wavelengths are directed to ports 1 and 2 of Demux 1, respectively. The frequency difference between channels 1 and 3 is $2\Delta f$. If ΔL is chosen using Eq. (9.48), channels 1 and 2 are directed to ports 1 and 2 of Demux 1, respectively. Since the power transmittance is periodic with period $2\Delta f$ (see Fig. 9.7), channels 1 and 3 have the maximum power transmittance at port 1 of Demux 1. Channels 1 and 3 are separated using Demux 2. Since the frequency difference between channels 1 and 3 is $2\Delta f$, ΔL of Demux 2 should be half that of Demux 1. In this analysis, we assume that the couplers are ideal 3-dB couplers and the MZ interferometer arms have no losses. As a result, we find that when the power output at port 1 is maximum that at port 2 is zero, and vice versa (see Fig. 9.7). This corresponds to zero cross-talk between channels. In practice, the power-coupling ratio deviates from 3 dB and the loss due to propagation in MZ cannot be ignored. When these effects are included, it is found that the power output at port 2 is not zero while that at port 1 is maximum, which leads to cross-talk between channels [5]. A 10 GHz-spaced silica-based integrated-optic 8-channel MZ multi/demultiplexer is fabricated with a cross-talk of -10 dB or less [5].

9.3.1.2 Diffraction-Based Multiplexer/Demultiplexers

Diffraction-based multi/demultiplexers make use of Bragg diffraction to isolate/combine the wavelength components [6, 7]. Fig. 9.9 shows a schematic of the bulk grating-based demultiplexer. The WDM signal consisting of multiple wavelength components is incident on the grating. Different wavelength components diffract at different angles and they are collected by output fibers. One of the problems with bulk grating-based demultiplexers is that the output fiber core must be much larger than the input fiber core in order to obtain the required flat pass band [6, 7]. Instead, an array of optical waveguides acting as a grating could be used. Such gratings are known as *arrayed-waveguide gratings* or *phased-array demultiplexers*.

9.3.1.3 Arrayed-Waveguide Gratings

The principle of wavelength multiplexing/demultiplexing using the AWG is discussed in Refs. [8–12]. Suppose the input consists of two channels centered around λ_1 and λ_2. The input field propagates in a uniform

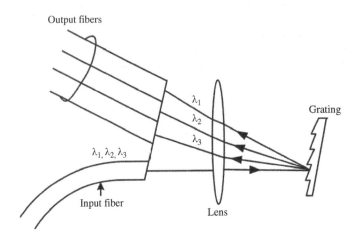

Figure 9.9 Bulk grating-based demultiplexer.

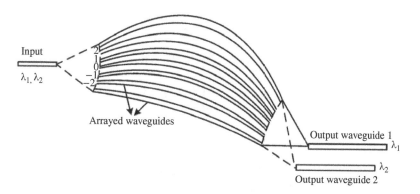

Figure 9.10   AWG demultiplexer.

medium and then it is incident on the waveguide array (or arrayed waveguides), as shown in Fig. 9.10. The lengths of waveguides $l_j, j = -N/2, \dots, -1, 0, 1, \dots, N/2 - 1$ are chosen so that the phase shifts introduced by each single-mode waveguide are an integral multiple of 2π at wavelength λ_1. The phase of the optical field arriving through waveguide j at the output is

$$\phi_j(\lambda_1) = \beta(\lambda_1)l_j + \theta_j(\lambda_1) = 2\pi m_j + \theta_j(\lambda_1), \tag{9.49}$$

where β is the propagation constant of the waveguide j, θ_j is the initial phase at the input of the waveguide j and m_j is an integer. For simplicity, let us assume that the input is a point source and the wave front at the input of the waveguides is spherical, so that $\theta_j = \theta$. When the lengths of the waveguides are chosen so that Eq. (9.49) is satisfied, the waveguide array behaves like a phased array with uniform phase distribution at the output. Because of the geometry, the diverging wave front at the input of the waveguide array becomes a converging wave front at its output. The part of the spectrum centered around $\lambda_j, j = 1, 2$ at the output of the waveguide array focuses on the output waveguide j. The reason for the spatial separation of the wavelengths can be understood as follows. From Eq. (9.49), the phase difference between the adjacent waveguides at λ_1 is

$$\delta\phi(\lambda_1) \equiv \phi_j(\lambda_1) - \phi_{j-1}(\lambda_1) = \beta(\lambda_1)\Delta l = 2m\pi, \tag{9.50}$$

where $\Delta l = l_j - l_{j-1}$ and $m = m_j - m_{j-1}$ is an integer. Let

$$\lambda_2 = \lambda_1 + \Delta\lambda. \tag{9.51}$$

The Taylor-series expansion of the propagation constant around λ_1 is

$$\beta(\lambda_2) \cong \beta(\lambda_1) + k\Delta\lambda, \tag{9.52}$$

$$k = \left.\frac{d\beta}{d\lambda}\right|_{\lambda=\lambda_1}. \tag{9.53}$$

From Eqs. (9.49) and (9.52), we have

$$\begin{aligned}\phi_j(\lambda_2) &= \beta(\lambda_2)l_j + \theta(\lambda_2) \\ &= \phi_j(\lambda_1) + \Delta\theta + k\Delta\lambda l_j,\end{aligned} \tag{9.54}$$

where

$$\Delta\theta = \theta(\lambda_2) - \theta(\lambda_1) = \theta_j(\lambda_2) - \theta_j(\lambda_1). \tag{9.55}$$

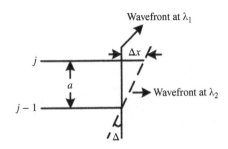

Figure 9.11 Wavefronts at λ_1 and λ_2.

From Eqs. (9.50) and (9.54), it follows that the phase difference between adjacent waveguides at λ_2 is

$$\delta\phi(\lambda_2) = \phi_j(\lambda_2) - \phi_{j-1}(\lambda_2) = k\Delta\lambda\Delta l. \tag{9.56}$$

Thus, although the phases of the adjacent waveguides are identical at λ_1, they are shifted by $k\Delta\lambda\Delta l$ at λ_2. Consider two adjacent waveguides j and $j-1$ separated by a, as shown in Fig. 9.11. Suppose the optical field in waveguide j propagates an additional distance Δx, then the phases of the output of waveguides j and $j-1$ become identical, i.e.,

$$k\Delta\lambda\Delta l = \beta(\lambda_2)\Delta x. \tag{9.57}$$

Thus, the wave front (the locus of all the points having the same phase) at λ_2 is tilted by an angle $\Delta = \Delta x/a$. Therefore, in Fig. 9.10, the output of the waveguide array corresponding to wavelength λ_2 focuses on a different port than the wavelength λ_1.

A waveguide grating demultiplexer on InP which resolves 16 channels with a channel spacing of 1.8 nm and with low polarization sensitivity was demonstrated in 1994 [11]. A 4-channel phased-array wavelength demultiplexer on InGaAsP/InP with a channel spacing of 1 nm was demonstrated in 1996 [12].

Example 9.3

A 1×2 AWG demultiplexer has to be designed. $\lambda_1 = 1550\,\text{nm}$, $\lambda_2 = 1550.8\,\text{nm}$, $\beta_0 = 5.87 \times 10^6\,\text{m}^{-1}$ at λ_1, $\beta_1 = 4.86 \times 10^{-9}$ s/m at λ_1. (a) Find the lengths of the adjacent waveguides l_1 and l_2 such that the phase shifts ϕ_1 and ϕ_2 at the fiber outputs at λ_1 are integral multiples of 2π. (b) Calculate ϕ_1 and ϕ_2 at λ_2. Assume $\theta_1 = \theta_2 = 0$.

Solution:
(a) Let

$$\beta(\lambda_1)l_1 = 2\pi m_1, \tag{9.58}$$

$$\beta(\lambda_1)l_2 = 2\pi m_2. \tag{9.59}$$

Here, m_1 and m_2 could be any integers. Let $m_1 = 100$ and $m_2 = 110$:

$$\beta(\lambda_1) = \beta_0 = 5.87 \times 10^6\,\text{m}^{-1}, \tag{9.60}$$

$$l_1 = \frac{2\pi \times 100}{\beta_0} = 107.04\,\mu\text{m}, \tag{9.61}$$

$$l_2 = \frac{2\pi \times 110}{\beta_0} = 117.74\,\mu\text{m}. \tag{9.62}$$

(b)

$$\frac{d\beta}{d\lambda} = \frac{d\beta}{d\omega}\frac{d\omega}{d\lambda} = 2\pi\frac{d\beta}{d\omega}\frac{df}{d\lambda}, \tag{9.63}$$

$$f = \frac{c}{\lambda}, \tag{9.64}$$

$$df = -\frac{c}{\lambda^2}d\lambda, \tag{9.65}$$

$$\left.\frac{df}{d\lambda}\right|_{\lambda_1=1550\,\text{nm}} = \frac{-3\times10^8}{(1550\times10^{-9})^2}\,\text{m}^{-1}\,\text{s}^{-1} = -1.24\times10^{20}\,\text{m}^{-1}\,\text{s}^{-1} \tag{9.66}$$

$$\beta_1 = \left.\frac{d\beta}{d\omega}\right|_{\lambda=1550\,\text{nm}} = 4.86\times10^{-9}\,\text{s/m}. \tag{9.67}$$

Eq. (9.63) may be written as

$$\left.\frac{d\beta}{d\lambda}\right|_{\lambda_1=1550\,\text{nm}} = 2\pi\beta_1\left.\frac{df}{d\lambda}\right|_{\lambda_1=1550\,\text{nm}}$$

$$= -3.81\times10^{12}\,\text{m}^{-2}. \tag{9.68}$$

From Eqs. (9.49) and (9.54), we have

$$\phi_1(\lambda_2) = \beta(\lambda_2)l_1$$

$$= \beta(\lambda_1)l_1 + (\lambda_2 - \lambda_1)l_1\left.\frac{d\beta}{d\lambda}\right|_{\lambda_1=1550\,\text{nm}}$$

$$= 2\pi \times 100 - 0.8\times10^{-9}\times107.04\times10^{-6}\times3.81\times10^{12}\,\text{rad}$$

$$= 6.279\times10^2\,\text{rad}, \tag{9.69}$$

$$\phi_2(\lambda_2) = \beta(\lambda_2)l_2$$

$$= \beta(\lambda_1)l_2 + (\lambda_2 - \lambda_1)l_2\left.\frac{d\beta}{d\lambda}\right|_{\lambda_1=1550\,\text{nm}}$$

$$= 2\pi \times 110 - 0.8\times10^{-9}\times117.74\times10^{-6}\times3.81\times10^{12}\,\text{rad}$$

$$= 6.907\times10^2\,\text{rad}. \tag{9.70}$$

9.3.2 WDM Experiments

WDM has been studied extensively and is used in commercial transmission systems. Early WDM experiments were carried out using direct detection [13]. With the advent of digital coherent receivers, the information capacity and spectral efficiency have increased significantly. A data rate of 7.2 Tb/s over a distance of 7040 km with a spectral efficiency of 2 bit/s/Hz is achieved using PDM and QPSK [14]. The baud rate of each WDM channel in this experiment is 25 GBaud, excluding the FEC overhead. Because of QPSK modulation formation, the bit rate of each PDM channel is 50 Gb/s. Since there are two PDM channels for each optical carrier, the total bit rate per WDM channel is 100 Gb/s. The total number of WDM channels is 72, leading to a total bit rate of 7.2 Tb/s. The optical bandwidth occupied by the WDM signal is 28 nm (or 3.5 THz), leading to a spectral efficiency of 7.2/3.5 ≅ 2 bit/s/Hz. The spectral efficiency can be increased further by using MPSK or QAM. A 17-Tb/s (161 × 114 Gb/s) polarization-multiplexed WDM signal is transmitted over 662 km using RZ-8PSK [15]. A spectral efficiency of 4.2 bit/s/Hz was achieved in this experiment [15]. A 32-Tb/s (320 × 114 Gb/s) polarization-multiplexed RZ-8QAM signal was transmitted over 580 km of SMF-28 with a spectral efficiency of 4 bit/s/Hz [16]. As the spectral efficiency increases, the transmission distance decreases because of fiber nonlinear effects (see Chapter 10).

9.4 OFDM

WDM is a FDM technique in which the carriers are typically not orthogonal. A special class of FDM in which the carriers (or subcarriers) are orthogonal is known as orthogonal frequency-division multiplexing (OFDM). In a WDM system, if the channel spacing is smaller than the bandwidth of the channels, this leads to cross-talk and performance degradation. However, in an OFDM system, if the separation between carriers is smaller than the band width of the data in each carrier, there is a significant spectral overlap between the neighboring channels and yet there would be no cross-talk or performance degradation because of carrier orthogonality conditions.

OFDM has drawn significant research interest in optical communications recently [17–33]. OFDM is widely used in wired and wireless communication systems because it is resilient to ISI caused by dispersive channels. It has been used for digital audio broadcasting, HDTV terrestrial broadcasting, and wireless LANS. The first proposal to use orthogonal frequencies for transmission appeared in a 1966 patent by Chang [34]. In 1969, Salz and Weinstein [35] introduced orthogonal carriers by using the discrete Fourier transform (DFT). In 1971, Weinstein and Ebert [36] applied the discrete cosine transform (DCT) to a multi-carrier transmission system as part of a modulation and demodulation process. The cyclic prefix, which is an important aspect of the OFDM system, was proposed in 1980 [37].

9.4.1 OFDM Principle

Consider a multi-carrier communication system as shown in Fig. 9.12. Suppose $d_1(t), d_2(t), \dots, d_N(t)$ are the complex data streams to be transmitted:

$$d_n(t) = \begin{cases} d_{n0} & \text{for } 0 < t < T_s \\ 0 & \text{otherwise,} \end{cases} \tag{9.71}$$

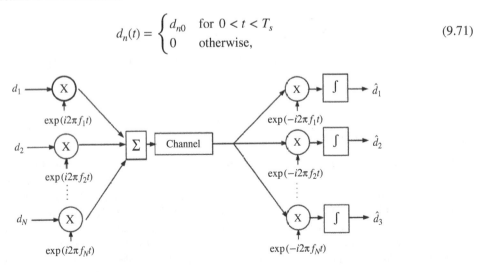

Figure 9.12 A multi-carrier communication system.

where T_s is the symbol period and d_{n0} is a constant. Let $f_1, f_2, \ldots, f_N$ be the carrier frequencies. The carrier $\exp(i2\pi f_n t)$ is modulated by the data stream $d_n, n = 1, 2, \ldots, N$ and then the modulated signals are combined to obtain the transmitted signal.

$$s_{tr}(t) = \sum_{n=1}^{N} d_n(t) \exp(i2\pi f_n t). \tag{9.72}$$

At the receiver, the received signal is multiplexed by the bank of local oscillators and integrators. In the absence of channel distortion and noise effects, the signal after the integrator can be written as

$$\widehat{d}_m = \frac{1}{T_s} \int_0^{T_s} s_{tr}(t) \exp(-i2\pi f_m t) dt, \quad m = 1, 2, \ldots, N \tag{9.73}$$

$$= \frac{1}{T_s} \int_0^{T_s} \sum_{n=1}^{N} d_n(t) \exp[i2\pi(f_n - f_m)] dt. \tag{9.74}$$

The integrator is nothing but a low-pass filter and the terms oscillating at frequencies $f_n - f_m, n \neq m$ do not contribute significantly if $f_n - f_m$ is larger than $1/T_s$. Therefore, a significant contribution comes only from the d.c. term corresponding to $n = m$ in Eq. (9.74), leading to

$$\widehat{d}_m \cong d_{m_0}. \tag{9.75}$$

One of the disadvantages of this approach is that frequency separation between carriers should be sufficiently large so that the contributions from the cross-terms at frequencies $f_n - f_m, n \neq m$ in Eq. (9.74) are small. This leads to excessive bandwidth requirements. Besides, the transmitter and receiver require a bank of analog oscillators, product modulators, and filters, increasing the complexity of the architecture.

The bandwidth can be utilized efficiently if the carriers are *orthogonal*. Suppose we choose the carrier frequencies such that

$$\frac{1}{T_s} \int_0^{T_s} \exp[i2\pi(f_m - f_n)t] = \begin{cases} 1 & \text{if } m = n \\ 0 & \text{otherwise.} \end{cases} \tag{9.76}$$

Now, the carriers are said to be orthogonal over the interval $[0, T_s]$. If

$$f_m = m/T_s, \quad m = 1, 2, \ldots, N, \tag{9.77}$$

it can easily be verified that Eq. (9.76) is satisfied. Therefore, the carrier frequencies should be integral multiples of the symbol rate $(= 1/T_s)$. For example, the first carrier is a sinusoid with period T_s, the second carrier is a sinusoid with period $T_s/2$, and so on. Using Eqs. (9.71) and (9.77), Eq. (9.72) may be rewritten as

$$s_{tr}(t) = \sum_{n=1}^{N} d_{n_0} \exp\left(\frac{i2\pi nt}{T_s}\right), \quad 0 < t < T_s. \tag{9.78}$$

If we discretize the time interval

$$t = k\Delta t, \quad k = 1, 2, \ldots, N, \tag{9.79}$$

where Δt is the sampling interval with

$$N\Delta t = T_s, \tag{9.80}$$

Eq. (9.78) is modified as

$$s_{tr}(k\Delta t) \equiv s_k = \sum_{n=1}^{N} d_{n_0} \exp\left[\frac{i2\pi kn}{N}\right], \quad k = 1, 2, \ldots, N. \tag{9.81}$$

s_k is the inverse discrete Fourier transform (IDFT) of the data sequence $\{d_{n_0}\}$, $n = 1, 2, \dots, N$. IDFT can be computed efficiently using the inverse fast Fourier transform (IFFT). In other words, the bank of oscillators, product modulators, and adders in the transmitter section of Fig. 9.12 can be replaced by an IDFT operation in the digital domain. Similarly, the bank of correlators at the receivers can be replaced by a DFT. In Eq. (9.74), because of the orthogonality condition given by Eq. (9.76), all the terms at frequencies $f_n - f_m$ vanish except for the d.c. term with $n = m$. Therefore,

$$\widehat{d}_m = d_{m_0}. \tag{9.82}$$

Replacing the integral in Eq. (9.73) by summation with $t = k\Delta t$ and using Eqs. (9.77) and (9.80), we obtain

$$\widehat{d}_m = \frac{1}{N} \sum_{k=1}^{N} s_{tr}(k\Delta t) \exp\left(-\frac{i2\pi mk}{N}\right), \quad m = 1, 2, \dots, N. \tag{9.83}$$

Thus, $\widehat{d}_m$ is the DFT of s_k, which can be computed using the FFT. Fig. 9.13 shows a simplified block diagram of an optical OFDM system. Consider the data sequence $\{d_1, d_2, \dots, d_N\}^k$ in a symbol interval $[kT_s, (k+1)T_s]$. The IDFT of this sequence is computed by means of IFFT. After parallel-to-serial conversion (P/S), the optical carrier (laser) is modulated by the electrical OFDM symbol and then it propagates through the fiber-optic link. At the receiver, a coherent receiver is used to retrieve the electrical OFDM symbol. After performing the serial-to-parallel (S/P) and DFT operations, the transmitted data sequence $\{d_1, d_2, \dots, d_N\}^k$ of the kth OFDM symbol can be recovered in the absence of noise and distortion. The IDFT operation at the transmitter and the DFT operation at the receiver are repeated for every OFDM symbol.

In a dispersive channel such as an optical fiber, different frequency components travel with different speeds. In a normal dispersion fiber, a higher-frequency subcarrier of a given OFDM symbol is delayed and it would interfere with the data in the neighboring symbol. Suppose, at the receiver, the timing offset is chosen so that the DFT window is synchronized with subcarrier 1. Now, subcarrier N is delayed by (see Eqs. (2.197) and (2.203))

$$\Delta T = |\beta_2| \Delta \omega L = D \Delta \lambda L, \tag{9.84}$$

where β_2 is the dispersion coefficient, D is the dispersion parameter, L is the fiber length, and $\Delta \omega$ is the angular frequency difference between subcarrier 1 and subcarrier N. From Fig. 9.14, it can be seen that subcarrier

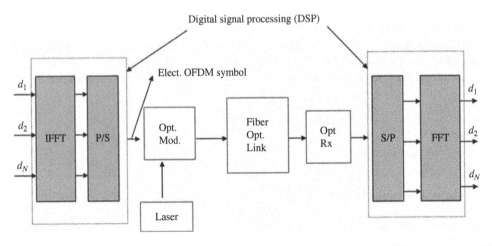

Figure 9.13 Block diagram of an optical OFDM system. Opt. mod. = optical modulator, P/S = parallel to serial, S/P = serial to parallel.

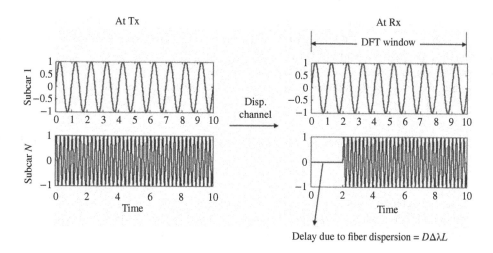

Figure 9.14   Subcarriers of the OFDM system with no cyclic prefix.

N has moved out of the DFT window. This leads to two issues. First, it would interfere with the data in the neighboring OFDM symbol if there is no guard interval between the two. This is known as intercarrier interference (ICI) and could lead to performance degradation. Second, as can be seen from Fig. 9.14, the first few cycles of the subcarrier N at the receiver are empty within the DFT window and, therefore, subcarriers 1 and N are no longer orthogonal over the symbol interval. The breakdown of orthogonality conditions also leads to performance degradations.

To preserve the orthogonality, a cyclic prefix is used [27, 28]. Instead of leaving the guard interval empty, the last few cycles of subcarrier N within a block are copied to the guard interval, as shown in Fig. 9.15.

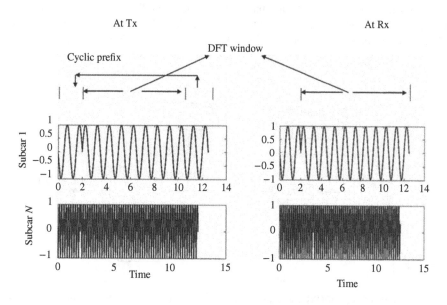

Figure 9.15   Subcarriers of the OFDM system with cyclic prefix.

Although subcarrier N moves out of the DFT due to fiber dispersion, because of its identical copy in the guard interval, the signal corresponding to subcarrier N at the receiver is the same as that at the transmitter except for a phase jump, as shown in Fig. 9.15. The phase jump can be removed after channel estimation at the receiver. As long as the maximum delay ΔT introduced by the dispersive channel is less than the guard interval T_g, the subcarriers are orthogonal at the receiver. The maximum delay ΔT is

$$\Delta T = |\beta_2|(2\pi N\Delta f)L, \tag{9.85}$$

where $\Delta f = 1/T_s$ is the frequency spacing between subcarriers. Therefore, the guard interval should be chosen so that

$$|\beta_2|(2\pi N\Delta fL) < T_g. \tag{9.86}$$

To preserve the orthogonality of all the subcarriers, the last few cycles of each subcarrier should be copied to the guard interval. Although the guard interval increases the tolerance against delay due to dispersion, it reduces the efficiency since the guard interval is discarded by the receiver.

9.4.2 Optical OFDM Transmitter

Fig. 9.16 shows a block diagram of an optical OFDM transmitter. First, the binary serial input is converted to parallel (S/P). For example, a bit sequence {00110111} is broken into {00}, {11}, {01}, and {11}. Each of these subsequences is mapped into QPSK data using a symbol mapper, i.e.,

$$\{00\} \rightarrow (-1-i)/\sqrt{2} = d_1$$
$$\{11\} \rightarrow (1+i)/\sqrt{2} = d_2$$

and so on. By breaking the bit sequence {00110111} into {0011} and {0111}, a symbol mapper could map it onto QAM-16 data. The output of the symbol mapper is complex data which passes through the IFFT block. After the parallel-to-serial conversion (P/S) and guard interval insertion, the digital signal is converted to an analog signal using a digital-to-analog converter (DAC). In fact, the IFFT output is complex in general and, therefore, two DACs are needed. The outputs of the DACs are used to modulate an optical IQ modulator.

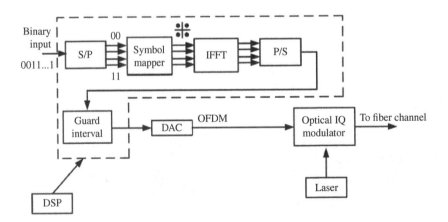

Figure 9.16 Block diagram of the OFDM transmitter. S/P = serial to parallel, P/S = parallel to serial, DAC = digital-to-analog converters.

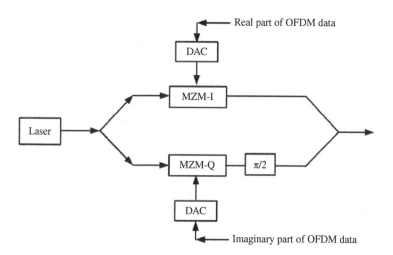

Figure 9.17   Block diagram of an IQ modulator. MZM = Mach–Zehnder modulator, DAC = digital-to-analog converter.

A description of the IQ modulator is provided in Fig. 9.17. Real and imaginary parts of the OFDM data modulate the laser light using a Mach–Zehnder modulator-I (MZM-I) and MZM-Q, respectively. Let the real and imaginary parts of the OFDM data be $m_r(t)$ and $m_i(t)$, respectively. Assuming that the MZMs operate in the linear regime, the MZM-I and MZM-Q outputs can be written as (see Section 4.6.2.2)

$$q_I = \frac{A_c}{\sqrt{2}} m_r(t) \exp(-i2\pi f_c t), \qquad (9.87)$$

$$q_Q = \frac{A_c}{\sqrt{2}} m_i(t) \exp(-i2\pi f_c t). \qquad (9.88)$$

The output of MZM-Q passes through a $\pi/2$ phase shifter, which is equivalent to multiplying by i. After the output y-branch in Fig. 9.17, the output is given by

$$q = (q_I + iq_Q)/\sqrt{2}$$
$$= \frac{A_c}{2} m(t) \exp(-i2\pi f_c t), \qquad (9.89)$$

where $m(t) = m_r(t) + im_i(t)$ is the complex OFDM data.

9.4.3   Optical OFDM Receiver

Fig. 9.18 shows a block diagram of an optical OFDM receiver with coherent detection. The output of the fiber-optic link passes through an optical IQ receiver (see Chapter 5) consisting of a 90° hybrid and an array of photodetectors. The I- and Q-branches of the IQ receiver output correspond to the real and imaginary parts of the OFDM data, respectively. After the analog-to-digital conversion (ADC), the I and Q signals pass through the DSP unit for further signal processing. Combining the real and imaginary parts, complex OFDM data is formed and the DFT of this data is computed using FFT, after serial-to-parallel conversion on each OFDM symbol. In the absence of laser phase noise, fiber propagation effects, and ASE, the output of the FFT

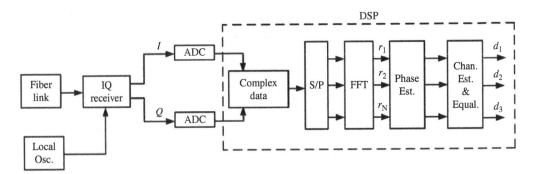

Figure 9.18 Block diagram of an optical OFDM receiver. S/P = serial to parallel, ADC = analog-to-digital converter, DSP = digital signal processing.

block would be the actual data in each subcarrier. However, because of the noise and fiber propagation effects, the signal is distorted. To undo these distortions, phase and channel estimation and equalization are carried out in the digital domain. More details can be found in Ref. [38].

9.4.4 Optical OFDM Experiments

Transmission of a 1-Tb/s OFDM superchannel over 8000 km of SSMF link at a spectral efficiency of 3.1 bit/s/Hz has been demonstrated experimentally [39]. Each channel of a WDM system can use OFDM as modulation format. Since OFDM has a compact spectrum, the guard band between WDM channels can be reduced. The experimental demonstration of an 8-channel dense WDM (DWDM) transmission with a spectral efficiency of 7 bit/s/Hz using a 65.1-Gb/s coherent PDM–OFDM signal in each channel with 8-GHz channel spacing utilizing 32-QAM data on each subcarrier of an OFDM over a 240-km SSMF link has been carried out [40].

Using a coded modulation scheme for coherent OFDM, a 231.5-Gb/s OFDM signal at a record spectral efficiency of 11.15 bit/s/Hz over an 800-km ultra-large-area fiber link has been demonstrated experimentally [41]. The spectral efficiency in this experiment is close to the Shannon limit. However, the signal reception in most of the experimental reports is processed off-line. This is because it is hard to implement DFT and IDFT in real time at higher data rates. Multi-band real-time coherent OFDM reception at a data rate of 110 Gb/s based on a field programmable gate array (FPGA) with an individual sub-band of 3.33 Gb/s over a 600-km fiber-optic link has been demonstrated [42].

Example 9.4

An optical OFDM system has 128 subcarriers with a frequency spacing of 78.125 MHz between subcarriers. The dispersion of the transmission fiber is $-22 \text{ ps}^2/\text{km}$. The guard interval is 1.28 ns. Calculate the maximum reach up to which the carrier orthogonality is preserved.

Solution:
The carrier orthogonality will not be preserved if

$$|\beta_2|(2\pi N \Delta f L) < T_g. \tag{9.90}$$

The maximum reach is given by

$$L_{max} = \frac{T_g}{|\beta_2|(2\pi N \Delta f)} = \frac{1.28 \times 10^{-9}}{22 \times 10^{-27} \times 2\pi \times 128 \times 78.125 \times 10^6}$$
$$= 925 \text{ km.} \tag{9.91}$$

9.5 Time-Division Multiplexing

In the case of frequency-division multiplexing, parallel streams of data are modulated by carriers with different frequencies so that the data spectra do not overlap (see Fig. 9.4). Instead, the parallel streams of data can be converted to serial data in such a way that the individual streams do not overlap in time. This type of multiplexing is known as *time-division multiplexing* (TDM). Fig. 9.19 illustrates the schematic of a two-channel TDM. Let T_s be the symbol interval of the individual data streams corresponding to a symbol rate of $B_s = 1/T_s$. TDM converts the parallel data streams into serial data with two symbols within the symbol interval T_s, as shown in Fig. 9.19. To avoid the overlap in time, the pulse widths should be less than $T_s/2$. In this example, the total symbol rate is $2B_s$. In general, for an N-channel TDM system, the pulse widths should be less than T_s/N and the total symbol rate is NB_s.

TDM can be performed in either an electrical or an optical domain. However, as the bit rate increases beyond 40 Gb/s, if becomes hard to do electrical TDM because of the limitations imposed by high-speed electronics. Instead, channels can be multiplexed in the optical domain and such a scheme is known as optical TDM (OTDM). Because of the wide bandwidth of optical devices, OTDM can be used to obtain a total bit rate of several terabits per seconds.

9.5.1 Multiplexing

To realize OTDM, ultra-short laser pulses and delay lines are required [7, 43]. Fig. 9.20 shows a schematic of a four-channel OTDM system. A train of ultra-short pulses is split into four branches. Each branch is modulated by the electrical data. To avoid the temporal overlap of channels, channel j, $j = 1, 2, 3, 4$, is delayed by $(j - 1)T$, where $T = T_s/4$, using a delay line. The delay lines can be realized using fiber segments (see Example 2.5). The output of all branches is combined to obtain a TDM signal.

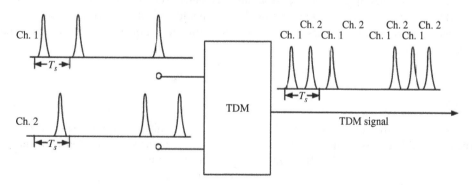

Figure 9.19 Schematic of a two-channel TDM.

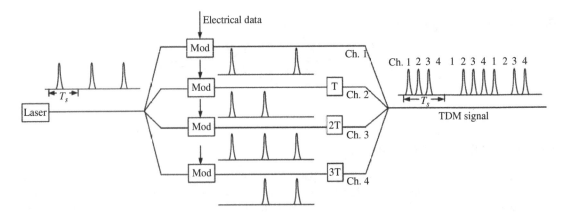

Figure 9.20   Schematic of a four-channel OTDM. Mod = optical modulator and T refers to a delay of $T_s/4$.

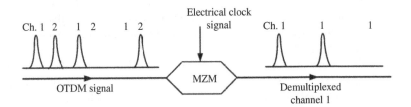

Figure 9.21   Schematic of an OTDM demultiplexer. MZM = Mach–Zehnder modulator.

9.5.2   *Demultiplexing*

Fig. 9.21 shows a schematic of a two-channel OTDM demultiplexer. The MZM is driven by the electrical signal at a clock rate of B_s. To demultiplex channel 1, the amplitude of the electrical driving voltage is chosen so that channel 1 is at the peak of the MZM transmittivity and channel 2 is at the null (see Section 4.6.2.2). Therefore, the modulator transmits channel 1 without significant attenuation while it rejects channel 2. A similar MZM with the appropriate delay transmits channel 2 while rejecting channel 1. MZMs can easily be cascaded to demultiplex a channel from an N-channel OTDM signal [7, 43]. To avoid cross-talk from other channels, the extinction ratio of the modulator should be high [43]. A moderate extinction ratio ($\sim 15\,\mathrm{dB}$) can be realized with MZMs. To enhance the extinction further, electroabsorption modulators can be used.

Example 9.5

A pulse incident on a 3-dB splitter as shown in Fig. 9.22 has a pulse width (FWHM) = 5 ps and a peak power = 5 mW. The length of fiber 1 is 1 mm. Find the length of fiber 2 so that the separation between pulses after the combiner is 25 ps. Assume $\beta_1 = 0.5 \times 10^{-8}$ s/m and $\beta_2 = 0$ for both fibers.

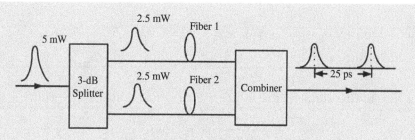

Figure 9.22   Generation of OTDM bit stream.

Solution:
The 3-dB splitter splits the optical pulse into two pulses, each with half power.
The delay occurring in fiber 1 is $\beta_1 L = 0.5 \times 10^{-8} \times 1 \times 10^{-3}$ s = 5 ps.
To have a separation of 25 ps between pulses, the delay in fiber 2 should be 25 ps + 5 ps = 30 ps. Therefore, the length of fiber 2 is given by

$$\frac{\text{delay}}{\beta_1} = \frac{30 \times 10^{-12}}{0.5 \times 10^{-8}} = 6 \text{ mm.} \tag{9.92}$$

Example 9.6

Develop an OTDM to multiplex four 10-Gb/s data streams into a single 40-Gb/s data stream. Assume the same parameters as in Example 9.5.

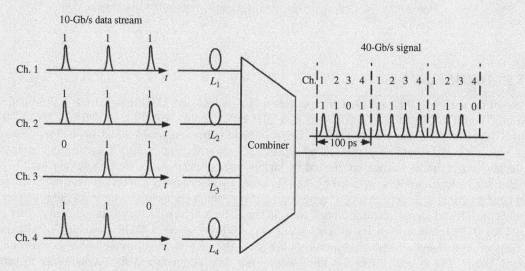

Figure 9.23   Multiplexing four 10-Gb/s data streams into a 40-Gb/s bit stream.

Solution:
The bit interval for a 10-Gb/s signal is equal to

$$\frac{1}{10 \times 10^9} \text{ ps} = 100 \text{ ps}. \tag{9.93}$$

The bit interval for a 40-Gb/s signal is equal to

$$\frac{1}{40 \times 10^9} \text{ ps} = 25 \text{ ps}. \tag{9.94}$$

Fig. 9.23 shows a schematic of an OTDM which multiplexes four 10-Gb/s bit streams into a 40-Gb/s bit stream. To have a delay of 25 ps between Ch. 1 and Ch. 2, $L_2 = L_1 + 25 \text{ ps}/\beta_1$. If $L_1 = 1 \text{ mm}$, $L_2 = 6 \text{ mm}$. Similarly, $L_3 = 11 \text{ mm}$ and $L_4 = 16 \text{ mm}$. Fig. 9.24 shows the pulses of 10-Gb/s channels within a bit interval of the 40-Gb/s signal.

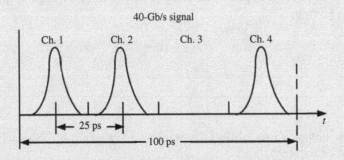

Figure 9.24   40-Gb/s signal obtained by multiplexing four 10-Gb/s signals.

9.5.3   OTDM Experiments

In one of the early OTDM transmission experiments [44], a 100-Gb/s OTDM signal was transmitted over 560 km. The OTDM signal was obtained by multiplexing 16 channels at a bit rate of 6.3 Gb/s. A 40-km normal dispersion fiber was used in the first half of an 80-km fiber span and a 40-km anomalous dispersion fiber was used in the other half so that second-order dispersion was compensated. The transmission distance in the above experiment was mainly limited by higher-order dispersion. As the bit rate increased beyond 40 Gb/s, the performance was degraded by the ISI caused by higher-order dispersion. This problem can be alleviated by using a dispersion slope compensation fiber which compensates for third-order dispersion. A 400-Gb/s OTDM signal was transmitted over 40 km using a dispersion slope compensation fiber [45]. A 1-Tb/s OTDM soliton signal was transmitted over 1000 km using a DMF consisting of alternating sections of normal and anomalous dispersion fiber [46]. The section length in the above experiment was around 10 km. The normal dispersion fiber section not only compensated for second-order dispersion of the anomalous dispersion fiber section, but also for the dispersion slope. OTDM can be combined with WDM to increase the capacity. A six-channel WDM system with each channel consisting of a 170.6-Gb/s OTDM signal was demonstrated over a 2000-km nonzero dispersion fiber using RZ-DPSK format [47].

9.6 Additional Examples

Example 9.7

In a polarization-multiplexed WDM system based on QAM-16, the number of channels = 24 and the symbol rate per polarization = 28 GBaud. Ideal Nyquist pulses are used in each channel. Assuming that the channel spacing Δf is equal to the signal bandwidth f_s in a channel, calculate (a) the total data rate and (b) the spectral efficiency.

Solution:
(a) The data rate of a channel per polarization, $B_s = 28$ GBaud. For QAM-16, we have

$$B = B_s \log_2 16$$
$$= 112 \,\text{Gb/s}. \tag{9.95}$$

Here, B is the data rate of a channel per polarization.

$$\text{Total data rate} = B \times \text{no. of polarizations} \times \text{no. of channels}$$
$$= 112 \times 2 \times 24 \,\text{Gb/s}$$
$$= 5.376 \,\text{Tb/s}. \tag{9.96}$$

(b) For a Nyquist pulse, the signal bandwidth $f_s = B_s = 28$ GHz. With $f_s = \Delta f$, the total WDM signal bandwidth is

$$N\Delta f = 28 \times 24 \,\text{GHz}$$
$$= 672 \,\text{GHz}, \tag{9.97}$$

$$\text{spectral efficiency} = \frac{\text{total data rate}}{\text{total bandwidth}}$$
$$= \frac{5376}{672} \,\text{b/s/Hz}$$
$$= 8 \,\text{b/s/Hz}. \tag{9.98}$$

Example 9.8

In an optical OFDM system, each subcarrier is modulated by QPSK data. The guard interval is 7% of the OFDM symbol period and the carrier orthogonality should be preserved over a transmission distance of at least 5000 km. Find the number of subcarriers required to transmit information at a data rate of about 10 Gb/s. Assume $\beta_2 = -22 \,\text{ps}^2/\text{km}$.

Solution:
Let the symbol rate of a subcarrier be B_s. For QPSK, we have

$$B = B_s \log_2 4 = 2B_s. \tag{9.99}$$

Therefore, the total bit rate is

$$NB = 2NB_s = 10\,\text{Gb/s}, \tag{9.100}$$

where N is the number of subcarriers. Since the frequency separation between the subcarriers $\Delta f = \frac{1}{T_s} = B_s$, Eq. (9.86) may be rewritten as

$$L_{\max} = \frac{T_g}{|\beta_2|(2\pi NB_s)}, \tag{9.101}$$

where $L_{\max}$ is the maximum reach up to which carrier orthogonality is preserved. From Eq. (9.101), we find

$$T_g = 5000 \times 10^3 \times 22 \times 10^{-27} \times 10 \times 10^9 \times \pi\,\text{s}$$

$$= 3.4558\,\text{ns}. \tag{9.102}$$

Since $T_g = 0.07T_s$, $T_s = 49.368$ ns. Using Eq. (9.100), with $T_s = 1/B_s$, we find

$$N = \text{floor}\left\{\frac{10 \times 10^9 \times 49.368 \times 10^{-9}}{2}\right\}$$

$$= 246. \tag{9.103}$$

Example 9.9

In a polarization-multiplexed OFDM system, there are 64 subcarriers and each carrier is modulated by QAM-64. OFDM symbol period = 12.8 ns, launch power to the fiber = 2 dBm, fiber loss = 0.19 dB/km, fiber length = 70 km. Calculate (a) the signal power/subcarrier/polarization at the fiber output, (b) the data rate and (c) the spectral efficiency.

Solution:
(a)

$$\text{Total loss} = 0.19 \times 70 = 13.3\,\text{dB}. \tag{9.104}$$

$$\text{Input power} = 2\,\text{dBm}. \tag{9.105}$$

$$\text{Output power}\ P_{\text{out}}(\text{dBm}) = 2 - 13.3\,\text{dBm}$$

$$= -11.3\,\text{dBm}. \tag{9.106}$$

$$P_{\text{out}}(\text{mW}) = 10^{-11.3/10}\,\text{mW}$$

$$= 0.0741\,\text{mW}. \tag{9.107}$$

$$\text{No. of subcarriers} = 64. \tag{9.108}$$

$$\text{No. of polarizations} = 2. \tag{9.109}$$

$$\text{Signal power/subcarrier/polarization} = \frac{0.0741}{2 \times 64}\,\text{mW}$$

$$= 5.789 \times 10^{-4}\,\text{mW}. \tag{9.110}$$

(b)
$$\text{OFDM symbol period } T_s = 12.8 \times 10^{-9} \text{ s.} \tag{9.111}$$

$$\text{Symbol rate/subcarrier } B_s = \frac{1}{T_s} = 78.125 \text{ MBaud.} \tag{9.112}$$

For QAM-64, we have
$$\text{bit rate } B = \log_2 64 B_s = 468.75 \text{ Mb/s.} \tag{9.113}$$

Since there are 64 subcarriers and two polarizations, the total data rate is

$$B_{\text{tot}} = 468.75 \times 10^6 \times 2 \times 64$$

$$= 60 \text{ Gb/s.} \tag{9.114}$$

(c)
$$\text{Separation between subcarriers } \Delta f = \frac{1}{T_s} = 78.125 \text{ MHz.} \tag{9.115}$$

$$\text{Total bandwidth = no. of subcarriers} \times \Delta f$$

$$= 5 \text{ GHz.} \tag{9.116}$$

$$\text{Spectral efficiency} = \frac{\text{total data rate}}{\text{total bandwidth}}$$

$$= 12 \text{ b/s/Hz.} \tag{9.117}$$

Exercises

9.1 In a polarization-multiplexed WDM system, number of channels = 20, total data rate = 2 Tb/s, and spectral efficiency = 4 b/s/Hz. Calculate the channel spacing.

(Ans: 25 GHz.)

9.2 In a polarization-multiplexed WDM system based on NRZ-OOK, the first null of the NRZ spectrum occurs at $f_0 = 40$ GHz. WDM signal bandwidth = 34.3 nm. The channel spacing = $2.5 f_0$ and spectral efficiency = 0.2 b/s/Hz. Calculate (a) the number of channels and (b) the total data rate.

(Ans: (a) 43, (b) 0.86 Tb/s.)

9.3 A polarization-multiplexed WDM signal is transmitted over a 60-km-long fiber. Number of channels = 20, fiber loss = 0.18 dB/km, channel spacing = 100 GHz, signal in each channel band-limited to 40 GHz. If the total power at the fiber output is −12.8 dBm, find the signal power/channel/polarization at the fiber output.

(Ans: 0.0158 mW.)

9.4 Explain the operating principles of an AWG multiplexer/demultiplexer.

9.5 In a WDM system, an AWG is used to demultiplex two channels. Find the length difference Δl of the adjacent single-mode waveguides such that the corresponding phase-shift difference is 10π. Assume $\beta_0 = 5.8 \times 10^6 \text{ m}^{-1}$.

(Ans: 5.416 μm.)

9.6 Explain the difference between WDM and OFDM systems.

9.7 In a polarization-multiplexed optical OFDM system, there are 256 subcarriers and each subcarrier is modulated by QAM-16 data. OFDM symbol period = 81.92 ns and fiber dispersion $\beta_2 = -22$ ps^2/km. The optical OFDM signal needs to be transmitted over a distance of 1000 km. Find (a) the minimum guard interval to ensure carrier orthogonality and (b) the total data rate.

(Ans: (a) 0.4319 ns, (b) 25 Gb/s.)

9.8 Discuss the significance of the cyclic prefix.

9.9 A polarization-multiplexed OFDM signal is transmitted over a 50-km-long fiber. Total OFDM bandwidth = 2.5 GHz, fiber loss = 0.19 dB/km, OFDM symbol period = 204.8 ns. If the total power at the fiber output is −13 dBm, find the signal power/subcarrier/polarization at the transmitter.

(Ans: 4.362×10^{-4} mW.)

9.10 In a polarization-multiplexed optical OFDM system, there are 128 subcarriers and each carrier is modulated by QAM-16 data. It is desired that the guard interval should not exceed 5% of the OFDM symbol period and the carrier orthogonality should be preserved over a distance of 500 km. Calculate (a) the OFDM symbol period and (b) the spectral efficiency. Assume $\beta_2 = -22$ ps^2/km.

(Ans: (a) 13.302 ns, (b) 8 b/s/Hz.)

9.11 Write a computer program to simulate the polarization-multiplexed OFDM system with the following parameters: total data rate = 28 Gb/s, modulation = QPSK, transmission distance = 1000 km, amplifier spacing = 100 km, fiber loss = 0.18 dB/km, fiber dispersion $\beta_2 = -22$ ps^2/km. The guard interval should not exceed 6% of the OFDM symbol period. Choose the OFDM symbol period such that carrier orthogonality is preserved. Assume that Mach−Zehnder modulators operating in the linear regime are used and ignore fiber nonlinearity and amplifier noise. Plot the OFDM symbol in the time and frequency domain at the fiber-optic link input and at the receiver after DFT.

9.12 Develop an optical TDM scheme to multiplex four 25-Gb/s data streams into a single 100-Gb/s data stream. Explain how the TDM signal can be demultiplexed at the receiver.

References

[1] ITU-T-694.1, Spectral grids for WDM applications: DWDM freq-grid. ITU-T website.
[2] H. Toba *et al.*, *Electron. Lett.*, vol. 23, p. 788, 1987.
[3] K. Oda *et al.*, *IEEE Photon. Technol. Lett.*, vol. 1, p. 137, 1989.
[4] B.H. Verbeek *et al.*, *J. Lightwave Technol.*, vol. 6, p. 1011, 1988.
[5] N. Takato *et al.*, *IEEE J. Select. Top. Comm.*, vol. 8, p. 1120, 1990.
[6] H. Ishio, J. Minowa, and K. Nosu, *J. Lightwave Technol.*, vol. 2, p. 448, 1984.
[7] G.P. Agrawal, *Fiber-Optic Communication System*, 4th edn. John Wiley & Sons, Hoboken, NJ, 2010.
[8] M.K. Smit, *Electron. Lett.*, vol. 29, p. 285, 1988.
[9] A.R. Vellekoop and M.K. Smit, *J. Lightwave Technol.*, vol. 9, p. 310, 1991.
[10] A.R. Vellekoop and M.K. Smit, *J. Lightwave Technol.*, vol. 8, p. 118, 1990.
[11] H. Bissessur *et al.*, *Electron. Lett.*, vol. 30, p. 336, 1994.
[12] L.H. Spiekman *et al.*, *J. Lightwave Technol.*, vol. 14, p. 991, 1996.
[13] N.S. Bergano and C.R. Davidson, *J. Lightwave Technol.*, vol. 14, p. 1299, 1996.

[14] G. Charlet *et al.*, Optical Fiber Communication Conference/National Fiber Optic Engineers Conference, PDP B6, 2009.

[15] J. Yu *et al.*, *European Conference on Optical Communications (ECOC)*, vol. 7–27, Th.3.E2, 2008.

[16] X. Zhou *et al.*, Optical Fiber Communication Conference (OFC), PDP B4, 2009.

[17] W. Shie and C. Athaudage, *Electron. Lett.*, vol. 42, p. 587, 2006.

[18] A.J. Lowery and J. Armstrong, *Opt. Expr.*, vol. 14, p. 2079, 2006.

[19] I.B. Djordjevic and B. Vasic, *Opt. Expr.*, vol. 14, p. 3767, 2006.

[20] A.J. Lowery, S. Wang, and M. Premaratne, *Opt. Expr.*, vol. 15, p. 13,282, 2007.

[21] A.J. Lowery, *Opt. Expr.*, vol. 15, p. 12,965, 2007.

[22] A.J. Lowery, *J. Lightwave Technol.*, vol. 25, p. 131, 2007.

[23] H. Bao and W. Shieh, *Opt. Expr.*, vol. 15, p. 4410, 2007.

[24] W. Shieh, H. Bao, and Y. Tang, *Opt. Expr.*, vol. 16, p. 841, 2008.

[25] S.L. Jansen, I. Morita, T.C.W. Schenk, N. Takeda, and H. Tanaka, *J. Lightwave Technol.*, vol. 26, p. 6, 2008.

[26] L.B. Du and A.J. Lowery, *Opt. Expr.*, vol. 16, p. 19,920, 2008.

[27] W. Shie, Q. Yang, and Y. Ma, *Opt. Expr.*, vol. 16, p. 6378, 2008.

[28] J. Armstrong, *J. Lightwave Technol.*, vol. 27, p. 189, 2009.

[29] Q. Yang, Y. Tang, Y. Ma, and W. Shieh, *J. Lightwave Technol.*, vol. 27, p. 168, 2009.

[30] Q. Yang, S. Chen, Y. Ma, and W. Shieh, *Opt. Expr.*, vol. 17, p. 7985, 2009.

[31] X. Yi, W. Shie, and Y. Ma, *J. Lightwave Technol.*, vol. 26, p. 1309, 2008.

[32] S.L. Jansen, I. Morita, T.C.W. Schenk, and H. Tanaka, *J. Lightwave Technol.*, vol. 27, p. 177, 2009.

[33] X. Yi, W. Shie, and Y. Tang, *IEEE Photon. Technol. Lett.*, vol. 19, p. 919, 2007.

[34] R.W. Chang, Orthogonal frequency multiplex data transmission, US patent 3,488,445, 1966.

[35] J. Salz and S.B. Weinstein, Fourier transform communication system. In Proceedings of ACM Symposium on Probability Optimization in Data Communication Systems, Pine Mountain, GA, 1969.

[36] S.B. Weinstein and P.M. Ebert, Data transmission by FDM using discrete Fourier transform. *IEEE Trans. Comm. Tech.*, vol. COM-19, 1971.

[37] A. Peled and A. Ruiz, Proc. ICASSP 80, Denver, CO, Vol. III, pp. 964–967, 1980.

[38] W. Shieh and I. Djordjevic, *OFDM for Optical Communications*. Academic Press, New York, 2010.

[39] A. Li, X. Chen, and W. Sheih, *J. Lightwave Technol.*, vol. 30, p. 3931, 2012.

[40] H. Takahashi *et al.*, *J. Lightwave Technol.*, vol. 28, p. 406, 2010.

[41] T. Lotz *et al.*, *J. Lightwave Technol.*, vol. 31, p. 538, 2013.

[42] S. Chen, Y. Ma, and W. Shieh, *IEEE Photon. J.*, vol. 2, p. 454, 2010.

[43] D.M. Spirit, A.D. Ellis, and P.E. Barnsley, *IEEE Commun.*, vol. 32, p. 56, 1994.

[44] S. Kawanishi *et al.*, *Electron. Lett.*, vol. 32, p. 470, 1996.

[45] S. Kawanishi *et al.*, *Electron. Lett.*, vol. 32, p. 916, 1996.

[46] H. Anis *et al.*, *European Conference on Optical Communication (ECOC)*, vol. I, p. 230, 1999.

[47] A.H. Gnauck *et al.*, *IEEE Photon. Technol. Lett.*, vol. 15, p. 1618, 2003.

10

Nonlinear Effects in Fibers

10.1 Introduction

So far, we have treated the fiber optic system as a linear system, but it is actually a nonlinear system because the refractive index of the fiber changes with the intensity of signal due to the Kerr and Raman effects. In Section 10.2, the origin of linear and nonlinear refractive indices and the Kerr effect are discussed. Since the change in refractive index due to the Kerr effect translates into a phase shift, the signal phase is modulated by its power distribution, which is known as self-phase modulation (SPM). SPM leads to spectral broadening and the exact balance between dispersion and SPM leads to soliton formation. A soliton is a pulse that propagates without any change in shape over long distances. Sections 10.3–10.6 present the effects of dispersion, SPM, and soliton formation. In WDM systems, several channels co-propagate down the fiber. The phase of a signal in a channel is modulated not only by its channel power, but also by other channels, which is known as cross-phase modulation (XPM). In addition, nonlinear interaction among two or more channels leads to four-wave mixing (FWM), which acts as noise on channels. The impact of XPM and FWM on the system performance of a WDM system is discussed in Section 10.7. In a high-bit-rate highly dispersive single-channel system, signal pulses overlap strongly in the time domain, leading to intra-channel four-wave mixing (IFWM) and intra-channel cross-phase modulation (IXPM). These intrachannel nonlinear effects are discussed in Sections 10.8–10.10. The propagation of a high-intensity optical pulse leads to an instantaneous as well as a delayed change in refractive index. The instantaneous response is responsible for the Kerr effect, while the delayed response is associated with the Raman effect. Section 10.11 is devoted to the stimulated Raman effect, which is responsible for the amplification of a low-frequency signal by a high-frequency intense pump.

10.2 Origin of Linear and Nonlinear Refractive Indices

In a dielectric medium, light travels at a speed lower than that in free space. This can be understood qualitatively as follows. The electric field of the light wave acts on an electron, making it oscillate in accordance with Coulomb's law. An oscillating charge acts as a tiny antenna which radiates electromagnetic radiation at a frequency the same as that of the incident wave in a linear approximation. The newly generated electromagnetic field is the same as the incident field, except for a phase shift. In other words, absorption of the incident field by a molecule and re-radiation delays the propagation of light compared with free-space propagation. The exact determination of the displacement of electrons due to the electric field of a light wave is a complicated problem of quantum mechanics. Instead, we use a classical electron oscillator model in which the electron is modeled as a charged cloud surrounding the nucleus, as shown in Fig. 10.1(a) [1].

Fiber Optic Communications: Fundamentals and Applications, First Edition. Shiva Kumar and M. Jamal Deen.
© 2014 John Wiley & Sons, Ltd. Published 2014 by John Wiley & Sons, Ltd.

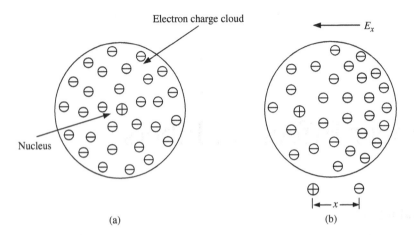

Figure 10.1   Classical electron oscillator model: (a) in equilibrium; (b) in the presence of an external field.

When an external electric field intensity E_x is applied to an atom, the electron charge cloud is displaced from its equilibrium position, as shown in Fig. 10.1(b). The equation of motion for the center of the electron charge cloud is given by Newton's law,

$$m\frac{d^2x}{dt^2} = F_{\text{ext}} = q_e E_x, \tag{10.1}$$

where $x(t)$ is the displacement of the center of the electron charge cloud, m is the electron mass, and q_e is the electron charge. When the center of the electron cloud moves away from the equilibrium position (Fig. 10.1(b)), there is a force of attraction between the nucleus and the electron charge cloud. If the displacement $x(t)$ is small, the restoration force can be approximated as

$$F_{\text{restoration}} = -Kx, \tag{10.2}$$

where K is a constant. The negative sign indicates that the restoration force acts in a direction opposite to the external force. The situation is similar to the case of a simple pendulum pushed away from the equilibrium position by an external force; there is a restoration force due to gravitation which pulls it back to the equilibrium position. The net force acting on the electron is given by

$$F_{\text{net}} = F_{\text{ext}} + F_{\text{restoration}} = q_e E_x - Kx. \tag{10.3}$$

Combining Eq. (10.3) with Newton's law, we obtain

$$m\frac{d^2x}{dt^2} = F_{\text{net}} = q_e E_x - Kx \tag{10.4}$$

or

$$\frac{d^2x}{dt^2} + \omega_0^2 x = \left(\frac{q_e}{m}\right)E_x, \tag{10.5}$$

where $\omega_0 = (K/m)^{1/2}$ is the natural frequency of oscillation. Suppose the applied field is of the form

$$E_x = E_0 \exp(-i\omega t). \tag{10.6}$$

We expect that the displacement $x(t)$ should also change harmonically in the steady state and try a trial solution

$$x(t) = B \exp(-i\omega t). \tag{10.7}$$

Substituting Eqs. (10.7) and (10.6) into Eq. (10.5), we obtain

$$B = \frac{E_0 q_e}{m(\omega_0^2 - \omega^2)} \tag{10.8}$$

and

$$x(t) = \frac{q_e}{m(\omega_0^2 - \omega^2)} E_x. \tag{10.9}$$

The *dipole moment* of an atom is defined as

$$p_x = q_e x = \frac{q_e^2}{m(\omega_0^2 - \omega^2)} E_x. \tag{10.10}$$

In general,

$$\mathbf{p} = \frac{q_e^2}{m(\omega_0^2 - \omega^2)} \mathbf{E}. \tag{10.11}$$

Our next step is to determine the electromagnetic field generated by the oscillating electron charge cloud. Each atom acts as a current source since the oscillating electron cloud can be imagined as a tiny current element. One of Maxwell's equations in the presence of a current source is (Eq. (1.47))

$$\nabla \times \mathbf{H} = \mathbf{J} + \epsilon_0 \frac{\partial \mathbf{E}}{\partial t}. \tag{10.12}$$

Consider an incremental volume $dV = A dx$ of an atomic system as shown in Fig. 10.2. Using Eq. (10.10), the current I is given by

$$I = \frac{dq}{dt} = \frac{dq}{dx}\frac{dx}{dt} = \frac{1}{q_e}\frac{dq}{dx}\frac{dp_x}{dt}. \tag{10.13}$$

Let N be the number of atoms per unit volume. The charge in volume dV is

$$dq = q_e N \, dV = q_e N \, A dx \tag{10.14}$$

or

$$\frac{dq}{dx} = q_e NA. \tag{10.15}$$

Using Eq. (10.15) in Eq. (10.13), we obtain

$$I = NA\frac{dp_x}{dt}. \tag{10.16}$$

Since $J_x = I/A$, we obtain

$$J_x = N\frac{dp_x}{dt}. \tag{10.17}$$

In general,

$$\mathbf{J} = N\frac{d\mathbf{p}}{dt}. \tag{10.18}$$

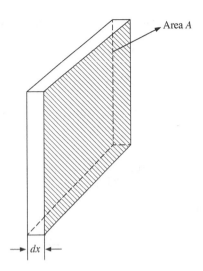

Figure 10.2   A slice of atomic system consisting of N atoms per unit volume.

If we define $\mathbf{P} = N\mathbf{p}$ as the *polarization*, Eq. (10.18) can be written as

$$\mathbf{J} = \frac{d\mathbf{P}}{dt}. \tag{10.19}$$

From Eq. (10.11), it follows that

$$\mathbf{P} = \frac{Nq_e^2}{m(\omega_0^2 - \omega^2)}\mathbf{E}. \tag{10.20}$$

Polarization is directly proportional to electric field intensity, and is often written as

$$\mathbf{P} = \epsilon_0 \chi^{(1)}\mathbf{E}, \tag{10.21}$$

where $\epsilon_0\chi^{(1)}$ is known as the *first-order susceptibilty* or linear susceptibility. Comparing Eqs. (10.20) and (10.21), we find

$$\chi^{(1)} = \frac{Nq_e^2}{m(\omega_0^2 - \omega^2)\epsilon_0}. \tag{10.22}$$

Substituting Eq. (10.19) in Eq. (10.12), we find

$$\nabla \times \mathbf{H} = \frac{\partial(\epsilon_0 \mathbf{E} + \mathbf{P})}{\partial t}. \tag{10.23}$$

If we define the electric flux density D as

$$\mathbf{D} = \epsilon_0 \mathbf{E} + \mathbf{P}, \tag{10.24}$$

Eq. (10.23) becomes

$$\nabla \times \mathbf{H} = \frac{\partial \mathbf{D}}{\partial t}, \tag{10.25}$$

which is the same as Maxwell's equation in a medium (Eq. (1.51)) in the absence of current source. In fact, the induced current and charge due to applied electromagnetic field are taken into account by using electric flux density $\mathbf{D}$ instead of $\epsilon_0 \mathbf{E}$ in Maxwell's equations for free space. Substituting Eq. (10.21) in Eq. (10.24), we obtain

$$\mathbf{D} = \epsilon_0 \left[1 + \chi^{(1)} \right] \mathbf{E}. \tag{10.26}$$

In Section 1.2, we defined the electric flux density as

$$\mathbf{D} = \epsilon_0 \epsilon_r \mathbf{E}. \tag{10.27}$$

Therefore, the relative permittivity is

$$\epsilon_r = 1 + \chi^{(1)}. \tag{10.28}$$

Since the relative permittivity and refractive index are related by $n^2 = \epsilon_r$, we obtain the result

$$n^2 = 1 + \chi^{(1)} = 1 + \frac{N q_e^2}{m \epsilon_0 (\omega_0^2 - \omega^2)}. \tag{10.29}$$

From Eq. (10.29), we see that the refractive index is dependent on the frequency of the incident electromagnetic signal. This is known as chromatic dispersion. In free space, $n = 1$ for all frequencies and the medium is not dispersive. In a dispersive medium, suppose $\omega < \omega_0$. As ω increases (with $\omega < \omega_0$), the denominator of Eq. (10.29) decreases and the refractive index increases with frequency. This explains why a prism bends light more at the violet end than the red end of the visible spectrum.

10.2.1 Absorption and Amplification

From Eq. (10.9), when $\omega_0 = \omega$, we see that the displacement becomes infinite, which is unphysical. This is because we ignored the loss effects. The situation is similar to that of an oscillating simple pendulum in which the oscillations decrease with time due to frictional forces that carry away the energy. In the case of an electron cloud, its vibration leads to the generation of electromagnetic waves that carry away the energy and as a result, the vibration is damped. There are also other reasons for the dissipation of energy, such as collision between atoms. Mathematical modeling of the vibrating electron cloud should include a damping force of the form

$$F_{\text{damp}} = -r \frac{dx}{dt}, \tag{10.30}$$

to account for energy dissipation. Here, r is the damping coefficient. When this force is included, the equation of motion (Eq. (10.5)) is modified as

$$m \frac{d^2 x}{dt^2} + r \frac{dx}{dt} + Kx = q_e E_x, \tag{10.31}$$

Assuming that the applied electric field intensity is of the form given by Eq. (10.6) and proceeding as before, the expression for the displacement is

$$x(t) = \frac{q_e E_x}{m(\omega_0^2 - \omega^2 - ir\omega/m)}. \tag{10.32}$$

Using Eq. (10.32), the polarization becomes

$$\mathbf{P} = \frac{N q_e^2 \mathbf{E}}{m(\omega_0^2 - \omega^2 - ir\omega/m)} \tag{10.33}$$

and

$$n^2 = 1 + \frac{Nq_e^2}{m(\omega_0^2 - \omega^2 - ir\omega/m)} \tag{10.34}$$

$$= (n_r + in_i)^2, \tag{10.35}$$

where $n_r = \text{Re}(n)$ and $n_i = \text{Im}(n)$. From Eq. (10.34), we see that the refractive index becomes complex in the presence of damping force. The forward-propagating plane wave solution takes the form (Eq. (1.95))

$$E_x = E_0 \exp\left[-\alpha z/2 - i(\omega t - \omega n_r z/c)\right], \tag{10.36}$$

where

$$\alpha = \frac{2\omega n_i}{c}. \tag{10.37}$$

When atoms absorb electromagnetic energy (which is used to increase the internal energy of the atomic system), $\alpha > 0$ and the incident electromagnetic signal is attenuated. If the atoms transfer energy to the electromagnetic signal due to a mechanism such as stimulated emission (see Chapter 3), α becomes negative and the incident electromagnetic signal is amplified.

10.2.2 Nonlinear Susceptibility

So far we have assumed that the restoration force is proportional to the displacement of the electron cloud (Eq. (10.2)), which leads to harmonic oscillation of the electron cloud when the incident electromagnetic field is harmonic (Eq. (10.6)). This holds true when the incident field is weak. As the incident electromagnetic field becomes intense, the assumption of the linear dependence of the restoration force on displacement breaks down and in this case, electron cloud oscillation is not harmonic. As a result, the generated electromagnetic field is also not harmonic. When the incident electromagnetic field is weak, we have found that polarization is directly proportional to incident electric field intensity (Eq. (10.21)),

$$\mathbf{P} = \epsilon_0 \chi^{(1)} \mathbf{E}. \tag{10.38}$$

If the medium is not isotropic, the susceptibility depends on direction as well and Eq. (10.38) is modified as

$$P_j = \chi_{jx}^{(1)} E_x + \chi_{jy}^{(1)} E_y + \chi_{jz}^{(1)} E_z, \quad j = x, y, z \tag{10.39}$$

or

$$\mathbf{P} = \epsilon_0 \chi^{(1)} . \mathbf{E}. \tag{10.40}$$

Here, $\chi^{(1)}$ is a 3×3 matrix and . denotes matrix multiplication. However, as the incident field becomes strong, the linear dependence does not hold true and $\mathbf{P}$ becomes a function of $\mathbf{E}$. In general, $\mathbf{P}$ can be expanded in terms of increasing powers of $\mathbf{E}$,

$$\mathbf{P} = \epsilon_0 \chi^{(1)} \cdot \mathbf{E} + \epsilon_0 \chi^{(2)} : \mathbf{EE} + \epsilon_0 \chi^{(3)} \vdots \mathbf{EEE} + \dots , \tag{10.41}$$

where $\chi^{(j)}$ is the jth-order susceptibility and is a tensor of rank $j + 1$. The first-order susceptibility $\chi^{(1)}$ is related to the linear refractive index (Eq. (10.29)). The second-order susceptibility $\chi^{(2)}$ is responsible for second harmonic generation; if the incident optical wave is sinusoidal of frequency ω, a new optical wave of frequency 2ω is generated. Anharmonic oscillation of the electron cloud due to intense electromagnetic field can be expanded as a Fourier series with frequency components $\omega, 2\omega, \dots , n\omega$, and electron cloud oscillations at frequency 2ω lead to the generation of an electromagnetic wave at 2ω. Crystals such as quarts have no center of symmetry and, therefore, they have a nonzero $\chi^{(2)}$ coefficient. In the case of optical glass fibers, SiO_2 (silica) is a symmetric molecule and $\chi^{(2)}$ is zero. Therefore, second harmonic generation does not normally occur in optical fibers. The third-order susceptibility $\chi^{(3)}$ is responsible for third harmonic generation and the Kerr effect.

Suppose the incident electromagnetic field has only E_x and H_y components. For a centrally symmetric dielectric material, the tensor equation (10.41) can be simplified to obtain

$$P_x = \epsilon_0 \chi_{xx}^{(1)} E_x + \epsilon_0 \chi_{xxxx}^{(3)} E_x^3, \tag{10.42}$$

where $\chi_{xxxx}^{(3)}$ is a component of the fourth-rank tensor $\chi^{(3)}$. Suppose the incident optical field is a monochromatic wave,

$$E_x = E_0 \exp(-i\omega t). \tag{10.43}$$

To find E_x^3, as pointed out in Section 1.6.2, we should first take the real part of E_x,

$$\text{Re}[E_x] = \frac{1}{2}[E_0 \exp(-i\omega t) + E_0^* \exp(i\omega t)], \tag{10.44}$$

$$\begin{aligned}\{\text{Re}[E_x]\}^3 = \frac{1}{8}\{E_0^3 \exp(-3i\omega t) + E_0^{*3} \exp(3i\omega t) \\ + 3|E_0|^2[E_0 \exp(-i\omega t) + E_0^* \exp(i\omega t)]\}.\end{aligned} \tag{10.45}$$

From Eqs. (10.42) and (10.45), we find that the incident field oscillating at frequency ω leads to a component of polarization oscillating at frequency 3ω, which is responsible for third harmonic generation. The electromagnetic wave at frequency 3ω becomes significant only when special phase-matching techniques are used. Otherwise, the component of polarization at frequency 3ω can be ignored. Hence, we ignore the first two terms on the right-hand side of Eq. (10.45).

Let the polarization at frequency ω be

$$P_x = P_0 \exp(-i\omega t), \tag{10.46}$$

$$\text{Re}[P_x] = \frac{1}{2}[P_0 \exp(-i\omega t) + P_0^* \exp(i\omega t)]. \tag{10.47}$$

From Eq. (10.42), we have

$$\text{Re}[P_x] = \epsilon_0 \chi_{xx}^{(1)} \text{Re}[E_x] + \epsilon_0 \chi_{xxxx}^{(3)} \text{Re}[E_x]^3, \tag{10.48}$$

where the imaginary parts of the susceptibility are ignored. Substituting Eqs. (10.44) and (10.45) into Eq. (10.48), collecting the terms that are proportional to $\exp(-i\omega t)$, and comparing it with Eq. (10.47), we obtain

$$P_0 = \epsilon_0 \left(\chi_{xx}^{(1)} + \frac{3|E_0|^2}{4} \chi_{xxxx}^{(3)} \right) E_0 = \epsilon_0 \chi_{\text{eff}} E_0, \tag{10.49}$$

where χ_{eff} is the effective susceptibility that includes both linear and nonlinear susceptibilities. From Eq. (10.29), in the absence of nonlinearity, we have $n^2 = 1 + \chi_{xx}^{(1)}$. Now, we modify it as

$$n^2 = 1 + \chi_{\text{eff}} = 1 + \chi_{xx}^{(1)} + \frac{3|E_0|^2}{4} \chi_{xxxx}^{(3)} \tag{10.50}$$

$$= n_0^2 + \frac{3|E_0|^2}{4} \chi_{xxxx}^{(3)}, \tag{10.51}$$

where n_0 is the linear refractive index and the second term of Eq. (10.51) represents the nonlinear contribution to the refractive index. Typically, the nonlinear part of the refractive index is much smaller than the linear part. From Eq. (10.51), we have

$$n = n_0 \left(1 + \frac{3|E_0|^2}{4n_0^2} \chi_{xxxx}^{(3)} \right)^{1/2}$$

$$\cong n_0 + n_2|E_0|^2, \tag{10.52}$$

where

$$n_2 = \frac{3\chi_{xxxx}^{(3)}}{8n_0} \tag{10.53}$$

is called the *Kerr coefficient*. In Eq. (10.52), we have used the following approximation:

$$(1+x)^{1/2} \approx 1 + x/2, \quad \text{if } x \ll 1, \tag{10.54}$$

which is valid since the nonlinear part of the refractive index is much smaller than its linear part. From Eq. (10.52), we see that the change in refractive index $(n - n_0)$ is directly proportional to the optical intensity $|E_0|^2$. This effect is called the *Kerr effect*.

For silica, $n_2 \approx 3 \times 10^{-20}\,\text{m}^2/\text{W}$. If a light beam of intensity $1\,\text{W/m}^2$ is incident on silica media of cross-sectional area $1\,\text{m}^2$, the change in refractive index is 3×10^{-20}, which is very small. However, silica fiber has an effective cross-sectional area of $100\,\mu\text{m}^2$ or less, and the change in refractive index due to the Kerr effect is comparable with the variations in refractive index due to dispersion, leading to interesting nonlinear phenomena such as soliton formation.

10.3 Fiber Dispersion

As mentioned in Chapter 2, a pulse propagating in a fiber broadens due to fiber dispersion. When the fiber nonlinear effects are ignored, the complex field envelope in a field is given by Eqs. (2.119) and (2.107) as

$$\tilde{q}(\omega, z) = \tilde{q}(\omega, 0)e^{-\alpha z/2 + i\beta_1\omega z + i\beta_2\omega^2 z/2}. \tag{10.55}$$

Differentiating Eq. (10.55) with respect to z, we find

$$\frac{\partial \tilde{q}(\omega, z)}{\partial z} = \tilde{q}(\omega, z)\left(-\frac{\alpha}{2} + i\beta_1\omega + i\beta_2\omega^2 \right). \tag{10.56}$$

Since

$$\mathcal{F}^{-1}\{(-i\omega)^n \tilde{q}(\omega, z)\} = \frac{\partial^n q(t, z)}{\partial t^n}, \tag{10.57}$$

taking the inverse Fourier transform of Eq. (10.56), we find

$$\frac{\partial q(t, z)}{\partial z} = -\frac{\alpha}{2}q(t, z) - \beta_1\frac{\partial q(t, z)}{\partial t} - \frac{i\beta_2}{2}\frac{\partial^2 q(t, z)}{\partial t^2}. \tag{10.58}$$

As mentioned in Chapter 2, the term $\exp(i\beta_1\omega z)$ of Eq. (10.55) introduces a constant time shift due to propagation and it could be dropped as we are primarily interested in assessing the quality of the signal at the fiber output. Let

$$Z = z, \tag{10.59}$$

$$T = t - \beta_1 z. \tag{10.60}$$

Note that (t, z) are the coordinates of an optical impulse with the origin $(0, 0)$ at the transmitter. If a receiver is placed at $Z = z$, T denotes the time at the receiver if the receiver clock is shifted from the transmitter clock by the time of flight $\beta_1 z$. Since Z and T are functions of z and t, we have

$$\frac{\partial T}{\partial t} = 1, \quad \frac{\partial T}{\partial z} = -\beta_1, \tag{10.61}$$

$$\frac{\partial Z}{\partial t} = 0, \quad \frac{\partial Z}{\partial z} = 1, \tag{10.62}$$

$$\frac{\partial q}{\partial z} = \frac{\partial q}{\partial Z}\frac{\partial Z}{\partial z} + \frac{\partial q}{\partial T}\frac{\partial T}{\partial z}$$

$$= \frac{\partial q}{\partial Z} \cdot 1 + \frac{\partial q}{\partial T}(-\beta_1), \tag{10.63}$$

$$\frac{\partial q}{\partial t} = \frac{\partial q}{\partial Z}\frac{\partial Z}{\partial t} + \frac{\partial q}{\partial T}\frac{\partial T}{\partial t}$$

$$= \frac{\partial q}{\partial T} \cdot 1, \tag{10.64}$$

$$\frac{\partial^2 q}{\partial t^2} = \frac{\partial}{\partial Z}\left(\frac{\partial q}{\partial t}\right)\frac{\partial Z}{\partial t} + \frac{\partial}{\partial T}\left(\frac{\partial q}{\partial T}\right)\frac{\partial T}{\partial t}$$

$$= \frac{\partial^2 q}{\partial T^2}. \tag{10.65}$$

Substituting Eqs. (10.63)–(10.65) in Eq. (10.58), we find

$$\frac{\partial q(T, Z)}{\partial Z} + \frac{\partial q(T, Z)}{\partial T}(-\beta_1) = \frac{-\alpha}{2}q(T, Z) - \beta_1\frac{\partial q(T, Z)}{\partial T} - i\frac{\beta_2}{2}\frac{\partial^2 q(T, Z)}{\partial T^2}$$

or

$$i\frac{\partial q}{\partial Z} - \frac{\beta_2}{2}\frac{\partial^2 q}{\partial T^2} = -\frac{i\alpha q}{2}. \tag{10.66}$$

Eq. (10.66) describes the propagation of the optical field envelope in a fiber when the nonlinear effects are ignored. Eq. (10.66) is equivalent to Eq. (10.56). For a Gaussian input, the output electric field envelope is given by Eq. (2.158),

$$q(T, Z) = \frac{\sqrt{P_0}T_0}{T_1}\exp\left(-\frac{T^2}{2T_1^2}\right), \tag{10.67}$$

$$T_1 = (T_0^2 - i\beta_2 Z)^{1/2}. \tag{10.68}$$

Let

$$T_1 = |T_1|\exp(i\theta_1), \tag{10.69}$$

where

$$|T_1|^2 = (T_0^4 + \beta_2^2 Z^2)^{1/2}, \tag{10.70}$$

$$\theta_1 = -\frac{1}{2}\tan^{-1}\left(\frac{\beta_2 Z}{T_0^2}\right). \tag{10.71}$$

Eq. (10.67) may be rewritten as

$$q(T, Z) = A(Z, T)e^{i\theta(T,Z)}$$

$$= \frac{\sqrt{P_0}T_0}{|T_1|e^{i\theta_1}} \exp\left[-\frac{T^2\left(T_0^2 + i\beta_2 Z\right)}{2(T_0^4 + \beta_2^2 Z^2)}\right], \tag{10.72}$$

$$A(T, Z) = \frac{\sqrt{P_0}T_0}{|T_1|} \exp\left[-\frac{T^2 T_0^2}{2\left(T_0^4 + \beta_2^2 Z^2\right)}\right], \tag{10.73}$$

$$\theta(T, Z) = -\theta_1(Z) - \frac{T^2 \beta_2 Z}{2(T_0^4 + \beta_2^2 Z^2)}, \tag{10.74}$$

The instantaneous power is

$$P(T, Z) = A^2(T, Z) = \frac{P_0 T_0^2}{|T_1|^2} \exp\left[-\frac{T^2 T_0^2}{|T_1|^4}\right], \tag{10.75}$$

and the instantaneous frequency is (see Eq. (2.165))

$$\delta\omega(T) = -\frac{\partial\theta}{\partial T} = \frac{T\beta_2 Z}{(T_0^4 + \beta_2^2 Z^2)}. \tag{10.76}$$

The negative sign is chosen because the carrier wave is $\exp(-i\omega_0 t)$. The actual instantaneous frequency is $\omega_0 + \delta\omega$.

The instantaneous power $P(T, Z)$ and instantaneous frequency $\delta\omega$ are plotted in Fig. 2.32 for an anomalous dispersion fiber ($\beta_2 < 0$). From Eq. (10.76) and Fig. 10.3, when $\beta_2 < 0$, we see that near the leading edge ($T < 0$) $\delta\omega$ is positive (blue shift) whereas it is negative (red shift) near the trailing edge ($T > 0$). These changes in frequency occur continuously as the signal propagates down the fiber. Since the blue components travel faster than the red components in the anomalous dispersion fiber, the frequency components near the leading edge arrive early and the frequency components near the trailing edge arrive late. This explains why the pulse is broadened at the fiber output. For a fiber with normal dispersion, the situation is exactly opposite.

10.4   Nonlinear Schrödinger Equation

Owing to the Kerr effect, an optical signal undergoes a phase shift that is proportional to the signal power as given by Eq. (10.52). If this nonlinear effect is included, Eq. (10.58) is modified as (see Appendix B)

$$i\left(\frac{\partial q}{\partial z} + \beta_1 \frac{\partial q}{\partial t}\right) - \frac{\beta_2}{2}\frac{\partial^2 q}{\partial t^2} + \gamma|q|^2 q = -i\frac{\alpha q}{2}, \tag{10.77}$$

where γ is the nonlinear coefficient related to the Kerr coefficient n_2 by (Appendix B)

$$\gamma = \frac{n_2 \omega_0}{cA_{\text{eff}}}, \tag{10.78}$$

where A_{eff} is the effective area of the fiber mode. Using the frame of reference that moves with the group speed of the pulse,

$$T = t - \beta_1 z, \tag{10.79}$$

$$Z = z, \tag{10.80}$$

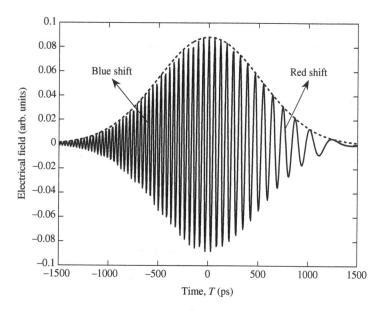

Figure 10.3 The electric field distribution at the fiber output. The broken line shows the field envelope and the rapid oscillation shows the actual field. $\beta_2 < 0$.

Eq. (10.77) becomes

$$i\frac{\partial q}{\partial Z} - \frac{\beta_2}{2}\frac{\partial^2 q}{\partial T^2} + \gamma|q|^2 q = -i\frac{\alpha q}{2}. \tag{10.81}$$

Eq. (10.81) is known as the *nonlinear Schrödinger equation* (NLSE); it is of significant importance in modeling fiber-optic transmission systems and can not be solved analytically for arbitrary inputs. Numerical techniques such as the split-step Fourier scheme (SSFS) are used to solve the NLSE (see Chapter 11).

Example 10.1

The Kerr coefficient of a single-mode fiber is 2.5×10^{-20} m^2/W. Its effective area is $80 \,\mu$m^2. Find the nonlinear coefficient γ at the wavelength 1550 nm.

Solution:
From Eq. (10.78), we have

$$\gamma = \frac{n_2 \omega_0}{cA_{\text{eff}}} = \frac{n_2 2\pi f_0}{cA_{\text{eff}}} = \frac{2\pi n_2}{\lambda_0 A_{\text{eff}}}, \tag{10.82}$$

$$\gamma = \frac{2\pi \times 2.5 \times 10^{-20}}{1550 \times 10^{-9} \times 80 \times 10^{-12}} \,\text{W}^{-1}\,\text{m}^{-1}$$

$$= 1.26 \times 10^{-3} \,\text{W}^{-1}\,\text{m}^{-1}. \tag{10.83}$$

10.5 Self-Phase Modulation

To find the impact of fiber nonlinearity acting alone, let us ignore β_2. To solve Eq. (10.81) under this condition, we separate the amplitude and phase,

$$q = Ae^{i\theta}. \tag{10.84}$$

Substituting Eq. (10.84) into Eq. (10.81), we find

$$i\left[\frac{\partial A}{\partial Z} + i\frac{\partial \theta}{\partial Z}A\right] = -\left[\gamma A^2 + i\frac{\alpha}{2}\right]A. \tag{10.85}$$

Separating real and imaginary parts, we obtain

$$\frac{\partial A}{\partial Z} = -\frac{\alpha}{2}A \tag{10.86}$$

or

$$A(T,Z) = A(T,0)\exp\left(-\alpha Z/2\right), \tag{10.87}$$

$$\frac{\partial \theta}{\partial Z} = \gamma A^2(T,Z) = \gamma A^2(T,0)e^{-\alpha Z}. \tag{10.88}$$

Let the fiber length be L. Integrating Eq. (10.88) from 0 to L, we obtain

$$\theta(T,L) = \theta(T,0) + \gamma A^2(T,0)\int_0^L e^{-\alpha Z}dZ$$

$$= \theta(T,0) + \gamma A^2(T,0)L_{\text{eff}}, \tag{10.89}$$

where

$$L_{\text{eff}} = \frac{1 - \exp\left(-\alpha L\right)}{\alpha}. \tag{10.90}$$

Substituting Eqs. (10.87) and (10.89) into Eq. (10.84), we find

$$q(T,L) = A(T,L)e^{i\theta(T,L)}$$

$$= A(T,0)e^{-\alpha L/2}e^{i[\theta(T,0)+\gamma A^2(T,0)L_{\text{eff}}]}$$

$$= q(T,0)e^{-\alpha L/2 + i\gamma|q(T,0)|^2 L_{\text{eff}}}, \tag{10.91}$$

where

$$q(T,0) = A(T,0)e^{i\theta(T,0)}. \tag{10.92}$$

Here, $|q(T,0)|^2$ represents the instantaneous power at the input. Since the phase of the optical signal is modulated by its own power distribution, this effect is known as *self-phase modulation* (SPM). From Eq. (10.91), we find that

$$|q(T,L)| = |q(T,0)|e^{-\alpha L/2}. \tag{10.93}$$

So, the amplitude of the signal decreases exponentially with distance, but the pulse width at the fiber output remains the same as that at the fiber input. However, the spectral width at the output is larger than that at the input. This is because the nonlinear mixing of the input frequency components due to SPM generates new frequency components. Using Eq. (10.89), the instantaneous frequency at L is

$$\delta\omega(T,L) = -\frac{d\theta(T,L)}{dT} = -\frac{d\theta(T,0)}{dT} - \gamma L_{\text{eff}}\frac{d|q(T,0)|^2}{dT}.$$

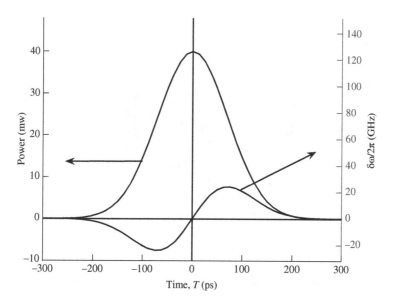

Figure 10.4   Power and instantaneous frequency at the fiber output.

Suppose the input field envelope is given by

$$q(T,0) = \sqrt{P}\exp\left(-T^2/2T_0^2\right). \tag{10.94}$$

The instantaneous frequency at L is

$$\delta\omega(T) = \frac{2\gamma P L_{\text{eff}} T}{T_0^2}\exp\left(-\frac{T^2}{T_0^2}\right). \tag{10.95}$$

The instantaneous power and frequency at the fiber output are shown in Fig. 10.4. The instantaneous frequency is negative (or less than the carrier frequency) near the leading edge, whereas it is positive near the trailing edge. In other words, it is down-shifted in frequency (red shift) near the leading edge and up-shifted (blue shift) near the trailing edge, as shown in Fig. 10.5.

Example 10.2

In a 1000-km fiber-optic link, it is desired that the peak nonlinear phase shift accumulated over the link should be less than 0.5 rad. The system has the following parameters: loss coefficient $\alpha = 0.046\,\text{km}^{-1}$, amplifier spacing $= 100\,\text{km}$, Kerr coefficient $n_2 = 2.5 \times 10^{-20}\,\text{m}^2/\text{W}$, $\lambda_0 = 1550\,\text{nm}$, and peak power at the fiber input $= 0\,\text{dBm}$. Find the lower limit on the effective area of the fiber. Ignore β_2.

Solution:

The peak nonlinear phase shift accumulated over a single span is given by Eq. (10.91),

$$\varphi_{\text{NL}} = \gamma L_{\text{eff}} P_{\text{peak}}, \tag{10.96}$$

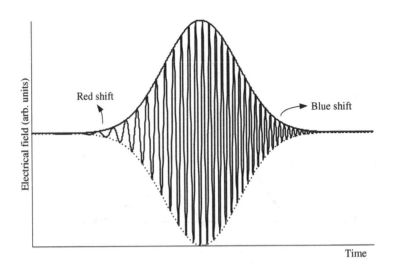

Figure 10.5 Instantaneous electric field at the fiber output.

where P_{peak} is the peak power and

$$L_{\text{eff}} = \frac{1 - \exp(-\alpha L)}{\alpha}, \tag{10.97}$$

$$\alpha = 0.046\,\text{km}^{-1}, \tag{10.98}$$

$$L = 100\,\text{km}. \tag{10.99}$$

$$L_{\text{eff}} = 21.52\,\text{km}. \tag{10.100}$$

$$\text{No. of spans} = \frac{\text{total length}}{\text{amp. spacing}}$$

$$= \frac{1000\,\text{km}}{100\,\text{km}}$$

$$= 10. \tag{10.101}$$

$$\text{Total nonlinear phase shift} = 10\,\varphi_{\text{NL}}. \tag{10.102}$$

$$P_{\text{peak}}(\text{dBm}) = 0\,\text{dBm}, \tag{10.103}$$

$$P_{\text{peak}} = 10^{P_{\text{peak}}(\text{dBm})/10}\,\text{mW}$$

$$= 1\,\text{mW}, \tag{10.104}$$

$$10\,\varphi_{\text{NL}} < 0.5, \tag{10.105}$$

$$10 \times \gamma \times 21.52 \times 10^3 \times 1 \times 10^{-3} < 0.5,$$

$$\gamma < 2.32 \times 10^{-3}\,\text{W}^{-1}\,\text{m}^{-1}.$$

From Eq. (10.78), we have

$$\gamma = \frac{n_2 \omega_0}{c A_{\text{eff}}} = \frac{2\pi n_2}{\lambda_0 A_{\text{eff}}} < 2.32 \times 10^{-3} \, \text{W}^{-1} \, \text{m}^{-1}, \tag{10.106}$$

$$A_{\text{eff}} > \frac{2\pi n_2}{\lambda_0 \times 2.32 \times 10^{-3}} \, \text{m}^2, \tag{10.107}$$

$$A_{\text{eff}} > 43.61 \, \mu\text{m}^2. \tag{10.108}$$

The effective area should be greater than 43.61 μm^2 to have the peak nonlinear phase shift less than or equal to 0.5 rad.

10.6   Combined Effect of Dispersion and SPM

First let us consider the case of a normal dispersion fiber. Fig. 10.6(a) shows the optical field at the fiber input. Owing to SPM acting down, the instantaneous frequency near the trailing edge is higher than that near the leading edge (see Fig. 10.5). Since the high-frequency (blue) components travel slower than the low-frequency (red) components in a normal dispersion fiber, the trailing edge arrives late while the leading edge arrives early at the fiber output. In other words, the combined effect of SPM and normal dispersion is to cause the pulse broadening as shown in Fig. 10.6(b).

Next, let us consider the case of anomalous dispersion. Owing to SPM acting alone (see Fig. 10.5), the leading edge is red-shifted (lower frequency) while the trailing edge is blue-shifted (higher frequency). Since the high-frequency components travel faster than the low-frequency components in an anomalous dispersion fiber, the trailing edge arrives early whereas the leading edge arrives later, causing pulse compression. Comparing Figs. 10.3 and 10.5, we see that the instantaneous frequency due to SPM and that due to anomalous dispersion are of opposite sign. For a specific pulse shape and power level, we might expect that these frequency shifts cancel exactly. Under this condition, the instantaneous frequency (relative to the carrier frequency) across the pulse is zero (or a constant) and therefore, all parts of the pulse travel at the same speed,

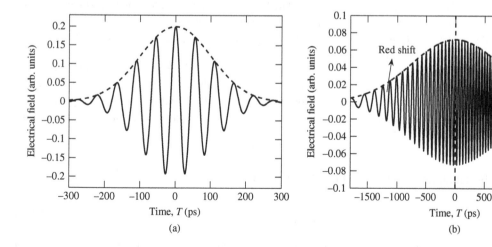

Figure 10.6   Electric field intensity at (a) fiber input and (b) fiber output. $\beta_2 > 0$.

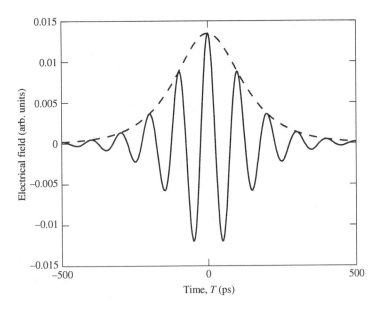

Figure 10.7   Electric field intensity of a soliton. Solid and broken lines show the carrier and the field envelope, respectively.

which implies that there is no change in pulse shape as it propagates down the fiber. Such a pulse is called a *soliton*. A pulse can be called a soliton if and only if (i) it preserves its shape and (ii) when it collides with another soliton or a pulse it comes out unscathed except for a phase shift. In this sense, the soliton mimics the massive particles. The broken line in Fig. 10.7 shows the field envelope corresponding to a soliton. This pulse shape does not change as the pulse propagates down a loss-less fiber. The instantaneous frequency across the pulse remains constant as a function of T as well as Z.

A soliton is a normal mode of a special class of nonlinear systems which can be integrated by means of an inverse scattering transform (IST) [2–6]. IST plays the role of a Fourier transform in a nonlinear system, and sometimes it is called the nonlinear Fourier transform. The NLS equation is solved using an inverse scattering transform to obtain soliton and breather solutions. The breathers or higher-order solitons undergo periodic compression and expansion with a period which is known as the soliton period, whereas the fundamental soliton propagates without any change in shape.

Even though the existence of optical solitons in fibers was theoretically predicted in 1973 [7], the experimental verification only appeared in the early 1980s [8–10]. The analytical expression for the pulse shape of a fundamental soliton can be calculated easily from the loss-less NLS equation without using an inverse scattering transform. The phase of an optical soliton can change with propagation distance. Therefore, we look for a solution of the form

$$q = g(T) \exp\left[i\theta(Z)\right]. \tag{10.109}$$

When $\alpha = 0$, substituting Eq. (10.109) into Eq. (10.81), we obtain

$$-kg - \frac{\beta_2}{2} \frac{d^2 g}{dT^2} + \gamma g^3 = 0, \tag{10.110}$$

where $k = d\theta/dZ$. If k changes with T, this would imply that the instantaneous frequency is not a constant and could lead to pulse broadening or compression. However, for the fundamental soliton, the pulse shape should not change as a function of Z and hence, we set k to be a constant. To solve Eq. (10.110), we multiply Eq. (10.110) by dg/dT and integrate from $-\infty$ to T to obtain

$$-k \int_{-\infty}^{T} g \frac{dg}{dT} dT - \frac{\beta_2}{2} \int_{-\infty}^{T} \frac{d^2g}{dT^2} \frac{dg}{dT} dT + \gamma \int_{-\infty}^{T} g^3 \frac{dg}{dT} dT = C,$$ (10.111)

$$-kg^2 - \frac{\beta_2}{4} \left(\frac{dg}{dT} \right)^2 + \frac{\gamma g^4}{4} = C,$$ (10.112)

where C is the constant of integration. To obtain Eq. (10.112), we have assumed that $g(\pm\infty) = 0$. When $\beta_2 < 0$, Eq. (10.112) can be rewritten as

$$\frac{dg}{dT} = \frac{2}{\sqrt{-\beta_2}} \left[C + kg^2 - \frac{\gamma g^4}{4} \right]^{1/2}$$ (10.113)

or

$$\int_{g_0}^{g} \frac{dg}{\left[C + kg^2 - \frac{\gamma g^4}{4} \right]^{1/2}} = \frac{2}{\sqrt{-\beta_2}} \int_{0}^{T} dT,$$ (10.114)

where $g_0 = g(0)$. Using the table of integrals [11], Eq. (10.114) can be solved to give

$$g(T) = \frac{\eta}{\sqrt{\gamma}} \text{sech} \left(\frac{\eta T}{\sqrt{-\beta_2}} \right),$$ (10.115)

where $\eta = \sqrt{2k}$. Therefore, the total solution is

$$q = \frac{\eta}{\sqrt{\gamma}} \text{sech} \left(\frac{\eta T}{\sqrt{-\beta_2}} \right) \exp\left(i\eta^2 Z/2 \right).$$ (10.116)

The above solution represents a fundamental soliton that propagates without any change in pulse shape. It acquires a phase shift due to propagation that is proportional to the square of the amplitude.

Example 10.3

The FWHM of a fundamental soliton is 50 ps. Fiber dispersion coefficient $\beta_2 = -21$ ps^2/km, and nonlinear coefficient $\gamma = 1.1$ W^{-1} km^{-1}. Calculate the peak power required to form a soliton. Ignore fiber loss.

Solution:
From Eq. (10.116), we have

$$P(t) = |q|^2 = \frac{\eta^2}{\gamma} \text{sech}^2 \left(\frac{\eta T}{\sqrt{-\beta_2}} \right).$$ (10.117)

Let

$$\frac{\eta^2}{\gamma} = P_{\text{peak}},$$ (10.118)

$$\frac{\sqrt{-\beta_2}}{\eta} = T_0,$$ (10.119)

$$P(t) = P_{\text{peak}} \; \text{sech}^2 \left(\frac{T}{T_0} \right). \tag{10.120}$$

From Fig. 10.8, at $t = T_h$, we find

$$P(T_h) = 0.5 P_{\text{peak}}$$

$$= P_{\text{peak}} \text{sech}^2 \left(\frac{T_h}{T_0} \right), \tag{10.121}$$

$$\text{sech} \left(\frac{T_h}{T_0} \right) = \sqrt{0.5}, \tag{10.122}$$

$$T_h = T_0 \; \text{sech}^{-1}(\sqrt{0.5})$$

$$= 0.8813 \, T_0, \tag{10.123}$$

$$T_{\text{FWHM}} = 2T_h = 1.763 T_0, \tag{10.124}$$

$$T_0 = \frac{50}{1.763} \text{ ps} = 28.37 \text{ ps}. \tag{10.125}$$

Note that the peak power is proportional to η^2 and the pulse width (FWHM) is inversely proportional to η. Thus, for a soliton, as the peak power increases, its pulse width decreases.

From Eq. (10.119), we have

$$\eta = \frac{\sqrt{-\beta_2}}{T_0} = \frac{\sqrt{21 \times 10^{-27}}}{28.37 \times 10^{-12}} = 5.1 \times 10^{-3} \text{ m}^{-1/2}. \tag{10.126}$$

From Eq. (10.118), we find

$$P_{\text{peak}} = \frac{\eta^2}{\gamma} = \frac{(5.1 \times 10^{-3})^2}{1.1 \times 10^{-3}} \text{ W}$$

$$= 4.6 \text{ mW}. \tag{10.127}$$

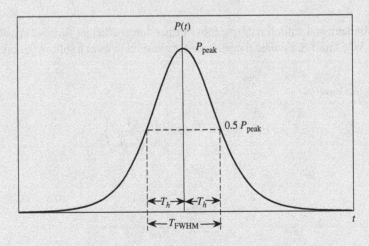

Figure 10.8   FWHM of a soliton pulse.

10.7   Interchannel Nonlinear Effects

So far we have considered the modulation of a single optical carrier. In a wavelength-division multiplexing system (see Chapter 9), multiple carriers are modulated by electrical data and, in this case, the nonlinear Schrödinger equation is valid if the total spectral width is much smaller than the reference carrier frequency. Typically, the spectral width of a WDM signal is about 3 to 4 THz, which is much smaller than the carrier frequency of 194 THz corresponding to the center of the erbium window. The total optical field may be written as

$$\Psi = \sum_{n=-N/2}^{N/2-1} q_n(t,z)e^{-i(\omega_n t - \beta_n z)}, \tag{10.128}$$

where ω_n and $q_n(t,z)$ are the carrier frequency and the slowly varying envelope of the nth channel, respectively, and N is the number of channels. Eq. (10.128) may be rewritten as

$$\Psi = q e^{-i(\omega_0 t - \beta_0 z)}, \tag{10.129}$$

where

$$q = \sum_{n=-N/2}^{N/2-1} q_n(t,z)e^{-i(\Omega_n t - \delta_n z)} \tag{10.130}$$

and $\Omega_n \equiv \omega_n - \omega_0$ is the relative center frequency of channel n with respect to a reference frequency ω_0, which is usually chosen equal to the center of the signal spectrum (see Fig. 10.9); $\delta_n \equiv \beta_n - \beta_0$ is the relative propagation constant.

When the total spectral width $\Delta\omega \ll \omega_0$, the slowly varying envelope can be described by the NLSE, Eq. (10.77). Substituting Eq. (10.130) in Eq. (10.77), we obtain

$$\sum_n \left\{ i\frac{\partial q_n}{\partial z} - \delta_n q_n + i\beta_1 \frac{\partial q_n}{\partial t} + \Omega_n \beta_1 q_n - \frac{\beta_2}{2}\left[-\Omega_n^2 - 2i\Omega_n \frac{\partial}{\partial t} + \frac{\partial^2}{\partial t^2} \right] q_n + i\frac{\alpha}{2}q_n \right\} e^{-i\theta_n}$$

$$+\gamma \sum_{j=-N/2}^{N/2-1} q_j e^{-i\theta_j} \sum_{k=-N/2}^{N/2-1} q_k e^{-i\theta_k} \sum_{l=-N/2}^{N/2-1} q_l^* e^{i\theta_l} = 0, \tag{10.131}$$

where

$$\theta_j(z,t) = \Omega_j t - \delta_j z. \tag{10.132}$$

To obtain the last term of Eq. (10.131), we have used $|q|^2 q = q q q^*$. When the spectral width $\Delta\omega$ is large and/or the dispersion slope is high, third- and higher-order dispersion terms may have to be included in

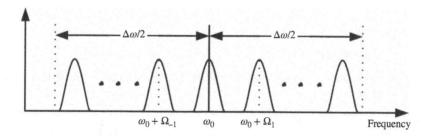

Figure 10.9   The spectrum of a WDM signal.

Eq. (10.131). From the last term on the left-hand side of Eq. (10.131), we see that nonlinear effects in the fiber generate frequency components of the form $\Omega_j + \Omega_k - \Omega_l$. For a frequency band centered around Ω_n, the only frequency components that are of importance are $\Omega_j + \Omega_k - \Omega_l = \Omega_n$. All the other frequency components generated through fiber nonlinearity have no effect on the frequency band centered around Ω_n. Therefore, collecting all the terms that oscillate at frequency Ω_n and noticing that

$$\beta_n = \beta_0 + \beta_1 \Omega_n + \frac{\beta_2}{2}\Omega_n^2, \tag{10.133}$$

we obtain

$$i\left[\frac{\partial q_n}{\partial z} + (\beta_1 + d_n)\frac{\partial q_n}{\partial t}\right] - \frac{\beta_2}{2}\frac{\partial^2 q_n}{\partial t^2} + \gamma \sum_{j=-N/2}^{N/2-1}\sum_{k=-N/2}^{N/2-1}\sum_{l=-N/2}^{N/2-1} q_j q_k q_l^* e^{i\Delta\beta_{jkln}z} = -i\frac{\alpha}{2}q_n \tag{10.134}$$

where

$$d_n = \beta_2 \Omega_n, \tag{10.135}$$

$$\Delta\beta_{jkln} = \beta_j + \beta_k - \beta_l - \beta_n, \tag{10.136}$$

$$\Omega_n = \Omega_j + \Omega_k - \Omega_l. \tag{10.137}$$

If $j = k = l = n$, the last term on the left-hand side is $|q_n|^2 q_n$, which represents SPM. If $j = n$ and $k = l \neq j$, the corresponding term in the summation is $|q_k|^2 q_n$, which represents cross-phase modulation (XPM). All other terms in the above summation represent four-wave mixing (FWM).

As before, using a reference frame that moves at the group speed of the reference channel at ω_0,

$$T = t - \beta_1 z, \tag{10.138}$$

$$Z = z, \tag{10.139}$$

we find

$$i\left(\frac{\partial q_n}{\partial Z} + d_n\frac{\partial q_n}{\partial T}\right) - \frac{\beta_2}{2}\frac{\partial^2 q_n}{\partial T^2} +$$

$$\gamma\left\{\underbrace{|q_n|^2 q_n}_{\text{SPM}} + \underbrace{2\sum_{\substack{k=-N/2\\k\neq n}}^{N/2}|q_k|^2 q_n}_{\text{XPM}} + \underbrace{\sum_{j=-N/2}^{N/2-1}\sum_{k=-N/2}^{N/2-1}\sum_{\substack{l=-N/2\\j+k-l=n}}^{N/2-1} q_j q_k q_l^* e^{i\Delta\beta_{jkln}Z}}_{\text{FWM}}\right\} \tag{10.140}$$

$$= -i\frac{\alpha}{2}q_n.$$

10.7.1 Cross-Phase Modulation

In this section, we focus on XPM by ignoring the FWM terms of Eq. (10.140). The nonlinear interaction due to SPM and XPM is described by

$$i\left(\frac{\partial q_n}{\partial Z} + d_n\frac{\partial q_n}{\partial T}\right) - \frac{\beta_2}{2}\frac{\partial^2 q_n}{\partial T^2} + \gamma\left\{|q_n|^2 q_n + 2\sum_{k=-N/2}^{N/2-1}|q_k|^2 q_n\right\} = -i\frac{\alpha}{2}q_n. \tag{10.141}$$

Here, d_n denotes the difference between the inverse group speed of channel n and β_1,

$$d_n = \beta_2 \Omega_n. \tag{10.142}$$

When the bandwidth of the WDM and/or the dispersion slope is large, d_n should be modified as

$$
\begin{aligned}
d_n &= \beta_1(\omega_0 + \Omega_n) - \beta_1(\omega_0) \\
&= \beta_1(\omega_0) + \frac{d\beta_1}{d\omega}\bigg|_{\omega=\omega_0} \Omega_n + \frac{1}{2}\frac{d^2\beta_1}{d\omega^2}\bigg|_{\omega=\omega_0} \Omega_n^2 - \beta_1(\omega_0) \\
&= \beta_2 \Omega_n + \frac{\beta_3}{2}\Omega_n^2.
\end{aligned}
\tag{10.143}
$$

In Section 10.7.1.1, the XPM-induced timing shift is discussed qualitatively. In Section 10.7.1.2, a simple analytical expression for the XPM efficiency is obtained and the impact of XPM on transmission performance is covered in Section 10.7.1.3.

10.7.1.1 Timing Shift Due to XPM

Let us consider a two-channel WDM system described by Eq. (10.141) with $N = 2$. Owing to the Kerr effect, two propagating pulses with different wavelengths induce a nonlinear phase shift on each other. This phase shift is time-dependent and, therefore, the instantaneous frequency across a pulse in a channel is modified. In a dispersive fiber, this frequency shift is translated into a timing shift since different frequency components propagate at different speeds.

Fig. 10.10 shows two WDM channel inputs. A single pulse is launched at symbol slot 0 in each channel, and initially the pulses are aligned. We assume that the fiber dispersion is anomalous ($\beta_2 = -10\,\mathrm{ps}^2/\mathrm{km}$) and the center wavelength of channel 2 is longer than that of channel 1, so channel 2 propagates slower than channel 1. Our reference frame is fixed to channel 1 and the pulse in channel 2 moves with the inverse walk-off speed of $d_2 = \beta_2 \Omega$ (ignoring β_3) relative to channel 1, where Ω is the channel spacing. Fig. 10.11 shows the pulses of channels at the end of a $L = 80\,\mathrm{km}$ span. As can be seen, due to the different channel speeds, channel 1 walked off channel 2. The pulse separation at the end of the span is given by

$$\Delta T = d_2 L \cong 250\,\mathrm{ps}. \tag{10.144}$$

Fig. 10.12 shows the pulse shape of channel 1 at the end of the fiber for three different cases. (1) Linear case (Lin): $\gamma = 0\,\mathrm{W}^{-1}\,\mathrm{km}^{-1}$, which shows the linear response of the fiber-optics system. (2) SPM case (Lin+SPM): channel 2 is turned off. (3) XPM case (Lin+SPM+XPM): both channels are present. Since anomalous dispersion was assumed, in the presence of nonlinearity (Lin+SPM), the output pulse width is narrower compared with the linear case (Lin). From Fig. 10.12, it can also be seen that when both channels are present (Lin+SPM+XPM), the center of the pulse has moved to the right. As the fast channel (channel 1) walks off, it induces the phase modulation on the slower channel (channel 2), and vice versa. In Fig. 10.10, the pulses are initially aligned and during the propagation, the leading edge of the slow channel overlaps with the trailing edge of the fast channel. The slope is positive at the leading edge of the pulse in channel 2. This leads to a negative instantaneous frequency shift of the pulse in channel 1 (see Eq. (10.167)) or, in other words, channel 1 is red-shifted. Since red-shifted components travel slowly in an anomalous dispersion fiber, the pulse of channel 1 arrives late or, in other words, there is a timing shift due to XPM (Fig. 10.12). Since this timing shift is bit-pattern dependent, it leads to timing jitter in soliton systems [12, 13]. The timing fluctuations leads to amplitude fluctuations of the signal samples used for decision and, hence, performance degradations (see Section 10.7.1.3).

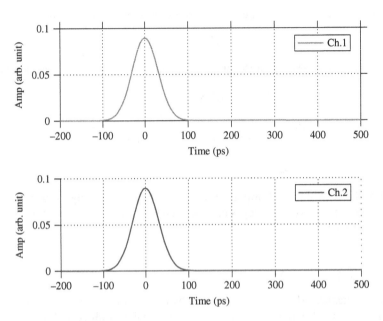

Figure 10.10   Input pulses for channels 1 and 2. The following parameters were assumed: $P_{\text{peak}} = 8\,\text{mW}$, $T_{\text{FWHM}} = 50\,\text{ps}$, channel spacing $= 50\,\text{GHz}$, $\beta_2 = -10\,\text{ps}^2/\text{km}$, $\beta_3 = 0\,\text{ps}^3/\text{km}$, $\gamma = 2.43\,\text{W}^{-1}\,\text{km}^{-1}$, fiber loss $= 0.2\,\text{dB/km}$, and span length $80\,\text{km}$.

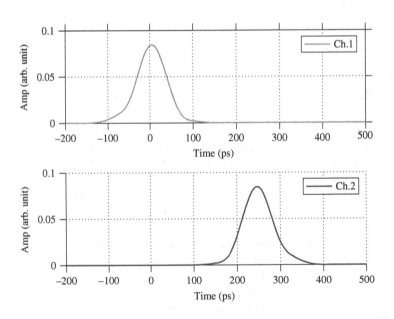

Figure 10.11   Output pulses for channels 1 and 2. Parameters are the same as those of Fig. 10.10.

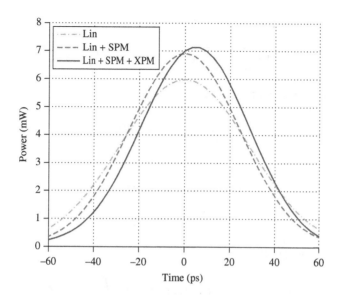

Figure 10.12 Channel 1 output pulse when (1) Lin: $\gamma = 0\,\mathrm{W^{-1}\,km^{-1}}$, (2) Lin+SPM: no pulse presents in channel 2, and (3) Lin+SPM+XPM: both channels are present. Parameters are the same as those of Fig. 10.10.

10.7.1.2 XPM Efficiency

To see the impact of XPM, let us first ignore the pulse broadening due to dispersion and consider only two channels–pump and signal. Let the central frequency of the signal channel be the same as the reference frequency ω_0, so that d_n for the signal is zero. We further assume that the pump is much stronger than the signal, i.e., $|q_p|^2 \gg |q_s|^2$. With these assumptions and approximations, Eq. (10.141) becomes

$$i\frac{\partial q_s}{\partial Z} = \left(-i\frac{\alpha}{2} - 2\gamma |q_p|^2\right) q_s, \tag{10.145}$$

$$i\left(\frac{\partial q_p}{\partial Z} + d_p\frac{\partial q_p}{\partial T}\right) = \left(-i\frac{\alpha}{2} - \gamma |q_p|^2\right) q_p, \tag{10.146}$$

where the subscripts p and s denote pump and signal, respectively, and the walk-off parameter is

$$d_p = \beta_2 \Omega_p + \frac{\beta_3}{2}\Omega_p^2; \tag{10.147}$$

Ω_p is the frequency separation between the pump and the signal. To solve Eq. (10.146), we use the following transformation:

$$T' = T - d_p Z, \tag{10.148}$$

$$Z' = Z, \tag{10.149}$$

to obtain

$$i\frac{\partial q_p}{\partial Z'} = \left(-i\frac{\alpha}{2} - \gamma |q_p|^2\right) q_p. \tag{10.150}$$

Let

$$q_p(T, Z) \equiv q_p(T', Z') = A_p(T', Z')e^{i\theta_p(T', Z')}. \tag{10.151}$$

Substituting Eq. (10.151) in Eq. (10.150), we find

$$\left(i\frac{\partial A_p}{\partial Z'} - \frac{\partial \theta_p}{\partial Z'}A_p\right) = \left(-i\frac{\alpha}{2} - \gamma A_p^2\right)A_p. \tag{10.152}$$

Separating the real and imaginary parts, we obtain

$$\frac{\partial A_p(T',Z')}{\partial Z'} = -\frac{\alpha}{2}A_p(T',Z') \tag{10.153}$$

or

$$A_p(T',Z') = A_p(T',0)e^{-\alpha Z'/2}, \tag{10.154}$$

$$\frac{\partial \theta_p(T',Z')}{\partial Z'} = \gamma A_p^2(T',0)e^{-\alpha Z'}. \tag{10.155}$$

Integrating Eq. (10.155) from 0 to Z', we find

$$\theta_p(T',Z') - \theta_p(T',0) = \gamma A_p^2(T',0)\int_0^{Z'} e^{-\alpha Z'}dZ'$$
$$= \gamma A_p^2(T',0)L_{\text{EFF}}(\alpha, Z') \tag{10.156}$$

where

$$L_{\text{EFF}}(\alpha, x) = \frac{1 - e^{-\alpha x}}{\alpha}. \tag{10.157}$$

Substituting Eqs. (10.154) and (10.156) in Eq. (10.151), we find

$$q_p(T',Z') = A_p(T',0)e^{-\alpha Z'/2}e^{i[\theta_p(T',0)+\gamma A_p^2(T',0)L_{\text{EFF}}(\alpha, Z')]}$$
$$= q_p(T',0)e^{i\gamma A_p^2(T',0)L_{\text{EFF}}(\alpha, Z')-\alpha Z'/2}. \tag{10.158}$$

Using Eqs. (10.148) and (10.149), Eq. (10.158) can be rewritten as

$$q_p(T,Z) \equiv q_p(T',Z') = q_p(T - d_pZ, 0)e^{-\alpha Z/2}e^{i\gamma|q_p(T-d_pZ,0)|^2 L_{\text{EFF}}(\alpha, Z)}. \tag{10.159}$$

Similarly, Eq. (10.145) can be solved by setting

$$q_s = A_s e^{i\theta_s}. \tag{10.160}$$

Substituting Eq. (10.160) in Eq. (10.145) and proceeding as before, we find

$$A_s(T,Z) = A_s(T,0)e^{-\alpha Z/2}, \tag{10.161}$$

$$\frac{d\theta_s}{dZ} = 2\gamma|q_p(T,Z)|^2$$
$$= 2\gamma|q_p(T - d_pZ, 0)|^2 e^{-\alpha Z}, \tag{10.162}$$

$$\theta_s(T,Z) = \theta_s(T,0) + 2\gamma\int_0^Z |q_p(T - d_pZ, 0)|^2 e^{-\alpha Z}dZ, \tag{10.163}$$

$$q_s(T,Z) = q_s(T,0)e^{-\alpha Z/2+i\phi_{\text{XPM}}(T,Z)}, \tag{10.164}$$

$$\phi_{\text{XPM}}(T, Z) = 2\gamma \int_0^Z |q_p(T - d_p Z, 0)|^2 e^{-\alpha Z} dZ, \tag{10.165}$$

$$P_s(T, Z) = |q_s(T, Z)|^2 = P_s(T, 0)e^{-\alpha Z}. \tag{10.166}$$

As in the case of SPM, the pulse width of the signal remains unchanged during propagation since we have ignored dispersion. However, as can be seen from Eq. (10.165), the phase of the signal is modulated by the pump. Hence, this is known as cross-phase modulation. The instantaneous frequency shift of the signal due to XPM is

$$\delta\omega_{\text{XPM}} = -\frac{\partial\phi_{\text{XPM}}}{\partial T} = -2\gamma \int_0^Z \frac{\partial|q_p(T - d_p Z, 0)|^2}{\partial T} e^{-\alpha Z} dZ. \tag{10.167}$$

When the pump is sinusoidally modulated, its field envelope at the fiber input may be written as

$$q_p(T, 0) = \sqrt{P_{p_0}} \cos(\Omega T), \tag{10.168}$$

$$|q_p(T - d_p Z, 0)|^2 = P_{p0}\cos^2[\Omega(T - d_p Z)]$$

$$= \frac{P_{p0}}{2}\{1 + \cos[2\Omega(T - d_p Z)]\}. \tag{10.169}$$

Substituting Eq. (10.169) in Eq. (10.165), we find

$$\phi_{\text{XPM}}(T, L) = \gamma P_{p0} \int_0^L \{1 + \cos[2\Omega(T - d_p Z)]\}e^{-\alpha Z} dZ$$

$$= \gamma P_{p0} L_{\text{eff}} + \gamma P_{p0} \text{Re}\left\{\int_0^L e^{-\alpha Z + i2\Omega(T - d_p Z)} dZ\right\}, \tag{10.170}$$

where

$$L_{\text{eff}} = \frac{1 - \exp(-\alpha L)}{\alpha}. \tag{10.171}$$

The first term on the right-hand side of Eq. (10.170) is the constant phase shift due to XPM, which is of no importance. The second term denotes the time-dependent phase shift, which could potentially degrade the performance. Ignoring the first term, Eq. (10.170) can be simplified as follows [14]:

$$\phi_{\text{XPM}}(T, L) = \gamma P_{p0} \text{Re}\left\{e^{i2\Omega T}\left[\frac{1 - \exp[-(\alpha + i2\Omega d_p)L]}{\alpha + i2\Omega d_p}\right]\right\}$$

$$= \gamma P_{p0} L_{\text{eff}} \sqrt{\eta_{\text{XPM}}} \cos(2\Omega T + \theta), \tag{10.172}$$

where η_{XPM} is the XPM efficiency given by

$$\eta_{\text{XPM}}(\Omega) = \frac{\alpha^2}{\alpha^2 + 4\Omega^2 d_p^2}\left[1 + \frac{4\sin^2(\Omega d_p L) e^{-\alpha L}}{(1 - e^{-\alpha L})^2}\right], \tag{10.173}$$

$$\theta = \tan^{-1}\left\{\frac{e^{-\alpha L}\sin(2\Omega d_p L)}{1 - e^{-\alpha L}\cos(2\Omega d_p L)}\right\} - \tan^{-1}\left\{\frac{2\Omega d_p}{\alpha}\right\}. \tag{10.174}$$

When the walk-off parameter $d_p = 0$ or the modulation frequency $\Omega = 0$, the XPM efficiency is maximum. From Eq. (10.173), we find that $\eta_{\text{XPM}} = 1$ for this case. As the walk-off increases, the interaction between the

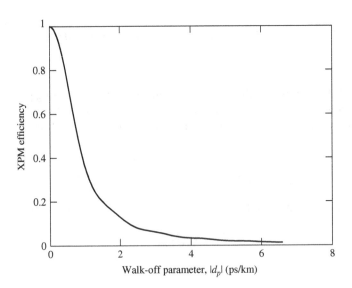

Figure 10.13 XPM efficiency versus absolute of the walk-off parameter. Parameters: $\alpha = 0.046\,\text{km}^{-1}$, fiber length = 80 km, modulating frequency $\Omega/2\pi = 5\,\text{GHz}$.

pump and the probe decreases and the XPM efficiency decreases. Fig. 10.13 shows the FWM efficiency as a function of the absolute walk-off parameter. From Eq. (10.147), we see that as the channel spacing increases, the walk-off increases and the XPM efficiency decreases. In other words, in a WDM system, the impact of the XPM due to the nearest-neighbor channels is the greatest. An arbitrary pump may be written as a superposition of sinusoids of the form given by Eq. (10.168), and the total XPM-induced phase shift can be calculated by adding terms of the form given by Eq. (10.172) due to each frequency component.

Example 10.4

A pump is sinusoidally modulated with modulating frequency 10 GHz. The fiber-optic system has the following parameters: loss coefficient $\alpha = 0.046\,\text{km}^{-1}$, length $L = 50\,\text{km}$, dispersion coefficient $D = 17\,\text{ps/nm·km}$, and dispersion slope $S = 0.06\,\text{ps/nm}^2/\text{km}$. The signal wavelength is 1550 nm and the pump wavelength is 1549.6 nm. Calculate the XPM efficiency.

Solution:
From Eqs. (2.216) and (2.202), we have

$$\beta_3 = S\left(\frac{\lambda^2}{2\pi c}\right)^2 + D\frac{\lambda^3}{2\pi^2 c^2}, \tag{10.175}$$

$$\beta_2 = -D\frac{\lambda^2}{2\pi c}, \tag{10.176}$$

$$D = 17 \times 10^{-6}\,\text{s/m}^2, \tag{10.177}$$

$$S = 0.06 \times 10^3\,\text{s/m}^3, \tag{10.178}$$

$$\lambda = \lambda_s = 1550 \times 10^{-9} \, \text{m}, \tag{10.179}$$

$$c = 3 \times 10^8 \, \text{m/s}, \tag{10.180}$$

$$\beta_3 = 1.33 \times 10^{-40} \, \text{s}^3/\text{m}, \tag{10.181}$$

$$\beta_2 = -2.166 \times 10^{-26} \, \text{s}^2/\text{m}. \tag{10.182}$$

Using Eq. (10.147), the walk-off parameter is given by

$$d_p = \beta_2 \Omega_p + \frac{\beta_3 \Omega_p^2}{2}, \tag{10.183}$$

$$\Omega_p = 2\pi(c/\lambda_p - c/\lambda_s)$$

$$= 3.14 \times 10^{11} \, \text{rad/s}, \tag{10.184}$$

$$d_p = -6.8 \times 10^{-15} \, \text{s/m}. \tag{10.185}$$

The XPM efficiency is given by Eq. (10.173),

$$\eta_{\text{XPM}} = \frac{\alpha^2}{\alpha^2 + 4\Omega^2 d_p^2} \left[1 + \frac{4\sin^2\left(\Omega d_p L\right) e^{-\alpha L}}{(1 - e^{-\alpha L})^2} \right]. \tag{10.186}$$

Modulating frequency $= 10^{10} \, \text{Hz}$. So, $\Omega = 2\pi \times 10^{10} \, \text{rad/s}$. Using this value in Eq. (10.186), we find

$$\eta_{\text{XPM}} = 3.34 \times 10^{-3}. \tag{10.187}$$

10.7.1.3 XPM Impact on System Performance

For intensity-modulated direct detection (IMDD) systems, the phase shift due to XPM does not degrade the system performance if dispersion is absent. In a dispersive fiber, the frequency components generated due to XPM travel at different speeds and arrive at different times at the fiber output, leading to amplitude distortion. In other words, dispersion translates phase modulation (PM) into amplitude modulation (AM). This is known as PM-to-AM conversion. The degradation due to XPM is one of the dominant impairments in WDM systems and, hence, it has drawn significant attention [15–21]. The amplitude fluctuations due to XPM can not be calculated analytically without approximations. In this section, we make a few approximations to find a closed-form approximation for the amplitude distortion due to XPM. As before, we assume that the pump is much stronger than the signal, so that the SPM of the probe can be ignored. Distortion of the pump due to dispersion and nonlinearity is also ignored. While calculating the phase shift due to XPM, fiber dispersion is ignored but its effect will be included later while converting PM to AM. Let the signal be CW,

$$q_s(T, 0) = \sqrt{P_{s0}}. \tag{10.188}$$

Let the pump be the modulated signal,

$$q_p(T, 0) = \sqrt{P_{p0}} \sum_n a_n f(t - nT_b), \tag{10.189}$$

where $\{a_n\}$ is the data sequence and $f(t)$ is the pulse shape function. The pump power at the fiber input is

$$P_p(T,0) = |q_p(T,0)|^2 = P_{p0} \left| \sum_n a_n f\left(t - nT_b\right) \right|^2, \tag{10.190}$$

$$|q_p(T - d_p Z, 0)|^2 = P_p(T - d_p Z, 0). \tag{10.191}$$

Taking the Fourier transform of Eq. (10.191), we find

$$\mathcal{F}\{|q_p(T - d_p Z, 0)|^2\} = \tilde{P}_p(\omega) e^{i\omega d_p Z}, \tag{10.192}$$

$$\tilde{P}_p(\omega) = \mathcal{F}[P_p(T, 0)]. \tag{10.193}$$

The phase shift of the signal due to XPM over a fiber length dZ can be found by differentiating Eq. (10.165) with respect to Z,

$$d\phi_{XPM}(T, Z) = 2\gamma |q_p(T - d_p Z, 0)|^2 e^{-\alpha Z} dZ. \tag{10.194}$$

Taking the Fourier transform of Eq. (10.194) and using Eq. (10.192), we obtain

$$d\tilde{\phi}_{XPM}(\omega, Z) = 2\gamma \tilde{P}_p(\omega) e^{-(\alpha - i\omega d_p)Z} dZ. \tag{10.195}$$

Let us deviate from XPM and consider a different problem. Suppose we have a linear dispersive fiber of length Z and let the fiber input in the frequency domain be

$$\tilde{q}_{in}(\omega) = A_{in} e^{i\tilde{\phi}_{in}(\omega)}. \tag{10.196}$$

The input phase $\tilde{\phi}_{in}(\omega)$ is assumed to be small, and A_{in} is a constant. After passing through the dispersive fiber, the phase fluctuations $\delta\tilde{\phi}_{in}(\omega)$ lead to amplitude fluctuations $\delta\tilde{A}(\omega)$ at the fiber output given by [22]

$$\tilde{A}_{out}(\omega) = A_{in} + \delta\tilde{A}(\omega), \tag{10.197}$$

$$\delta\tilde{A}(\omega) = -A_{in}\tilde{\phi}_{in}(\omega)\sin\left(\frac{\beta_2\omega^2 Z}{2}\right). \tag{10.198}$$

Now let us return to the phase shift due to XPM. Let $d\tilde{\phi}_{XPM}(\omega, Z_0)$ be the phase shift of the signal due to XPM at Z_0. After passing through the dispersive fiber of length $L - Z_0$ where L is the fiber length, this phase shift leads to an amplitude shift, as shown in Fig. 10.14, [16, 18],

$$d\tilde{A}_s(\omega) = -\sqrt{P_{s0}} \; d\tilde{\phi}_{XPM}(\omega, Z_0)\sin\left[\frac{\beta_2\omega^2(L - Z_0)}{2}\right]. \tag{10.199}$$

The nonlinear phase shift due to XPM is distributed over the fiber length, with each infinitesimal phase shift leading to an infinitesimal amplitude shift at the fiber output. Substituting Eq. (10.195) into Eq. (10.199) and integrating the XPM contributions originating from 0 to L, we obtain [16–18]

$$\Delta\tilde{A}_s(\omega) = -2\gamma\sqrt{P_{s0}} \int_0^L P_p(\omega) e^{-(\alpha - i\omega d_p)Z_0}\sin\left(\frac{\beta_2\omega^2(L - Z_0)}{2}\right) dZ_0$$

$$= -\frac{\sqrt{P_{s0}}\gamma P_p(\omega)}{i} \int_0^L [e^{-[\alpha - i\omega d_p + ix]Z_0 + ixL} - e^{-ixL - [\alpha - i\omega d_p - ix]Z_0}] dZ_0$$

$$= i\gamma\sqrt{P_{s0}}P_p(\omega)\{e^{ixL}L_{EFF}[(\alpha - i\omega d_p + ix), L] - e^{-ixL}L_{EFF}[(\alpha - i\omega d_p - ix), L]\}, \tag{10.200}$$

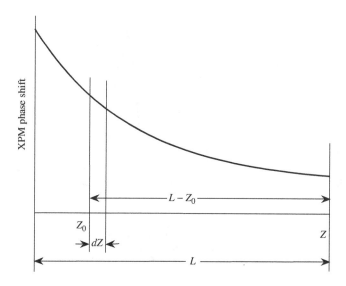

Figure 10.14   Conversion of XPM-induced phase shift into amplitude shift.

where

$$x = \beta_2 \omega^2 / 2, \tag{10.201}$$

$$L_{\text{EFF}}(a, x) = \frac{1 - \exp(-ax)}{a}. \tag{10.202}$$

Since the power modulation of the pump fluctuates as the bit pattern changes, the distortion of the signal due to XPM, $\Delta \tilde{A}_s(\omega)$, changes as a function of the bit pattern. To quantify the magnitude of the XPM distortion, let us calculate the PSD of the XPM distortion as

$$\rho_{\text{XPM}}(\omega) = \lim_{T \to \infty} \frac{\langle |\Delta \tilde{A}_s^{(T)}(\omega)|^2 \rangle}{T} = \lim_{T \to \infty} \frac{\langle |P_p^{(T)}(\omega)|^2 \rangle}{T}$$
$$\times \gamma^2 P_{s0} |e^{ixL} L_{\text{EFF}}[(\alpha - i\omega d_p + ix), L] - e^{-ixL} L_{\text{EFF}}[(\alpha - i\omega d_p - ix), L]|^2, \tag{10.203}$$

where T is the time interval of the bit pattern and

$$\Delta \tilde{A}_s^{(T)}(\omega) = \int_{-T/2}^{T/2} \Delta A_s(t) e^{i\omega t} dt. \tag{10.204}$$

As an example, consider an OOK system that uses unipolar NRZ pulses. The pump field envelope may be written as

$$q_p(t) = \sqrt{P_{p0}} \sum_n a_n \text{rect} \left(\frac{t - n T_b}{T_b} \right), \tag{10.205}$$

where a_n is a random variable which takes the values 0 or 1 with equal probability,

$$P_p(t) = |q_p(t)|^2 = P_{p0} \sum_n a_n^2 \text{rect} \left(\frac{t - n T_b}{T_b} \right). \tag{10.206}$$

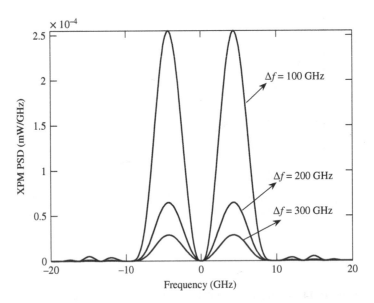

Figure 10.15 The power spectral density of the XPM distortion vs. frequency. Parameters: $\alpha = 0.046\,\mathrm{km}^{-1}$, $P_{p0} = 4\,\mathrm{mW}$, $P_{s0} = 0.1\,\mathrm{mW}$, $L = 80\,\mathrm{km}$, $\gamma = 1.1 \times 10^{-3}$, $\beta_2 = -21\,\mathrm{ps}^2/\mathrm{km}$, and $\beta_3 = 0$. Bit rate $= 10\,\mathrm{Gb/s}$.

For the OOK–NRZ signal, $a_n^2 = a_n$. Therefore, the PSD of $P_p(t)$ is given by Eq. (4.20), i.e.,

$$\lim_{T \to \infty} \frac{\langle |P_p(\omega)|^2 \rangle}{T} = \frac{P_{p0}{}^2 T_b}{4}\mathrm{sinc}^2\,(\omega T_b/2)\left[1 + \frac{2\pi\delta\,(\omega)}{T_b}\right]. \tag{10.207}$$

Fig. 10.15 shows the signal distortion due to XPM using Eqs. (10.203) and (10.207), and ignoring the discrete part of the spectrum (second term in Eq. (10.207)). As can be seen, the PSD of XPM distortion decreases as the channel spacing $\Delta f(= \Omega_p/(2\pi))$ increases.

This analysis can be modified by taking into account the pump envelope change due to dispersion [18]. The amplitude and phase fluctuations of the modulated signal due to XPM can also be calculated using a first- or second-order perturbation theory [20].

10.7.2   Four-Wave Mixing

Four-wave mixing refers to the generation of a fourth wave at the frequency Ω_n due to the nonlinear interaction of three waves at frequencies Ω_j, Ω_k, and Ω_l. To study the impact of FWM alone, let us ignore the SPM and XPM terms in Eq. (10.140). In addition, in order to simplify the analysis, let us assume that the signals in each channel are CW (constant envelope), so that the second and third terms on the left-hand side of Eq. (10.140) can also be ignored. With these simplifications, Eq. (10.140) becomes

$$\frac{\partial q_n}{\partial Z} + \frac{\alpha}{2}q_n = i\gamma \sum_{\substack{j=-N/2 \\ j+k-l=n}}^{N/2-1} \sum_{k=-N/2}^{N/2-1} \sum_{l=-N/2}^{N/2-1} q_j q_k q_l^* e^{i\Delta\beta_{jkln}Z}. \tag{10.208}$$

No SPM, no XPM

Let us first consider a single triplet $\{jkl\}$ corresponding to channels at frequencies Ω_j, Ω_k, and Ω_l satisfying

$$\Omega_j + \Omega_k - \Omega_l = \Omega_n. \tag{10.209}$$

Let the FWM field generated at Ω_n be ϵ_n. Now, q_n may be written as

$$q_n = q_n^{(0)} + \epsilon_n, \tag{10.210}$$

where $q_n^{(0)}$ is the signal field in the absence of nonlinearity. We assume that $|\epsilon_n| \ll |q_n|$. Considering only the triplet $\{jkl\}$, Eq. (10.208) becomes

$$\left(\frac{dq_n^{(0)}}{dZ} + \frac{\alpha}{2} q_n^{(0)} \right) + \left(\frac{d\epsilon_n}{dZ} + \frac{\alpha}{2} \epsilon_n \right) = i\gamma q_j^{(0)} q_k^{(0)} q_l^{*(0)} e^{i\Delta\beta_{jkln}Z}, \tag{10.211}$$

where

$$\Delta\beta_{jkln} = \beta_j + \beta_k - \beta_l - \beta_n \tag{10.212}$$

is the phase mismatch. Note that $q_j q_k q_l^*$ of Eq. (10.208) is replaced by $q_j^{(0)} q_k^{(0)} q_l^{*(0)}$ in Eq. (10.211), which is known as the undepleted pump approximation. When the FWM power is much smaller than the signal power, depletion of signal terms (i.e., FWM pumps) appearing in Eq. (10.211) may be ignored. Since $q_n^{(0)}$ is the signal field in the absence of nonlinearity, it can be written as

$$q_n^{(0)} = A_n e^{-\frac{\alpha}{2}Z + i\theta_n}, \quad n = -\frac{N}{2}, -\frac{N}{2} + 1, \ldots, \frac{N}{2} - 1, \tag{10.213}$$

where A_n and θ_n are amplitude and phase at $Z = 0$, respectively. It can easily be seen that

$$\frac{dq_n^{(0)}}{dZ} + \frac{\alpha}{2} q_n^{(0)} = 0. \tag{10.214}$$

So, Eq. (10.211) becomes

$$\frac{d\epsilon_n}{dZ} + \frac{\alpha}{2} \epsilon_n = i\gamma q_j^{(0)} q_k^{(0)} q_l^{*(0)} e^{i\Delta\beta_{jkln}Z}. \tag{10.215}$$

When the third-order dispersion is ignored, the propagation constant is given by Eq. (10.133). Using Eqs. (10.133), and Eq. (10.209) in Eq. (10.212), we find

$$\Delta\beta_{jkln} = \beta_1(\Omega_j + \Omega_k - \Omega_l - \Omega_n) + \frac{\beta_2}{2}[\Omega_j^2 + \Omega_k^2 - \Omega_l^2 - (\Omega_j + \Omega_k - \Omega_l)^2]$$

$$= \beta_2[\Omega_l\Omega_n - \Omega_j\Omega_k]. \tag{10.216}$$

When the bandwidth of the WDM signal and/or the dispersion slope is large, the third-order dispersion coefficient can not be ignored. In this case, Eq. (10.216) is modified as (see Example 10.9)

$$\Delta\beta_{jkln} = (\Omega_l\Omega_n - \Omega_j\Omega_k)\left[\beta_2 + \frac{\beta_3}{2}(\Omega_j + \Omega_k)\right]. \tag{10.217}$$

Substituting Eq. (10.213) in Eq. (10.215), we find

$$\frac{d\epsilon_n}{dZ} + \frac{\alpha}{2} \epsilon_n = i\gamma A_j A_k A_l e^{-(3\alpha/2 - i\Delta\beta_{jkln})Z + i\Delta\theta_{jkl}}, \tag{10.218}$$

where

$$\Delta\theta_{jkl} = \theta_j + \theta_k - \theta_l. \tag{10.219}$$

Eq. (10.218) is a first-order ordinary differential equation. The integrating factor is $e^{\alpha Z/2}$. So, multiplying Eq. (10.218) by $e^{\alpha Z/2}$, we find

$$\frac{d(\epsilon_n e^{\alpha Z/2})}{dZ} = i\gamma A_j A_k A_l e^{-(\alpha - i\Delta\beta_{jkln})Z + i\Delta\theta_{jkl}}. \tag{10.220}$$

Integrating Eq. (10.220) from 0 to L with the condition $\epsilon_n(0) = 0$, we obtain

$$\epsilon_n(L) = i\gamma A_j A_k A_l e^{-\alpha L/2 + i\Delta\theta_{jkl}} \int_0^L e^{-(\alpha - i\Delta\beta_{jkln})Z} dZ$$

$$= K_{jkl} \frac{[1 - e^{-\delta_{jkln}L}]}{\delta_{jkln}}, \tag{10.221}$$

where

$$K_{jkl} = i\gamma A_j A_k A_l e^{i\Delta\theta_{jkl} - \alpha L/2} \tag{10.222}$$

and

$$\delta_{jkln} = \alpha - i\Delta\beta_{jkln}. \tag{10.223}$$

The power of the FWM component is [23, 24]

$$P_{\text{FWM},n} = |\epsilon_n|^2 = \frac{|K_{jkl}|^2 |1 - e^{-\delta_{jkln}L}|^2}{|\delta_{jkln}|^2} e^{-\alpha L}$$

$$= \gamma^2 P_j P_k P_l L_{\text{eff}}^2 \eta_{jkln} e^{-\alpha L}, \tag{10.224}$$

$$\eta_{jkln} = \frac{\alpha^2 + 4e^{-\alpha L}\sin^2(\Delta\beta_{jkln}L/2)/L_{\text{eff}}^2}{\alpha^2 + (\Delta\beta_{jkln})^2}, \tag{10.225}$$

$$P_j = A_j^2, \tag{10.226}$$

$$L_{\text{eff}} = \frac{1 - \exp(-\alpha L)}{\alpha}. \tag{10.227}$$

Here, η_{jkln} represents the *FWM efficiency*. Fig. 10.16 shows the dependence of the efficiency on the dispersion coefficient β_2 when $j = 1, k = 2$, and $l = 3$. When $\beta_2 = 0$, the efficiency is maximum and this is known as *phase matching*. As $|\beta_2|$ increases, the FWM efficiency decreases and it becomes significantly smaller when $|\beta_2| > 6\,\text{ps}^2/\text{km}$. When the fiber is sufficiently long, the second term in Eq. (10.225) may be ignored and Eq. (10.225) may be approximated as

$$\eta_{jkln} \cong \frac{\alpha^2}{\alpha^2 + (\Delta\beta_{jkln})^2}. \tag{10.228}$$

Let the channel spacing be Δf and $\Omega_j = j2\pi\Delta f, j = -N/2, -N/2 + 1, \ldots, N/2 - 1$. Now, Eqs. (10.209) and (10.216) become

$$j + k - l = n, \tag{10.229}$$

$$\Delta\beta_{jkln} = (2\pi\Delta f)^2 \beta_2 [nl - jk]. \tag{10.230}$$

With $j = 1, k = 2$, and $l = 3$, we find $n = 0$ and

$$\eta_{1230} \cong \frac{\alpha^2}{\alpha^2 + 4\beta_2^2(2\pi\Delta f)^4}. \tag{10.231}$$

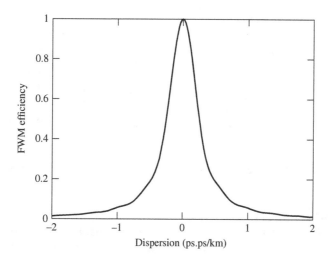

Figure 10.16 FWM efficiency vs. β_2 for $j = 1, k = 2$, and $l = 3$. Channel spacing, $\Delta f = 100\,\text{GHz}$, $\Omega_1 = 2\pi\Delta f$, $\Omega_2 = 4\pi\Delta f$, $L = 80\,\text{km}$, loss $= 0.2\,\text{dB/km}$, and $\Omega_3 = 6\pi\Delta f$.

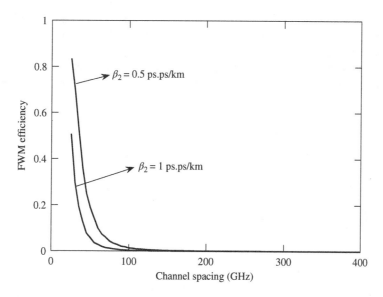

Figure 10.17 FWM efficiency vs. channel spacing, Δf. $L = 80\,\text{km}$, loss $= 0.2\,\text{dB/km}$, $j = 1, k = 2$, and $l = 3$.

Fig. 10.17 shows the dependence of the FWM efficiency on the channel spacing. As can be seen, the efficiency decreases as the channel spacing and/or $|\beta_2|$ increases. So far we have considered FWM generation due to a single triple $\{jkl\}$. Considering all the triplets in Eq. (10.208), Eq. (10.221) should be modified as

$$\epsilon_n(L) = \underset{\substack{j+k-l=n \\ \text{No SPM, no XPM}}}{\sum_j \sum_k \sum_L} K_{jkl} \frac{\left[1 - e^{-\delta_{jkln}L}\right]}{\delta_{jkln}} \tag{10.232}$$

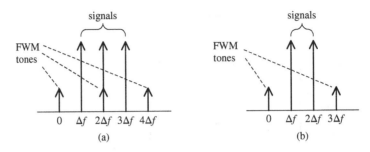

Figure 10.18 Two types of FWM: (a) non-degenerate FWM; (b) degenerate FWM.

and the FWM power is $P_n = |\epsilon_n|^2$. Consider three channels of a WDM system centered at $l\Delta f$, $l = 1, 2, 3$. The nonlinear interaction between these channels leads to a FWM field at $4\Delta f$ and 0, as shown in Fig. 10.18. If we choose $j = 1$, $k = 2$, and $l = 3$, the FWM tone falls on the channel at 0 since $j + k - l = 0$. Choosing $j = 2$, $k = 3$, and $l = 1$, we find $j + k - l = 4$ and, therefore, the FWM tone is generated at $4\Delta f$ as well, as shown in Fig. 10.18. These types of FWM are known as *non-degenerate FWM* as j, k, and l are distinct. When $j = k = 1$ and $l = 2$, $j + k - l = 0$ and the FWM tone falls on the channel at 0, as shown in Fig. 10.18(b). The other possibility is $j = k = 2$ and $l = 1$, $j + k - l = 3$ and the FWM tone falls on the channel at $3\Delta f$. These types of FWM are known as *degenerate FWM*. Eq. (10.232) includes both types of FWM. Adding all the possible FWM tones that satisfy the condition $j + k - l = n$, the total FWM field on the channel n can be calculated using Eq. (10.232).

Fig. 10.19 shows the mean FWM power on the middle channel as a function of the number of channels in a WDM system. Initial phases of channels (θ_j) are assumed to be random and the mean FWM power

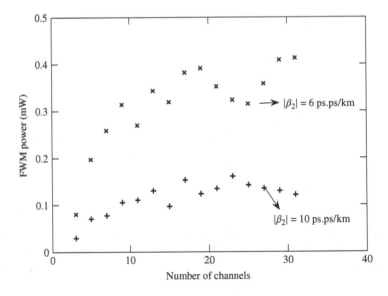

Figure 10.19 Mean FWM power on the middle channel vs. number of channels in a WDM system. Parameters: channel spacing $= 50$ GHz, power/channel $= 3$ mW, number of spans $= 20$, amplifier spacing $= 80$ km, loss $= 0.2$ dB/km, and $\gamma = 1.1$ W^{-1} km^{-1}.

is calculated by averaging over the random phases. The FWM field acts as noise on the channel, leading to performance degradations. The FWM impairment becomes smaller as the dispersion increases, since the phase matching becomes more difficult, and it increases as the channel spacing decreases. Therefore, FWM is one of the dominant impairments in OFDM systems in which the subcarriers are closely spaced [25, 26].

There are a number of approximations in this FWM model. Firstly, the modulation of the channels is ignored. When the channel modulation is included, the signal field q_n is not independent of time and, therefore, the second term on the left-hand side of Eq. (10.140) can not be ignored. Owing to dispersion, different channels propagate at different speeds, which is not taken into account in this simple model. The FWM model may be modified taking into account pump depletion [27], pump modulation [28–30], pump modulation and walk-off [31–33]. Experimental validation of the CW FWM model can be found in Ref.[34].

Example 10.5

A WDM system consists of three channels centered at $-\Delta f$, 0, and Δf, with $\Delta f = 50\,\text{GHz}$. The fiber loss coefficient $\alpha = 0.046\,\text{km}^{-1}$ and the fiber length $L = 40\,\text{km}$. Calculate the efficiency of the non-degenerate FWM tone at $-2\Delta f$ if (a) $\beta_2 = -4\,\text{ps}^2/\text{km}$, (b) $\beta_2 = 0\,\text{ps}^2/\text{km}$. Ignore β_3.

Solution:
From Eq. (10.90), we have

$$L_{\text{eff}} = \frac{1 - \exp(-\alpha L)}{\alpha} = \frac{1 - \exp(-0.046 \times 40)}{0.046}$$

$$= 18.28\,\text{km}. \tag{10.233}$$

From Eq. (10.225), the FWM efficiency is

$$\eta_{jkln} = \frac{\alpha^2 + 4e^{-\alpha L}\sin^2(\Delta\beta_{jkln}L/2)/L_{\text{eff}}^2}{\alpha^2 + (\Delta\beta_{jkln})^2}. \tag{10.234}$$

Let $j = -1$, $k = 0$, and $l = 1$ so that $n = j + k - l = -2$ corresponding to the FWM tone at $-2\Delta f$. From Eq. (10.216), we have

$$\Delta\beta_{jkln} = \beta_2[\Omega_l\Omega_n - \Omega_j\Omega_k]$$

$$= \beta_2(2\pi\Delta f)^2[1 \cdot (-2) - (-1) \cdot 0]. \tag{10.235}$$

(a) $\beta_2 = -4\,\text{ps}^2/\text{km}$:

$$\Delta\beta_{-101-2} = -4 \times 10^{-27} \times (2\pi \times 50 \times 10^9)^2 \times (-2)\,\text{m}^{-1}$$

$$= 7.89 \times 10^{-4}\,\text{m}^{-1}; \tag{10.236}$$

$$\eta_{-101-2} = \frac{(0.046 \times 10^{-3})^2 + 4\exp(-0.046 \times 40)\sin^2(7.89 \times 10^{-1} \times 40/2)/(18.28 \times 10^3)^2}{(0.046 \times 10^{-3})^2 + (7.89 \times 10^{-4})^2}$$

$$= 3.4 \times 10^{-3}. \tag{10.237}$$

(b) $\beta_2 = 0\,\text{ps}^2/\text{km}$:

$$\Delta\beta_{-101-2} = 0\,\text{m}^{-1}. \tag{10.238}$$

Now, from Eq. (10.225), we have $\eta = 1$.

10.8 Intrachannel Nonlinear Impairments

In quasi-linear systems, dispersive effects are much stronger than nonlinear effects and the fiber nonlinearity can be considered as a small perturbation in the linear system. Since the dispersive effects are dominant in quasi-linear systems, neighboring pulses overlap and this system is also known as a *strongly pulse-overlapped system* [35] or *pseudo-linear system* [36]. In contrast, in classical soliton systems, dispersion is balanced by nonlinearity, the pulses are well confined within the bit period. In quasi-linear systems, the pulses that are separated by several bit periods could interact nonlinearly because of the strong pulse overlap among the pulses. In this section, we consider single-channel nonlinear impairments such as intrachannel four-wave mixing (IFWM) [35–46] and intrachannel cross-phase modulation (IXPM) [47–49]. The variance in signal distortion due to nonlinear effects is used as a measure to compare different fiber-optic systems. IXPM and IFWM can be considered as deterministic signal–signal nonlinear impairments because if we know the bit pattern, these effects can be undone using digital back propagation (DBP) at the transmitter or receiver (see Chapter 11). In contrast, the nonlinear signal–ASE interaction such as Gordon–Mollenauer phase noise is stochastic and the DBP can not compensate for it. In single-channel systems, the nonlinear interaction can be divided into three types: (i) intrapulse SPM; (ii) IXPM; and (iii) IFWM. SPM has already been discussed in Section 10.5. Here we discuss IXPM and IFWM.

10.8.1 Intrachannel Cross-Phase Modulation

IXPM is the phase modulation of a pulse by another pulse of the same channel. Consider the interaction between two pulses $q_1(T, Z)$ and $q_2(T, Z)$ separated by T_b at the fiber input. Let the total field envelope be

$$q(T, Z) = q_1(T, Z) + q_2(T, Z). \tag{10.239}$$

Substituting Eq. (10.239) in Eq. (10.81), we find

$$i\frac{\partial(q_1 + q_2)}{\partial Z} - \frac{\beta_2}{2}\frac{\partial^2(q_1 + q_2)}{\partial T^2} + \gamma|q_1 + q_2|^2(q_1 + q_2) = -i\alpha(q_1 + q_2)/2. \tag{10.240}$$

The last term on the left-hand side can be written as

$$|q_1 + q_2|^2(q_1 + q_2) = (|q_1|^2 + 2|q_2|^2)q_1 + (|q_2|^2 + 2|q_1|^2)q_2 + q_1^2 q_2^* + q_2^2 q_1^*. \tag{10.241}$$

The last two terms in Eq. (10.241) represent the intrachannel four-wave mixing, and this will be considered in the next section. The term $2|q_2|^2 q_1$ represents the phase modulation of q_1 due to q_2. If the IXPM terms $2|q_2|^2 q_1$ and $2|q_1|^2 q_2$ and the IFWM term $q_1^2 q_2^* + q_2^2 q_1^*$ were to be absent, the pulses would experience intra-pulse SPM only and there would be no change in the temporal position of the pulses as a function of the propagation distance. However, due to IXPM, pulses could attract or repel each other. Figs. 10.20 and 10.21 show the nonlinear interaction between adjacent pulses. In this example, pulses repel each other, leading to performance degradation. The timing jitter due to IXPM can be calculated using a variation approach [42, 49] or a perturbation technique [48]. The repulsion between pulses can be explained as follows. The phase modulation caused by the IXPM leads to instantaneous frequency change of a pulse. In a dispersive fiber, different frequency components travel at different speeds and, therefore, the frequency change due to IXPM translates into group speed changes. Therefore, the first pulse moves faster than the second pulse and it arrives at the fiber output earlier than the second one, leading to a temporal separation longer than the bit interval. In the absence of IXPM, pulses would have the same group speed and the separation between pulses would be equal to the bit period. For systems based on OOK, '1' and '0' occur randomly and the timing shift caused by IXPM is random, leading to time jitters and performance degradation.

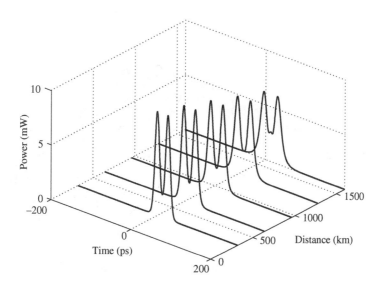

Figure 10.20 Nonlinear interaction between adjacent pulses. The pulses are separated by 25 ps at the input. Transmission fiber is a standard single-mode fiber with $\beta_2 = -22 \, \mathrm{ps}^2/\mathrm{km}$. The dispersion is fully compensated in each span.

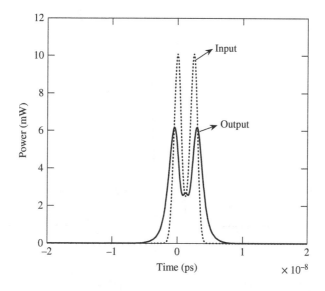

Figure 10.21 Optical power distribution at the fiber input and output. Fiber length = 1600 km. The parameters are the same as in Fig. 10.20.

10.8.2   Intrachannel Four-Wave Mixing

When two or more pulses of the same channel interact nonlinearly, *echo* or *ghost* pulses are generated, as shown in Figs. 10.22 and 10.23. This is called *intrachannel four-wave* mixing. In the case of interchannel FWM shown in Fig. 10.24(a), the nonlinear interaction between the frequency components f_1, f_2, and f_3 leads

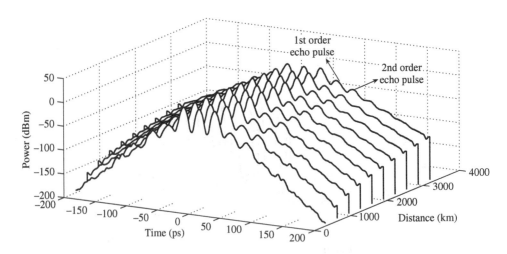

Figure 10.22   Interaction of three signal pulses leading to echo pulses. The parameters are the same as in Fig. 10.20.

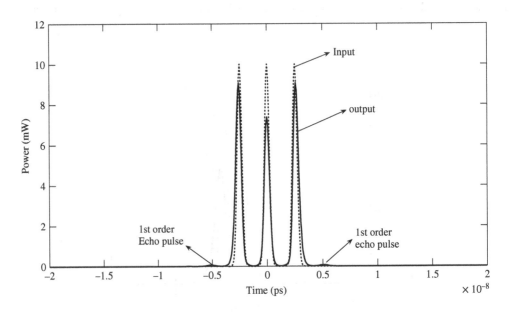

Figure 10.23   Optical power distributions at the fiber input and output. Fiber length $= 3200$ km.

to FWM sidebands at $f_1 + f_2 - f_3$ and $f_2 + f_3 - f_1$. Similarly, in the case of IFWM, the nonlinear interaction between pulses centered at t_1, t_2, and t_3 leads to echo pulses at $t_1 + t_2 - t_3$ and $t_2 + t_3 - t_1$. The difference between FWM and IFWM is that echo pulses appear in the time domain instead of in the frequency domain, as shown in Fig. 10.24(b). Hence, it is also known as time-domain FWM. The nonlinear interaction between signal pulses leads to first-order echo pulses, as shown in Figs. 10.22 and 10.23, and the nonlinear interaction of signal pulses and first-order echo pulses leads to second-order echo pulses. However, the amplitudes of the

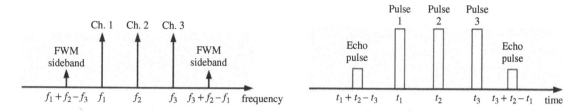

Figure 10.24   The analogy between interchannel FWM and intrachannel FWM.

second-order echo pulses are very small, and they are not visible in the linear plot shown in Fig. 10.23. We ignore second-order echo pulses in the analysis of IFWM.

Suppose the pulse centered at mT_b is q_m. The nonlinear interaction between q_l, q_m, and q_n due to IFWM is described by $q_l q_m q_n^*$ and the resulting echo pulse is centered near $(l + m - n)T_b$. For example, in Figs. 10.22 and 10.23, the first-order echo pulses centered around -50 ps and 50 ps and generated due to the nonlinear interaction of signal pulses centered at -25 ps, 0 ps, and 25 ps. The echo pulse at 50 ps is generated by a nonlinear interaction of the form $q_l q_m q_n^*$ where $l = 1$ (25 ps), $m = 0$ (0 ps), $n = -1$ (-25 ps), and $l + m - n = 2$ (50 ps). The nonlinear interaction of the first-order echo pulses and signal pulses leads to second-order echo pulses at -75 ps and 75 ps (see Fig. 10.22).

When $l = m$, it is known as *degenerate IFWM*, similar to the degenerate interchannel FWM. Otherwise, it is called *non-degenerate IFWM*. The echo pulse centered around 50 ps is generated not only by signal pulses centered at -25 ps, 0 ps, and 25 ps ($q_1 q_0 q_{-1}^*$) due to non-degenerate IFWM, but also by pulses centered at 0 ps and 25 ps ($q_1^2 q_0^*$) due to degenerate IFWM. Note that the echo pulses are generated in the locations of signal pulses as well. In Fig. 10.23, the nonlinear interaction of the signal pulses centered at -25 ps, 0 ps, and 25 ps ($q_1 q_{-1} q_0^*$) leads to an echo pulse around 0 ps ($l = 1$, $m = -1$, $n = 0$, $l + m - n = 0$). The coherent superposition of the signal pulse and echo pulse around $T = 0$ ps leads to the distortion of the signal pulse at $T = 0$ ps, as shown in Fig. 10.23. Section 10.9 provides the mathematical description of IFWM.

10.8.3   Intra- versus Interchannel Nonlinear Effects

Fig. 10.25 illustrates the difference between intrachannel and interchannel nonlinear impairments. The pulse located at the center interacts nonlinearly with the pulses within the trapezoids. This interaction includes both intrachannel and interchannel nonlinear effects (SPM, IXPM, IFWM, XPM, FWM). The area of the trapezoids depends on the system parameters such as fiber dispersion, nonlinear coefficient, launch power, and transmission distance. For example, if the launch power is higher and/or the transmission distance is longer, the area of the trapezoids would be larger. The nonlinear interaction of the pulse located at the center with the pulses within the ellipse corresponds to intrachannel impairments (SPM, IXPM, IFWM).

10.9   Theory of Intrachannel Nonlinear Effects

The optical field envelope in a fiber-optic transmission system can be described by the nonlinear Schrödinger equation (NLS) (see Eq. (10.81))

$$i\frac{\partial q}{\partial Z} - \frac{\beta_2(Z)}{2}\frac{\partial^2 q}{\partial T^2} + \gamma|q|^2 q = -i\frac{\alpha(Z)}{2}q, \tag{10.242}$$

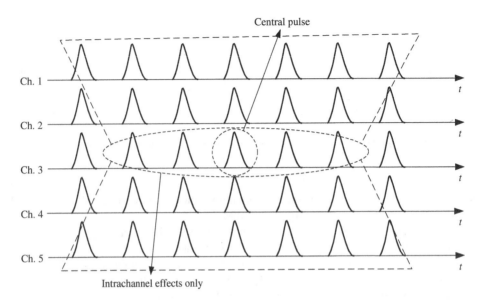

Figure 10.25 Illustration of the difference between intrachannel and inter channel nonlinear effects.

where $\alpha(Z)$ is the loss/gain profile which includes fiber loss as well as amplifier gain, β_2 is the second-order dispersion profile, and γ is the fiber nonlinear coefficient. Fig. 10.26(a) shows a typical fiber-optic transmission system. We assume that the amplifier compensates for the fiber loss. To separate the fast variation of optical power due to fiber loss/gain, we use the following transformation [50]:

$$q(T, Z) = a(Z)u(T, Z), \tag{10.243}$$

where $a(Z)$ is real. Differentiating Eq. (10.243), we find

$$\frac{\partial \dot{q}}{\partial Z} = \dot{a}u + a\frac{\partial u}{\partial Z}, \tag{10.244}$$

where $\cdot$ denotes differentiation with regard to Z. Let

$$\dot{a} = -\frac{\alpha(Z)}{2}a. \tag{10.245}$$

Substituting Eqs. (10.244) and (10.245) in Eq. (10.242), we obtain the NLS equation in the lossless form as

$$i\frac{\partial u}{\partial Z} - \frac{\beta_2(Z)}{2}\frac{\partial^2 u}{\partial T^2} = -\gamma a^2(Z)|u|^2 u. \tag{10.246}$$

Solving Eq. (10.245) with the initial condition $a(0) = 1$, we obtain

$$a(Z) = \exp\left[-\int_0^Z \frac{\alpha(s)}{2}ds\right]. \tag{10.247}$$

The choice of this initial condition is arbitrary. The sole purpose of introducing $a(Z)$ is to separate the variations of the optical field due to loss/gain from that due to dispersion and nonlinear effects. Between amplifiers, when the fiber loss is constant, $\alpha(Z) = \alpha_0$, it becomes

$$a(Z) = \exp\left(-\alpha_0 Z'/2\right), \tag{10.248}$$

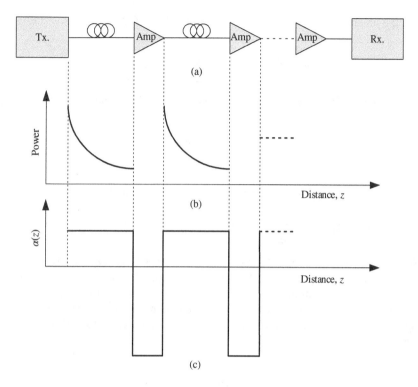

Figure 10.26 Typical fiber-optic transmission system: (a) block diagram, (b) power variation, (c) loss/gain profile.

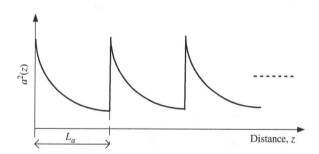

Figure 10.27 Plot of $a^2(Z)$ v distance. L_a = amplifier spacing.

where $Z' = \text{mod}(Z, L_a)$, where L_a = amplifier spacing. Fig. 10.27 shows $a^2(Z)$ for a fiber-optic link with fiber loss exactly compensated by the amplifier gain (see Example 10.12 for more details). The mean optical power $< |q|^2 >$ fluctuates as a function of distance due to fiber loss and amplifier gain, but $< |u|^2 >$ is independent of distance since the variations due to loss/gain are separated out using Eq. (10.243). Note that the nonlinear coefficient is constant in Eq. (10.242), but the effective nonlinear coefficient $\gamma a^2(Z)$ changes as a function of distance in Eq. (10.246). Eq. (10.246) can be solved using perturbation theory. The solution of Eq. (10.246) can be written as

$$u(T,Z) = u_0(T,Z) + \gamma u_1(T,Z) + \gamma^2 u_2(T,Z) + \cdots, \tag{10.249}$$

where $u_j(T, Z)$ is the jth order solution. The nonlinear term in Eq. (10.246) may be written as

$$|u(T,Z)|^2 u(T,Z) = \left| \sum_{n=0}^{\infty} \gamma^n u_n \right|^2 \sum_{n=0}^{\infty} \gamma^n u_n. \qquad (10.250)$$

Using Eqs. (10.249) and (10.250) in Eq. (10.246) and separating terms proportional to γ^n, $n = 0, 1, 2, \ldots$, we obtain [20]

$$(\gamma^0): \qquad i\frac{\partial u_0}{\partial Z} - \frac{\beta_2}{2}\frac{\partial^2 u_0}{\partial T^2} = 0, \qquad (10.251)$$

$$(\gamma^1): \qquad i\frac{\partial u_1}{\partial Z} - \frac{\beta_2}{2}\frac{\partial^2 u_1}{\partial T^2} = -a^2(Z)|u_0|^2 u_0, \qquad (10.252)$$

$$(\gamma^2): \qquad i\frac{\partial u_2}{\partial Z} - \frac{\beta_2}{2}\frac{\partial^2 u_2}{\partial T^2} = -a^2(Z)(2|u_0|^2 u_1 + u_0^2 u_1^*). \qquad (10.253)$$

Eq. (10.251) is the linear Schrödinger equation. The linear solution $u_0(T, Z)$ can be obtained using the linear fiber transfer function as discussed in Chapter 2. Eqs. (10.252) and (10.253) represent the first- and second-order corrections due to nonlinear effects. The first (second)-order term u_1 (u_2) corresponds to the first (second)-order echo pulses shown in Fig. 10.22. When the nonlinear effects are small, the terms of order γ^n, $n > 1$, can be ignored. As an example, let us consider a single-span lossless ($\alpha = 0$) zero-dispersion ($\beta_2 = 0$) fiber. Eq. (10.251) becomes

$$\frac{du_0}{dZ} = 0, \qquad (10.254)$$

$$u_0 = k(\text{const.}). \qquad (10.255)$$

Note that k may be a function of t. Since $a^2(Z) = 1$ in this example, Eq. (10.252) becomes

$$i\frac{\partial u_1}{\partial Z} = -|k|^2 k, \qquad (10.256)$$

$$u_1(T,Z) = u_1(T,0) + i|k|^2 kZ. \qquad (10.257)$$

The initial condition is

$$u(T,0) = u_0(T,0) + \gamma u_1(T,0) + \gamma^2 u_2(T,0) + \cdots . \qquad (10.258)$$

Since $u_j(T, 0)$ are arbitrary functions, one convenient choice would be

$$u_0(T,0) = u(T,0) = k,$$

$$u_j(T,0) = 0 \quad \text{for} \quad j > 1. \qquad (10.259)$$

Using Eqs. (10.255)–(10.259) and ignoring terms of order γ^n, $n > 1$ in Eq. (10.249), we obtain

$$u(T,Z) = (1 + i\gamma |k|^2 Z)\, k. \qquad (10.260)$$

In this simple example, Eq. (10.246) can easily be solved directly:

$$i\frac{du}{dZ} = -\gamma |u|^2 u,$$

$$u(T,Z) = \exp\left[i\gamma |u(T,0)|^2\right] u(T,0). \qquad (10.261)$$

Taylor series expansion of the exponential function in Eq. (10.261) yields

$$u(T, Z) = \left(1 + i\gamma |k|^2 Z - \frac{\gamma^2 Z^2}{2!} |k|^4 + \cdots\right) k. \tag{10.262}$$

Comparing Eqs. (10.260) and (10.262), we see that they match up to the first-order term in γ. If we solve Eq. (10.253) and add the second-order correction in Eq. (10.260), we find that Eqs. (10.260) and (10.262) would match up to second order in γ (see Example 10.11).

Now, let us consider a more general problem in which neither β_2 nor α is zero. Let the optical field envelope at the fiber input be

$$q(T, 0) = a(0)u(T, 0) = \sum_{n=-\infty}^{\infty} b_n f(T - nT_s), \tag{10.263}$$

where T_s is the symbol interval and $f(T)$ is the pulse shape. For systems based on OOK,

$$b_n = \begin{cases} 1 & \text{for '1',} \\ 0 & \text{for '0'.} \end{cases} \tag{10.264}$$

For systems based on PSK or DPSK,

$$b_n = \begin{cases} 1 & \text{for '1',} \\ -1 & \text{for '0'.} \end{cases} \tag{10.265}$$

Eq. (10.251) can be solved using the Fourier transform technique (see Chapter 2). The solution is

$$u_0(T, Z) = \mathcal{F}^{-1}\{\tilde{u}_0(\omega, 0) \exp[i\omega^2 S(Z)/2]\}, \tag{10.266}$$

where $\tilde{u}_0(\omega, 0) = \mathcal{F}[u_0(T, 0)]$, $u_0(T, 0) = u(T, 0)$, and $S(Z)$ is the accumulated dispersion

$$S(Z) = \int_0^Z \beta_2(x)dx. \tag{10.267}$$

Taking the Fourier transform of Eq. (10.252), we obtain

$$i\frac{d\tilde{u}_1}{dZ} + \frac{\beta_2(Z)}{2}\omega^2 \tilde{u}_1(\omega, Z) = -a^2(Z)\tilde{F}(\omega, Z), \tag{10.268}$$

where $\tilde{u}_1(\omega, Z) = \mathcal{F}[u_1(T, Z)]$ and $\tilde{F}(\omega, Z) = \mathcal{F}[|u_0(T, Z)|^2 u_0(T, Z)]$. Eq. (10.268) is a first-order ordinary differential equation which can be solved to yield

$$\tilde{u}_1(\omega, L_{\text{tot}})IF(L_{\text{tot}}) = i\int_0^{L_{\text{tot}}} a^2(x)\tilde{F}(\omega, x)IF(x)dx, \tag{10.269}$$

where the integrating factor is

$$IF(Z) = \exp\left[iS(Z)\omega^2/2\right], \tag{10.270}$$

and L_{tot} is the total transmission distance. We assume that the dispersion is fully compensated either in the optical or the electrical domain before the decision device. So, $S(L_{\text{tot}}) = 0$ and Eq. (10.269) becomes

$$\tilde{u}_1(\omega, L_{\text{tot}}) = i\int_0^{L_{\text{tot}}} a^2(x)\tilde{F}(\omega, x)IF(x)dx. \tag{10.271}$$

The first-order correction $u_1(T, Z)$ is obtained by performing the inverse Fourier transformation of $\tilde{u}_1(\omega, Z)$. Typically, in quasi-linear systems, the nonlinear effects are smaller than the dispersive effects and the first-order correction $u_1(T, Z)$ is often adequate to describe the nonlinear propagation. However, when the transmission distance is long and/or the launch power is large, a second-order perturbation theory is needed [20].

A closed-form expression for $u_0(T, Z)$ and $u_1(T, Z)$ can be obtained if we assume that the pulse shape $f(T)$ is Gaussian, i.e.,

$$f(T) = \sqrt{P_0} \exp\left(-\frac{T^2}{2T_0^2}\right), \tag{10.272}$$

where P_0 is the peak power. The linear propagation of this pulse is described by (see Eq. (2.158))

$$\frac{T_0\sqrt{P_0}}{T_1(Z)} \exp\left[-\frac{T^2}{2T_1^2(Z)}\right], \tag{10.273}$$

where $T_1^2 = T_0^2 - iS(Z)$. When a long bit sequence is launched to the fiber, $u_0(T, 0)$ is given by Eq. (10.263). In this case, the linear solution is

$$u_0(T, Z) = \frac{T_0\sqrt{P_0}}{T_1} \sum_{n=-\infty}^{\infty} b_n \exp\left[-\frac{(T - nT_s)^2}{2T_1^2(Z)}\right], \tag{10.274}$$

$$F(T, Z) = |u_0(T, Z)|^2 u_0(T, Z) = P_0^{3/2} \frac{T_0^3}{|T_1|^2 T_1} \sum_{l=-\infty}^{\infty} \sum_{m=-\infty}^{\infty} \sum_{n=-\infty}^{\infty} b_l b_m b_n$$

$$\times \exp\left[-\frac{(T - lT_s)^2}{2T_1^2} - \frac{(T - mT_s)^2}{2T_1^2} - \frac{(T - nT_s)^2}{2(T_1^2)^*}\right]. \tag{10.275}$$

The Fourier transform of $F(T, Z)$ is (Example 10.13)

$$\tilde{F}(\omega, Z) = \frac{P_0^{3/2} T_0^3}{|T_1|^2 T_1} \sqrt{\frac{\pi}{C(Z)}} \sum_{lmn} b_l b_m b_n \exp\left[-g(Z) + [i\omega - d(Z)]^2/4C(Z)\right], \tag{10.276}$$

where

$$C(Z) = \frac{3T_0^2 + iS}{2(T_0^4 + S^2)}, \tag{10.277}$$

$$d(Z) = \frac{T_s[(l + m + n)T_0^2 + i(l + m - n)S]}{T_0^4 + S^2}, \tag{10.278}$$

$$g(Z) = \frac{T_s^2[(l^2 + m^2 + n^2)T_0^2 + i(l^2 + m^2 - n^2)S]}{2(T_0^4 + S^2)}. \tag{10.279}$$

Substituting Eq. (10.276) in Eq. (10.269), and after performing the inverse Fourier transformation, we find

$$u_1(T, L_{\text{tot}}) = \sum_{lmn} \delta u_{lmn}(T, L_{\text{tot}}) b_l b_m b_n, \tag{10.280}$$

$$\delta u_{lmn}(T, L_{\text{tot}}) = iP_0^{3/2} T_0^3 \int_0^{L_{\text{tot}}} \frac{a^2(x) \exp\left[-(2CT - d)^2/(4C(1 + i2SC) - C)\right]}{\sqrt{(1 + i2SC)(T_0^2 - iS)(T_0^4 + S^2)}} dx. \tag{10.281}$$

For a single-span system with constant loss α and dispersion coefficient β_2, Eq. (10.281) reduces to [38, 43]

$$\delta u_{lmn}(T, L) = iP_0^{3/2} T_0^3 \int_0^L \frac{\exp\left[-\alpha Z - (2CT - d)^2 / (4C(1 + i2\beta_2 CZ) - C)\right]}{\sqrt{(1 + i2\beta_2 CZ)(T_0^2 - i\beta_2 Z)(T_0^4 + \beta_2^2 Z^2)}} dZ, \tag{10.282}$$

where L is the fiber length.

10.9.1   Variance Calculations

Without loss of generality, we consider the nonlinear distortion on the pulse located at $T = 0$. The total field at the end of the transmission line is

$$u(T = 0, L_{\text{tot}}) = u_0(T = 0, L_{\text{tot}}) + \gamma u_1(T = 0, L_{\text{tot}})$$

$$= \frac{\sqrt{P_0} T_0}{T_1(L_{\text{tot}})} \left[b_0 + \sum_{n=-\infty, n\neq 0}^{\infty} b_n \exp\left(-\frac{n^2 T_s^2}{2T_1^2}\right) \right]$$

$$+ \gamma \sum_{l=-\infty}^{\infty} \sum_{m=-\infty}^{\infty} \sum_{n=-\infty}^{\infty} b_l b_m b_n \delta u_{lmn}. \tag{10.283}$$

The second term on the right-hand side of Eq. (10.283) represents the ISI from the neighboring symbols and the last term on the right-hand side represents the nonlinear distortion due to SPM, IXPM, and IFWM. The nonlinear interaction between pulses centered at lT_s, mT_s, and nT_s results in an echo pulse centered approximately at $(l + m - n)T_s$. Therefore, the dominant contributions to the nonlinear distortion at $T = 0$ come from the symbol slots that satisfy $l + m - n = 0$ and all the other triplets in Eq. (10.283) can be ignored. When $l = m = n = 0$, δu_{000} corresponds to SPM. When $l = 0$ and $m = n$, δu_{0mm} corresponds to intrachannel XPM (IXPM). All the other triplets satisfying $l + m - n = 0$ represent the echo pulses due to intrachannel FWM (IFWM). Let us first calculate the variance of '1' in BPSK systems. Let us assume that the bit in the symbol slot is '1', i.e., $b_0 = 1$. Considering only the triplets that satisfy $l + m - n = 0$, Eq. (10.283) can be written as

$$u(T = 0, L_{\text{tot}}) = \frac{\sqrt{P_0} T_0}{T_1(L_{\text{tot}})} \left[1 + \sum_{n=-\infty, n\neq 0}^{\infty} b_n \exp\left(-\frac{n^2 T_s^2}{2T_1^2}\right) \right]$$

$$+ \gamma \left[\underbrace{\delta u_{000}}_{\text{SPM}} + \underbrace{2 \sum_{m=-\infty}^{\infty} \delta u_{0mm}}_{\text{IXPM}} + \underbrace{\sum_{l+m-n=0, l\neq 0, m\neq 0} \delta u_{lmn} b_l b_m b_n}_{\text{IFWM}} \right]. \tag{10.284}$$

The last term in Eq. (10.284) excludes SPM and IXPM. As can be seen from Eq. (10.284), the contribution from SPM and IXPM leads to deterministic amplitude and phase changes. At the receiver, the dispersion is fully compensated either in the optical domain or using the DSP (see Chapter 11). So, we assume that $T_1(L_{\text{tot}}) = T_0$ in Eq. (10.284). For BPSK systems, the photocurrent is proportional to the real part of $u(0, L_{\text{tot}})$. Setting the constant proportionality to be unity, the current at $T = 0$ can be written as

$$I = I_0 + \delta I, \tag{10.285}$$

where I_0 is the mean current given by

$$I_0 = \sqrt{P_0}\,\mathrm{Re}\left\{ 1 + \gamma \left(\delta u_{000} + 2 \sum_{m=-\infty}^{\infty} \delta u_{0mm} \right) \right\} \qquad (10.286)$$

and

$$\delta I = \sum_{n=-\infty, n\neq 0}^{\infty} b_n \delta u_{\mathrm{lin},n} + \gamma \sum_{l+m-n=0, l\neq 0, m\neq 0} \mathrm{Re}(\delta u_{lmn}) b_l b_m b_n, \qquad (10.287)$$

where

$$\delta u_{\mathrm{lin},n} = \sqrt{P_0}\,\exp\left(-n^2 T_s^2 / 2 T_0^2\right). \qquad (10.288)$$

For BPSK, we have

$$\langle b_n \rangle = 0, \qquad (10.289)$$

$$\langle b_n b_m \rangle = \delta_{nm}, \qquad (10.290)$$

where δ_{nm} is the Kronecker delta function. To calculate the variance, Eq. (10.287) is rewritten as

$$\delta I = \delta I_{\mathrm{lin}} + \delta I_{\mathrm{IFWM},d} + \delta I_{\mathrm{IFWM},nd}, \qquad (10.291)$$

where δI_{lin}, $\delta I_{\mathrm{IFWM},d}$, and $\delta I_{\mathrm{IFWM},nd}$ represent random currents due to linear ISI, degenerate IFWM, and non-degenerate IFWM, respectively. An IFWM triplet is degenerate if $l = m$. From Eq. (10.287), we have

$$\delta I_{\mathrm{lin}} = \sum_{n=-\infty, n\neq 0}^{\infty} b_n \delta u_{\mathrm{lin},n}, \qquad (10.292)$$

$$\delta I_{\mathrm{IFWM},d} = \gamma \sum_{l+m-n=0, l=m} \mathrm{Re}\{\delta u_{lln}\} b_n, \qquad (10.293)$$

$$\delta I_{\mathrm{IFWM},nd} = 2\gamma \sum_{l+m-n=0, l<m, l\neq m\neq n} \mathrm{Re}\{\delta u_{lmn}\} b_l b_m b_n. \qquad (10.294)$$

The factor 2 is introduced to account for the fact that the summation is carried out only over the region of $l < m$. In Eqs. (10.293) and (10.294), the terms corresponding to intra-pulse SPM and IXPM are excluded. The variance of δI_{lin} is

$$< \delta I_{\mathrm{lin}}^2 > = \sum_{m\neq 0} \sum_{n\neq 0} < b_m b_n > \delta u_{\mathrm{lin},m} \delta u_{\mathrm{lin},n}, \qquad (10.295)$$

Using Eq. (10.290), Eq. (10.295) simplifies to

$$< \delta I_{\mathrm{lin}}^2 > = \sum_{m\neq 0} \delta u_{\mathrm{lin},m}^2 = P_0 \sum_{m\neq 0} \exp\left(\frac{-m^2 T_s^2}{T_0^2} \right), \qquad (10.296)$$

$$< \delta I_{\mathrm{IFWM},d}^2 > = \gamma^2 \sum_{l+m-n=0, l=m} \sum_{l'+m'-n'=0, l'=m'} \mathrm{Re}[\delta u_{lln}]\mathrm{Re}[\delta u_{l'l'n'}] < b_n b_{n'} >$$

$$= \gamma^2 \sum_{l+m-n=0, l=m} (\mathrm{Re}[\delta u_{lln}])^2. \qquad (10.297)$$

In Eq. (10.297), we have used Eq. (10.290) and when $n = n'$, l has to be equal to l' to satisfy $l + m - n = 0$ and $l' + m' - n' = 0$.

Next consider the correlation between linear and degenerate IFWM,

$$< \delta I_{\text{lin}} \delta I_{\text{IFWM},d} > = \gamma \sum_{l+m-n=0,l=m} \sum_{n',n'\neq 0} \text{Re}[\delta u_{lln}] \delta u_{\text{lin},n'} < b_n b_{n'} >$$

$$= \gamma \sqrt{P_0} \sum_{l+m-n=0,l=m,n\neq 0} \text{Re}[\delta u_{lln}] \exp\left(\frac{-n^2 T_s^2}{T_0^2}\right). \tag{10.298}$$

The variance of $\delta I_{\text{IFWM},nd}$ is

$$< \delta I_{\text{IFWM},nd}^2 > = 4\gamma^2 \sum_{\substack{l+m-n=0 \\ l<m,l\neq m\neq n}} \sum_{\substack{l'+m'-n'=0 \\ l'<m',l'\neq m'\neq n'}} \text{Re}[\delta u_{lmn}]$$

$$\text{Re}[\delta u_{l'm'n'}] < b_l b_m b_n b_{l'} b_{m'} b_{n'} > . \tag{10.299}$$

Since $< b_l b_m b_n b_{l'} b_{m'} b_{n'} > = \delta_{ll'} \delta_{mm'} \delta_{nn'}$, Eq. (10.299) is simplified as

$$< \delta I_{\text{IFWM},nd}^2 > = 4\gamma^2 \sum_{l+m-n=0,l<m,l\neq m\neq n} (\text{Re}\{\delta u_{lmn}\})^2. \tag{10.300}$$

Using Eqs. (10.289) and (10.290) in Eq. (10.287), it is easy to show that

$$< \delta I > = 0, \tag{10.301}$$

$$\sigma_{\text{PSK}}^2 = < \delta I^2 > = < \delta I_{\text{lin}}^2 > + < \delta I_{\text{IFWM},d}^2 > + < \delta I_{\text{IFWM},nd}^2 > + 2 < \delta I_{\text{lin}} \delta I_{\text{IFWM},d} >$$
$$+ 2 < \delta I_{\text{lin}} \delta I_{\text{IFWM},nd} > + 2 < \delta I_{\text{IFWM},nd} \delta I_{\text{IFWM},d} > . \tag{10.302}$$

It can be shown that the correlation between degenerate IFWM and non-degenerate IFWM is zero and that between linear ISI and non-degenerate IFWM is also zero. Hence, Eq. (10.302) becomes

$$\sigma_{\text{PSK}}^2 = < \delta I_{\text{lin}}^2 > + < \delta I_{\text{IFWM},d}^2 > + < \delta I_{\text{IFWM},nd}^2 > + 2 < \delta I_{\text{lin}} \delta I_{\text{IFWM},d} > . \tag{10.303}$$

Next, let us consider a direct detection OOK system. The photocurrent is

$$I \propto P = |u(T, L_{\text{tot}})|^2. \tag{10.304}$$

Setting the constant of proportionality to be unity and using Eq. (10.283),

$$I(T=0) = |u_0(T=0, L_{\text{tot}}) + \gamma u_1(T=0, L_{\text{tot}})|^2$$
$$= |u_0(0, L_{\text{tot}})|^2 + 2\gamma \text{Re}\{u_0(0, L_{\text{tot}}) u_1^*(0, L_{\text{tot}})\} + \gamma^2 |u_1(0, L_{\text{tot}})|^2. \tag{10.305}$$

In Eq. (10.305), the first, second, and last terms on the right-hand side represent the currents due to the linear transmission, signal–nonlinear distortion beating, and nonlinear distortion–nonlinear distortion beating. When the nonlinear distortion is small, the last term can be ignored. Eq. (10.305) can be written as

$$I = I_0 + \delta I_{\text{lin}} + \delta I_{nl}, \tag{10.306}$$

where

$$I_0 = P_0, \tag{10.307}$$

$$\delta I_{\text{lin}} \approx 2\sqrt{P_0} \sum_{n=-\infty, n\neq0}^{\infty} b_n \delta u_{\text{lin},n}, \tag{10.308}$$

$$\delta I_{\text{nl}} \approx 2\gamma\sqrt{P_0} \sum_{l+m-n=0} b_l b_m b_n \text{Re}(\delta u_{\text{lmn}}). \tag{10.309}$$

The variance is calculated as (see Example 10.14)

$$\sigma_{\text{OOK}}^2 = <I^2> - <I>^2 \tag{10.310}$$

$$= \sigma_{\text{lin}}^2 + \sigma_{nl}^2, \tag{10.311}$$

where

$$\sigma_{\text{lin}}^2 = P_0 \sum_{m=-\infty}^{\infty} \exp\left(\frac{-m^2 T_s^2}{T_0^2}\right), \tag{10.312}$$

$$\sigma_{nl}^2 = 4\gamma^2 P_0 \sum_{l+m-n=0} \sum_{l'+m'-n'=0} \left(\frac{1}{2^{x(l,m,n,l',m',n')}} - \frac{1}{2^{r(l,m,n)-r(l',m',n')}}\right) \text{Re}(\delta u_{lmn}) \text{Re}(\delta u_{l'm'n'}). \tag{10.313}$$

$r(l, m, n)$ is the number of non-degenerate indices in the set $\{l, m, n\}$ and $x(l, m, n, l', m', n')$ is the number of non-degenerate indices in the set $\{l, m, n, l', m', n'\}$.

10.9.2   Numerical Simulations

To test the accuracy of the semi-analytical expressions for the variance, numerical simulation of the NLSE is carried out using the symmetric split-step Fourier scheme (see Chapter 11). The fiber-optic link is shown in Fig. 10.28. A dispersion-compensating fiber (DCF) is used for pre-, inline, and post-compensation. The parameters of the transmission fiber (TF) and DCF are shown in Table 10.1. Two-stage EDFA is used with a DCF between the amplifiers. Let the accumulated dispersions of the pre- and post-compensating fibers be $\mathcal{D}_{\text{pre}}$

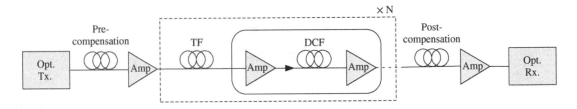

Figure 10.28   A typical fiber-optic transmission system. TF = transmission fiber, DCF = dispersion compensating fiber.

Table 10.1   Parameters of the transmission fiber and DCF.

Fiber type	D (ps/km/nm)	γ (W^{-1} km^{-1})	Loss (dB/km)
TF	17	1.1	0.2
DCF	−120	4.86	0.45

and $\mathcal{D}_{\text{post}}$, respectively. Dispersion of the TF is partially compensated by the in-line dispersion compensation. Let $\mathcal{D}_{\text{res}}$ be the residual accumulated dispersion of a single span, i.e.,

$$\mathcal{D}_{\text{res}} = D_{\text{TF}}L_{\text{TF}} + D_{\text{inline}}L_{\text{inline}}, \tag{10.314}$$

where D and L denote the dispersion parameter (see Chapter 2) and length, respectively, and the subscripts TF and inline correspond to the transmission fiber and inline DCF, respectively. We have assumed that the total accumulated dispersion from transmitter to receiver is zero, i.e.,

$$\mathcal{D}_{\text{pre}} + N\mathcal{D}_{\text{res}} + \mathcal{D}_{\text{post}} = 0, \tag{10.315}$$

where N is the number of TF spans. The following parameters are used in the numerical simulation of the direct detection OOK system. Bit rate = 40 Gb/s, pulsewidth (FWHM) = 5 ps, $N = 10$, peak powers launched to TF and DCF are 10 dBm and 0 dBm, respectively. The lengths of pre-, inline, and post-compensating fibers are chosen so that $\mathcal{D}_{\text{res}} = 100$ ps/nm and $\mathcal{D}_{\text{post}} = \mathcal{D}_{\text{pre}} = -500$ ps/nm. The amplifier noise is turned off. Two pulses centered at 25 ps and 50 ps are launched to the fiber-optic link. Owing to IFWM, echo pulses are generated around 0 ps and 75 ps. The solid and broken lines in Fig. 10.29 show the echo pulses after 10 spans obtained by the analytical expression (Eq. (10.281)) and numerical simulations, respectively. In this example, a small pulse width is chosen so that the echo pulse is not affected by the ISI from the pulse centered at 25 ps. In practice, short-duty-cycle pulses are rarely used because of the large bandwidth which leads to cross-talk in WDM systems.

The percentage pre-compensation ratio is defined as

$$\%\text{pre-compensation ratio} = \frac{\mathcal{D}_{\text{pre}} \times 100}{\mathcal{D}_{\text{pre}} + \mathcal{D}_{\text{post}}}. \tag{10.316}$$

Fig. 10.30 shows the variance as a function of pre-compensation ratio. The following parameters are used for Fig. 10.30: pulsewidth (FWHM) = 12.5 ps, $D_{\text{res}} = 100$ ps/nm, peak powers launched to TF and DCF are 0 dBm

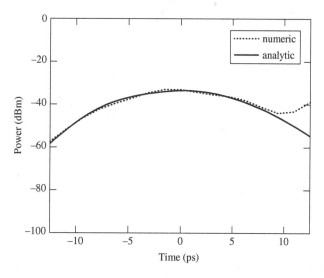

Figure 10.29 Comparison of the echo pulse power at the output obtained by the analytical expression (Eq. (10.281)) and numerical simulations. Two signal pulses centered around 25 ps and 50 ps are launched to the fiber (not shown in the figure).

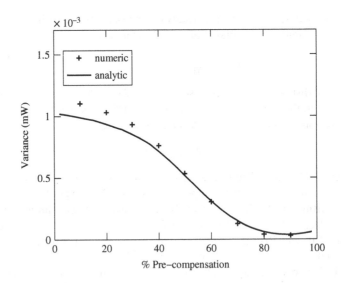

Figure 10.30 Variance of '1' at the fiber output for an OOK direct detection system. $D_{res}=100\,\text{ps/nm}$.

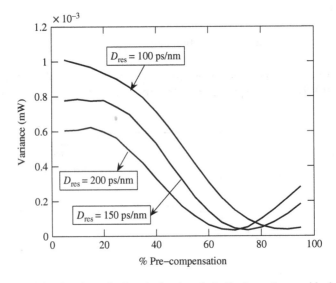

Figure 10.31 The variance of the signal amplitude calculated analytically for various residual accumulated dispersions per span. The other parameters are the same as in Fig. 10.30.

and $-3\,\text{dBm}$, respectively. The rest of the parameters are the same as in Fig. 10.29. To calculate the variance of '1', a 13-bit-long random bit sequence with the center bit being '1' is transmitted over the fiber-optic link in simulations. The bits are randomly varied, keeping the center bit fixed as '1' and exhausting all possible bit patterns. From Fig. 10.30, we see that the optimum pre-compensation ratio is about 90% when $D_{res} = 100\,\text{ps/nm}$. However, if the residual dispersion per span is increased, the pre-compensation ratio is lowered, as shown in Fig. 10.31.

Example 10.6

A rectangular pulse of peak power 6 mW is transmitted over a dispersion-free fiber of length 40 km. Find the nonlinear phase shift at the center of the pulse. Compare the exact results with those obtained using first- and second-order perturbation theory. Assume fiber loss = 0.2 dB/km and $\gamma = 1.1 \, \text{W}^{-1} \, \text{km}^{-1}$.

Solution:

In the absence of dispersion, the field envelope evolution is given by the NLSE in lossless form (Eq. 10.246),

$$i\frac{\partial u}{\partial Z} = -\gamma e^{-\alpha Z}|u|^2 u. \tag{10.317}$$

Let

$$u = Ae^{i\theta}. \tag{10.318}$$

Substituting Eq. (10.318) in Eq. (10.317), we find

$$i\frac{dA}{dZ}e^{i\theta} - \frac{d\theta}{dZ}Ae^{i\theta} = -\gamma e^{-\alpha Z}|A|^2 Ae^{i\theta}. \tag{10.319}$$

Comparing the real and imaginary parts of Eq. (10.319), we find

$$\frac{dA}{dZ} = 0, \quad A = \text{constant}, \tag{10.320}$$

$$\frac{d\theta}{dZ} = \gamma e^{-\alpha Z}|A|^2, \tag{10.321}$$

$$\theta(T,Z) - \theta(T,0) = \gamma|A|^2 \int_0^Z e^{-\alpha Z} dZ$$

$$= \gamma|A|^2 Z_{\text{eff}}, \tag{10.322}$$

$$Z_{\text{eff}} = \frac{1 - e^{-\alpha Z}}{\alpha}. \tag{10.323}$$

Note that A does not change as a function of Z, but it may depend on T. Since $|u(T,Z)| = A$ does not change with Z, we have

$$|u(T,0)| = |u(T,Z)| = A. \tag{10.324}$$

Using Eq. (10.324) in Eqs. (10.322) and (10.318), we find

$$\theta(T,Z) = \theta(T,0) + \gamma|u(T,0)|^2 Z_{\text{eff}}, \tag{10.325}$$

$$u(T,Z) = |u(T,0)|e^{i[\theta(T,0)+\gamma|u(T,0)|^2 Z_{\text{eff}}]} \tag{10.326}$$

$$= u(T,0)e^{i\gamma|u(T,0)|^2 Z_{\text{eff}}}. \tag{10.327}$$

From Eqs. (2.177) and (10.323), we have

$$\alpha = \frac{0.2}{4.343} \, \text{km}^{-1} = 4.605 \times 10^{-2} \, \text{km}^{-1}, \tag{10.328}$$

$$Z_{\text{eff}} = \frac{1 - \exp(-4.605 \times 10^{-2} \times 40)}{4.605 \times 10^{-2}} = 18.27 \, \text{km}. \tag{10.329}$$

The nonlinear phase shift is

$$\phi_{NL}(T) = \gamma |u(T,0)|^2 Z_{\text{eff}}, \tag{10.330}$$

$$= \gamma P(T,0) Z_{\text{eff}}. \tag{10.331}$$

At $T = 0$, we have

$$\phi_{NL}(0) = 1.1 \times 10^{-3} \times 6 \times 10^{-3} \times 18.27 \times 10^3 = 0.1206 \,\text{rad}. \tag{10.332}$$

The Taylor expansion of Eq. (10.326) yields

$$u(T,Z) = u(T,0) \left\{ 1 + i\gamma |u(T,0)|^2 Z_{\text{eff}} - \frac{\gamma^2}{2!} |u(T,0)|^4 Z_{\text{eff}}^2 + \cdots \right\}. \tag{10.333}$$

Here, the second and third terms on the right-hand sied of Eq. (10.333) represent the first-order and second-order corrections due to nonlinear effects. First, let us consider only the first-order term

$$u(T,Z) = u(T,0)\{1 + i\gamma |u(T,0)|^2 Z_{\text{eff}}\} \tag{10.334}$$

$$= u(T,0)B(T)e^{i\xi(T)}, \tag{10.335}$$

where

$$B(T) = \sqrt{1 + \gamma^2 |u(T,0)|^4 Z_{\text{eff}}^2}, \tag{10.336}$$

$$\xi(T) = \tan^{-1}(\gamma |u(T,0)|^2 Z_{\text{eff}}). \tag{10.337}$$

At $T = 0$, we have

$$B(0) = 1.007, \tag{10.338}$$

$$\xi(0) = 0.12002 \,\text{rad}. \tag{10.339}$$

B and ξ represent the amplitude shift and nonlinear phase shift using the first-order theory. Note that from the exact solution given by Eq. (10.326), we see that there is no change in amplitude due to fiber nonlinearity; the first-order approximation shows that the amplitude is shifted by a factor of 1.007.

Next, consider the terms up to second order in Eq. (10.333),

$$u(T,Z) = u(T,0)\{x + iy\}, \tag{10.340}$$

$$x = 1 - \frac{\gamma^2}{2!} |u(T,0)|^4 Z_{\text{eff}}^2, \tag{10.341}$$

$$y = \gamma |u(T,0)|^2 Z_{\text{eff}}. \tag{10.342}$$

Let $x + iy = Be^{i\xi}$. Proceeding as before, we find

$$B(0) = 1.00002, \tag{10.343}$$

$$\xi(0) = 0.12089 \,\text{rad}. \tag{10.344}$$

Comparing Eqs. (10.338) and (10.339) with Eqs. (10.343) and (10.344), we see that the second-order theory is closer to the exact result.

10.10   Nonlinear Phase Noise

So far we have ignored the nonlinear interaction between the signal and the ASE of inline amplifiers. Owing to ASE, the amplitude or power of the optical signal fluctuates randomly about a mean value. Since the nonlinear phase shift due to the SPM is proportional to power, the phase of the signal fluctuates randomly. This type of noise was first studied by Gordon and Mollenauer [51] and, hence, this noise is also known as *Gordon-Mollenauer phase noise*. The nonlinear phase-noise leads to performance degradations in phase-modulated systems such as DPSK or QPSK systems. The analysis of nonlinear phase noise in phase-modulated fiber-optic transmission systems has drawn significant attention [51–76]. In the following section, we consider the impact of ASE when the nonlinear effects are absent and in Section 10.10.2, an expression for the variance of phase noise including the SPM is derived.

10.10.1   Linear Phase Noise

Consider the output of the optical transmitter, $s_{in}(T)$, which is confined to the bit interval $-T_b/2 < T < T_b/2$. Let

$$s_{\text{in}}(T) = a_0 \sqrt{E} p(T), \tag{10.345}$$

where a_0 is the symbol in the interval $-T_b/2 < T < T_b/2$, $p(T)$ is the pulse shape, E is the energy of the pulse, and

$$\int_{-\infty}^{\infty} |p(T)|^2 dT = 1. \tag{10.346}$$

For BPSK, a_0 takes values 1 and -1 with equal probability. In this section, we ignore the fiber dispersion and nonlinearity and include only fiber loss. To compensate for fiber loss, amplifiers are introduced periodically along the transmission line with a spacing of L_a. The amplifier compensates for the loss exactly and introduces ASE noise. Let us consider a single-span fiber-optic system with a single amplifier at the fiber output. Let the amplifier compensate for the fiber loss exactly. The output of the amplifier may be written as

$$s_{\text{out}}(T) = s_{\text{in}}(T) + n(T), \tag{10.347}$$

where $n(T)$ is the ASE noise which can be treated as white,

$$< n(T) >= 0, \tag{10.348}$$

$$< n(T)n^*(T') >= \rho\delta(T - T'), \tag{10.349}$$

$$< n(T)n(T') >= 0, \tag{10.350}$$

where ρ is the ASE power spectral density per polarization given by (Eq. (6.17))

$$\rho = n_{sp}h\bar{f}(G - 1). \tag{10.351}$$

Here, G is the gain of the amplifier, n_{sp} is a spontaneous noise factor, h is Planck's constant, and $\bar{f}$ is the mean optical carrier frequency.

A signal of bandwidth B and duration T_b has $2J = 2BT_b$ degrees of freedom (DOF). From the Nyquist sampling theorem, it follows that if the highest frequency component of a signal is $B/2$, the signal is described completely by specifying the values of the signal at instants of time separated by $1/B$. Therefore, in the interval T_b, there are BT_b complex samples which fully describe the signal. Equivalently, the signal can be described by J complex coefficients (or $2J$ real coefficients) of the expansion in a set of orthonormal basis functions. Let us represent the signal and noise fields using an orthonormal set of basis functions as

$$s_{\text{in}}(T) = \sum_{j=0}^{J-1} s_j F_j(T), \tag{10.352}$$

$$n(T) = \sum_{j=0}^{J-1} n_j F_j(T), \tag{10.353}$$

where $\{F_j(T)\}$ is a set of orthonormal functions,

$$\int_{-\infty}^{\infty} F_j(T) F_k^*(T) dT = 1 \ \text{ if } j = k$$

$$= 0 \ \text{ otherwise.} \tag{10.354}$$

Because of the orthogonality of the basis functions, it follows that

$$n_j = \int_{-\infty}^{\infty} n(T) F_j^*(T) dT. \tag{10.355}$$

Using Eqs. (10.355) and (10.348)–(10.350), we obtain

$$< n_j > = 0, \tag{10.356}$$

$$< n_j n_k^* > = \rho \ \text{ if } j = k$$

$$= 0 \ \text{ otherwise,} \tag{10.357}$$

$$< n_j n_k) > = 0. \tag{10.358}$$

Using Eqs. (10.352) and (10.353) in Eq. (10.347), we find

$$s_{\text{out}}(T) = \sum_{j=0}^{J-1} (s_j + n_j) F_j(T). \tag{10.359}$$

Suppose '1' is transmitted ($a_0 = 1$). We choose $F_0(T) = p(T)$ so that

$$s_j = \sqrt{E} \ \text{ if } j = 0$$

$$= 0 \ \text{ otherwise.} \tag{10.360}$$

Eq. (10.359) can be written as

$$s_{\text{out}}(T) = (\sqrt{E} + n_0) p(T) + \sum_{j=1}^{J-1} n_j F_j(T). \tag{10.361}$$

Let us assume that the signal power is much larger than the noise power and $s_{\text{in}}(T)$ is real. Let

$$n(T) = n_r(T) + i n_i(T), \tag{10.362}$$

where $n_r = \text{Re}\{n(T)\}$ and $n_i = \text{Im}\{n(T)\}$. Eq. (10.347) can be written as

$$s_{\text{out}}(T) = A(T)\exp[i\phi(T)], \tag{10.363}$$

where

$$A(T) = \{[s_{\text{in}}(T) + n_r(T)]^2 + n_i^2(T)\}^{1/2}, \tag{10.364}$$

$$\phi(T) = \tan^{-1}\left\{\frac{n_i(T)}{s_{\text{in}}(T) + n_r(T)}.\right\}$$

$$\approx \frac{n_i(T)}{s_{\text{in}}(T)}. \tag{10.365}$$

In Eq. (10.365), we have ignored higher-order terms such as n_i^2 and n_r^2. Using Eqs. (10.352), (10.353), (10.360), and (10.361) in Eq. (10.365), we obtain

$$\phi(T) = \frac{n_{0i}}{\sqrt{E}} + \sum_{j=1}^{J-1} \frac{n_{ji}F_j(T)}{p(T)\sqrt{E}}, \tag{10.366}$$

where $n_{jr} = \text{Re}\{n_j\}$ and $n_{ji} = \text{Im}\{n_j\}$. From Eqs. (10.366) and (10.356), it follows that

$$< \phi(T) >= 0. \tag{10.367}$$

Squaring and averaging Eq. (10.366) and using Eqs. (10.357) and (10.358), we obtain the variance of phase noise as

$$\sigma_{\text{lin}}^2 =< \phi^2 >= \frac{\rho}{2E} + \frac{\rho}{2E} \sum_{j=1}^{J-1} \frac{F_m^2(T)}{F_0^2(T)}. \tag{10.368}$$

Next, let us consider the impact of a matched filter on the phase noise. When a matched filter is used, the received signal is

$$r = \int_{-\infty}^{\infty} s_{\text{out}}(T)F_0^*(T)dT. \tag{10.369}$$

Substituting Eq. (10.361) in Eq. (10.369) and using Eq. (10.354), we obtain

$$r = (\sqrt{E} + n_0). \tag{10.370}$$

Note that the higher-order noise components given by the second term on the right-hand side of Eq. (10.361) do not contribute because of the orthogonality of basis functions. Now, Eq. (10.368) reduces to

$$\sigma_{\text{lin}}^2 = \frac{< n_{0i}^2 >}{E} = \frac{\rho}{2E}. \tag{10.371}$$

From Eq. (10.370), we see that when a matched filter is used, the noise field is fully described by two degrees of freedom, namely, the in-phase component n_{0r} and the quadrature component n_{0i}. The other degrees of freedom are orthogonal to the signal and do not contribute after the matched filter. From Eq. (10.371), we see that the quadrature component n_{0i} is responsible for the linear phase noise.

10.10.2   Gordon–Mollenauer Phase Noise

The optical field envelope in a fiber-optic transmission system can be described by the NLSE in the lossless form (Eq. (10.246)),

$$i\frac{\partial u}{\partial Z} - \frac{\beta_2(Z)}{2}\frac{\partial^2 u}{\partial T^2} = -\gamma a^2(Z)|u|^2 u. \tag{10.372}$$

Amplifier noise effects can be introduced in Eq. (10.372) by adding a source term on the right-hand side, which leads to

$$i\frac{\partial u}{\partial Z} - \frac{\beta_2(Z)}{2}\frac{\partial^2 u}{\partial T^2} = -\gamma a^2(Z)|u|^2 u + iR(Z,T), \tag{10.373}$$

where

$$R(Z,T) = \sum_{m=1}^{N_a} \delta(Z - mL_a)n(T). \tag{10.374}$$

Here, N_a is the number of amplifiers and $n(T)$ is the noise field due to ASE, with statistical properties defined in Eqs. (10.348)–(10.350).

In this section, we first consider the case in which the fiber dispersion is zero. Let us consider the solution of Eq. (10.373) in the absence of noise. Let

$$u(Z,T) = A(Z,T)\exp[i\phi(Z,T)], \tag{10.375}$$

and

$$u(0,T) = \sqrt{E}p(T). \tag{10.376}$$

Substituting Eq. (10.375) in Eq. (10.372), we find

$$\frac{dA}{dZ} = 0 \rightarrow A(Z,T) = A(0,T) = \sqrt{E}|p(T)|, \tag{10.377}$$

$$\frac{d\phi}{dZ} = \gamma a^2(Z)|u(0,T)|^2$$

$$= \gamma a^2(Z)E|p(T)|^2. \tag{10.378}$$

Solving Eq. (10.378), we find

$$\phi(Z,T) = \gamma E|p(T)|^2 \int_0^Z a^2(s)ds, \tag{10.379}$$

$$u(Z,T) = u(0,T)\exp\left[i\gamma|u(0,T)|^2 \int_0^Z a^2(s)ds\right]. \tag{10.380}$$

We assume that the signal pulse shape is rectangular, with pulse width T_b. From Eq. (10.346), it follows that $|p(T)|^2 = 1/T_b$. Since $a^2(Z) = \exp(-\alpha_0 Z)$ between amplifiers, we have

$$\int_0^{mL_a-} a^2(Z)dZ = mL_{\text{eff}}, \tag{10.381}$$

where

$$L_{\text{eff}} = \frac{1 - \exp(-\alpha_0 L_a)}{\alpha_0}. \tag{10.382}$$

Substituting Eq. (10.381) in Eqs. (10.379) and (10.380), we find

$$\phi(mL_a-, T) = \frac{\gamma E m L_{\text{eff}}}{T_b},$$ (10.383)

$$u(mL_a-, T) = \sqrt{E}p(T) \exp [i\phi(mL_a-)].$$ (10.384)

Next, let us consider the case when there is only one amplifier located at mL_a that introduces ASE noise. The optical field envelope after the amplifier is

$$u(mL_a+, T) = u(mL_a-, T) + n(T).$$ (10.385)

We assume that two degrees of freedom in the noise field are of importance. They are the in-phase component n_{0r} and the quadrature component n_{0i}; we ignore other noise components. As mentioned in Section 10.10.1, the noise field is fully described by these two degrees of freedom for a linear system. Gordon and Mollenauer [51] assumed that these two degrees of freedom are adequate to describe the noise field even for a nonlinear system. Using Eqs. (10.384) and (10.353) in Eq. (10.385), we find

$$u(mL_a+, T) = \sqrt{E}p(T) \exp [i\phi(mL_a-)] + n_0 p(T)$$

$$= (\sqrt{E} + n_0')p(T) \exp [i\phi(mL_a-)],$$ (10.386)

where

$$n_0' = n_0 \exp [-i\phi(mL_a-)].$$ (10.387)

n_0' is the same as n_0, except for a deterministic phase shift which does not alter the statistical properties, i.e.,

$$< n_0' >= 0,$$ (10.388)

$$< n_0' n_0'^* >= \rho,$$ (10.389)

$$< n_0' n_0' >= 0.$$ (10.390)

From Eq. (10.386), we see that the complex amplitude of the field envelope has changed because of the amplifier noise. Using $u(mL_a+, T)$ as the initial condition, the NLSE (10.372) is solved to obtain the field at the end of the transmission line as

$$u(L_{\text{tot}}, T) = u(mL_a+, T) \exp \left\{ i\gamma |u(mL_a+, T)|^2 \int_{mL_a+}^{L_{\text{tot}}} a^2(Z)dZ \right\}$$

$$= (\sqrt{E} + n_0')p(T) \exp [i\phi(mL_a-) + i\gamma |\sqrt{E} + n_0'|^2 (N_a - m)L_{\text{eff}}/T_b],$$ (10.391)

where $L_{\text{tot}} = N_a L_a$ is the total transmission distance. The phase at L_{tot} is

$$\phi = \tan^{-1} \left\{ \frac{n_{0i}'}{\sqrt{E} + n_{0r}'} \right\} + \frac{\gamma |\sqrt{E} + n_0'|^2 (N_a - m)L_{\text{eff}}}{T_b} + \frac{\gamma E m L_{\text{eff}}}{T_b}$$

$$\approx \frac{n_{0i}'}{\sqrt{E}} + \gamma (E + 2\sqrt{E}n_{0r}')(N_a - m)L_{\text{eff}}/T_b + \gamma E m L_{\text{eff}}/T_b.$$ (10.392)

The total phase given by Eq. (10.392) can be separated into two parts:

$$\phi = \phi_d + \delta\phi, \tag{10.393}$$

where ϕ_d is the deterministic nonlinear phase shift given by

$$\phi_d = \gamma E N_a L_{\text{eff}}/T_b \tag{10.394}$$

and $\delta\phi$ represents the phase noise,

$$\delta\phi = \frac{n'_{0i}}{\sqrt{E}} + \frac{2\gamma\sqrt{E}n'_{0r}(N_a - m)L_{\text{eff}}}{T_b}. \tag{10.395}$$

The first and second terms in Eq. (10.395) represent the linear and nonlinear phase noise, respectively. As can be seen, the in-phase component n'_{0r} and the quadrature component, n'_{0i} are responsible for nonlinear and linear phase noise, respectively. From Eq. (10.388), it follows that

$$< \delta\phi > = 0. \tag{10.396}$$

Squaring and averaging Eq. (10.395) and using Eqs. (10.389) and (10.390), we find the variance of the phase noise as

$$\sigma_m^2 = \frac{\rho}{2E} + 2\rho E\left[\frac{\gamma(N_a - m)L_{\text{eff}}}{T_b}\right]^2. \tag{10.397}$$

So far, we have ignored the impact of ASE due to other amplifiers. In the presence of ASE due to other amplifiers, the expression for the optical field envelope at mL_a- given by Eq. (10.384) is inaccurate since it ignores the noise field added by the amplifiers preceding the mth amplifier. However, when the signal power is much larger than the noise power, second-order terms such as n_{0r}^2 and n_{0i}^2 can be ignored. At the end of the transmission line, the dominant contribution would come from the linear terms n_{0i} and n_{0i} of each amplifier. Since the noise fields of amplifiers are statistically independent, the total variance is the sum of the variance due to each amplifier,

$$\sigma^2 = \sum_{m=1}^{N_a} \sigma_m^2$$

$$= \frac{\rho N_a}{2E} + 2\rho E\left[\frac{\gamma L_{\text{eff}}}{T_b}\right]^2 \sum_{m=1}^{N_a-1}(N_a - m)^2$$

$$= \frac{\rho N_a}{2E} + \frac{(N_a - 1)N_a(2N_a - 1)\rho E\gamma^2 L_{\text{eff}}^2}{3T_b^2}. \tag{10.398}$$

Refs. [55–58] provide a more rigorous treatment of the nonlinear phase noise without ignoring the higher-order noise terms. From Eq. (10.398), we see that the variance of the linear phase noise (the first term on the right-hand side) increases linearly with the number of amplifiers, whereas the variance of the nonlinear phase noise (the second term) increases cubically with the number of amplifiers when N_a is large, indicating that nonlinear phase noise could be the dominant penalty for ultra-long-haul fiber-optic transmission systems. In addition, the variance of linear phase noise is inversely proportional to the energy of the pulse, whereas the variance of nonlinear phase noise is directly proportional to the energy. This implies that there exists an optimum energy at which the total phase variance is minimum. By setting $d\sigma^2/dE$ to zero, the optimum energy is calculated as

$$E_{\text{opt}} = \frac{T_b}{\gamma L_{\text{eff}}} \sqrt{\frac{3}{2(N_a - 1)(2N_a - 1)}}. \tag{10.399}$$

When N_a is large, $(N_a - 1)(2N_a - 1) \approx 2N_a^2$ and using Eq. (10.394), we find that the phase variance is minimum when the deterministic nonlinear phase shift $\phi_d \approx 0.87$ rad. Eqs. (10.397) and (10.398) are derived under the assumption that dispersion is zero. In the presence of dispersion, Eq. (10.397) is modified as [60]

$$\sigma_m^2 = \frac{\rho}{2E} + 2\rho E[\gamma g_{fr}(mL_a)^2], \tag{10.400}$$

where

$$g_{fr}(x) = \frac{T_0}{\sqrt{\pi}} \text{Re} \left\{ \int_x^{L_{\text{tot}}} G(s)ds \right\}, \tag{10.401}$$

$$G(s) = \frac{a^2(s)}{\sqrt{(1 + T_0^2 \Delta(s))(T_0^4 + 3S^2(s) + 2iT_0^2 S(s))}}, \tag{10.402}$$

$$\Delta(s) = \frac{T_0^2 - iS(s)}{T_0^2[T_0^2 + i3S(s)]}. \tag{10.403}$$

T_0 and $S(z)$ are defined in Section 10.9. If a dispersion-managed fiber with zero mean dispersion per span is used, the total variance can be expressed in a form similar to Eq. (10.398) [60],

$$\sigma^2 = \frac{\rho N_a}{2E} + \frac{(N_a - 1)N_a(2N_a - 1)\rho E(\gamma h_{fr})^2}{3}, \tag{10.404}$$

$$h_{fr} = \frac{T_0}{\sqrt{\pi}} \text{Re} \left\{ \int_0^{L_a} G(s)ds \right\}, \tag{10.405}$$

Comparing Eqs. (10.398) and (10.404), we see that these two expressions are the same except that L_{eff}/T_0 is replaced by h_{fr}. For a highly dispersive system, h_{fr} is much smaller than L_{eff}/T_0 and, hence, the variance of nonlinear phase noise due to SPM is much smaller in a highly dispersive system. Eq. (10.404) does not include contributions due to IXPM. Even if IXPM contributions are included, numerical simulations have shown that for highly dispersive systems, the variance of nonlinear phase noise (signal–noise interaction) is much smaller than that due to IFWM and IXPM (signal–signal interactions). In a WDM system, interaction between ASE and XPM leads to nonlinear phase noise as well [76]. Using the digital back propagation technique discussed in Chapter 11, it is possible to compensate for deterministic (symbol pattern-dependent signal–signal interactions) nonlinear effects, but not for nonlinear phase noise (signal–ASE interactions). So, when the DBP is used to compensate for intra- and interchannel nonlinear impairments, nonlinear phase noise is likely to be one of the dominant impairments.

Example 10.7

A rectangular pulse of peak power 2 mW and pulse width 25 ps is transmitted over a periodically amplified dispersion-free fiber-optic transmission system operating at 1550 nm. The fiber-optic link consists of 20 amplifiers with an amplifier spacing of 80 km. The parameters of the link are as follows: nonlinear coefficient

$\gamma = 1.1\,\mathrm{W}^{-1}\,\mathrm{km}^{-1}$, loss coefficient $\alpha = 0.0461\,\mathrm{km}^{-1}$, spontaneous emission factor $n_{sp} = 1.5$. Find the variance of (a) linear phase noise, (b) nonlinear phase noise at the receiver.

Solution:

The PSD of ASE is

$$\rho = h\bar{f}(G-1)n_{sp},$$

$$G = \exp\left(\alpha L_a\right) = \exp\left(0.0461 \times 80\right) = 39.96,$$

$$\bar{f} = \frac{c}{\lambda} = \frac{3 \times 10^8}{1550 \times 10^{-9}}\,\mathrm{Hz} = 193.54\,\mathrm{THz},$$

$$\rho = 6.626 \times 10^{-34} \times 193.54(39.96 - 1) \times 1.5\,\mathrm{J} = 7.495 \times 10^{-18}\,\mathrm{J}.$$

The pulse energy is

$$E = PT_b = 2 \times 10^{-3} \times 25 \times 10^{-12}\,\mathrm{J} = 5 \times 10^{-14}\,\mathrm{J}.$$

The variance of linear phase noise is

$$\sigma_{\mathrm{lin}}^2 = \frac{N_a\rho}{2E} = \frac{20 \times 7.495 \times 10^{-18}}{2 \times 5 \times 10^{-14}}\,\mathrm{rad}^2 = 1.499 \times 10^{-3}\,\mathrm{rad}^2.$$

The effective length is

$$L_{\mathrm{eff}} = \frac{1 - \exp\left(-\alpha L_a\right)}{\alpha} = \frac{1 - \exp\left(-0.0461 \times 80\right)}{0.0461}\,\mathrm{km} = 21.14\,\mathrm{km}.$$

The variance of the nonlinear phase noise is

$$\sigma_{\mathrm{nl}}^2 = \frac{(N_a - 1)N_a(2N_a - 1)\rho E\gamma^2 L_{\mathrm{eff}}^2}{3T_b^2}$$

$$= \frac{19 \times 20 \times 39 \times 7.495 \times 10^{-18} \times 5 \times 10^{-14} \times (1.1 \times 10^{-3} \times 21.14 \times 10^3)^2}{3 \times (25 \times 10^{-12})^2}\,\mathrm{rad}^2$$

$$= 1.603 \times 10^{-3}\,\mathrm{rad}^2.$$

The total variance is

$$\sigma_{\mathrm{tot}}^2 = \sigma_{\mathrm{lin}}^2 + \sigma_{\mathrm{nl}}^2$$

$$= 1.499 \times 10^{-3} + 1.603 \times 10^{-3}\,\mathrm{rad}^2$$

$$= 3.102 \times 10^{-3}\,\mathrm{rad}^2.$$

10.11   Stimulated Raman Scattering

When a light wave propagates in a medium, a small fraction of light emerges in directions other than that of the incident wave. Most of the scattered light has the same frequency as the incident light, a small part has frequencies different from the incident light. This phenomenon was first observed by Raman [77] (and

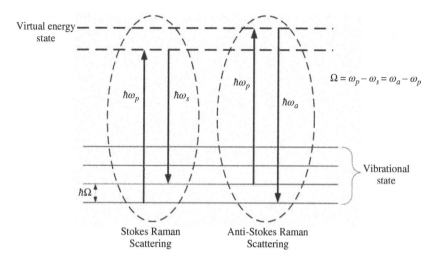

Figure 10.32   Stokes and anti-Stokes Raman scattering.

independently by Landsberg and Manderlstam [78]) in 1928. The molecules in the medium have several vibrational states (or phonon modes). When a light wave (photons) interacts with sound waves (phonons), the frequency of the light wave is shifted up or down. The shift in frequency gives information about the phonon modes of the molecules. When the scattered photon has a frequency lower than the incident photon, it is known as *Stokes shift*. Stokes Raman scattering can be described quantum mechanically as the annihilation of a pump photon of energy $\hbar\omega_p$ and the creation of a Stokes photon of lower energy $\hbar\omega_s$, and absorption of the energy $\hbar(\omega_p - \omega_s)$ by the molecules by making transition from a low-energy vibrational state to a high-energy vibrational state (see Fig. 10.32). A lower-energy photon has lower frequency and, therefore, Stokes Raman scattering leads to a red shift of the incident light wave. If the molecule makes transition from a high-energy vibrational state to a low-energy vibrational state in the presence of an incident pump of energy $\hbar\omega_p$, the difference in energy is added to the incident photon, leading to a photon of higher energy $\hbar\omega_a$ (which is of higher frequency). This is known as anti-Stokes Raman scattering. Raman scattering is quite useful in chemistry, since vibrational information is specific to the chemical bonds and symmetry of molecules.

Spontaneous Raman scattering is typically very weak. In 1962, it was found that an intense-pump optical wave can excite molecular vibrations and, thereby, stimulate molecules to emit photons of reduced energy (a Stokes wave), to which most of the pump energy is transferred [79]. This is known as *stimulated Raman scattering*. The interaction between the pump and the Stokes wave is described by the following coupled equations under CW conditions [80]:

$$\frac{d\mathcal{I}_s}{dZ} = g_R \mathcal{I}_p \mathcal{I}_s - \alpha_s \mathcal{I}_s, \tag{10.406}$$

$$\frac{d\mathcal{I}_p}{dZ} = -\frac{\omega_p}{\omega_s} g_R \mathcal{I}_p \mathcal{I}_s - \alpha_p \mathcal{I}_p, \tag{10.407}$$

where $\mathcal{I}_p$ and $\mathcal{I}_s$ are the optical intensities of the pump and the Stokes waves, respectively, α_p and α_s are the fiber loss coefficients at the pump and the Stokes frequencies, respectively, and $g_R(\Omega)$, $\Omega = \omega_p - \omega_s$ is the Raman gain coefficient. The amplification of the Stokes wave by the pump wave can be understood from Eqs. (10.406) and (10.407). To simplify the analysis, we assume that the pump intensity is much larger than the Stokes intensity so that the depletion of the pump due to Stokes wave amplification can be ignored. Eq. (10.407) may be approximated as

$$\frac{d\mathcal{I}_p}{dZ} \cong -\alpha_p \mathcal{I}_p, \tag{10.408}$$

$$\mathcal{I}_p(Z) = \mathcal{I}_p(0)e^{-\alpha_p Z}. \tag{10.409}$$

From Eq. (10.406), we have

$$\frac{d\mathcal{I}_s}{\mathcal{I}_s} = g_R \mathcal{I}_p(0)e^{-\alpha_p Z} - \alpha_s. \tag{10.410}$$

Integrating Eq. (10.410) from 0 to L, we obtain

$$\ln\left(\frac{\mathcal{I}_s(L)}{\mathcal{I}_s(0)}\right) = g_R \mathcal{I}_p(0)L_{\text{eff},p} - \alpha_s L, \tag{10.411}$$

where the effective length of the pump, $L_{\text{eff},p}$, is given by

$$L_{\text{eff},p} = \frac{1 - \exp(-\alpha_p L)}{\alpha_p}. \tag{10.412}$$

Rearranging Eq. (10.411), we find

$$\mathcal{I}_s(L) = \mathcal{I}_s(0)e^{-\alpha_s L + g_R \mathcal{I}_p(0)L_{\text{eff},p}}. \tag{10.413}$$

The Stokes wave is amplified if $g_R \mathcal{I}_p(0)L_{\text{eff},p} > \alpha_s L$. If a signal (Stokes wave) is down-shifted in frequency by about 14 THz, it would have the highest amplification since $g_R(\Omega)$ is maximum when the frequency shift is about 14 THz (see Fig. 6.21).

Example 10.8

Stokes and pump beams co-propagate in a fiber of length 2 km. The Raman coefficient of the fiber $g_R = 1 \times 10^{-13}$ m/W, input Stokes's signal power $= -10$ dBm, input pump power $= 20$ dBm, $\alpha_s = 0.046$ km^{-1}, $\alpha_p = 0.08$ km^{-1}, and effective area of the fiber $= 40$ μm^2. Calculate the Stokes signal power at the fiber output.

Solution:
The Stokes and pump powers in a fiber may be approximated as

$$\mathcal{I}_{p,s} \cong \frac{P_{p,s}}{A_{\text{eff}}}. \tag{10.414}$$

So, Eq. (10.413) can be rewritten as

$$P_s(L) = P_s(0)e^{-\alpha_s L + g_R P_p(0)L_{\text{eff},p}/A_{\text{eff}}}, \tag{10.415}$$

$$P_p(0)\,(\text{dBm}) = 20\,\text{dBm}. \tag{10.416}$$

$$P_p(0) = 10^{P_p(0)\,(\text{dBm})/10}\,\text{mW}$$

$$= 100\,\text{mW}, \tag{10.417}$$

$$P_s(0)\,(\text{dBm}) = -10\,\text{dBm}, \tag{10.418}$$

$$P_s(0) = 10^{P_s(0)\,(\mathrm{dBm})/10}\,\mathrm{mW}$$

$$= 0.1\,\mathrm{mW}, \tag{10.419}$$

$$L_{\mathrm{eff},p} = \frac{1 - \exp(-\alpha_p L)}{\alpha_p}$$

$$= \frac{1 - \exp(-0.08 \times 2)}{0.08 \times 10^{-3}}$$

$$= 1.848\,\mathrm{km}, \tag{10.420}$$

$$\frac{g_R P_p(0) L_{\mathrm{eff},p}}{A_{\mathrm{eff}}} = \frac{1 \times 10^{-13} \times 100 \times 10^{-3} \times 1.848 \times 10^3}{40 \times 10^{-12}}$$

$$= 0.462, \tag{10.421}$$

$$\alpha_s L = 0.092. \tag{10.422}$$

So, the gain dominates the loss. From Eq. (10.413), we find

$$P_s(L) = P_s(0) e^{-0.092 + 0.462}$$

$$= 0.1447\,\mathrm{mW}. \tag{10.423}$$

10.11.1   Time Domain Description

In the time domain, stimulated Raman scattering can be explained as follows [81]. When a high-intensity optical pulse interacts with a molecule, it perturbs the electronic structure of the molecule and results in intensity-dependent polarizability of the molecule (see Section 10.2). This electronic effect occurs on a time scale shorter than the pulse width of the optical pulse, and it can be considered instantaneous. However, perturbation of the electronic structure by the optical pulse also perturbs the Coloumb interaction between the nuclei and the electronic structure, which can excite molecular vibrations. These vibrations, in turn, perturb the electronic structure, leading to a delayed change in polarizability. The time- and intensity-dependent change in polarizability (or equivalent refractive index) associated with the excitation of a molecular vibration is the Raman effect, whereas the instantaneous intensity-dependent change in polarizability is the Kerr effect.

As mentioned in Section 10.4, the Kerr effect is taken into account by the term $\gamma|q|^2 q$ in the nonlinear Schrödinger equation. In the presence of the Raman effect, it can be modified as [82]

$$\gamma|q(T,Z)|^2 \to \gamma(1-\alpha)|q(T,Z)|^2 + \gamma\alpha \int_{-\infty}^{\infty} h(s)|q(T-s,Z)|^2 ds, \tag{10.424}$$

where α is the fraction of the nonlinearity resulting from the Raman contributions and $h(T)$ is the normalized Raman response function, with

$$\int_{-\infty}^{\infty} h(T)dT = 1 \tag{10.425}$$

and $h(-|t|) = 0$ to ensure causality. The response function $h(T)$ is specific to the medium and the imaginary part of its Fourier transform is related to the Raman gain coefficient $g(\Omega)$ [81, 83]. Let

$$P(T,Z) = |q(T,Z)|^2, \tag{10.426}$$

$$P(T-s,Z) = |q(T-s,Z)|^2. \tag{10.427}$$

The Taylor series expansion of $P(T - s, Z)$ around T is

$$P(T - s, Z) = P(T, Z) - s\frac{\partial P}{\partial T} + \frac{s^2}{2}\frac{\partial^2 P}{\partial T^2} + \cdots \tag{10.428}$$

If the spectral width of the signal is sufficiently small, terms that are proportional to s^2 and beyond may be ignored. Under this condition, Eq. (10.424) becomes

$$\gamma|q(T, Z)|^2 \to \gamma(1 - \alpha)|q(T, Z)|^2 + \gamma\alpha \int_{-\infty}^{\infty}\left[|q(T, Z)|^2 - s\frac{\partial|q(T, Z)|^2}{\partial T}\right]h(s)ds. \tag{10.429}$$

Using Eq. (10.425), Eq. (10.429) is rewritten as

$$\gamma|q(T, Z)|^2 \to \gamma|q(T, Z)|^2 - \kappa\frac{\partial|q(T, Z)|^2}{\partial T}, \tag{10.430}$$

where

$$\kappa = \gamma\alpha \int_{-\infty}^{\infty} sh(s)ds. \tag{10.431}$$

Substituting Eq. (10.430) in Eq. (10.81), we obtain

$$i\frac{\partial q}{\partial Z} - \frac{\beta_2}{2}\frac{\partial^2 q}{\partial T^2} + \gamma|q|^2 q + i\frac{\alpha q}{2} = \kappa\frac{\partial|q(T, Z)|^2}{\partial T}q. \tag{10.432}$$

Eq. (10.432) is the modified nonlinear Schrodinger equation and the term on the right-hand side denotes the Raman contributions. The energy exchange between the pump and the Stokes waves can be understood from Eq. (10.432) by considering the pump and Stokes waves as CW:

$$q = q_p + q_s, \tag{10.433}$$

$$q_p = A_p e^{-i\Omega_p T}, \tag{10.434}$$

$$q_s = A_s e^{-i\Omega_s T}, \tag{10.435}$$

where A_p and A_s denote the complex amplitudes of pump and Stokes's waves, respectively, and Ω_p and Ω_s are the corresponding angular frequency offset from the reference. Let us first consider

$$|q|^2 = |A_p|^2 + |A_s|^2 + A_p A_s^* e^{-i\Omega T} + A_p^* A_s e^{i\Omega T}, \tag{10.436}$$

$$\frac{\partial|q|^2}{\partial T}q = (-i\Omega)A_p|A_s|^2 e^{-i\Omega_p T} + (i\Omega)A_s|A_p|^2 e^{-i\Omega_s T} + \text{terms at } 2\Omega_p - \Omega_s \quad \text{and} \quad 2\Omega_s - \Omega_p, \tag{10.437}$$

where

$$\Omega = \Omega_p - \Omega_s. \tag{10.438}$$

Substituting Eq. (10.437) in Eq. (10.432) and collecting the terms that are proportional to $e^{-i\Omega_p T}$ and $e^{-i\Omega_s T}$, we find

$$i\frac{dA_p}{dZ} + \frac{\beta_2}{2}\Omega_p^2 A_p + \gamma\{|A_p|^2 + 2|A_s|^2\}A_p = -i\kappa\Omega|A_s|^2 A_p - \alpha A_p, \tag{10.439}$$

$$i\frac{dA_s}{dZ} + \frac{\beta_2}{2}\Omega_s^2 A_s + \gamma\{|A_s|^2 + 2|A_p|^2\}A_s = i\kappa\Omega|A_p|^2 A_s - \alpha A_s. \tag{10.440}$$

Eqs. (10.439) and (10.440) represent the evolution of the complex amplitudes of the pump and the Stokes waves. To obtain the expression for the energy exchange between the pump and the Stokes waves, we multiply Eq. (10.439) by A_p^* and subtract its complex conjugate to obtain

$$\frac{dP_p}{dZ} = -gP_pP_s - \alpha P_p, \tag{10.441}$$

where $P_p = |A_p|^2$ and $g = 2\kappa\Omega$. Similar operations on Eq. (10.440) leads to

$$\frac{dP_s}{dZ} = gP_pP_s - \alpha P_s. \tag{10.442}$$

From Eqs. (10.441) and (10.442), we see that the gain coefficients at the pump and the Stokes frequencies are identical, but they are not the same in Eqs. (10.406) and (10.407). This is because of our linear approximation (first term in the Taylor expansion) to the Raman time response function.

SRS has a number of applications. If an intense Raman pump is launched to the fiber, it can amplify a weak signal if the frequency difference lies within the bandwidth of the Raman gain spectrum. SRS can also be used to construct Raman fiber lasers which can be tuned over a wide frequency range (~ 10 THz) [84–88]. In other types of fiber amplifiers, the SRS process can be detrimental since the pump energy is used to amplify the range of wavelengths over which amplification is not desired. In WDM systems, a channel of higher frequency transfers energy to a channel of lower frequency, leading to Raman cross-talk and performance degradations [89–91].

10.12 Additional Examples

Example 10.9

When the bandwidth of the WDM signal and/or the dispersion slope are large, β_3 can not be ignored. In this case, show that the phase mismatch factor given by Eq. (10.216) should be modified as

$$\Delta\beta_{jkln} = (\Omega_l\Omega_n - \Omega_j\Omega_k)\left[\beta_2 + \frac{\beta_3}{2}(\Omega_j + \Omega_k)\right]. \tag{10.443}$$

Solution:

$$\beta_j = \beta_0 + \beta_1\Omega_j + \frac{\beta_2}{2}\Omega_j^2 + \frac{\beta_3}{6}\Omega_j^3, \tag{10.444}$$

$$\Omega_n = \Omega_j + \Omega_k - \Omega_l. \tag{10.445}$$

Consider the contribution to the phase mismatch $\Delta\beta_{jkln}$ due to the last term of Eq. (10.444),

$$\frac{\beta_3}{6}[\Omega_j^3 + \Omega_k^3 - \Omega_l^3 - \Omega_n^3] = \frac{\beta_3}{6}[\Omega_j^3 + \Omega_k^3 - \Omega_l^3 - (\Omega_j + \Omega_k - \Omega_l)^3]. \tag{10.446}$$

Using the formula

$$(a+b+c)^3 = a^3 + b^3 + 3a^2b + 3ab^2 + c^3 + 3(a+b)^2c + 3(a+b)c^2, \tag{10.447}$$

Eq. (10.446) is simplified as

$$\frac{\beta_3}{6}\{\Omega_j^3 + \Omega_k^3 - \Omega_l^3 - [\Omega_j^3 + \Omega_k^3 + 3\Omega_j^2\Omega_k + 3\Omega_j\Omega_k^2 - \Omega_l^3 - 3(\Omega_j + \Omega_k)^2\Omega_l + 3(\Omega_j + \Omega_k)\Omega_l^2]\}$$

$$= -\frac{\beta_3}{6}\{3\Omega_j\Omega_k(\Omega_j + \Omega_k) + 3(\Omega_j + \Omega_k)\Omega_l(\Omega_l - \Omega_j - \Omega_k)\}$$

$$= -\frac{\beta_3}{2}(\Omega_j + \Omega_k)\{\Omega_j\Omega_k - \Omega_l\Omega_n\}. \tag{10.448}$$

Adding the contributions due to β_3 and β_2, Eq. (10.216) is modified as

$$\Delta\beta_{jkln} = [\Omega_l\Omega_n - \Omega_j\Omega_k]\left[\beta_2 + \frac{\beta_3}{2}(\Omega_j + \Omega_k)\right]. \tag{10.449}$$

Example 10.10

A WDM system has five channels centered at $l\Delta f$, $l = -2, -1, 0, 1, 2$, $\Delta f = 50\,\text{GHz}$. The launch power per channel is 3 dBm, and the channels are CW. The transmission fiber has the following parameters: $\alpha = 0.046\,\text{km}^{-1}$, $L = 20\,\text{km}$, $\beta_2 = -4\,\text{ps}^2/\text{km}$, and $\gamma = 1.8\,\text{W}^{-1}\,\text{km}^{-1}$. The initial phases of the channels are $\theta_{-2} = 0.5\,\text{rad}$, $\theta_{-1} = -0.7\,\text{rad}$, $\theta_0 = 1.2\,\text{rad}$, $\theta_1 = 0.8\,\text{rad}$, and $\theta_2 = -1\,\text{rad}$. Find the FWM power on the central channel ($l = 0$). Ignore β_3.

Solution:

The FWM tones falling on the central channel should satisfy the condition

$$(j + k - l)\Delta f = n\Delta f = 0. \tag{10.450}$$

The possible triplets are shown in Table 10.2. Here, ND and D refer to non-degenerate and degenerate FWM tones, respectively. From Eq. (10.216), we see that $\Delta\beta_{jkln}$ is invariant under the exchange of j and k, i.e.,

$$\Delta\beta_{jkln} = \Delta\beta_{kjln}. \tag{10.451}$$

Hence, the FWM tones corresponding to $\{j, k, l\}$ and $\{k, j, l\}$ should be identical. For example, the triplets $\{-2, 1, -1\}$ and $\{1, -2, -1\}$ produce identical FWM tones. So, we need to consider only the tones listed in Table 10.3. First consider the triplet $\{-2, 1, -1\}$. The FWM field for this triplet is given by Eq. (10.221),

$$\epsilon_0^{(-2,1,-1,0)}(L) = i\gamma P^{3/2} e^{-\frac{\alpha L}{2} + i\Delta\theta_{-2,1,-1}} \frac{(1 - e^{-\delta_{-2,1,-1,0}L})}{\delta_{-2,1,-1,0}}, \tag{10.452}$$

$$\delta_{-2,1,-1,0} = \alpha - i\Delta\beta_{-2,1,-1,0}$$

$$= \alpha + i\beta_2\Omega_{-2}\Omega_1$$

$$= 0.046 \times 10^{-3} - i4 \times 10^{-27} \times (2\pi \times 50 \times 10^9)^2 \times (-2)\,\text{m}^{-1}$$

$$= 4.6 \times 10^{-5} + 7.89 \times 10^{-4}i\,\text{m}^{-1}, \tag{10.453}$$

$$\Delta\theta_{-2,1,-1} = 0.5 + 0.8 - (-0.7) \quad \text{rad}$$

$$= 2\,\text{rad}, \tag{10.454}$$

$$P(\text{dBm}) = 3\,\text{dBm}, \tag{10.455}$$

$$P(\text{mW}) = 10^{P(\text{dBm})/10}\,\text{mW}$$

$$= 2\,\text{mW}. \tag{10.456}$$

Substituting Eqs. (10.453), (10.454), and (10.456) in Eq. (10.452), we find

$$\epsilon_0^{(-2,1,-1,0)}(L) = (-0.8 + 1.6 \text{ i}) \times 10^{-4}\,\sqrt{\text{W}}. \tag{10.457}$$

Similarly, the FWM fields due to other triplets are

$$\epsilon_0^{(-2,2,0,0)}(L) = (3.33 \times 10^{-7} - 3.92 \times 10^{-5} \text{ i}) \quad \sqrt{\text{W}}, \tag{10.458}$$

$$\epsilon_0^{(-1,-1,-2,0)}(L) = (2.03 + 1.9 \text{ i}) \times 10^{-4}\,\sqrt{\text{W}}, \tag{10.459}$$

$$\epsilon_0^{(-1,1,0,0)}(L) = (2.28 - 1.6 \text{ i}) \times 10^{-4}\,\sqrt{\text{W}}, \tag{10.460}$$

$$\epsilon_0^{(-1,2,1,0)}(L) = (-1.4 - 1.12 \text{ i}) \times 10^{-4}\,\sqrt{\text{W}}, \tag{10.461}$$

$$\epsilon_0^{(1,1,2,0)}(L) = (1.435 - 2.394 \text{ i}) \times 10^{-4}\,\sqrt{\text{W}}. \tag{10.462}$$

The total FWM field is

$$\epsilon_0 = 2\epsilon^{(-2,1,-1,0)} + 2\epsilon^{(-2,2,0,0)} + 2\epsilon^{(-1,1,0,0)} + 2\epsilon^{(-1,2,1,0)} + \epsilon^{(1,1,2,0)} + \epsilon^{(-1,-1,-2)}$$

$$= (3.64 - 3.509 \text{ i}) \times 10^{-4}\,\sqrt{\text{W}}. \tag{10.463}$$

The FWM power at the fiber output is

$$P_{\text{FWM}} = |\epsilon_0|^2 = 2.56 \times 10^{-4}\,\text{mW}. \tag{10.464}$$

Table 10.2 FWM tones on the central channel.

j	k	l	Type
−2	1	−1	ND
−2	2	0	ND
−1	−1	−2	D
−1	1	0	ND
−1	2	1	ND
1	−2	−1	ND
1	−1	0	ND
1	1	2	D
2	−2	0	ND
2	−1	1	ND

Table 10.3 FWM tones on the central channel with the degeneracy factor.

Tone number	Number of tones	j	k	l	Type
1	2	-2	1	-1	ND
2	2	-2	2	0	ND
3	2	-1	1	0	ND
4	2	-1	2	1	ND
5	1	1	1	2	D
6	1	-1	-1	-2	D

Example 10.11

For a single-span dispersion-free fiber, find the nonlinear distortion up to second order using the perturbation theory.

Solution:
When $\beta_2 = 0$, from Eq. (10.252), we have

$$i\frac{du_1}{dZ} = -a^2(Z)|u_0|^2 u_0. \tag{10.465}$$

For a single-span system, $a^2(Z) = \exp(-\alpha Z)$. From Eq. (10.255), we have $u_0 = k$. Integrating Eq. (10.465) and using $u_1(T, 0) = 0$, we obtain

$$u_1(T, Z) = iZ_{\text{eff}}|k|^2 k, \tag{10.466}$$

where

$$Z_{\text{eff}} = \frac{1 - \exp(-\alpha Z)}{\alpha}. \tag{10.467}$$

From Eq. (10.253), we have

$$\frac{du_2}{dZ} = i\exp(-\alpha Z)(2iZ_{\text{eff}}|k|^4 k - iZ_{\text{eff}}|k|^4 k)$$

$$= \frac{\exp(-\alpha Z) - \exp(-2\alpha Z)}{\alpha}(-|k|^4 k). \tag{10.468}$$

Integrating Eq. (10.468), we obtain

$$u_2(T, Z) = -\left[\frac{1 - \exp(-\alpha Z)}{\alpha^2} - \frac{1 - \exp(-2\alpha Z)}{2\alpha^2}\right]|k|^4 k$$

$$= -\frac{|k|^4 k}{2}Z_{\text{eff}}^2. \tag{10.469}$$

The total solution up to second order is

$$u = u_0 + \gamma u_1 + \gamma^2 u_2$$

$$= k\left(1 + i\gamma|k|^2 Z_{\text{eff}} - \frac{\gamma^2|k|^4 Z_{\text{eff}}^2}{2}\right). \tag{10.470}$$

Example 10.12

The evolution of the complex field envelope in a periodically amplified fiber-optic system is governed by

$$i\frac{\partial q}{\partial Z} - \frac{\beta_2(Z)}{2}\frac{\partial^2 q}{\partial T^2} + \gamma(Z)|q|^2 q = -i\frac{\alpha(Z)}{2}q + iA\sum_{n=1}^{N}\delta(Z - nL_a)q(T, nL_{a-}), \tag{10.471}$$

where

$$\beta_2(Z) = \beta_{20} \quad \text{for } 0 < \text{mod}(Z, L_a) < L_a$$
$$= 0 \qquad \text{otherwise,} \tag{10.472}$$

$$\gamma(Z) = \gamma_0 \quad \text{for } 0 < \text{mod}(Z, L_a) < L_a$$
$$= 0 \qquad \text{otherwise} \tag{10.473}$$

$$\alpha(Z) = \alpha_0 \quad \text{for } 0 < \text{mod}(Z, L_a) < L_a$$
$$= 0 \qquad \text{otherwise} \tag{10.474}$$

Using the transformation

$$q(Z, T) = a(Z)u(Z, T), \tag{10.475}$$

show that

$$i\frac{\partial u}{\partial Z} - \frac{\beta_2(Z)}{2}\frac{\partial^2 u}{\partial T^2} + \frac{\gamma}{2}a^2(Z)|u|^2 u = 0, \tag{10.476}$$

$$a(Z) = e^{-\alpha_0 Z/2} \quad \text{for } 0 < \text{mod}(Z, L_a) < L_a$$
$$= 1 \qquad \text{otherwise.} \tag{10.477}$$

Assume that the fiber loss is exactly compensated by the amplifier gain.

Solution:
Consider the propagation over a short length from nL_{a-} to $nL_{a-} + \Delta Z$ corresponding to the amplifier located at nL_a. In this short length, $\beta_2(Z) = \gamma(Z) = \alpha(Z) = 0$. Integrating Eq. (10.471) from nL_{a-} to $nL_{a-} + \Delta Z$, we obtain

$$i\int_{nL_{a-}}^{nL_{a-}+\Delta Z}\frac{dq}{dZ}dZ = iA\int_{nL_{a-}}^{nL_{a-}+\Delta Z}\delta(Z - nL_a)q(T, nL_{a-}), \tag{10.478}$$

$$q(T, nL_{a-} + \Delta Z) - q(T, nL_{a-}) = Aq(T, nL_{a-}), \tag{10.479}$$

$$\frac{q(T, nL_{a-} + \Delta Z)}{q(T, nL_{a-})} = A + 1. \tag{10.480}$$

Since $q(T, nL_{a-} + \Delta Z)$ and $q(T, nL_{a-})$ represent the amplifier output and input, respectively, we have

$$q(T, nL_{a-} + \Delta Z) = \sqrt{G}q(T, nL_{a-}), \tag{10.481}$$

where $G = \exp(\alpha_0 L_a)$ is the power gain of the amplifier. Comparing Eqs. (10.480) and (10.481), we find

$$A = \sqrt{G} - 1. \tag{10.482}$$

Substituting Eq. (10.475) in Eq. (10.471), we find

$$iu\frac{da}{dZ} + ia\frac{\partial u}{\partial Z} - \frac{\beta_2(Z)}{2}a\frac{\partial^2 u}{\partial T^2} + \gamma a^2|q|^2 q = -i\frac{\alpha(Z)}{2}au$$

$$+ i(\sqrt{G} - 1)\sum_{n=1}^{N}\delta(Z - nL_a)a(nL_{a-})u(T, nL_{a-}). \tag{10.483}$$

Let

$$u\frac{da}{dZ} = \frac{-\alpha a u}{2} + (\sqrt{G} - 1)\sum_{n=1}^{N}\delta(Z - nL_a)a(nL_{a-})u(T, nL_{a-}), \tag{10.484}$$

so that Eq. (10.483) becomes

$$i\frac{\partial u}{\partial Z} - \frac{\beta_2}{2}\frac{\partial^2 u}{\partial T^2} + \gamma a^2(Z)|u|^2 u = 0. \tag{10.485}$$

Note that the optical field q increases abruptly at amplifier locations. Using the transformation of Eq. (10.475), the amplitude fluctuations due to fiber loss and amplifier gain are separated out so that $u(T, Z)$ changes smoothly as a function of Z, i.e.,

$$u(T, nL_{a-}) = u(T, nL_{a+}). \tag{10.486}$$

Consider the region $0 < \text{mod}(Z, L_a) < L_a$. In this region, $\alpha(Z) = \alpha_0$ and the second term on the right-hand side of Eq. (10.484) is zero. Solving Eq. (10.484), we find

$$a(Z) = a(0)e^{-\alpha_0 Z/2}. \tag{10.487}$$

Here $a(0)$ could be chosen arbitrarily. For convenience, let $a(0) = 1$. Next, consider the length from nL_{a-} to $nL_{a-} + \Delta Z$. Integrating Eq. (10.484) from nL_{a-} to $nL_{a-} + \Delta Z$, we find

$$a(nL_{a-} + \Delta Z) - a(nL_{a-}) = (\sqrt{G} - 1)a(nL_{a-}), \tag{10.488}$$

$$a(nL_{a-} + \Delta Z) = \sqrt{G}a(nL_{a-}) = \sqrt{G}e^{-\alpha_0 L_a/2} = 1. \tag{10.489}$$

Combining Eqs. (10.489) and (10.487), we find

$$a(Z) = e^{-\alpha_0 Z/2} \quad \text{for } 0 < \text{mod}(Z, L_a) < L_a$$

$$= 1 \quad \text{otherwise}. \tag{10.490}$$

Fig. 10.33 shows a plot of $a(Z)$ as a function of Z. Note that $a(Z)$ jumps by $\sqrt{G}$ at the amplifier locations.

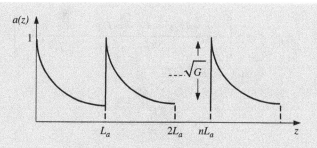

Figure 10.33   Plot of $a(Z)$ vs. distance Z.

Example 10.13

Show that the Fourier transformation of $|u_0|^2 u_0$ is

$$\mathcal{F}\{|u_0|^2 u_0\} = \frac{P_0^{3/2} T_0^3}{|T_1|^2 T_1} \sqrt{\frac{\pi}{C}} \sum_{lmn} b_l b_m b_n \exp\left[g + (i\omega - d)^2/4C\right],$$

where

$$u_0 = \frac{\sqrt{P_0} T_0}{T_1} \sum_{n=-\infty}^{\infty} b_n \exp\left[-\frac{(T - nT_s)^2}{2T_1^2}\right],$$

$$C = \frac{3T_0^2 + iS}{2(T_0^4 + S^2)},$$

$$d = \frac{[(l + m + n)T_0^2 + i(l + m - n)S]T_s}{T_0^4 + S^2},$$

$$g = \frac{[(l^2 + m^2 + n^2)T_0^2 + (l^2 + m^2 - n^2)iS]T_s^2}{2(T_0^4 + S^2)}.$$

Solution:

Let

$$r_l = \exp\left[-\frac{(T - lT_s)^2}{2T_1^2}\right], \tag{10.491}$$

$$|u_0|^2 u_0 = u_0 u_0 u_0^* = \frac{P_0^{3/2} T_0^3}{|T_1|^2 T_1} \sum_{l=-\infty}^{\infty} b_l r_l \sum_{m=-\infty}^{\infty} b_m r_m \sum_{n=-\infty}^{\infty} b_n r_n^*$$

$$= \frac{P_0^{3/2} T_0^3}{|T_1|^2 T_1} \sum_{lmn} b_l b_m b_n r_l r_m r_n^*, \tag{10.492}$$

$$r_l r_m r_n^* = \exp\left[-\frac{(T - lT_s)^2 - (T - mT_s)^2}{2T_1^2} - \frac{(T - nT_s)^2}{2(T_1^2)^*}\right]$$

$$= \exp\left[-(CT^2 - dT) - g\right],\tag{10.493}$$

where

$$C = \frac{3T_0^2 + iS}{2(T_0^4 + S^2)},\tag{10.494}$$

$$d = \frac{[(l + m + n)T_0^2 + i(l + m - n)S]T_s}{T_0^4 + S^2},\tag{10.495}$$

$$g = \frac{[(l^2 + m^2 + n^2)T_0^2 + (l^2 + m^2 - n^2)iS]T_s^2}{2(T_0^4 + S^2)}.\tag{10.496}$$

$$\mathcal{F}\{r_l r_m r_n^*\} = \exp(-g)\int_{-\infty}^{\infty}\exp\left[-(CT^2 - dT) - i\omega T\right]dT$$

$$= \exp(-g)\int_{-\infty}^{\infty}\exp\left[-C(T^2 + 2xT + x^2) + Cx^2\right]dt,\tag{10.497}$$

where

$$x = \frac{i\omega - d}{2C}.\tag{10.498}$$

$$\mathcal{F}\{r_l r_m r_n^*\} = \exp(-g + Cx^2)\int_{-\infty}^{\infty}\exp\left[-C(T + x)^2\right]dT.\tag{10.499}$$

Let

$$u = \sqrt{C}(T + x) \quad \text{and} \quad du = \sqrt{C}dT,\tag{10.500}$$

$$\mathcal{F}\{r_l r_m r_n^*\} = \exp(-g + Cx^2)\frac{1}{\sqrt{C}}\int_{-\infty}^{\infty}\exp(-u^2)du$$

$$= \sqrt{\frac{\pi}{C}}\exp\left[-g + (i\omega - d)^2/4C\right],\tag{10.501}$$

$$\mathcal{F}\{|u_0|^2 u_0\} = \frac{P_0^{3/2}T_0^3}{|T_1|^2 T_1}\sqrt{\frac{\pi}{C}}\sum_{lmn}b_l b_m b_n \exp\left[g + (i\omega - d)^2/4C\right].\tag{10.502}$$

Example 10.14

Find the variance of '1' in a direct detection OOK system due to linear and nonlinear distortion.

Solution:
From Eq. (10.308), we obtain

$$< \delta I_{\text{lin}}^2 > = 4P_0 \sum_{\substack{n=-\infty \\ n\neq 0}}^{\infty} \sum_{\substack{m=-\infty \\ m\neq 0}}^{\infty} < b_n b_m > \delta u_{\text{lin},n}\delta u_{\text{lin},m}.\tag{10.503}$$

For OOK, we have

$$< b_n >= 1/2, \tag{10.504}$$

$$< b_n b_m >= \begin{cases} 1/4 & \text{if } n \neq m \\ 1/2 & \text{if } n = m, \end{cases} \tag{10.505}$$

$$< \delta I_{\text{lin}}^2 >= P_0 \left[\sum_{\substack{n=-\infty \\ n\neq 0}}^{\infty} 2\delta u_{\text{lin},n}^2 + \sum_{\substack{m=-\infty \\ m\neq 0 \\ m\neq n}}^{\infty} \sum_{\substack{n=-\infty \\ n\neq 0}}^{\infty} \delta u_{\text{lin},m} \delta u_{\text{lin},n} \right], \tag{10.506}$$

$$< \delta I_{\text{lin}} >= \sqrt{P_0} \sum_{\substack{n=-\infty \\ n\neq 0}}^{\infty} \delta u_{\text{lin},n}, \tag{10.507}$$

$$\sigma_{\text{lin}}^2 =< \delta I_{\text{lin}}^2 > - < \delta I_{\text{lin}}>^2 = P_0 \sum_{\substack{n=-\infty \\ n\neq 0}}^{\infty} \delta u_{\text{lin},n}^2 = P_0 \sum_{\substack{n=-\infty \\ n\neq 0}}^{\infty} \exp\left[\frac{-m^2 T_s^2}{T_0^2} \right]. \tag{10.508}$$

From Eq. (10.309), we have

$$< \delta I_{nl} >= 2\gamma \sqrt{P_0} \text{Re} \left[\sum_{l+m-n=0} < b_l b_m b_n > \delta u_{lmn} \right], \tag{10.509}$$

$$< b_l b_m b_n >= \frac{1}{2^{r(l,m,n)}}, \tag{10.510}$$

where $r(l, m, n)$ is the number of non-degenerate indices in the set $\{l, m, n\}$. For example, if $\{l, m, n\} = \{2, 5, 7\}$, none of the indices are equal $(l \neq m \neq n)$ and hence $r(l, m, n) = 3$; in a set $\{l, m, n\} = \{0, 3, 3\}$ $(l \neq m = n)$, $r(l, m, n) = 2$; if $l = m = n$, $r(l, m, n) = 1$.

$$< \delta I_{nl}^2 >= 4\gamma^2 P_0 \sum_{l+m-n=0} \sum_{l'+m'-n'=0} < b_l b_m b_n b_l' b_m' b_n' > \text{Re}(\delta u_{lmn}) \text{Re}(\delta u_{l'm'n'}), \tag{10.511}$$

$$< b_l b_m b_n b_l' b_m' b_n' >= \frac{1}{2^{x(l,m,n,l',m',n')}}, \tag{10.512}$$

where $x(l, m, n, l', m', n')$ is the number of non-degenerate indices in a set $\{l, m, n, l', m', n'\}$. For example, if $\{l, m, n, l', m', n'\} = \{1, 2, 3, 2, 3, 5\}$, x is 4. Using Eqs. (10.509) and (10.511), the variance is calculated as

$$\sigma_{nl}^2 =< \delta I_{nl}^2 > - < \delta I_{nl}>^2$$

$$= 4\gamma^2 P_0 \sum_{l+m-n=0} \sum_{l'+m'-n'=0} \left(\frac{1}{2^{x(l,m,n,l',m',n')}} - \frac{1}{2^{r(l,m,n)+r(l',m',n')}} \right) \text{Re}(\delta u_{lmn}) \text{Re}(\delta u_{l'm'n'}). \tag{10.513}$$

Exercises

10.1 Discuss the origin of the nonlinear refractive index.

10.2 A Kerr medium has a cross-sectional area of $500\,\mu m^2$. Calculate the optical power required to change the refractive index by 10^{-8}. Assume $n_2 = 3 \times 10^{-20}\,m^2/W$.

(Ans: 166.66 W.)

10.3 The nonlinear coefficient of a single-mode fiber is $1.2\,W^{-1}\,km^{-1}$. Calculate the effective area. Assume $n_2 = 2.5 \times 10^{-20}\,m^2/W$ and wavelength = 1530 nm.

(Ans: $85\,\mu m^2$.)

10.4 A fiber-optic system has the following parameters: span length = 75 km, number of spans = 20, fiber loss = 0.21 dB/km, Kerr coefficient $n_2 = 2.6 \times 10^{-20}\,m^2/W$, wavelength $\lambda_0 = 1540$ nm, effective area = $50\,\mu m^2$. Find the upper limit on the transmitter output power so that the nonlinear phase shift accumulated over 20 spans is less than 0.5 rad.

(Ans: 0.585 mW.)

10.5 The effective length of a fiber L_{eff} is 18 km and fiber loss = 0.17 dB/km. Find the fiber length.

(Ans: 31.15 km.)

10.6 A single Gaussian pulse of width (FWHM) 20 ps and a peak power of 10 mW is transmitted in a dispersion-free fiber over 80 km. Find the nonlinear phase shift at the center of the pulse at the fiber output. Assume $\gamma = 2.2\,W^{-1}\,km^{-1}$ and fiber loss = 0.2 dB/km.

(Ans: 0.465 rad.)

10.7 Solve the previous exercise numerically using the split-step Fourier scheme (see Chapter 11) and verify the analytical calculations.

10.8 Repeat Exercise 10.7 if $\beta_2 = -2\,ps^2/km$ (instead of $0\,ps^2/km$). Is the nonlinear phase shift at the center of the pulse smaller? Explain.

10.9 Explain the differences between the instantaneous frequency of a pulse due to (i) SPM, (ii) anomalous dispersion, and (iii) normal dispersion.

10.10 Discuss the properties of a soliton in single-mode fibers.

10.11 A modulated pump with a modulating frequency of 8 GHz co-propagated with a weak CW signal. Fiber loss = 0.18 dB/km, length $L = 80$ km, walk-off parameter $d = 13.2$ ps/km, signal wavelength = 1530 nm, and pump wavelength = 1530.78 nm. Calculate the XPM efficiency.

(Ans: 9.72×10^{-4}.)

10.12 Explain the differences between XPM and FWM.

10.13 A WDM system consists of three channels centered at Δf, $2\Delta f$, and $3\Delta f$ with $\Delta f = 100$ GHz. Fiber loss coefficient $\alpha = 0.0461\,km^{-1}$, fiber length $L = 60$ km, and $\beta_2 = -4\,ps^2/km$. Calculate the efficiency of non-degenerate as well as degenerate FWM tones at $4\Delta f$. Ignore β_3.

(Ans: Non-degenerate, 2.27×10^{-4}; degenerate, 8.66×10^{-4}.)

10.14 Two Gaussian pulses of width 40 ps separated by 25 ps are transmitted in a dispersion-free fiber over 80 km. Find the nonlinear phase shift at the center of either of the pulses at the fiber output. Assume $\gamma = 2.2\,\text{W}^{-1}\,\text{km}^{-1}$, fiber loss $= 0.2\,\text{dB/km}$, and peak power $= 4\,\text{mW}$.

(Ans: 0.466 rad.)

10.15 In a 10-Gb/s fiber-optic system based on BPSK operating at 1530 nm there are N_a in-line amplifiers with a noise figure of 4.5 dB. NRZ rectangular pulses are used with a mean power of 0 dBm. The fiber parameters are as follows: $\gamma = 2.2\,\text{W}^{-1}\,\text{km}^{-1}$ and $\alpha = 0.0461\,\text{km}^{-1}$. The variances of linear and nonlinear phase noises at the fiber output are found to be $2.27 \times 10^{-3}\,\text{rad}^2$ and $3.98 \times 10^{-3}\,\text{rad}^2$, respectively. Calculate the amplifier spacing. Ignore dispersion.

(Ans: 100 km.)

10.16 Discuss stimulated Raman scattering in optical fibers.

10.17 In a partially Raman amplified fiber-optic system, it is desired that the gain provided by the Raman pump is 10 dB. The Raman coefficient of the fiber $= 1 \times 10^{-13}\,\text{m/W}$, signal loss $\alpha_s = 0.046\,\text{km}^{-1}$, pump loss $\alpha_p = 0.09\,\text{km}^{-1}$, length $= 80\,\text{km}$, and effective area of the fiber $= 80\,\mu\text{m}^2$. Assuming that the pump co-propagates with the signal, calculate the input pump power.

(Ans: 431 mW.)

Further Reading

G.P. Agrawal, *Nonlinear Fiber Optics, 3rd edn.* Academic Press, San Diego, CA, 2001.
R.W. Boyd, *Nonlinear Optics, 3rd edn.* Academic Press, San Diego, CA, 2007.
Y.R. Shen, *Principles of Nonlinear Optics.* John Wiley & sons, Hoboken, NJ, 2003.
A. Hasegawa and M. Matsumoto, *Optical Solitons in Fibers, 3rd edn.* Springer-Verlag, Berlin, 2003.
N. Bloembergen, *Nonlinear Optics, 4th edn.* World Scientific, Singapore, 1996.
J.V. Moloney and A.C. Newell, *Nonlinear Optics.* Westview Press, Boulder, CO, 2004.

References

[1] A.E. Siegman, *Lasers.* University Science Books, Sausalito, CA, 1986.
[2] V.E. Zakharov and A.B. Shabat, *Sov. Phys. JETP*, vol. **34**, p. 62, 1972.
[3] V.E. Zakharov and F. Calogero, *What is Integrability?* Springer-Verlag, Berlin, 1991.
[4] S. Novikov, S.V. Manakov, L.P. Pitaevskii, and V.E. Zakharov, *Theory of Solitons: The inverse scattering method.* Consultants Bureau, New York, 1984.
[5] M.J. Ablowitz, B. Fuchssteiner, and M. Kruskal, *Topics in Soliton Theory and Exactly Solvable Nonlinear Equations.* World Scientific, Singapore, 1987.
[6] G.L. Lamb Jr., *Elements of Soliton Theory.* John Wiley & Sons, New York, 1980.
[7] A. Hasegawa and F. Tappert, *Appl. Phys. Lett.*, vol. **23**, p. 171, 1973.
[8] L.F. Mollenauer, R.H. Stolen, and J.P. Gordon, *Phys. Rev. Lett.*, vol. **45**(13), p. 1095, 1980.
[9] L.F. Mollenauer, R.H. Stolen, J.P. Gordon, and W.J. Tomlinson, *Opt. Lett.*, vol. **8**(5), p. 289, 1983.
[10] R.H. Stolen, L.F. Mollenauer, and W.J. Tomlinson, *Opt. Lett.*, vol. **8**(3), p. 186, 1983.
[11] I.S. Gradshteyn and I.M. Ryzhik, *Table of Integrals, Series and Products*, 6th edn. Academic Press, San Diego, 2000.

[12] L.F. Mollenauer, S.G. Evangelides, and J.P. Gordon, *J. Lightwave Technol.*, vol. **9**, p. 362, 1991.

[13] A. Hasegawa, S. Kumar, and Y. Kodama, *Opt. Lett.*, vol. **21**, p. 39, 1996.

[14] T.K. Chiang, N. Kagi, M.E. Marhic, and L.G. Kazovsky, *J. Lightwave Technol.*, vol. **14**(3), p. 249, 1996.

[15] D. Marcuse, A.R. Chraplyvy, and R.W. Tkach, *J. Lightwave Technol.*, vol. **12**(5), p. 885, 1994.

[16] R. Hui, Y. Wang, K. Demarest, and C. Allen, *IEEE Photon. Technol. Lett.*, vol. **10**(9), p. 1271, 1998.

[17] R. Hui, K. Demarest, and C. Allen, *J. Lightwave Technol.*, vol. **17**(6), p. 1018, 1999.

[18] A.T. Cartaxo, *J. Lightwave Technol.*, vol. **17**(2), p. 178, 1999.

[19] Z. Jiang and C. Fan, *J. Lightwave Technol.*, vol. **21**(4), p. 953, 2003.

[20] S. Kumar and D. Yang, *J. Lightwave Technol.*, vol.**23**(6), p. 2073, 2005.

[21] Z. Tao *et al.*, *J. Lightwave Technol.*, vol. **29**(7), p. 974, 2011.

[22] J. Wang and K. Petermann, *J. Lightwave Technol.*, vol. **10**(1), p. 96, 1992.

[23] K. Inoue, *IEEE Photon. Technol. Lett.*, vol. **10**(11), p. 1553, 1992.

[24] R.W. Tkach, *J. Lightwave Technol.*, vol. **13**(5), p. 841, 1995.

[25] M. Nazarathy *et al.*, *Opt. Expr.*, vol. **16**, p. 15,777, 2008.

[26] M. Nazarathy and R. Weidenfeld. In S. Kumar (ed.), *Impact of Nonlinearities on Fiber Optic Communications.* Springer-Verlag, New York, 2011, chapter 3.

[27] Y. Chen and A.W. Snyder, *Opt. Lett.*, vol. **14**(1), p. 87, 1989.

[28] K. Inoue, *IEEE Photon. Technol. Lett.*, vol. **8**(2), p. 293, 1996.

[29] S. Burtsev *et al.*, European Conference on Optical Communication, Nice, France, 1999.

[30] F. Matera *et al.*, *Opt. Commun.*, vol. **181**(4–6), p. 407, 2000.

[31] S. Kumar, *J. Lightwave Technol.*, vol. **23**(1), p. 310, 2005.

[32] A. Akhtar, L. Pavel, and S. Kumar, *J. Lightwave Technol.*, vol. **24**(11), p. 4269, 2006.

[33] J. Du, *Opt. Commun.*, vol. **282**(14), p. 2983, 2009.

[34] K. Inoue, *J. Lightwave Technol.*, vol. **12**(6), p. 1023, 1994.

[35] P.V. Mamyshev and N.A. Mamysheva, *Opt. Lett.*, vol. **24**, p. 1454, 1999.

[36] R.J. Essiambre, B. Mikkelsen, and G. Raybon, *Electron. Lett.*, vol. **35**, p. 1576, 1999.

[37] I. Shake *et al.*, *Electron. Lett.*, vol. **34**, p. 1600, 1998.

[38] A. Mecozzi, C.B. Clausen, and M. Shtaif, *IEEE Photon. Technol. Lett.*, vol. **12**, p. 292, 2000.

[39] M.J. Ablowitz and T. Hirooka, *Opt. Lett.*, vol. **25**, p. 1750, 2000.

[40] P. Killey *et al.*, *IEEE Photon. Technol. Lett.*, vol. **12**, p. 1264, 2000.

[41] S. Kumar, *IEEE Photon. Technol. Lett.*, vol. **13**, p. 800, 2001.

[42] S. Kumar *et al.*, *IEEE J. Quant. Electron.*, vol. **8**, p. 626, 2002.

[43] R.J. Essiambre, G. Raybon, and B. Mikkelsen. In I.P. Kaminov and T. Li (eds), *Optical Fiber Telecommunications IVB*. Academic Press, New York, 2002, chapter 6.

[44] D. Yang and S. Kumar, *J. Lightwave Technol.*, vol. **27**, p. 2916, 2009.

[45] S. Turitsyn, M. Sorokina, and S. Derevyanko, *Opt. Lett.*, vol. **37**, p. 2931, 2012.

[46] A. Bononi *et al.*, *Opt. Exp.*, vol. **20**, p. 7777, 2012.

[47] T. Yu *et al.*, *Opt. Lett.*, vol. **22**, p. 793, 1997.

[48] A. Mecozzi *et al.*, *IEEE Photon. Technol. Lett.*, vol. **13**, p. 445, 2001.

[49] M. Matsumoto, *IEEE Photon. Technol. Lett.*, vol. **10**, p. 373, 1998.

[50] A. Hasegawa and Y. Kodama, *Opt. Lett.*, vol. **15**, p. 1443, 1990; vol. **66**, p. 161, 1991.

[51] J.P. Gordon and L.F. Mollenauer, *Opt. Lett.*, vol. **15**(23), p. 1351, 1990.

[52] H. Kim and A.H. Gnauck, *IEEE Photon. Technol. Lett.*, vol. **15**, p. 320, 2003.

[53] P.J. Winzer and R.-J. Essiambre, *J. Lightwave Technol.*, vol. **24**, no. 12, p. 4711, 2006.

[54] S.L. Jansen *et al.*, *IEEE J. Lightwave Technol.*, vol. **24**, p. 54–64, 2006.

[55] A. Mecozzi, *J. Lightwave Technol.*, vol. **12**(11), p. 1993, 1994.

[56] K.-P. Ho, *J. Opt. Soc. Am. B*, vol. **20**(9), p. 1875, 2003.

[57] K.-P. Ho, *Opt. Lett.*, vol. **28**(15), p. 1350, 2003.

[58] A. Mecozzi, *Opt. Lett.*, vol. **29**(7), p. 673, 2004.

[59] A.G. Green, P.P. Mitra, and L.G.L. Wegener, *Opt. Lett.*, vol. **28**, p. 2455, 2003.

[60] S. Kumar, *Opt. Lett.*, vol. **30**, p. 3278, 2005.

[61] C.J. McKinstrie, C. Xie, and T. Lakoba, *Opt. Lett.*, vol. **27**, p. 1887, 2002.

[62] C.J. McKinstrie and C. Xie, *IEEE J. Select. Top. Quantum Electron.*, vol. **8**, p. 616, 2002.

[63] M. Hanna, D. Boivin, P.-A. Lacourt, and J.-P. Goedgebuer, *J. Opt. Soc. Am. B*, vol. **21**, p. 24, 2004.

[64] K.-P. Ho and H.-C. Wang, *IEEE Photon. Technol. Lett.*, vol. **17**, p. 1426, 2005.

[65] K.-P. Ho and H.-C. Wang, *Opt. Lett.*, vol. **31**, p. 2109, 2006.

[66] F. Zhang, C.-A. Bunge, and K. Petermann, *Opt. Lett.*, vol. **31**(8), p. 1038, 2006.

[67] P. Serena, A. Orlandini, and A. Bononi, *J. Lightwave Technol.*, vol. **24**(5), p. 2026, 2006.

[68] X. Zhu, S. Kumar, and X. Li, *Appl. Opt.*, vol. **45**, p. 6812, 2006.

[69] A. Demir, *J. Lightwave Technol.*, vol. **25**(8), p. 2002, 2007.

[70] S. Kumar and L. Liu, *Opt. Exp.*, vol. **15**, p. 2166, 2007.

[71] M. Faisal and A. Maruta, *Opt. Commun.*, vol. **282**, p. 1893, 2009.

[72] S. Kumar, *J. Lightwave Technol.*, vol. **27**(21), p. 4722, 2009.

[73] A. Bononi, P. Serena, and N. Rossi, *Optical Fiber Tech.*, vol. **16**, p. 73, 2010.

[74] X. Zhu and S. Kumar, *Opt. Expr.*, vol. **18**(7), p. 7347, 2010.

[75] S. Kumar and X. Zhu. In S. Kumar (ed.), *Impact of Nonlinearities on Fiber Optic Communications*. Springer-Verlag, New York, 2011, chapter 7.

[76] K.-P. Ho. In S. Kumar (ed.), *Impact of Nonlinearities on Fiber Optic Communications*. Springer-Verlag, New York, 2011, chapter 8.

[77] C.V. Raman, *Ind. J. Phys.*, vol. **2**, p. 387, 1928.

[78] G. Landsberg and L. Mandelstam, *Naturwiss.*, vol. **16**(28), p. 557, 1928.

[79] E.J. Woodbury and W.K. Ng, Proceedings of IRE50, 1962, p. 2347.

[80] G.P. Agrawal, *Nonlinear Fiber Optics* 3rd edn. Academic Press, San Diego, CA, 2001, chapter 8.

[81] R.H. Stolen *et al.*, *J. Opt. Soc. Am. B*, vol. **6**(6), p. 1159, 1989.

[82] R.H. Stolen and W.J. Tomlinson, *J. Opt. Soc. Am. B*, vol. **9**(4), p. 565, 1992.

[83] J.P. Gordon, *Opt. Lett.*, vol. **11**, p. 662, 1986.

[84] K.O. Hill, B.S. Kawasaki, and D.C. Johnson, *Appl. Phys. Lett.*, vol. **28**(10), p. 608, 1976.

[85] R.H. Stolen, C. Lin, and R.K. Jain, *Appl. Phys. Lett.*, vol. **30**(7), p. 340, 1977.

[86] M. Nakazawa, T. Masamitsu, and N. Uchida, *J. Opt. Soc. Am. B*, vol. **1**(1), p. 86, 1984.

[87] A.J. Stentz, Proceedings of SPIE 3263, Nonlinear Optical Engineering, 1998, p. 91.

[88] Y. Li *et al.*, *J. Lightwave Technol.*, vol. **23**(5), p. 1907, 2005.

[89] F. Forghieri *et al.* In I. Kaminov and T.L. Koch (eds), *Optical Fiber Telecommunications IIIA*. Academic Press, San Diego, 1997.

[90] J. Wang, X. Sun, and M. Zhang *et al.*, *IEEE Photon. Technol. Lett.*, vol. **10**, p. 540, 1998.

[91] M. Muktoyuk and S. Kumar, *IEEE Photon. Technol. Lett.*, vol. **15**, p. 1222, 2003.

11

Digital Signal Processing

11.1 Introduction

The key component that revived coherent fiber communications in the mid-2000 was high-speed digital signal processing. In the 1990s, coherent receivers used optical phase-locked loops (OPLL) to align the phases and dynamic polarization controllers to match the polarization of the received signal with that of the LO. However, dynamic polarization controllers are bulky and expensive [1], and each channel of a WDM system needs a separate polarization controller. Phase locking in the optical domain using OPLL is difficult as well. With the advances in high-speed DSP, phase alignment and polarization management can be done in the electrical domain, as discussed in Sections 11.5 and 11.7, respectively. Linear impairments such as chromatic dispersion (CD) and polarization mode dispersion can be compensated using equalizers, as discussed in Sections 11.6 and 11.7, respectively. It is also possible to compensate for the interplay between dispersion and nonlinearity by using digital back propagation (DBP), in which the nonlinear Schrödinger equation is solved for a virtual fiber whose signs of dispersion, loss, and nonlinear coefficients are opposite to those of the transmission fiber. DBP is discussed in Section 11.8.

11.2 Coherent Receiver

Fig. 11.1 shows a schematic of the coherent IQ receiver with digital signal processing. The in-phase and quadrature components of the received signal can be written as (see Chapter 5, Eqs. (5.114) and (5.118))

$$y_I = RA_r A_{\mathrm{LO}} \mathrm{Re}\{s(t)\exp\left[-i(\omega_{\mathrm{IF}}t + \Delta\phi)\right]\}/2, \tag{11.1}$$

$$y_Q = RA_r A_{\mathrm{LO}} \mathrm{Im}\{s(t)\exp\left[-i(\omega_{\mathrm{IF}}t + \Delta\phi)\right]\}/2, \tag{11.2}$$

where $s(t)$ is the transmitted data:

$$s(t) = \sum_m a_m g(t - mT_s), \tag{11.3}$$

T_s is the symbol period, and $g(t)$ represents the pulse shape. Eqs. (11.1) and (11.2) have to be modified to take into account the noise and delays due to 90° hybrids:

$$y_I = K\mathrm{Re}\{s(t - \delta_I)\exp\left[-i(2\pi f_{\mathrm{IF}}(t - \delta_I) + \Delta\phi)\right] + n(t)\}, \tag{11.4}$$

$$y_Q = K\mathrm{Im}\{s(t - \delta_Q)\exp\left[-i(2\pi f_{\mathrm{IF}}(t - \delta_Q) + \Delta\phi)\right] + n(t)\}, \tag{11.5}$$

Fiber Optic Communications: Fundamentals and Applications, First Edition. Shiva Kumar and M. Jamal Deen.
© 2014 John Wiley & Sons, Ltd. Published 2014 by John Wiley & Sons, Ltd.

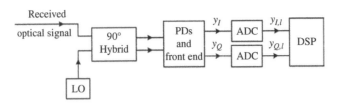

Figure 11.1 Block diagram of a coherent IQ receiver. LO = local oscillator, PDs = photodiodes, ADC = analog-to-digital converter, DSP = digital signal processing.

where $K = R\sqrt{P_r P_{LO}}/2$, $n(t)$ represents noise due to ASE and shot noise, and δ_I and δ_Q are the delays introduced by 90° hybrids and other parts of the coherent receiver. The constant K has no impact on the performance. So, from now on, we set it to unity. An ADC discretizes the analog signal at a sampling rate $R_{samp} \geq B_s$, where $B_s = 1/T_s$ is the symbol rate. Typically, two samples per symbol are required. The samples are combined into a complex number. The outputs of the ADC are written as

$$y_{I,l} = \text{Re}\{s_{I,l} \exp[-i(2\pi f_{IF}(t_l - \delta_{I,l}) + \Delta\phi_l)] + n_l\}, \tag{11.6}$$

$$y_{Q,l} = \text{Im}\{s_{Q,l} \exp[-i(2\pi f_{IF}(t_l - \delta_{Q,l}) + \Delta\phi_l)] + n_l\}, \quad l = 1, 2, \dots, \tag{11.7}$$

where $s_{I,l}$ and $s_{Q,l}$ are the samples of $s(t - \delta_I)$ and $s(t - \delta_Q)$, respectively, at $t = lT_{samp}$, $T_{samp} = 1/R_{samp}$. n_l is the sample of the noise at $t = lT_{samp}$. DSP performs the complex addition to obtain the received signal as

$$y_l = y_{I,l} + iy_{Q,l}. \tag{11.8}$$

In general, δ_I could be different from δ_Q. Therefore, $s_{I,l}$ and $s_{Q,l}$ may be different, and the real and imaginary parts of $\tilde{y}_l$ may not correspond to the same symbol, which could lead to symbol errors. However, this is a systematic error and can be corrected easily. Using the DSP, the delays experienced by I- and Q-channels can be removed. After correcting for δ_I and δ_Q, we have

$$y_l = x_l \exp[-i(2\pi f_{IF} t_l + \Delta\phi_l)] + n_l, \quad l = 1, 2, \dots, \tag{11.9}$$

where $x_l = s_l \equiv s(lT_{samp})$.

11.3 Laser Phase Noise

The output of a single-frequency laser is not strictly monochromatic but rather has frequency deviations that change randomly. The output field of a fiber-optic transmitter may be written as

$$q_T(t) = A_T s(t) \exp\{-i[2\pi f_c t - \phi(t)]\}, \tag{11.10}$$

where $s(t)$ is the data, f_c is the laser mean frequency, and $\phi(t)$ is the laser phase noise. The instantaneous frequency deviation can be written as (see Eq. (2.165))

$$f_i = -\frac{1}{2\pi}\frac{d\phi}{dt}. \tag{11.11}$$

The instantaneous frequency deviation is a zero-mean Gaussian noise process with standard deviation σ_f. Integrating Eq. (11.11), it follows that

$$\phi(t) = \phi(t_0) - 2\pi \int_{t_0}^{t} f_i(\tau)d\tau \tag{11.12}$$

is a Wiener process. If the interval $(t - t_0)$ is sufficiently small, the integration can be replaced by the rectangular rule. With $(t - t_0) = T_{\text{samp}}$,

$$\phi(t) = \phi(t - T_{\text{samp}}) - 2\pi f_i(t - T_{\text{samp}})T_{\text{samp}}. \tag{11.13}$$

After discretization, Eq. (11.13) becomes

$$\phi_l = \phi_{l-1} - 2\pi f_{i,l-1}T_{\text{samp}}, \tag{11.14}$$

where $t = lT_{\text{samp}}$. The phase noise can be interpreted as a one-dimensional random walk [2]. As an example, consider a drunken man walking randomly on the road. Suppose that at every step there is a 50% chance that he moves either forward or backward. After two steps, there is a 25% chance that he has moved two steps forward, a 25% chance that he has moved two steps backward, and 50% chance that he is at his initial position. After many steps, the mean distance traversed would be close to zero and there would be a large number of different paths he could have traversed. Since the chance of moving forward or backward at a given step is independent of the decision at the previous steps, the variance of distance traversed is proportional to the number of steps. Similarly, in the case of laser phase noise, the phase of the sample n is incremented by $-2\pi T_{\text{samp}} f_{i,l-1}$, where $f_{i,l-1}$ is a value of instantaneous frequency picked from the Gaussian distribution. From Eq. (11.14), we have

$$\delta\phi_l \equiv \phi(l) - \phi(0) = -2\pi T_{\text{samp}} \sum_{m=0}^{l-1} f_{i,m}. \tag{11.15}$$

Squaring Eq. (11.15), averaging, and noting that the frequency deviations at each step are independent, we find

$$< \delta\phi_l^2 > = 4\pi^2 T_{\text{samp}}^2 l\sigma_f^2. \tag{11.16}$$

Note that the phase variance is proportional to l. Solving the laser rate equations with Langevin noise terms, we find [3]

$$< \delta\phi_l^2 > = 2\pi \Delta v l T_{\text{samp}}, \tag{11.17}$$

where Δv is the laser linewidth (FWHM). Comparing Eqs. (11.16) and (11.17), we find

$$\sigma_f^2 = \frac{\Delta v}{2\pi T_{\text{samp}}}. \tag{11.18}$$

Fig. 11.2 shows a few possible evolutions of the laser phase $\phi(t)$ when the linewidth Δv is 5 MHz. Fig. 11.3 shows the evolutions of the phase for two different linewidths. It can be seen that the phase fluctuation is larger as the linewidth increases.

The phase noise is present in the LO output as well, and the LO output field may be written as

$$q_{\text{LO}}(t) = A_{\text{LO}} \exp\{-i[2\pi f_{\text{LO}}t + \phi_{\text{LO}}(t)]\}, \tag{11.19}$$

with

$$< \delta\phi_{\text{LO},l}^2 > = 2\pi \Delta v_{\text{LO}} l T_{\text{samp}}, \tag{11.20}$$

where Δv_{LO} is the linewidth of LO. The received signal after discretization is given by Eq. (11.9), with

$$\Delta\phi_l = \phi_{\text{TX},l} + \phi_p - \phi_{\text{LO},l}, \tag{11.21}$$

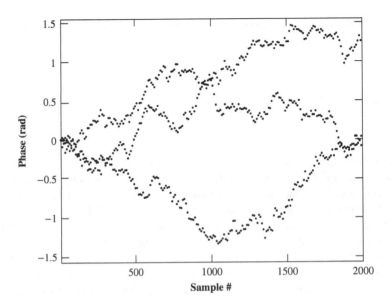

Figure 11.2 Evolution of laser phase noise. Each curve corresponds to a different realization of the phase noise. The laser linewidth = 1 MHz for all curves.

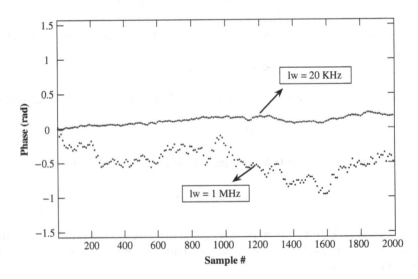

Figure 11.3 Laser phase noise evolution for two different linewidths; lw = linewidth. The fluctuations increase with linewidth and time.

where $\phi_{\text{TX},l}$ is the phase noise due to the transmitter laser and ϕ_p is the constant phase shift due to propagation. The variance of $\Delta\phi_l$ is

$$\sigma_{\Delta\phi}^2 = \; < \delta\phi_{\text{TX},l}^2 > + < \delta\phi_{\text{LO},l}^2 >$$

$$= 2\pi\{\Delta\nu_{\text{TX}} + \Delta\nu_{\text{LO}}\}lT_{\text{samp}}. \tag{11.22}$$

Here, we have assumed $\phi_{TX,0} = \phi_{LO,0} = 0$ and ignored dispersion and nonlinear effects in the fiber-optic channel.

11.4   IF Estimation and Compensation

Nowadays, coherent optical communication systems use free-running LO lasers without any optical/digital phase-locked loop (PLL). A typical temperature-stabilized DFB laser has a frequency fluctuation of about ± 1.25 GHz [4]. External cavity lasers (ECLs) with linewidths < 100 kHz are also available. Typically, the symbol rate is ≥ 10 GBaud and, therefore, coherent receivers with free-running LO lasers may be considered as intradyne receivers [5]. A constant IF offset causes the absolute value of the phase to increase with time, which leads to erroneous phase decisions. IF offset should be removed before channel synchronization if the intermediate frequency $f_{IF} > 12.5\%$ of symbol rate B_s, or it can be removed after channel synchronization if $f_{IF} < 12.5\%$ of B_s [6]. After channel synchronization, the complex signal is given by Eq. (11.9). There are various techniques to estimate f_{IF}, such as the phase increment algorithm [6,7], Tratter IF estimation algorithm, and Kay IF estimation algorithm [8]. In this book, we consider the phase increment algorithm because of its simplicity. In the absence of laser phase noise ($\Delta\phi_l = 0$) and phase modulation ($x_l = 1$), the phase shift between two consecutive samples y_l, y_{l+1} is

$$\Delta\theta = 2\pi f_{IF} T_{samp}. \tag{11.23}$$

The objective of the frequency estimator is to estimate the phase shift $\Delta\theta$ between two consecutive samples.

Fig. 11.4 shows a block diagram of the phase increment frequency estimator and compensator. First, the current sample is multiplied by the complex conjugate of the previous sample. Using Eq. (11.9), we find

$$y_l y_{l-1}^* = x_l x_{l-1}^* \exp\left[-i(2\pi f_{IF} T_{samp} + \Delta\phi_l - \Delta\phi_{l-1})\right] + n_l', \tag{11.24}$$

where $n_l' = x_l n_{l-1}^* + x_{l-1}^* n_l + n_l n_{l-1}^*$ is the effective noise. First consider the case $n_l = 0$ and $\Delta\phi_l = 0$. Eq. (11.24) may be rewritten as

$$y_l y_{l-1}^* = |x_l||x_{l-1}| \exp\left[-i(2\pi f_{IF} T_{samp} + \theta_{x,l})\right], \tag{11.25}$$

where $\theta_{x,l} = \text{Arg}(x_l x_{l-1}^*)$. For M-PSK systems, $\theta_{x,l}$ takes values $2\pi(m-n)/M$, $m,n = 0, 1, \ldots, M-1$. The IF estimation is complicated by the presence of phase modulation and laser phase noise. For an M-PSK system, if we take the Mth power of $y_l y_{l-1}^*$, $\theta_{x,l}$ is multiplied by M, resulting in a phase that is an integral multiple of 2π and hence it can be ignored. From Eq. (11.25), we have

$$\text{Arg}\{(y_l y_{l-1}^*)^M\} = -(2\pi f_{IF} T_{samp})M \tag{11.26}$$

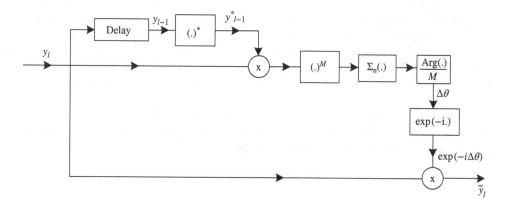

Figure 11.4   Block diagram of IF estimation and compensation.

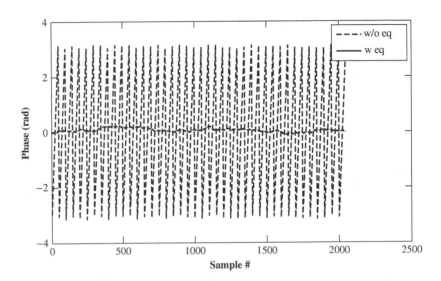

Figure 11.5 Phase vs. sample number for the back-to-back case with no phase modulation. Parameters: Tx laser linewidth = LO linewidth = 50 kHz, $f_{IF} = 200$ MHz.

or

$$\hat{f}_{IF} = \frac{-1}{2\pi T_{samp} M} \text{Arg}[(y_l y_{l-1}^*)^M].$$
(11.27)

However, in the presence of n_l and $\Delta\phi_l$, the estimated frequency offset $\hat{f}_{IF}$ fluctuates from symbol to symbol. Since n_l and $\Delta\phi_l - \Delta\phi_{l-1}$ are zero-mean random variables, the fluctuations can be minimized if we average over N samples,

$$\hat{f}_{IF} = \frac{-1}{2\pi T_{samp} M} \text{Arg} \left[\sum_{l=1}^{N} (y_l y_{l-1}^*)^M \right].$$
(11.28)

The frequency estimate $\hat{f}_{IF}$ gets better as the block size N increases, as long as $\hat{f}_{IF}$ remains constant over the block size. The IF is removed by multiplying y_l by $\exp(-i\Delta\theta)$, where $\Delta\theta = -2\pi\hat{f}_{IF}T_{samp}$. The solid and broken lines in Fig. 11.5 show the phases with and without IF equalization, respectively, when the phase modulation is turned off. In this example, we consider the back-to-back case with no fiber-optic channel between transmitter and receiver. When IF equalization is not used, the phase increases constantly because of the term $2\pi f_{IF}t_l$ in Eq. (11.9). However, the Arg($\cdot$) function can not distinguish phases that differ by 2π and produces results in the $[-\pi, \pi]$ interval. When the IF equalization is used, from Fig. 11.5, we see that the phase fluctuations are quite small, indicating that the equalizer is effective in removing IF. These phase fluctuations after the IF removal are due to laser phase noise. Equalization of the phase noise is discussed in the next section.

Fig. 11.6(a) and (b) shows the constellation diagrams before and after the IF removal, respectively, for the QPSK signal. Before the IF removal, the phase varies almost uniformly over the range of 0 to 2π. After the IF removal, the phase is close to one of the transmitted phases 0, $\pi/2$, π, $3\pi/2$.

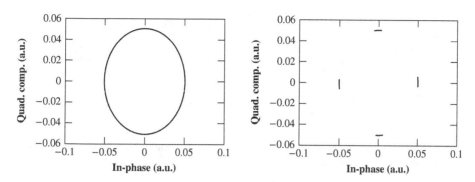

Figure 11.6   Constellation diagrams: (a) before IF removal, (b) after IF removal. Parameters: symbol rate = 10 GSym/s, NRZ-QPSK, other parameters the same as in Fig. 11.5.

11.5   Phase Estimation and Compensation

The linewidth of ECL/DFB lasers used at the transmitter and receiver (as LO) ranges from 10 kHz to 10 MHz, and the symbol rates are usually $\geq$ 10 GSym/s. Therefore, the phase $\Delta\phi_k$ of Eq. (11.9) varies much more slowly than the rate of phase modulation. By averaging the phase $\Delta\phi_k$ over many symbol intervals, it is possible to obtain an accurate phase estimate [9].

There exist a number of techniques for phase estimation and compensating [9–13]. Here, we describe the commonly used technique known as the block phase noise estimation or Viterbi–Viterbi algorithm [9, 10]. The block diagrams of the phase estimation technique are shown in Figs. 11.7 to 11.9. After removal of the IF, the signal input to the phase estimator is

$$\tilde{y}_l = x_l \exp\left(-i\Delta\phi_l\right) + n_l. \tag{11.29}$$

For M-PSK systems, the phase modulation effect is removed by taking the Mth power of the signal as before,

$$(\tilde{y}_l)^M = \left[x_l \exp\left(-i\Delta\phi_l\right) + n_l\right]^M. \tag{11.30}$$

Using the binomial theorem,

$$(A + B)^M = A^M + \binom{M}{1} A^{M-1}B + \binom{M}{2} A^{M-2}B^2 + \cdots + B^M, \tag{11.31}$$

Eq. (11.30) may be written as

$$(\tilde{y}_l)^M = x_l^M \exp\left(-iM\Delta\phi_l\right) + n_l', \tag{11.32}$$

where

$$n_l' = \binom{M}{1} x_l^{M-1} \exp\left[-i(M-1)\Delta\phi_l\right]n_l + \binom{M}{2} x_l^{M-2} \exp\left[-i(M-2)\Delta\phi_l\right]n_l^2 + \cdots n_l^M. \tag{11.33}$$

In Eq. (11.32), the first term is the desired term and n_l' is the sum of unwanted cross-terms due to signal–noise and noise–noise beating. It can be shown that n_l' is a zero-mean complex random variable (see Example 11.2)

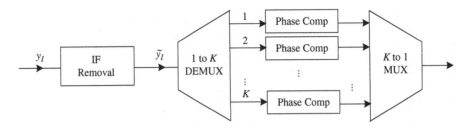

Figure 11.7   Block diagram of an IF and phase compensator. Demux = demultiplexer, Phase Comp = block phase estimator and compensator, Mux = multiplexer.

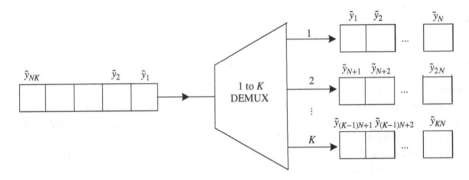

Figure 11.8   Demultiplexing of the data into K blocks with each block consisting of N samples.

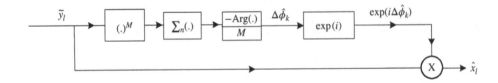

Figure 11.9   Block diagram of a block phase estimator and compensator for the kth block.

and, therefore, if we average $(\tilde{y}_l)^M$ over N samples, the impact of n'_l can be minimized. First, the signal is divided into K blocks with each block consisting of N samples, as shown in Fig. 11.7. In block k, $k = 1, 2, \ldots, K$, the signal is raised to the Mth power and summed over N samples to obtain

$$\sum_{l=(k-1)N+1}^{kN} (\tilde{y}_l)^M = \sum_{l=(k-1)N+1}^{kN} |x_l|^M \exp\left[-iM(\theta_l + \Delta\phi_l)\right] + \sum_{l=(k-1)N+1}^{kN} n'_l, \qquad (11.34)$$

where $\theta_l = \text{Arg}(x_l)$. For M-PSK systems, $|x_l|^M = A_0$ is a constant independent of modulation. In Eq. (11.34), we assumed that the $\Delta\phi_l$ is approximately constant within the block. $M\theta_l$ is an integral multiple of 2π and, hence, it can be ignored. If N is sufficiently large, the last term in Eq. (11.34) is close to zero. So, we have

$$\sum_{l=(k-1)N+1}^{kN} (\tilde{y}_l)^M \cong A_0 N \exp\left(-iM\Delta\phi_l\right). \qquad (11.35)$$

From Eq. (11.35), we find

$$\frac{-1}{M}\text{Arg}\left\{\sum_{l=(k-1)N+1}^{kN}(\tilde{y}_l)^M\right\} \equiv \Delta\hat{\phi}_k \cong \Delta\phi_l. \qquad (11.36)$$

Here, $\Delta\hat{\phi}_k$ is the phase estimate of the kth block. The signal sample $\tilde{y}_l$ is multiplied by $\exp(i\Delta\hat{\phi}_k)$ to obtain the estimate of the transmitted signal,

$$\hat{x}_l = \tilde{y}_l \exp(i\Delta\hat{\phi}_k). \qquad (11.37)$$

The computation in each block can be carried out using separate signal processors. Finally, the signal samples in each block are combined using a multiplexer to obtain the serial data. The block size should be chosen carefully. If N is too small, the impact of the noise term n'_l in Eq. (11.32) can not be ignored. If N is too large, the laser phase may drift and $\Delta\phi_l$ may not remain constant within each block. The block size should be optimized based on the laser linewidth.

11.5.1 Phase Unwrapping

The function Arg() in Eq. (11.36) can not distinguish between phases that differ by 2π and it returns the results in the interval $[-\pi, \pi]$. If the phase is $\pi + \epsilon$, $\epsilon > 0$, the function Arg() returns a phase of $-\pi + \epsilon$. This is known as *phase wrapping* and it could lead to symbol errors. Special techniques have to be used to unwrap phases. Consider the following example: suppose $\Delta\phi_l$ in the current block k is $\epsilon \ll \pi$, $\epsilon > 0$, and let $\Delta\phi_l$ be roughly constant over the block. From Eq. (11.36), it follows that $\Delta\hat{\phi}_k = \epsilon$. Now, let $\Delta\phi_l$ of the next block, $k + 1$ jump by π/M, i.e., $\Delta\phi_l$ of the $(k+1)$th block is $\epsilon + \pi/M$. From Eq. (11.36) for the $(k+1)$th block, we find

$$\frac{-1}{M}\text{Arg}\left\{\sum_{l=kN+1}^{(k+1)N}A_l\exp\left[-i\left(M\theta_l + M\epsilon + \pi\right)\right]\right\} = \frac{-1}{M}\text{Arg}\{\exp\left[-i(M\epsilon + \pi)\right]\}$$

$$= \frac{M\epsilon - \pi}{M}$$

$$= \Delta\hat{\phi}_{k+1}. \qquad (11.38)$$

Clearly, the estimated phase $\epsilon - \pi/M$ is different from the actual phase, $\epsilon + \pi/M$. This is because of the phase wrapping done by the function of Arg(). Phase wrapping in the context of coherent optical communication has been studied in Refs. [14, 15]. Let the carrier phase prior to the unwrapping be $\Delta\hat{\phi}_k$. If we add $2\pi/M$ to $\Delta\hat{\phi}_{k+1}$, the phase for the $(k + 1)$th block after the phase unwrapping is

$$\Delta\phi_{k+1} = \Delta\hat{\phi}_{k+1} + \frac{2\pi}{M} = \epsilon + \pi/M, \qquad (11.39)$$

which is actually the phase of the $(k + 1)$th block. In general, an integral multiple of $2\pi/M$ is added to the carrier phase. The carrier phase after the phase unwrapping can be written as

$$\Delta\phi_k = \Delta\hat{\phi}_k + m2\pi/M, \quad k = 1, 2, ..., K, \qquad (11.40)$$

where

$$m = \text{Floor}\left(0.5 + \frac{\Delta\phi_{k-1} - \Delta\hat{\phi}_k}{2\pi/M}\right). \qquad (11.41)$$

Here, Floor() returns the nearest integer toward $-\infty$. Suppose the phases of the kth and $(k - 1)$th block are both ϵ. In this case, $m = 0$ and the phase unwrapping block of the phase estimator Eq. (11.40) does not add $2\pi/M$.

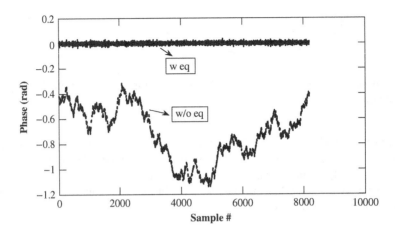

Figure 11.10 Plot of phase vs. sample number with and without phase equalizer. Parameters: no phase modulation, TX laser linewidth = LO linewidth = 125 kHz, f_{IF} = 200 MHz, block size N = 10. The signal passes through an IF equalizer prior to the phase equalizer.

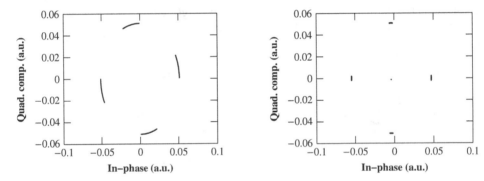

Figure 11.11 Constellation diagrams (a) before phase equalization, (b) after phase equalization. Parameters: symbol rate = 10 GSym/s, NRZ-QPSK, and other parameters the same as in Fig. 11.10.

A back-to-back case (no fiber-optic channel between transmitter and receiver) is simulated. Fig. 11.10 shows the phases of the signals after IF equalization when the phase modulation is turned off. In the absence of the phase equalizer, the phase varies randomly in the interval $[-\pi, \pi]$. After using the phase equalizer, the phase fluctuations are significantly reduced. Fig. 11.11(a) and (b) shows the constellation diagrams before and after phase noise removal for the QPSK signal, respectively. As can be seen, the phase equalizer is quite effective in removing the phase fluctuations introduced by the transmitter and LO.

11.6 *CD* Equalization

In this section, we ignore laser phase noise, fiber nonlinear effects, ASE, and other noise sources, and consider only the impact of fiber dispersion. The output field envelope of the fiber can be written as

$$y(t) = \mathcal{F}^{-1}[\tilde{y}(f)], \tag{11.42}$$

$$\tilde{y}(f) = \tilde{H}(f)\tilde{x}(f), \tag{11.43}$$

$$\tilde{x}(f) = \mathcal{F}[x(t)], \tag{11.44}$$

where $x(t)$ is the input field envelope of the fiber and $\tilde{H}(f)$ is the fiber transfer function. The dispersion-compensating filter (DCF) should have the transfer function

$$\tilde{W}(f) = \frac{1}{\tilde{H}(f)}, \qquad (11.45)$$

so that the output of the DCF is the same as the fiber input, as shown in Fig. 11.12:

$$\hat{x}(f) = \tilde{W}(f)\tilde{y}(f) \qquad (11.46)$$

$$= \tilde{W}(f)\tilde{H}(f)\tilde{x}(f) = \tilde{x}(f). \qquad (11.47)$$

Inverse Fourier transforming Eq. (11.46) and noting that a product in the frequency domain becomes a convolution in the time domain, we obtain

$$x(t) = \int_{-\infty}^{\infty} y(t - t')W(t')\,dt, \qquad (11.48)$$

where

$$W(t) = \mathcal{F}^{-1}[\tilde{W}(f)] \qquad (11.49)$$

is the impulse response of the dispersion-compensating filter. The DCF discussed in Chapter 2 is a dispersion-compensating filter in the optical domain. Owing to the linearity of coherent detection, a dispersion-compensating filter can be realized in the electrical domain as well. For digital implementation, Eq. (11.48) is discretized to obtain

$$x[n] = \sum_{k=-\infty}^{\infty} W[k]y[n-k]. \qquad (11.50)$$

Here, the time t is discretized as $t = kT_{\text{samp}}$, where $1/T_{\text{samp}}$ is the sampling rate, k is an integer,

$$x[n] = x(nT_{\text{samp}}), \qquad (11.51)$$

$$W[n] = T_{\text{samp}}W(nT_{\text{samp}}), \qquad (11.52)$$

$$y[n] = y(nT_{\text{samp}}). \qquad (11.53)$$

Thus, if we know the impulse response of the dispersion-compensating filter, convolving it with the fiber output field envelope could undo the distortions caused by fiber dispersion. As an example, consider the fiber transfer function given by Eq. (2.107) (with no loss and no delay),

$$H(f) = \exp{(i2\pi^2 f^2 \beta_2 L)}. \qquad (11.54)$$

Using Eq. (11.45), the transfer function of the dispersion-compensating filter is

$$\tilde{W}(f) = \exp{(-i2\pi^2 f^2 \beta_2 L)}. \qquad (11.55)$$

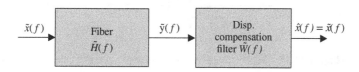

Figure 11.12 CD equalizer using a digital DCF.

The impulse response function of the dispersion-compensating filter is given by [16] (see Example 11.3)

$$W(t) = \sqrt{\frac{1}{2\pi i \beta_2 L}} \exp{[i\phi(t)]}, \qquad (11.56)$$

$$\phi(t) = t^2 / 2\beta_2 L. \qquad (11.57)$$

A dispersion-compensating filter is an all-pass filter and its impulse response $W(t)$ is infinite in duration. The summation in Eq. (11.50) may be truncated to a finite number of terms, known as a finite impulse response (FIR) filter. Now, Eq. (11.50) becomes

$$x[n] = \sum_{k=-K}^{K} W[k] y[n-k], \qquad (11.58)$$

$$W[k] = T_{\text{samp}} \sqrt{\frac{1}{2\pi i \beta_2 L}} \exp{\left[\frac{i k^2 T_{\text{samp}}^2}{2\beta_2 L}\right]}. \qquad (11.59)$$

Fig. 11.13 shows a schematic of the FIR filter. The number of taps, $2K + 1$, has to be decided based on the Nyquist sampling theorem, which states that if the signal is band-limited to B, the sampling rate, R_{samp}, has to be greater than or equal to $2B$. Otherwise, aliasing could occur. From Eq. (11.57), the instantaneous frequency of $W(t)$ is

$$f_i = \frac{-1}{2\pi} \frac{d\phi}{dt} = \frac{-t}{2\pi \beta_2 L}. \qquad (11.60)$$

From Eq. (11.60), we see that the magnitude of instantaneous frequency increases with t. When the summation in Eq. (11.50) is truncated to $2K + 1$ terms (see Eq. (11.58)), the highest-frequency component occurs at $t = K T_{\text{samp}}$:

$$B = |f_{i,\text{max}}| = \frac{K T_{\text{samp}}}{2\pi |\beta_2| L}. \qquad (11.61)$$

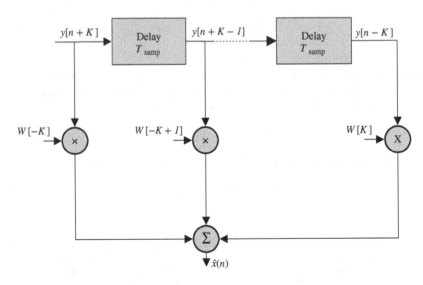

Figure 11.13   Schematic of the FIR dispersion-compensating filter.

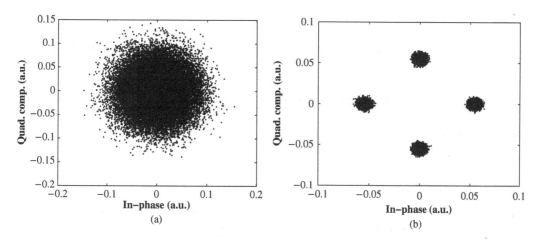

Figure 11.14 Constellation diagrams for NRZ-QPSK system: (a) before dispersion equalizer, (b) after dispersion equalizer. Parameters: accumulated dispersion = 13,600 ps/nm, number of samples/symbol = 2, number of taps = 47.

Using the Nyquist theorem, the sampling rate, R_{samp}, should be at least equal to $2B$,

$$R_{samp} = \frac{1}{T_{samp}} \geq \frac{K T_{samp}}{\pi |\beta_2| L} \tag{11.62}$$

or

$$K \geq \frac{\pi |\beta_2| L}{T_{samp}^2}. \tag{11.63}$$

Since K has to be an integer, we choose

$$K = \text{ceil}(\pi |\beta_2| L / T_{samp}^2), \tag{11.64}$$

where ceil() gives the nearest integer toward ∞. From Eq. (11.64), we see that the number of taps increases as $|\beta_2| L$. This can be understood from the fact that the pulse broadening increases with $|\beta_2| L$. To undo the distortion due to dispersion at $t = k T_{samp}$, samples of $y(t)$ extending from $(k - K)T_{samp}$ to $(k + K)T_{samp}$ are required.

Fig. 11.14(a) and (b) shows the constellation diagrams of a system based on QPSK before and after the dispersion-compensating filter, respectively. As can be seen, the distortions caused by fiber dispersion can be mitigated using the dispersion-compensating filter. Alternatively, the dispersion-compensating filter can be realized by using an IIR filter, which is computationally efficient but requires buffering [17]. When the accumulated dispersion is large, it would be more efficient to compensate dispersion in the frequency domain using FFTs, as discussed in Section 11.8.

Example 11.1

A 10-GSym/s fiber-optic system has the following parameters: $\beta_2 = -22$ ps^2/km and transmission distance = 800 km. Assuming two samples per symbol, calculate the minimum number of taps needed to compensate for the fiber dispersion.

Solution:
For a 10-GSym/s system, the symbol period is 100 ps. Since there are two samples per symbol, $T_{\text{samp}} = 50$ ps. Using Eq. (11.64), we find

$$K = \text{ceil}\left(\frac{\pi \times 22 \times 10^{-27} \times 800 \times 10^3}{(50 \times 10^{-12})^2}\right) = 23. \tag{11.65}$$

Therefore, the number of taps $2K + 1 = 47$.

11.6.1 Adaptive Equalizers

The fiber dispersion could fluctuate due to environmental conditions. However, these fluctuations occur at a rate that is much slower than the transmission data rate and, therefore, the tap weights of the FIR filter shown in Fig. 11.13 can be adjusted adaptively. There exist a number of techniques to realize the tunable dispersion-compensating filters [18–20]. In this section, we focus on two types of adaptive equalizer: least mean squares (LMS) and constant modulus algorithm (CMA) equalizers.

Fig. 11.15 shows a schematic of a fiber-optic system with adaptive equalizer in the digital domain. Let the input to the fiber-optic channel be $x[k]$. The channel output is

$$y[m] = \sum_{k=-N}^{N} H[k]x[m-k] + n[m], \tag{11.66}$$

where $H[k]$ is the channel impulse response and $n[m]$ is the noise added by the channel. In Eq. (11.66), we have assumed that the ISI at $t = mT_{\text{samp}}$ could occur due to the samples of the optical signal ranging from $m - N$ to $m + N$. In other words, $H[k]$ is assumed to be zero for $|k| > N$. The adaptive equalizer is a transversal filter with tap weights $W[k]$ and the output of the equalizer is

$$\hat{x}[n] = \sum_{k=-K}^{K} W[k]y[n-k]. \tag{11.67}$$

Here, $2K + 1$ is the number of taps. If the equalizer compensates for the channel effects, $\hat{x}[n]$ should be equal to $x[n]$ in the absence of noise. The error between the desired response $x[n]$ and the output of the equalizer $\hat{x}[n]$ is

$$e[n] = x[n] - \hat{x}[n]. \tag{11.68}$$

The mean square error is

$$
\begin{aligned}
J(W[-K], W[-K+1],...,W[K], W^*[-K],...,W^*[K]) &= <|e[n]^2|> \\
&= <|x[n]|^2 - x[n]\hat{x}^*[n] \\
&\quad -\hat{x}[n]x^*[n] + \hat{x}[n]\hat{x}^*[n]>.
\end{aligned}
\tag{11.69}
$$

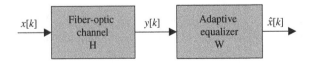

Figure 11.15 Adaptive equalization of the fiber-optic channel.

The adaptive equalizer has $2K + 1$ adjustable complex coefficients. The coefficients $W[k]$ can be adjusted so that the mean square error is minimum,

$$\frac{\partial J}{\partial W[k]} = 0 \tag{11.70}$$

and

$$\frac{\partial J}{\partial W^*[k]} = 0. \tag{11.71}$$

Using Eqs. (11.67) and (11.69) in Eq. (11.70), we find

$$\frac{\partial J}{\partial W[k]} = < -x^*[n]y[n-k] + y[n-k]\hat{x}^*[n] >$$

$$= - < y[n-k]e^*[n] >= 0. \tag{11.72}$$

Note that $W[k]$ and $W^*[k]$ are independent variables and, therefore, $\partial \hat{x}^*[n]/\partial W[k] = 0$. From Eq. (11.71), we obtain

$$\frac{\partial J}{\partial W^*[k]} = - < y^*[n-k]e[n] >= 0, \tag{11.73}$$

which is nothing but the complex conjugate of Eq. (11.72).

The tap weights $W[-K], W[-K+1], \ldots, W[K]$ are optimum when the cost function J is minimum. To find the optimum tap weights, we follow an iterative procedure. Initially, tap weights are chosen arbitrarily as

$$\mathbf{W}^{(0)} = [W^{(0)}[-K], W^{(0)}[-K+1], \ldots, W^{(0)}[K]], \tag{11.74}$$

where '(0)' stands for the zeroth iteration. To update the tap weights for the next iteration, we need to move in a vector space of $2K + 1$ dimensions such that we are closer to a minimum of the cost function J. The gradient vector is defined as

$$\mathbf{G} = [g[-K], g[-K+1], \ldots, g[K]], \tag{11.75}$$

$$g[k] = 2\frac{\partial J}{\partial W^*[k]} = -2 < y^*[n-k]e[n] > . \tag{11.76}$$

At the starting point, we have the tap weight vector $\mathbf{W}^{(0)}$ and the gradient vector $\mathbf{G}^{(0)}$. From Eq. (11.71), we see that J is minimum when $g[k]$ is zero. But at the starting point, $g[k]$ may not be zero. Iteratively, we need to find $W[k]$ such that $g[k]$ is close to zero. The tap weight vector for the next iteration should be chosen in a direction opposite to $\mathbf{G}^{(0)}$. This is because, if we move in the direction of $\mathbf{G}^{(0)}$, J would be maximized. So, the tap weights for the next iteration are chosen as

$$\mathbf{W}^{(1)} = \mathbf{W}^{(0)} - \frac{\Delta}{2}\mathbf{G}^{(0)} \tag{11.77}$$

or

$$W[k]^{(1)} = W[k]^{(0)} - \frac{\Delta}{2}g[k]^{(0)}$$

$$= W[k]^{(0)} + \Delta < y^*[n-k]e[n] >, \tag{11.78}$$

where Δ is a step-size parameter and the factor $1/2$ in Eq. (11.77) is introduced for convenience. The convergence of the iterative procedure depends on the value of Δ chosen. In practice, it is difficult to evaluate the expectation operator of Eq. (11.78), which requires knowledge of the channel response $H[n]$. Instead, the gradient vector is approximated by the instantaneous value or an estimate of the gradient vector. Ignoring the expectation operator in Eq. (11.78), the tap weights are altered at the $(n+1)$th iteration as [18–20]

$$W[k]^{(n+1)} = W[k]^{(n)} + y^*[n-k]e[n]\Delta, \quad k = -K, \dots, 0, \dots, K \tag{11.79}$$

$$e[n] = x[n] - \hat{x}[n]. \tag{11.80}$$

Eqs. (11.79) and (11.80) constitute the LMS algorithm for adaptive equalization. After a few iterations, $e[n] \cong 0$ and, thereafter, the tap weights remain roughly the same. Fig. 11.16 shows a schematic of the adaptive equalizer. Initially, the transmitter sends a training sequence $x[n]$, $n = 1, 2, 3, \dots$ which is known to the receiver. This is received as $y[n]$. The purpose of sending a training sequence is to let the receiver find the tap weights adaptively. The equalizer is switched to training mode, initially in Fig. 11.16. The period of training is pre-decided between the transmitter and receiver, and the receiver has full information on the information sequence $x[n]$. After the tap weights $W[k]$ have reached their optimum values, it may be assumed that the output of the decision device $\hat{x}[n]$ is a reliable estimate of the information sequence $x[n]$. At the end of the training period, actual data is transmitted. Since the receiver has no information on the transmitted data, the output of the decision device $\tilde{x}[n]$ is used to calculate the error signal $e[n]$ instead of the actual information sequence $x[n]$, as shown in Fig. 11.16. This is known as a *decision-directed mode* of adaption. In this mode, an error signal is obtained as

$$e[n] = \tilde{x}[n] - \hat{x}[n]. \tag{11.81}$$

The fiber dispersion varies slowly due to environmental fluctuations and the tap weights are adjusted adaptively to compensate for the slow variations in dispersion.

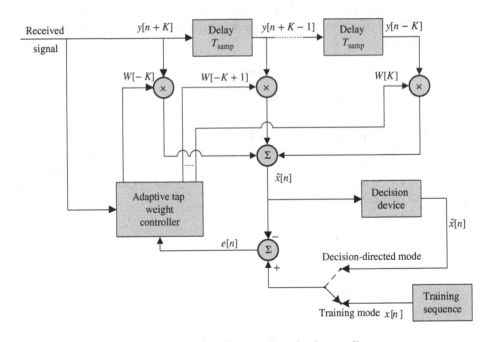

Figure 11.16 Block diagram of an adaptive equalizer.

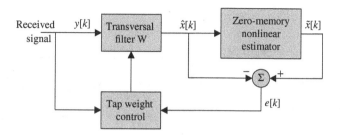

Figure 11.17   Block diagram of a blind equalizer.

11.6.1.1   Blind Equalizers

In some applications, it is desirable for receivers to undo the distortions without using the training sequences. Such equalizers are known as blind equalizers. Fig. 11.17 shows a schematic of a blind equalizer. The blind equalizer is similar to the decision-directed equalizer except that the error signal is obtained using the zero-memory nonlinear estimator instead of the decision device. Once the blind equalizer converges, it will be switched to a decision-directed mode of operation. Godard proposed a family of blind equalization algorithms [21]. In this section, we consider a special case of Godard's algorithms, known as the *constant-modulus algorithm* (CMA). In this case, the output of the zero-memory nonlinear estimator is [20, 21]

$$\tilde{x}[k] = \hat{x}[k](1 + R_2 - |\hat{x}[k]|^2), \tag{11.82}$$

where

$$R_2 = \frac{< |x[k]|^4 >}{< |x[k]|^2 >}. \tag{11.83}$$

The error signal is

$$e[k] = \tilde{x}[k] - \hat{x}[k]$$
$$= \hat{x}[k](R_2 - |\hat{x}[k]|^2). \tag{11.84}$$

For constant-intensity modulation formats such as QPSK-NRZ, $\langle |x[n]|^4 \rangle = \langle |x[n]|^2 \rangle = 1$ assuming that the transmitter power is normalized to unity. For these formats, Eq. (11.84) reduces to

$$e[k] = \hat{x}[k](1 - |\hat{x}[k]|^2). \tag{11.85}$$

If the tap weights are optimum, $|\hat{x}[k]|^2$ should be unity for constant-intensity formats and, therefore, the error signal $e[k]$ that is proportional to the deviation of $|\hat{x}[k]|^2$ from unity is used to adjust the tap weights. The tap weights are adjusted in accordance with the stochastic gradient algorithm as discussed previously,

$$w[k]^{(n+1)} = w[k]^{(n)} + y^*[n - k]e[n]\Delta. \tag{11.86}$$

11.7   Polarization Mode Dispersion Equalization

Consider a polarization-multiplexed fiber-optic system as shown in Fig. 11.18. Let $\psi_{x,\text{in}}$ and $\psi_{y,\text{in}}$ be the field envelopes of the *x*- and *y*- polarization components at the input of the fiber-optic channel. Ignoring the noise

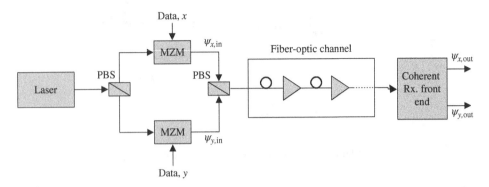

Figure 11.18   Polarization-multiplexed fiber-optic system. PBS = polarization beam splitter, MZM = Mach–Zehnder modulator.

effects, the output of the coherent receiver front end can be written as (see Section 5.6.5)

$$\begin{bmatrix} \tilde{\psi}_{x,\text{out}}(\omega) \\ \tilde{\psi}_{y,\text{out}}(\omega) \end{bmatrix} = F \begin{bmatrix} \tilde{H}_{xx}(\omega) & \tilde{H}_{xy}(\omega) \\ \tilde{H}_{yx}(\omega) & \tilde{H}_{yy}(\omega) \end{bmatrix} \begin{bmatrix} \tilde{\psi}_{x,\text{in}}(\omega) \\ \tilde{\psi}_{y,\text{in}}(\omega) \end{bmatrix}, \tag{11.87}$$

where F is a scalar that represents the loss in the fiber-optic channel. In the absence of the polarization-dependent loss (PDL) or polarization-dependent gain (PDG), total power should be conserved, which implies that the determinant of the matrix in Eq. (11.87) should be unity. Eq. (11.87) may be written as

$$\tilde{\psi}_{x,\text{out}}(\omega) = F[\tilde{H}_{xx}(\omega)\tilde{\psi}_{x,\text{in}}(\omega) + \tilde{H}_{xy}(\omega)\tilde{\psi}_{y,\text{in}}(\omega)], \tag{11.88}$$

$$\tilde{\psi}_{y,\text{out}}(\omega) = F[\tilde{H}_{yx}(\omega)\tilde{\psi}_{x,\text{in}}(\omega) + \tilde{H}_{yy}(\omega)\tilde{\psi}_{y,\text{in}}(\omega)]. \tag{11.89}$$

After taking the inverse Fourier transform and discretizing, Eqs. (11.88) and (11.89) become

$$\psi_{x,\text{out}}[m] = F \sum_{k=-N}^{N} \{H_{xx}[k]\psi_{x,\text{in}}[m-k] + H_{xy}[k]\psi_{y,\text{in}}[m-k]\}, \tag{11.90}$$

$$\psi_{y,\text{out}}[m] = F \sum_{k=-N}^{N} \{H_{yx}[k]\psi_{x,\text{in}}[m-k] + H_{yy}[k]\psi_{y,\text{in}}[m-k]\}. \tag{11.91}$$

Let

$$\boldsymbol{\Psi}_{\text{in}}[k] = \begin{bmatrix} \psi_{x,\text{in}}[k] \\ \psi_{y,\text{in}}[k] \end{bmatrix}, \tag{11.92}$$

$$\boldsymbol{\Psi}_{\text{out}}[k] = \begin{bmatrix} \psi_{x,\text{out}}[k] \\ \psi_{y,\text{out}}[k] \end{bmatrix}. \tag{11.93}$$

$$\mathbf{H}[k] = F \begin{bmatrix} H_{xx}[k] & H_{xy}[k] \\ H_{yx}[k] & H_{yy}[k] \end{bmatrix}. \tag{11.94}$$

Now, Eqs. (11.90) and (11.91) may be rewritten as

$$\boldsymbol{\Psi}_{\text{out}}[m] = \sum_{k=-N}^{N} \mathbf{H}[k]\boldsymbol{\Psi}_{\text{in}}[m-k]. \tag{11.95}$$

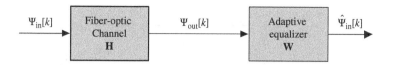

$\Psi_{\text{in}}[k]$ → Fiber-optic Channel **H** → $\Psi_{\text{out}}[k]$ → Adaptive equalizer **W** → $\hat{\Psi}_{\text{in}}[k]$

Figure 11.19   Adaptive equalization of the polarization-multiplexed fiber-optic channel.

Fig. 11.19 shows a schematic of the fiber-optic channel with adaptive equalizer in the digital domain. The output of the equalizer is

$$\hat{\psi}_{x,\text{in}}[n] = F \sum_{k=-K}^{K} \{W_{xx}[k]\psi_{x,\text{out}}[n-k] + W_{xy}[k]\psi_{y,\text{out}}[n-k]\}, \tag{11.96}$$

$$\hat{\psi}_{y,\text{in}}[n] = F \sum_{k=-K}^{K} \{W_{yx}[k]\psi_{x,\text{out}}[n-k] + W_{yy}[k]\psi_{y,\text{out}}[n-k]\}. \tag{11.97}$$

Let

$$\hat{\Psi}_{\text{in}}[k] = \begin{bmatrix} \hat{\psi}_{x,\text{in}}[k] \\ \hat{\psi}_{y,\text{in}}[k] \end{bmatrix}, \tag{11.98}$$

$$\mathbf{W}[k] = \begin{bmatrix} W_{xx}[k] & W_{xy}[k] \\ W_{yx}[k] & W_{yy}[k]. \end{bmatrix} \tag{11.99}$$

Eqs. (11.96) and (11.97) may be written as

$$\hat{\Psi}_{\text{in}}[n] = \sum_{k=-K}^{K} \mathbf{W}[k]\Psi_{\text{out}}[x-k]. \tag{11.100}$$

The adaptive equalizer for polarization mode dispersion consists of four transversal filters, W_{xx}, W_{xy}, W_{yx}, and W_{yy}, as shown in Fig. 11.20. The tap weights of the equalizer can be updated using the training sequence or blind equalization techniques, as described previously. Let us first consider an adaptive equalizer that uses a LMS algorithm and training sequences. The weights are updated as (see Example 11.4)

$$W_{xx}[k]^{(n+1)} = W_{xx}[k]^{(n)} + \psi_{x,\text{out}}^{*}[n-k]e_{x}[n]\Delta, \tag{11.101}$$

$$W_{xy}[k]^{(n+1)} = W_{xy}[k]^{(n)} + \psi_{y,\text{out}}^{*}[n-k]e_{x}[n]\Delta, \tag{11.102}$$

$$W_{yy}[k]^{(n+1)} = W_{yy}[k]^{(n)} + \psi_{y,\text{out}}^{*}[n-k]e_{y}[n]\Delta, \tag{11.103}$$

$$W_{yx}[k]^{(n+1)} = W_{yx}[k]^{(n)} + \psi_{x,\text{out}}^{*}[n-k]e_{y}[n]\Delta, \tag{11.104}$$

where

$$e_{r}[n] = \psi_{r,\text{in}}[n] - \hat{\psi}_{r,\text{in}}[n], \quad r = x, y. \tag{11.105}$$

For a blind equalizer that uses CMA, the error signals are given by Eq. (11.84),

$$e_{r}'[k] = \hat{\psi}_{r,\text{in}}[k](1 - |\hat{\psi}_{r,\text{in}}[n]|^{2}), \quad r = x, y. \tag{11.106}$$

The tap weights are adjusted in accordance with the stochastic gradient algorithm,

$$W_{xx}[k]^{(n+1)} = W_{xx}[k]^{(n)} + \psi_{x,\text{out}}^{*}[n-k]e_{x}'[n]\Delta, \tag{11.107}$$

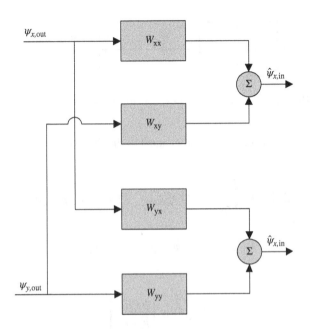

Figure 11.20   Polarization mode dispersion compensation using four transversal filters.

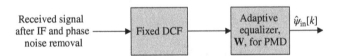

Figure 11.21   Block diagram of the digital equalizer for a polarization-multiplexed fiber-optic system.

$$W_{xy}[k]^{(n+1)} = W_{xy}[k]^{(n)} + \psi_{y,\text{out}}^*[n-k]e_x'[n]\Delta, \qquad (11.108)$$

$$W_{yy}[k]^{(n+1)} = W_{yy}[k]^{(n)} + \psi_{y,\text{out}}^*[n-k]e_y'[n]\Delta, \qquad (11.109)$$

$$W_{yx}[k]^{(n+1)} = W_{yx}[k]^{(n)} + \psi_{x,out}^*[n-k]e_y'[n]\Delta \qquad (11.110)$$

When a PMD equalizer is used, it is not necessary to have a separate adaptive equalizer for CD, as the diagonal elements of the matrix $\mathbf{H}$ have contributions from CD. Typically, the fixed dispersion-compensating filter compensates for the mean (non-time-varying) CD and the residual CD is compensated by the transversal filters W_{xx} and W_{yy}. W_{rr} compensates for the residual CD of the r-polarization component, $r = x, y$. Fig. 11.21 shows block diagram of the digital equalizer that compensates for CD and PMD.

11.8   Digital Back Propagation

So far in this chapter, we have assumed that the fiber-optic system is a linear system and focused on the mitigation of the linear impairments such as chromatic dispersion and polarization mode dispersion. In this section, we consider the mitigation of fiber nonlinear effects. Fiber dispersion and nonlinear effects can be compensated using the digital back-propagation techniques [22, 23]. Let us first consider a single-span system with constant fiber dispersion, nonlinear, and loss coefficients. The evolution of the field envelope in a fiber is described by the NLSE (see Chapter 10),

$$\frac{\partial q}{\partial z} = (N + D)q,$$ (11.111)

where D denotes the fiber dispersion effect,

$$D = -i\frac{\beta_2}{2}\frac{\partial^2}{\partial t^2}$$ (11.112)

and N denotes the nonlinear and loss effects,

$$N(t,z) = i\gamma|q(t,z)|^2 - \frac{\alpha}{2}.$$ (11.113)

The formal solution of Eq. (11.111) can be obtained as follows:

$$\frac{dq}{q} = (N + D),$$ (11.114)

$$\ln[q(t,z)]|_0^L = \int_0^L (N + D)dz,$$ (11.115)

$$q(t,L) = Mq(t,0),$$ (11.116)

where

$$M = \exp\left\{\int_0^L [N(t,z) + D(t)]dz\right\}$$ (11.117)

and L is the fiber length. In general, $q(t,L)$ can not be obtained in a closed form since $N(t,z)$ has a term proportional to $|q(t,z)|^2$ which is unknown for $z > 0$. Eq. (11.116) is just another way of writing Eq. (11.111), and numerical techniques have to be used to find $q(t,L)$ [24]. Multiplying Eq. (11.116) by M^{-1} on both sides, we find

$$q(t,0) = M^{-1}q(t,L).$$ (11.118)

In Eq. (11.118), $q(t,L)$ represents the received field envelope which is distorted due to fiber dispersion and nonlinear effects. If we multiply the received field by the inverse fiber operator, M^{-1}, distortions due to fiber dispersion and nonlinear effects can be completely undone. Since

$$\exp(\hat{x})\exp(-\hat{x}) = I,$$ (11.119)

where I is an identity operator (Example 11.4), taking

$$\hat{x} = \int_0^L [N(t,z) + D(t)]dz,$$ (11.120)

we find

$$M^{-1} = \exp\left[-\int_0^L [N(t,z) + D(t)]dz\right].$$ (11.121)

Eq. (11.118) with M^{-1} given by Eq. (11.121) is equivalent to solving the following partial differential equation:

$$\frac{\partial q_b}{\partial z} = -[N + D]q_b,$$ (11.122)

or

$$\frac{\partial q_b}{\partial(-z)} = [N + D]q_b, \tag{11.123}$$

with the initial condition $q_b(t,0) = q(t,L)$. From Eq. (11.118), it follows that

$$M^{-1}q_b(t,0) \equiv q_b(t,L) = q(t,0). \tag{11.124}$$

Thus, by solving Eq. (11.123), $q_b(t,L)$ can be found, which should be equal to the fiber input $q(t,0)$. In other words, if the fiber link inverse operator M^{-1} can be realized in the digital domain, by operating it on the fiber link output, we can retrieve the fiber input $q(t,0)$. Since Eq. (11.123) is nothing but Eq. (11.111) with $z \to -z$, this technique is referred to as *back propagation*. Eq. (11.122) may be rewritten as

$$\frac{\partial q_b}{\partial z} = [N_b + D_b]q_b, \tag{11.125}$$

with $q_b(t,0) \equiv q(t,L)$

$$D_b = -D = i\beta_2\frac{\partial^2}{\partial t^2}, \tag{11.126}$$

$$N_b = -N = -i\gamma|q_b|^2 + \frac{\alpha}{2}. \tag{11.127}$$

The NLSE with reversed signs of dispersion, loss, and nonlinear coefficients is solved in the digital domain to undo the distortion caused by the transmission fiber. Figs. 11.22 and 11.23 illustrate the forward and backward propagation. Eq. (11.125) can be solved numerically using the split-step Fourier scheme [24]. In Eq. (11.125), the operators N_b and D_b act simultaneously and N_b changes with z, which makes it harder to realize the operator M^{-1} numerically. However, over a small propagation step, $\Delta z, D_b$, and N_b may be approximated to act one after the other. Hence, this technique is known as the split-step technique. This is an approximation, and this technique becomes more accurate as $\Delta z \to 0$. First let us consider the unsymmetric split-step

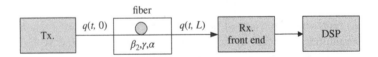

Figure 11.22   Propagation in a single-span fiber (forward propagation).

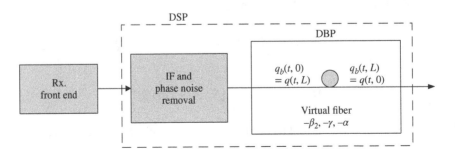

Figure 11.23   Backward propagation in the virtual fiber.

scheme. The received field $q(t, L) = q_b(t, 0)$. We wish to find $q_b(t, \Delta z)$, which corresponds to $q(t, L - \Delta z)$. The operator M^{-1} in this propagation step can be approximated as

$$M^{-1} = \exp\left[\int_0^{\Delta z} [N_b(t, z) + D_b(t)]dz\right] \cong \exp\left[\int_0^{\Delta z} [N_b(t, z)]dz\right] \exp[D_b(t)\Delta z], \quad (11.128)$$

$$q_b(t, \Delta z) = M^{-1}q_b(t, 0) = \exp\left[\int_0^{\Delta z} N_b(t, z)dz\right] q_b^l(t, \Delta z), \quad (11.129)$$

where

$$q_b^l(t, \Delta z) = \exp[D_b(t)\Delta z]q_b(t, 0). \quad (11.130)$$

Eq. (11.130) is equivalent to solving the following equation:

$$\frac{\partial q_b^l}{\partial z} = D_b q_b^l = \frac{i\beta_2}{2}\frac{\partial^2 q_b^l}{\partial t^2}, \quad (11.131)$$

with $q_b^l(t, 0) = q_b(t, 0)$. To solve Eq. (11.131), we take the Fourier transform on both sides:

$$\frac{d\tilde{q}_b^l(\omega, z)}{dz} = \frac{-i\beta_2\omega^2}{2}\tilde{q}_b^l(\omega, z), \quad (11.132)$$

$$\tilde{q}_b^l(\omega, \Delta z) = \frac{-i\beta_2\omega^2\Delta z}{2}\tilde{q}_b^l(\omega, 0). \quad (11.133)$$

The signal $q_b^l(t, \Delta z)$ can be obtained by the inverse Fourier transformation

$$q_b^l(t, \Delta z) = \mathcal{F}^{-1}[\tilde{q}_b^l(\omega, \Delta z)]. \quad (11.134)$$

In other words, the initial spectrum $\tilde{q}_b(\omega, 0)$ is multiplied by the inverse fiber linear transfer function to obtain $q_b(t, \Delta z)$ and, therefore, it represents the inverse linear response of the fiber. Computation of the Fourier transform/inverse Fourier transform takes N^2 complex additions/multiplications, where N is the number of samples. To facilitate fast computation of the Fourier transform, fast Fourier transform (FFT) is used which takes $\sim N\log_2 N$ complex additions/multiplications. Eq. (11.131) may also be solved using the FIR filter approach [22], as discussed in Section 11.6. Next, let us consider the nonlinear operator in Eq. (11.129). Eq. (11.129) is formally equivalent to the following equation:

$$\frac{\partial q_b}{\partial z} = N_b q_b = \left(-i\gamma|q_b|^2 + \frac{\alpha}{2}\right)q_b, \quad (11.135)$$

with $q_b(t, 0) = q_b^l(t, \Delta z)$. Let

$$q_b = A\exp(i\theta). \quad (11.136)$$

Substituting Eq. (11.136) into Eq. (11.135) and separating the real and imaginary parts, we find

$$\frac{dA}{dz} = \frac{\alpha}{2}A, \quad (11.137)$$

$$\frac{d\theta}{dz} = -\gamma|A|^2. \quad (11.138)$$

Integrating Eq. (11.137), we obtain

$$A(t, z) = \exp\left(\frac{\alpha z}{2}\right)A(t, 0), \quad (11.139)$$

and from Eq. (11.138), we have

$$\theta(t, \Delta z) = \theta(t, 0) - \gamma \int_0^{\Delta z} |A(t, z)|^2 dz = \theta(t, 0) - \gamma \Delta z_{\text{eff}} |A(t, 0)|^2, \tag{11.140}$$

where

$$\Delta z_{\text{eff}} = \frac{\exp(\alpha \Delta z) - 1}{\alpha}. \tag{11.141}$$

Substituting Eqs. (11.140) and (11.139) in Eq. (11.136), we find

$$q_b(t, \Delta z) = q_b(t, 0) \exp(-i\gamma \Delta z_{\text{eff}} |q_b(t, 0)|^2 + \alpha \Delta z), \tag{11.142}$$

where

$$q_b(t, 0) = A(t, 0) \exp[i\theta(t, 0)]. \tag{11.143}$$

With $q_b(t, 0) = q_b^l(t, \Delta z)$, Eq. (11.142) becomes

$$q_b(t, \Delta z) = q_b^l(t, \Delta z) \exp(-i\gamma \Delta z_{\text{eff}} |q_b^l(t, \Delta z)|^2 + \alpha \Delta z). \tag{11.144}$$

Fig. 11.24 illustrates the unsymmetric SSFS. This technique can be summarized as follows:

(i) Initial field $q_b(t, 0)$ is known. First, the nonlinear and loss effects ($\hat{N}$) are ignored and the output of a lossless, linear fiber $q_b^l(t, \Delta z)$ is calculated using the Fourier transformation technique.
(ii) Next, fiber dispersion ($\hat{D}$) is ignored. The NLSE is analytically solved with the initial condition $q_b(t, 0) = q_b^l(t, \Delta z)$ and the field envelope at Δz, $q_b(t, \Delta z)$ is calculated using Eq. (11.144).
(iii) $q_b(t, 2\Delta z)$ is calculated with $q_b(t, \Delta z)$ as the initial condition by repeating (i) and (ii). This process is repeated until $z = L$. The step size Δz should be chosen sufficiently small that the absolute value of the nonlinear phase shift $\Delta \theta$ accumulated over a distance Δz should be much smaller than π. From Eq. (11.140), it follows that

$$|\Delta \theta| = |\theta(t, \Delta z) - \theta(t, 0)| = \gamma \Delta z_{\text{eff}} |A(t, 0)|^2 \ll \pi. \tag{11.145}$$

A disadvantage of the unsymmetric SSFS is that the step size has to be really small since the error scales since as Δz^2 [24]. The step size can be made significantly larger using the symmetric SSFS, which is described as follows. From Eq. (11.125), we have

$$q_b(t, \Delta z) = \exp\left[\int_0^{\Delta z} [N_b(t, z) + D_b(t)]dz\right] q_b(t, 0). \tag{11.146}$$

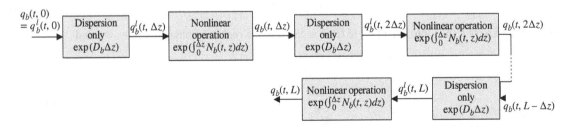

Figure 11.24 Unsymmetric split-step Fourier scheme for backward propagation.

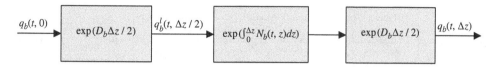

Figure 11.25   Symmetric split-step Fourier scheme for a single-step Δz.

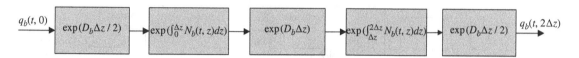

Figure 11.26   Symmetric split-step Fourier scheme for the propagation from 0 to $2\Delta z$.

$q_b(t, \Delta z)$ can be approximated as

$$q_b(t, \Delta z) = \left[\exp\left\{ \frac{D_b \Delta z}{2} \right\} \exp\left\{ \int_0^{\Delta z} N_b(t, z)dz \right\} \exp\left\{ \frac{D_b \Delta z}{2} \right\} \right] q_b(t, 0). \tag{11.147}$$

The above scheme is known as symmetric SSFS. Fig. 11.25 illustrates the symmetric SSFS. First, the NLSE is solved with $\hat{N}_b = 0$ over a distance $\Delta z/2$. The linear field $q_b^l(t, \Delta z/2)$ is multiplied by the nonlinear phase shift and amplified. The resulting field is propagated over a distance $\Delta z/2$ with $\hat{N}_b = 0$. It may appear that the computational effort for the symmetric SSFS is twice that of the unsymmetric SSFS. However, the computational efforts are roughly the same when the step size is much smaller than the fiber length. This can be understood from the propagation of the field from 0 to $2\Delta z$, as shown in Fig. 11.26 . The linear propagation operator, $e^{D_b \Delta z/2}$ shown in the last block of Fig. 11.25 can be combined with $e^{D_b \Delta z/2}$ corresponding to the first block of the propagation from Δz to $2\Delta z$, leading to a linear propagation operator $e^{D_b \Delta z}$, as indicated by the third block in Fig. 11.26. Since the evaluation of $e^{D_b \Delta z}$ or $e^{D_b \Delta z/2}$ requires $\sim N\log_2 N$ complex multiplications, the computational cost for the symmetric SSFS shown in Fig. 11.26 is roughly $3N\log_2 N$ complex multiplications, whereas that for the unsymmetric SSFS is roughly $2N\log_2 N$ for propagation up to $2\Delta z$. Over M propagation steps the computational overhead for the symmetric SSFS increases as $(M + 1)/M$. Thus, the overhead is insignificant when $M \gg 1$. For the given step size, the symmetric SSFS gives a more accurate result than the unsymmetric SSFS. This is because the error in the case of symmetric SSFS scales as Δz^3, whereas it scales as Δz^2 for unsymmetric SSFS [24]. Alternatively, for the given accuracy, a larger step size could be chosen in the case of symmetric SSFS.

11.8.1   Multi-Span DBP

Fig. 11.27 shows the propagation in an N-span fiber-optic system. To undo the propagation effect, amplifiers with gain G_n are substituted by loss elements $1/G_n$ in the digital domain and a real fiber with parameters $(\beta_{2n}, \gamma_n, \alpha_n)$, $n = 1, 2, \dots, N$ is replaced by a virtual fiber with parameters $(-\beta_{2n}, -\gamma_n, -\alpha_n)$, as shown in Fig. 11.28. Note that the signal distortions due to the last fiber in the fiber-optic link are compensated first in the digital domain. Although the digital back propagation can compensate for deterministic (and bit-pattern-dependent) nonlinear effects, it can not undo the impact of ASE and nonlinearity–ASE coupling, such as Gordon–Mollenauer phase noise.

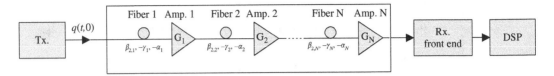

Figure 11.27   Propagation in an N-span fiber-optic system.

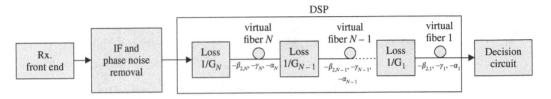

Figure 11.28   Digital back propagation for a N-span fiber-optic system.

11.9   Additional Examples

Example 11.2

The noise n_l is a zero-mean complex random variable with Gaussian distribution. Show that the mean of the effective noise n'_l given by Eq. (11.33) is zero.

Solution:
Let

$$n_l = A_l \exp(i\theta_l) = A_l \cos(\theta_l) + iA_l \sin(\theta_l). \tag{11.148}$$

Since n_l is a Gaussian random variable, it follows that θ_l is a random variable with uniform distribution in the interval $[0, 2\pi]$:

$$<n_l> = <A_l><\cos(\theta_l)> + i<A_l><\sin(\theta_l)> = 0. \tag{11.149}$$

Consider

$$n_l^k = A_l^k \exp(ik\theta_l) = A_l^k \cos(k\theta_l) + iA_l^k \sin(k\theta_l), \quad k = 1, 2, \ldots, M \tag{11.150}$$

$$<n_l^k> = <A_l^k><\cos(k\theta_l)> + i<A_l^k><\sin(k\theta_l)> = 0. \tag{11.151}$$

Since θ_l is a uniformly distributed random variable in the interval $[0, 2\pi]$, $k\theta_l$ is also a uniformly distributed random variable in the interval $[0, 2\pi k]$ and therefore $<\cos(k\theta_l)> = <\sin(k\theta_l)> = 0$. Eq. (11.33) may be rewritten as

$$n'_l = K_1 n_l + K_2 n_l^2 + \cdots + K_M n_l^M, \tag{11.152}$$

where K_m, $m = 1, 2, 3, \ldots M$, are complex constants:

$$<n'_l> = K_1 <n_l> + K_2 <n_l^2> + \cdots + K_M <n_l^M>. \tag{11.153}$$

Since $<n_l^k>$ is zero, it follows that

$$<n'_l> = 0. \tag{11.154}$$

Example 11.3

Show that the impulse response of a dispersion-compensating filter is

$$W(t) = \sqrt{\frac{1}{2\pi i \beta_2 L}} \exp\left[\frac{it^2}{2\beta_2 L}\right]. \tag{11.155}$$

Solution:

For a Gaussian pulse, we have the following relations:

$$\mathcal{F}[\exp(-\pi t^2)] = \exp[-\pi f^2], \tag{11.156}$$

$$\mathcal{F}[W(at)] = \frac{1}{a}\tilde{W}(f/a), \quad \text{Re}(a) > 0. \tag{11.157}$$

The transfer function of a dispersion-compensating filter is

$$\tilde{W}(f) = \exp(-i2\pi^2 f^2 \beta_2 L). \tag{11.158}$$

Choosing $a^2 = 1/(2\pi\beta_2 Li)$, Eq. (11.158) can be written as

$$\tilde{W}(f) = \exp(-\pi f^2/a^2). \tag{11.159}$$

Using Eqs. (11.156), (11.157), and (11.158), we find

$$W(t) = a\exp[-\pi(at)^2] = \frac{1}{\sqrt{2\pi\beta_2 Li}}\exp\left[\frac{it^2}{2\beta_2 L}\right]. \tag{11.160}$$

Example 11.4

Find the LMS for the updated weights of the PMD equalizer.

Solution:

Following the notation of Section 11.7, let

$$e_r[n] = \psi_{r,\text{in}}[n] - \hat{\psi}_{r,\text{in}}[n], \quad r = x, y, \tag{11.161}$$

$$J_r = < |e_r[n]|^2 > = < |\psi_{r,\text{in}}[n]|^2 - \psi_{r,\text{in}}[n]\hat{\psi}^*_{r,\text{in}}[n] - \psi^*_{r,\text{in}}[n]\hat{\psi}_{r,\text{in}}[n] + \hat{\psi}_{r,\text{in}}[n]\hat{\psi}^*_{r,\text{in}}[n] > . \tag{11.162}$$

Using Eq. (11.76), the gradient vectors are

$$2\frac{\partial J_x}{\partial W^*_{xx}[n]} = 2 < -\psi_{x,\text{in}}[n]\psi^*_{x,\text{out}}[n-k] + \psi^*_{x,\text{out}}[n-k]\hat{\psi}_{x,\text{in}}[n] > = -2 < \psi^*_{x,\text{out}}[n-k]e_x[n] >, \tag{11.163}$$

$$2\frac{\partial J_x}{\partial W^*_{xy}[n]} = -2 < \psi^*_{y,\text{out}}[n-k]e_x[n] > . \tag{11.164}$$

As discussed in Section 11.6.1, the tap weights for the next iteration should be chosen in a direction opposite to the gradient vector,

$$W_{xx}[k]^{(n+1)} = W_{xx}[k]^{(n)} + \psi^*_{x,\text{out}}[n-k]e_x[n]\Delta, \qquad (11.165)$$

$$W_{xy}[k]^{(n+1)} = W_{xy}[k]^{(n)} + \psi^*_{y,\text{out}}[n-k]e_x[n]\Delta. \qquad (11.166)$$

Similarly, the tap weights $W_{yy}[k]$ and $W_{yx}[k]$ are altered as

$$W_{yy}[k]^{(n+1)} = W_{yy}[k]^{(n)} + \psi^*_{y,\text{out}}[n-k]e_y[n]\Delta, \qquad (11.167)$$

$$W_{yx}[k]^{(n+1)} = W_{yx}[k]^{(n)} + \psi^*_{x,\text{out}}[n-k]e_y[n]\Delta. \qquad (11.168)$$

Example 11.5

Show that

$$\exp(\hat{x}) \cdot \exp(-\hat{x}) = I, \qquad (11.169)$$

where $\hat{x}$ is any operator and I is an identity operator.

Solution:

Expanding $\exp(\pm\hat{x})$ in a Taylor series, we find

$$\exp(\hat{x}) = I + \hat{x} + \frac{\hat{x} \cdot \hat{x}}{2!} + \cdots \qquad (11.170)$$

$$\exp(-\hat{x}) = I - \hat{x} + \frac{\hat{x} \cdot \hat{x}}{2!} + \cdots \qquad (11.171)$$

Now consider the product

$$\exp(\hat{x}) \cdot \exp(-\hat{x}) = \left(I + \hat{x} + \frac{\hat{x} \cdot \hat{x}}{2} + \cdots \right) \cdot \left(I - \hat{x} + \frac{\hat{x} \cdot \hat{x}}{2} + \cdots \right)$$

$$= I + (\hat{x} \cdot I - I \cdot \hat{x}) + \left(\frac{I \cdot \hat{x} \cdot \hat{x}}{2} - \hat{x} \cdot \hat{x} + \frac{\hat{x} \cdot \hat{x} \cdot I}{2} \right) + \cdots$$

$$= I. \qquad (11.172)$$

Exercises

11.1 Explain the phase increment algorithm for IF estimation.

11.2 Discuss the phase-unwrapping techniques used in phase compensation.

11.3 Write a computer program to compensate for IF and laser phase noise in a back-to-back configuration with the following parameters: transmitter laser linewidth = 5 MHz, LO linewidth = 10 MHz, $f_{IF} = 200$ MHz, symbol rate = 25 GSym/s, modulation = NRZ-QPSK. Determine the optimum block size.

11.4 Discuss the advantages and disadvantages of CD compensation in the time domain and the frequency domain.

11.5 Write a computer program to compensate for CD of a fiber-optic system with the following parameters: $\beta_2 = -22\,\text{ps}^2/\text{km}$, transmission distance = 1000 km, symbol rate = 25 GSym/s, modulation = NRZ-QPSK. Determine the number of taps required for the time domain technique using a FIR filter. Also, write a program to compensate for CD in the frequency domain using FFTs. Compare the computational costs associated with time and frequency domain techniques.

11.6 Provide the algorithms of an adaptive equalizer based on the constant-modulus algorithm (CMA).

11.7 Explain the principles of digital back propagation (DBP). Can DBP compensate for the impairments due to interaction between fiber nonlinearity and ASE?

11.8 Explain the differences between symmetric and asymmetric split-step Fourier schemes.

11.9 Write a computer program to simulate a single-span fiber-optic system using the symmetric split-step Fourier scheme. The parameters of the system are as follows: symbol rate = 25 GSym/s, NRZ-QPSK, span length $L = 80$ km, $\beta_2 = -22\,\text{ps}^2/\text{km}$, loss = 0.2 dB/km, $\gamma = 1.1\,\text{W}^{-1}\,\text{km}^{-1}$, launch power = 10 dBm. At the receiver, introduce DBP with (a) step size = L, (b) step size = L/2. Compare the constellation diagrams with and without DBP. Ignore laser phase noise.

Further Reading

R. Chassaing and D. Reay, *Digital Signal Processing*. John Wiley & Sons, New York 2008.

S. Haykin, *Adaptive Filter Theory*, 4th edn. Prentice-Hall, Englewood ciffs, NJ, 2001.

S. Haykin, *Communication System*, 4th edn. John Wiley & Sons, New York 2001.

J.G. Proakis, *Digital Communications*, 4th edn., McGraw-Hill, New York, 2001.

H. Meyr, M. Molenclaey, and S. Fechtel, *Digital Communication Receivers, Synchronization, Channel Estimation, and Signal Processing*. John Wiley & Sons, New York, 1998.

References

[1] G. Li, *Adv. Opt. Photon.*, vol. **1**, p. 279, 2009.

[2] A. Einstein, *Annal. Phys.*, vol. **17**, p. 549, 1905.

[3] C.H. Henry, *J. Lightwave Technol.*, vol. **LT-4**, p. 298, 1986.

[4] F. Funabashi *et al.*, *IEEE J. Select. Top. Quant. Electron.*, vol. **10**(2), p. 312, 2004.

[5] F. Derr, *Electron. Lett.*, vol. **23**, p. 2177, 1991.

[6] H. Meyr, M. Molenclaey, and S. Fechtel, *Digital Communication Receivers, Synchronization, Channel Estimation, and Signal Processing*. John Wiley & Sons, New York, 1998, chapter 8.

[7] A. Leven *et al*, *IEEE Photon. Technol. Lett.*, vol. **19**, p. 366, 2007.

[8] M. Morelli and U. Mengali, *Eur. Trans. Telecommun.*, vol. **2**, p. 103, 1998.

[9] D.S. Ly-Gagnon *et al*, *J. Lightwave Technol.*, vol. **24**, p. 12, 2006.

[10] A.J. Viterbi and A.M. Viterbi, *IEEE Trans. Inform. Theory*, vol. **IT-29**, p. 543, 1983.

[11] E. Ip and M. Kahn, *J. Lightwave Technol.*, vol. **25**, p. 2675, 2007.

[12] D.E. Crivelli, H.S. Cortnr, and M.L. Hunda, Proceedings of IEEE Global Telecommunications Conference (GLOBE-COM), Dallas, TX, Vol. **4**, p. 2545, 2004.

[13] T. Pfan, S. Hoffmann, and R. Nor, *J. Lightwave Technol.*, vol. **27**, p. 989, 2009.

[14] M.G. Taylor, Proceedings of European Conference on Optical Communication (ECOC), Vol. **2**, p. 263, 2005.

[15] E. Ip and J.M. Kahn, *J. Lightwave Technol.*, vol. **25**, p. 2765, 2007.

[16] S.J. Savory, *Opt. Expr.*, vol. **16**, p. 804, 2008.

[17] G. Goldfarb and G. Li, *IEEE Photon. Technol. Lett.*, vol. **19**, p. 969, 2007.

[18] J.G. Proakis, *Digital Communications*, 4th edn. McGraw-Hill, New York, 2001, chapter 11.

[19] S. Haykin, *Communication Systems*, 4th edn. John Wiley & Sons, New York, 2001.

[20] S. Haykin, *Adaptive Filter Theory*, 4th edn. Prentice-Hall, Englewood Cliffs, NJ, 2001, chapter 5.

[21] D.N. Godard, *IEEE Trans. Commun.*, vol. **com-28**, p. 1867, 1980.

[22] Y. Li *et al*, *Opt. Expr.*, vol. **16**, p. 880, 2008.

[23] E. Ip and J. Kahn, *J. Lightwave Technol.*, vol. **26**, p. 3416, 2008.

[24] G.P. Agrawal, *Nonlinear Fiber Optics*, 3rd edn. Academic Press, San Diego, 2001.

Appendix A

From Eq. (3.15), we find that the Einstein coefficients A and B are related by

$$A = \gamma \hbar \omega B,$$ (A.1)

where

$$\gamma = \frac{\omega^2 n_0^3}{\pi^2 c^3}.$$ (A.2)

The spontaneous emission rate per unit volume is given by Eq. (3.4),

$$R_{\text{spont}} = -\left(\frac{dN_2}{dt}\right)_{\text{spont}} = \gamma \hbar \omega B N_2.$$ (A.3)

In Eq. (A.3), the medium is assumed to be homogeneous with refractive index n_0 and this emission rate takes into account all the modes of the homogeneous medium in the frequency interval $[\omega, \ \omega + d\omega]$. Typically, amplifiers or lasers make use of single mode or multi-mode devices such as single/multi-mode fibers or channel waveguides. In a single-mode fiber amplifier, the ASE coupled to a radiation mode escapes to the cladding and does not contribute to the fiber amplifier output. Only the ASE coupled to the guided mode is of practical interest. Therefore, we modify Eq. (A.3) such that the spontaneous emission rate corresponds to ASE coupled to the guided mode. In fact, γ of Eq. (A.2) represents the number of modes of a homogeneous medium per unit volume per unit frequency interval. To see that, consider an electromagnetic wave in a homogeneous medium confined to a cube of volume L^3. The plane wave inside this cube is

$$\psi = A \cos(\omega t - k_x x - k_y y - k_z z),$$ (A.4)

with

$$\omega = kc/n_0,$$ (A.5)

$$k^2 = k_x^2 + k_y^2 + k_z^2.$$ (A.6)

If L is infinite, k_x, k_y, and k_z can take arbitrary values satisfying Eq. (A.6). The propagation of the plane wave is in the direction of $\mathbf{k} = k_x \hat{x} + k_y \hat{y} + k_z \hat{z}$. Therefore, spontaneous emission occurs uniformly in all directions.

Fiber Optic Communications: Fundamentals and Applications, First Edition. Shiva Kumar and M. Jamal Deen.
© 2014 John Wiley & Sons, Ltd. Published 2014 by John Wiley & Sons, Ltd.

When L is finite, and if we assume that the walls of the cube are perfectly conducting, the field should vanish at the walls. In this case k_x, k_y, and k_z take discrete values given by

$$k_x = \frac{2\pi n_x}{L}, k_y = \frac{2\pi n_y}{L}, \quad \text{and} \quad k_z = \frac{2\pi n_z}{L}, \tag{A.7}$$

where n_x, n_y, and n_z are integers. In other words, they are the standing waves formed by the superposition of plane waves propagating in opposite directions ($\cos(\omega t - k_x x - k_y y - k_z z)$ and $\cos(\omega t + k_x x + k_y y + k_z z)$). In this case, spontaneous emission occurs at discrete angles in the direction of $\mathbf{k} = k_x \hat{x} + k_y \hat{y} + k_z, \hat{z}$ with k_x, k_y, and k_z given by Eq. (A.7). We wish to find the number of modes per unit volume, with angular frequencies ranging from ω to $\omega + d\omega$. This corresponds to wave numbers ranging from $k (= |\mathbf{k}|)$ to $k + dk$. For the given value of k, there can be a number of modes with different values of k_x, k_y, and k_z satisfying Eq. (A.6). For example, $k_x = k, k_y = k_z = 0$ is a mode propagating in the x-direction and $k_x = k/\sqrt{2}, k_y = k/\sqrt{2}, k_z = 0$ is another mode propagating at angle $45°$ to the x-axis and $45°$ to the y-axis, and so on. The wave numbers ranging from k to $k + dk$ correspond to modes in the intervals $[k_x, k_x + dk_x],[k_y, k_y + dk_y]$, and $[k_z, k_z + dk_z]$ with

$$k^2 = k_x^2 + k_y^2 + k_z^2 \tag{A.8}$$

and

$$(k + dk)^2 = (k_x + dk_x)^2 + (k_y + dk_y)^2 + (k_z + dk_z)^2. \tag{A.9}$$

From Eq. (A.7), we have

$$dk_x = \frac{2\pi}{L} dn_x, \tag{A.10}$$

where dn_x is the number of modes in the interval $[k_x, \ k_x + dk_x]$. The total number of modes with the x-component of the wave vector ranging from k_x to $k_x + dk_x$, the y-component ranging from k_y to $k_y + dk_y$, and the z-component ranging from k_z to $k_z + dk_z$ is

$$dn_x dn_y dn_z = \frac{L^3}{(2\pi)^3} dk_x dk_y dk_z, \tag{A.11}$$

where $dk_x dk_y dk_z$ represents the volume of the spherical shell enclosed between two spheres with radii k and $k + dk$, as shown in Fig. A.1. Therefore,

$$dk_x dk_y dk_z = (\text{area of the sphere with radius } k) \times dk$$

$$= 4\pi k^2 dk. \tag{A.12}$$

Substituting Eq. (A.12) into Eq. (A.11), we find that the total number of modes per unit volume with angular frequency ranging from ω to $\omega + d\omega$ is

$$\frac{dn_x dn_y dn_z}{L^3} = \frac{4\pi k^2 dk}{(2\pi)^3} = \frac{\omega^2 n_0^3 d\omega}{2\pi^2 c^3}, \tag{A.13}$$

where

$$\omega n_0 / c = k. \tag{A.14}$$

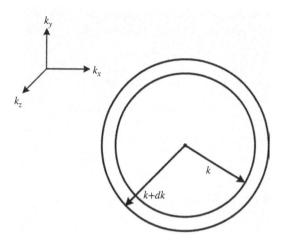

Figure A.1   Number of modes in the volume of a spherical shell enclosed between two spheres with radii k and $k + dk$.

For each mode defined by $(k_x \hat{x} + k_y \hat{y} + k_z \hat{z})$, there could be two polarizations (see Section 1.11). Therefore, each mode can be considered as two polarization modes. The total number of modes per unit volume per unit frequency interval, taking into account two polarization modes, is then

$$\gamma = \frac{2 dn_x dn_y dn_z}{L^3 d\omega} = \frac{N_m}{L^3} \qquad\qquad (A.15)$$

$$= \frac{\omega^2 n_0^3}{\pi^2 c^3}, \qquad\qquad (A.16)$$

where $N_m d\omega$ is the number of modes in the frequency interval $[\omega, \ \omega + d\omega]$. Eq. (A.16) is valid only for a homogeneous medium. In the case of an optical fiber, the general expression Eq. (A.15) should be used. From Eq. (A.3), the photon gain rate per unit volume due to spontaneous emission is

$$R_{\text{spont}} = \left(\frac{dN_{ph}}{dt}\right)_{\text{spont}} = \gamma \hbar \omega B N_2 = \frac{N_m \hbar \omega B N_2}{L^3}, \qquad\qquad (A.17)$$

where N_{ph} is the photon density. The spontaneous emission occurs over all the spatial and polarization modes of an optical fiber, and Eq. (A.17) represents the total spontaneous emission rate over all the modes. However, all the modes do not contribute to the spontaneous emission at the amplifier output. In a single-mode fiber, only the spontaneous emission coupled to the guided mode is of interest. For a single mode fiber with a single polarization mode, $N_m = 1$ and Eq. (A.17) becomes

$$R_{\text{spont}} = \left(\frac{dN_{ph}}{dt}\right)_{\text{spont}} = AN_2, \qquad\qquad (A.18)$$

where

$$A = \hbar \omega B / L^3. \qquad\qquad (A.19)$$

Next we consider the photon net gain rate due to absorption and spontaneous/stimulated emission. In this analysis, we ignore the loss of photons due to scattering and other possible mechanisms. From Eqs. (3.74) and (3.81), we have

$$\frac{dN_{ph}}{dt} = R_{stim} + R_{abs} + R_{spont}$$

$$= \hbar\omega B N_{ph}(N_2 - N_1) + A N_2. \tag{A.20}$$

Let n_{ph} be the number of photons in the volume L^3,

$$n_{ph} = N_{ph}L^3. \tag{A.21}$$

Using Eqs. (A.21) and (A.19), Eq. (A.20) may be rewritten as

$$\frac{dn_{ph}}{dt} = \hbar\omega B n_{ph}(N_2 - N_1) + \hbar\omega B N_2. \tag{A.22}$$

Using $v = dz/dt$ and simplifying Eq. (A.22), we obtain

$$\frac{dn_{ph}}{(N_2 - N_1)n_{ph} + N_2} = \frac{\hbar\omega B dz}{v}. \tag{A.23}$$

Eq. (A.23) can be rewritten as

$$\frac{dn_{ph}}{n_{ph} + n_{sp}} = \frac{\hbar\omega B(N_2 - N_1)dz}{v}, \tag{A.24}$$

where

$$n_{sp} = \frac{N_2}{N_2 - N_1} \tag{A.25}$$

is known as the *spontaneous emission factor* or *population-inversion factor*. For an amplifier, $N_2 > N_1$ and, therefore, $n_{sp} \geq 1$. Integrating Eq. (A.24) from 0 to L, we find

$$\ln\ [n_{ph}(L) + n_{sp}] - \ln\ [n_{ph}(0) + n_{sp}] = gL, \tag{A.26}$$

$$g = \frac{\hbar\omega B (N_2 - N_1)}{v} \tag{A.27}$$

or

$$n_{ph}(L) = n_{ph}(0)\exp(gL) + n_{sp}[\exp\ (gL) - 1]. \tag{A.28}$$

Since $G = \exp\ (gL)$, Eq. (A.28) can be written as

$$n_{ph}(L) = n_{ph}(0)G + n_{sp}(G - 1). \tag{A.29}$$

Eq. (A.29) is of fundamental significance. The first and second terms on the right hand side represent the photon gain due to stimulated emission and spontaneous emission, respectively.

Next, let us consider the average noise power due to spontaneous emission. A photon of energy $\hbar\omega_0$ is assumed to occupy a length L or equivalently time L/v[1, 2]. The noise power of a photon is

$$P_{\omega_0} = \frac{\hbar\omega_0}{L/v}. \tag{A.30}$$

In a single-mode fiber, the propagation part of the guided mode is of the form exp $[i(\omega t - kz)]$ with

$$\frac{\omega}{k} = v = \frac{c}{n_{\text{eff}}}, \tag{A.31}$$

where n_{eff} is the effective refractive index of the mode (see Chapter 2). The spontaneous emission occurs in both the forward and the backward direction. The forward propagating wave exp $[i(\omega t - kz)]$ and the backward propagating wave exp $[i(\omega t + kz)]$ form a standing wave. In Section 3.3, we have found that the frequency for the longitudinal modes is given by

$$\omega = \frac{\pi m v}{L}, \quad m = 0, \pm 1, \pm 2, \ldots \tag{A.32}$$

and

$$\Delta\omega = \frac{\pi r v}{L}, \tag{A.33}$$

where r is the number of longitudinal modes in the frequency interval $[\omega_0, \quad \omega_0 + \Delta\omega]$. The total noise power of photons in this frequency interval is

$$P_{\text{total}} = \sum_{j=0}^{r} \frac{\hbar(\omega_0 + j\delta\omega)v}{L}, \tag{A.34}$$

where $r\delta\omega = \Delta\omega$. Assuming $\Delta\omega \ll \omega_0$,

$$\hbar(\omega_0 + j\delta\omega) \cong \hbar\omega_0. \tag{A.35}$$

Now, using Eqs. (A.35) and (A.33), Eq. (A.34) reduces to

$$P_{\text{total}} = \frac{\hbar\omega_0 v r}{L} = \frac{\hbar\omega_0 \Delta\omega}{\pi}. \tag{A.36}$$

The noise power given by Eq. (A.36) includes the power emitted in the forward and backward directions. We are mainly interested in the noise power accompanying the amplified signal in the forward direction, which is half of that given by Eq. (A.36). Using $\Delta\omega = 2\pi\Delta f$, we have

$$P_{\text{total}}^{\text{forward}} = \hbar\omega_0 \Delta f. \tag{A.37}$$

So far we have assumed a single photon of energy $\hbar\omega_0$. If there are $n_{sp}(G-1)$ photons, Eq. (A.37) is modified as

$$P_{\text{ASE}} = n_{sp}(G-1)hf_0\Delta f, \tag{A.38}$$

where P_{ASE} is the mean noise power in the frequency interval $[f_0, f_0 + \Delta f]$.

References

[1] P.C. Becker, N.A. Olsson, and J.R. Simpson, *Erbium-Doped Fiber Amplifiers, Fundamentals and Technology*, Academic Press, San Diego, 1999.
[2] M.L. Dakss and P. Melman, *J. Lightwave Technol*, vol. LT-3, p. 806, 1985.

Appendix B

From Maxwell's equations, we have

$$\nabla^2 \mathbf{E} = \mu_0 \frac{\partial^2 \mathbf{D}}{\partial t^2}. \tag{B.1}$$

From Eq. (10.24), we have

$$\mathbf{D} = \epsilon_0 \mathbf{E} + \mathbf{P}. \tag{B.2}$$

Let us consider the case of a single polarization:

$$\mathbf{E} = E_x \mathbf{x},$$
$$\mathbf{P} = P_x \mathbf{x}. \tag{B.3}$$

From Eq. (10.42), we have

$$P_x(\mathbf{r},t) = P_L(\mathbf{r},t) + P_{NL}(\mathbf{r},t), \tag{B.4}$$

where

$$P_L(\mathbf{r},t) = \epsilon_0 \chi^{(1)} E_x(\mathbf{r},t), \tag{B.5}$$

$$P_{NL}(\mathbf{r},t) = \epsilon_0 \chi^{(3)} E_x^3(\mathbf{r},t). \tag{B.6}$$

Here, we have ignored the subscripts 'xx' and 'xxxx'. For a dispersive medium, the first-order susceptibility $\chi^{(1)}$ is a function of frequency (see Eq. (10.22)). Since the product in the frequency domain becomes a convolution in the time domain, for a dispersive medium, Eq. (B.5) should be modified as

$$P_L(\mathbf{r},t) = \epsilon \chi^{(1)}(\mathbf{r},t) \otimes E_x(\mathbf{r},t) \tag{B.7}$$

or

$$\tilde{P}_L(\mathbf{r},\omega) = \epsilon \tilde{\chi}^{(1)}(\mathbf{r},\omega) \tilde{E}_x(\mathbf{r},\omega), \tag{B.8}$$

where $\otimes$ denotes convolution. An optical pulse propagating down the fiber has rapidly varying oscillations at the carrier frequency and a slowly varying envelope corresponding to the pulse shape. Therefore, the electric field may be written in the following form:

$$E_x(\mathbf{r}, t) = \frac{1}{2}[E_0(\mathbf{r}, t) \exp(-i\omega_0 t) + \text{c.c.}], \tag{B.9}$$

where $E_0(r, t)$ is the slowly varying function of time and c.c stands for complex conjugate. Substituting Eq. (B.9) in Eq. (B.6), we find

$$P_{NL}(\mathbf{r}, t) = \frac{\epsilon_0 \chi^{(3)}}{8}[3|E(\mathbf{r}, t)|^2 E(\mathbf{r}, t)\exp(-i\omega_0 t) + E^3(\mathbf{r}, t)\exp(-i3\omega_0 t)] + \text{c.c.} \qquad \text{(B.10)}$$

The first term in the square bracket corresponds to oscillations at ω_0 and the second term corresponds to third harmonic frequency $3\omega_0$. The efficiency of third harmonic generation in fibers is very small unless special phase-matching techniques are used. Therefore, ignoring the second term and substituting Eqs. (B.7)–(B.10) in Eq. (B.1), we obtain

$$\nabla^2\Psi(\mathbf{r}, t) - \frac{1}{c^2}\frac{\partial^2\Psi(\mathbf{r}, t)}{\partial t^2} = \frac{1}{c^2}\frac{\partial^2}{\partial t^2}[\chi^{(1)}(\mathbf{r}, t) \otimes \Psi(\mathbf{r}, t)] + \frac{3\chi^{(3)}}{4c^2}\frac{\partial^2}{\partial t^2}[|\Psi(\mathbf{r}, t)|^2\Psi(\mathbf{r}, t)], \qquad \text{(B.11)}$$

where

$$\Psi(\mathbf{r}, t) = E_0(\mathbf{r}, t)\exp(-i\omega_0 t),$$

$$c^2 = \frac{1}{\mu_0 \epsilon_0}.$$

The electric field intensity in a single-mode fiber may be written as (see Chapter 2)

$$\Psi(\mathbf{r}, t) = q(z, t)\phi(x, y)e^{-i(\omega_0 t - \beta_0 z)}, \qquad \text{(B.12)}$$

where $\beta_0 = \beta_0(\omega_0)$ is the propagation constant, $\phi(x, y)$ is the transverse field distribution, and $q(z, t)$ is the field envelope which is a slowly varying function of t and z. Substituting Eq. (B.12) in Eq. (B.11) and taking the Fourier transform, we obtain

$$\left[\phi\frac{\partial^2\tilde{q}(z, \Omega)}{\partial z^2} + 2i\beta_0\phi\frac{\partial\tilde{q}(z, \Omega)}{\partial z} - \beta_0^2\tilde{q}(z, \Omega)\phi\right]$$

$$+ \left\{\frac{\partial^2\phi}{\partial x^2} + \frac{\partial^2\phi}{\partial y^2} + \frac{\omega^2\phi}{c^2}\left[1 + \tilde{\chi}^{(1)}(\mathbf{r}, \omega)\right]\right\}\tilde{q}(z, \Omega)$$

$$= -\frac{3\omega^2\chi^{(3)}}{4c^2}\{\phi^3(x, y)[\tilde{q}(z, \Omega) \otimes \tilde{q}^*(z, -\Omega) \otimes \tilde{q}(z, \Omega)]\}, \qquad \text{(B.13)}$$

where $\Omega = \omega - \omega_0$. To obtain Eq. (B.13), we have used the Fourier transform relations

$$\mathcal{F}\left(\frac{\partial^2 A}{\partial t^2}\right) = -\omega^2\tilde{A}(\omega) \qquad \text{(B.14)}$$

and

$$\mathcal{F}[A(t)B(t)] = \tilde{A}(\omega) \otimes \tilde{B}(\omega)$$

$$= \int \tilde{A}\left(\frac{\omega'}{2\pi}\right)\tilde{B}\left(\frac{\omega - \omega'}{2\pi}\right)\frac{d\omega}{2\pi}. \qquad \text{(B.15)}$$

Under the slowly varying envelope approximation, the first term in Eq. (B.13) can be ignored, which is a good approximation for pulse widths that are much longer than the period $2\pi/\omega_0$. From Eq. (10.29), we have

$$1 + \tilde{\chi}^{(1)}(\mathbf{r}, \omega) = n^2(\mathbf{r}, \omega), \qquad \text{(B.16)}$$

where n is the linear refractive index of the fiber. For a single-mode fiber we have

$$\frac{\partial^2 \phi}{\partial x^2} + \frac{\partial^2 \phi}{\partial y^2} + \frac{\omega^2 n^2(\mathbf{r}, \omega)}{c^2}\phi = \beta^2(\omega)\phi, \tag{B.17}$$

where $\beta(\omega)$ is the propagation constant. Substituting Eqs. (B.16) and (B.17) in Eq. (B.13), we obtain

$$2i\beta_0\frac{\partial \tilde{q}(z, \Omega)}{\partial z}\phi + [\beta^2(\omega) - \beta_0^2]\phi\tilde{q} = -\frac{3\omega^2\chi^{(3)}}{4c^2}\{\phi^3(x, y)[\tilde{q}(z, \Omega) \otimes \tilde{q}^*(z, -\Omega) \otimes \tilde{q}(z, \Omega)]\}. \tag{B.18}$$

To remove the dependence of transverse field distributions, we multiply Eq. (B.18) by $\phi(x, y)$ and integrate from $-\infty$ to ∞ in the x–y plane to obtain

$$i\frac{\partial \tilde{q}}{\partial z} + \frac{[\beta^2(\omega) - \beta_0^2]\tilde{q}}{2\beta_0} = -\frac{3\omega^2\chi^{(3)}}{8c^2 A_{\text{eff}}\beta_0}[\tilde{q}(z, \Omega) \otimes \tilde{q}^*(z, -\Omega) \otimes \tilde{q}(z, \Omega)], \tag{B.19}$$

where

$$A_{\text{eff}} = \frac{\int_{-\infty}^{\infty} \int_{-\infty}^{\infty} \phi^2(x, y) dx dy}{\int_{-\infty}^{\infty} \int_{-\infty}^{\infty} \phi^4(x, y) dx dy}. \tag{B.20}$$

The second term on the left-hand side of Eq. (B.19) can be approximated as

$$\frac{[\beta^2(\omega) - \beta_0^2]\tilde{q}}{2\beta_0} = \frac{[\beta(\omega) + \beta_0][\beta(\omega) - \beta_0]\tilde{q}}{2\beta_0} \cong [\beta(\omega) - \beta_0]\tilde{q}. \tag{B.21}$$

The above approximation is valid if the difference between $\beta(\omega)$ and β_0 is quite small. If the spectral width of the optical signal is comparable with or larger than ω_0, the above approximation could be incorrect. When the spectral width $\Delta\omega \ll \omega_0$, we can approximate $\beta(\omega)$ as a Taylor series around ω_0 and retain the first three terms,

$$\beta(\omega) = \beta_0 + \beta_1(\omega - \omega_0) + \frac{\beta_2}{2}(\omega - \omega_0)^2 + \frac{\beta_3}{6}(\omega - \omega_0)^3, \tag{B.22}$$

where

$$\beta_n = \frac{d^n\beta}{d\omega^n}\bigg|_{\omega = \omega_0} \tag{B.23}$$

is known as the nth-order dispersion coefficient (see Chapter 2). Substituting Eqs. (B.21) and (B.22) in Eq. (B.19), we obtain

$$i\frac{\partial \tilde{q}}{\partial z} + \left(\beta_1\Omega + \frac{\beta_2}{2}\Omega^2 + \frac{\beta_3}{6}\Omega^3\right)\tilde{q} = -\frac{3(\omega_0 + \Omega)^2\chi^{(3)}}{8c^2 A_{\text{eff}}\beta_0}[\tilde{q}(z, \Omega) \otimes \tilde{q}^*(z, -\Omega) \otimes \tilde{q}(z, \Omega)]. \tag{B.24}$$

If we include the fiber losses by treating the refractive index n as complex with its imaginary part being frequency independent, Eq. (B.24) is modified as

$$i\frac{\partial \tilde{q}}{\partial z} + \left(\beta_1\Omega + \frac{\beta_2}{2}\Omega^2 + \frac{\beta_3}{6}\Omega^3\right)\tilde{q} = -i\frac{\alpha\tilde{q}}{2} - \frac{3(\omega_0 + \Omega)^2\chi^{(3)}}{8c^2 A_{\text{eff}}\beta_0}[\tilde{q}(z, \Omega) \otimes \tilde{q}^*(z, -\Omega) \otimes \tilde{q}(z, \Omega)], \tag{B.25}$$

where α is the fiber loss coefficient related to the imaginary part of the refractive index through Eq. (10.37). We have assumed α to be independent of frequency. Since $\Omega \ll \omega_0$, $(\omega_0 + \Omega)^2 \simeq \omega_0^2 + 2\omega_0\Omega$. Now, performing the inverse Fourier transform, we obtain

$$i\left(\frac{\partial q}{\partial z} + \beta_1\frac{\partial q}{\partial t}\right) - \beta_2\frac{\partial^2 q}{\partial t^2} + i\gamma|q|^2 q = \frac{i\beta_3}{6}\frac{\partial^3 q}{\partial t^3} - \frac{i2\gamma}{\omega_0}\frac{\partial(|q|^2 q)}{\partial t} + \frac{i\alpha}{2}q, \tag{B.26}$$

where

$$\gamma = \frac{3\omega_0^2 \chi^{(3)}}{8c^2 A_{\text{eff}} \beta_0} \tag{B.27}$$

is the nonlinear coefficient. Note that multiplication by Ω in the frequency domain leads to the operator $i\frac{\partial}{\partial t}$ in the time domain. Eq. (B.26) in the absence of the right-hand side terms is called the *nonlinear Schrödinger equation* (NLSE). The third term on the left-hand side represents self-phase modulation, which is discussed in Section 10.5. The second term on the right-hand side of Eq. (B.26) is responsible for self-steepening. The terms on the right-hand side of Eq. (B.26) become important for ultra-short pulses (pulse width < 1ps). Eq. (B.26) in the presence of the right-hand-side terms is called the modified nonlinear Schrödinger equation (MNLSE). From Eq. (10.53), we have

$$n_2 = \frac{3\chi^{(3)}}{8n_0}. \tag{B.28}$$

Using Eq. (B.28) in Eq. (B.27) and noting that $\beta_0 \cong \omega_0 n_0/c$, we obtain

$$\gamma = \frac{n_2 \omega_0}{c A_{\text{eff}}}. \tag{B.29}$$

Index

Fiber Optic Communications: Fundamentals and Applications, First Edition. Shiva Kumar and M. Jamal Deen.
© 2014 John Wiley & Sons, Ltd. Published 2014 by John Wiley & Sons, Ltd.